*James E. Turner, Darryl J. Downing,
and James S. Bogard*

**Statistical Methods
in Radiation Physics**

Related Titles

Turner, J.E.

Atoms, Radiation, and Radiation Protection

Third Edition

2007

ISBN 978-3-527-40606-7

Lieser, K. H., J.V. Kratz

Nuclear and Radiochemistry

Fundamentals and Applications, Third Edition

2013

ISBN: 978-3-527-32901-4

Martin, J. E.

Physics for Radiation Protection

Second Edition

2003

ISBN: 978-3-527-40611-1

Bevelacqua, J. J.

Basic Health Physics

Problems and Solutions, Second Edition

2010

ISBN: 978-3-527-40823-8

Bevelacqua, J. J.

Contemporary Health Physics

Problems and Solutions, Second Edition

2009

ISBN: 978-3-527-40824-5

*James E. Turner, Darryl J. Downing,
and James S. Bogard*

Statistical Methods
in Radiation Physics

WILEY-
VCH

WILEY-VCH Verlag GmbH & Co. KGaA

The Authors

Dr. James E. Turner †
Oak Ridge, TN
USA

Dr. Darryl J. Downing
Loudon, TN 37774

Dr. James S. Bogard
Dade Moeller
704 S. Illinois Ave.
Oak Ridge, TN 37830
USA

All books published by **Wiley-VCH** are carefully produced. Nevertheless, authors, editors, and publisher do not warrant the information contained in these books, including this book, to be free of errors. Readers are advised to keep in mind that statements, data, illustrations, procedural details or other items may inadvertently be inaccurate.

Library of Congress Card No.: applied for

British Library Cataloguing-in-Publication Data
A catalogue record for this book is available from the British Library.

Bibliographic information published by the Deutsche Nationalbibliothek
The Deutsche Nationalbibliothek lists this publication in the Deutsche Nationalbibliografie; detailed bibliographic data are available on the Internet at http://dnb.d-nb.de.

© 2012 Wiley-VCH Verlag & Co. KGaA, Boschstr. 12, 69469 Weinheim, Germany

All rights reserved (including those of translation into other languages). No part of this book may be reproduced in any form – by photoprinting, microfilm, or any other means – nor transmitted or translated into a machine language without written permission from the publishers. Registered names, trademarks, etc. used in this book, even when not specifically marked as such, are not to be considered unprotected by law.

Print ISBN: 978-3-527-41107-8
ePDF ISBN: 978-3-527-64657-9
ePub ISBN: 978-3-527-64656-2
mobi ISBN: 978-3-527-64655-5
oBook ISBN: 978-3-527-64654-8

Cover Design   Adam-Design, Weinheim
Composition   Thomson Digital, Noida, India
Printing and Binding   Markono Print Media Pte Ltd, Singapore

Printed in Singapore
Printed on acid-free paper

James E. Turner, 1930–2008

Dedicated to the memory of James E. (Jim) Turner – scholar, scientist, mentor, friend.

Contents

Preface *XIII*

1   **The Statistical Nature of Radiation, Emission, and Interaction** *1*
1.1   Introduction and Scope *1*
1.2   Classical and Modern Physics – Determinism and Probabilities *1*
1.3   Semiclassical Atomic Theory *3*
1.4   Quantum Mechanics and the Uncertainty Principle *5*
1.5   Quantum Mechanics and Radioactive Decay *8*
       Problems *11*

2   **Radioactive Decay** *15*
2.1   Scope of Chapter *15*
2.2   Radioactive Disintegration – Exponential Decay *16*
2.3   Activity and Number of Atoms *18*
2.4   Survival and Decay Probabilities of Atoms *20*
2.5   Number of Disintegrations – The Binomial Distribution *22*
2.6   Critique *26*
       Problems *27*

3   **Sample Space, Events, and Probability** *29*
3.1   Sample Space *29*
3.2   Events *33*
3.3   Random Variables *36*
3.4   Probability of an Event *36*
3.5   Conditional and Independent Events *38*
       Problems *45*

4   **Probability Distributions and Transformations** *51*
4.1   Probability Distributions *51*
4.2   Expected Value *59*
4.3   Variance *63*

4.4	Joint Distributions	65
4.5	Covariance	71
4.6	Chebyshev's Inequality	76
4.7	Transformations of Random Variables	77
4.8	Bayes' Theorem	82
	Problems	84

5 Discrete Distributions 91

5.1	Introduction	91
5.2	Discrete Uniform Distribution	91
5.3	Bernoulli Distribution	92
5.4	Binomial Distribution	93
5.5	Poisson Distribution	98
5.6	Hypergeometric Distribution	106
5.7	Geometric Distribution	110
5.8	Negative Binomial Distribution	112
	Problems	113

6 Continuous Distributions 119

6.1	Introduction	119
6.2	Continuous Uniform Distribution	119
6.3	Normal Distribution	124
6.4	Central Limit Theorem	132
6.5	Normal Approximation to the Binomial Distribution	135
6.6	Gamma Distribution	142
6.7	Exponential Distribution	142
6.8	Chi-Square Distribution	145
6.9	Student's t-Distribution	149
6.10	F Distribution	151
6.11	Lognormal Distribution	153
6.12	Beta Distribution	154
	Problems	156

7 Parameter and Interval Estimation 163

7.1	Introduction	163
7.2	Random and Systematic Errors	163
7.3	Terminology and Notation	164
7.4	Estimator Properties	165
7.5	Interval Estimation of Parameters	168
7.5.1	Interval Estimation for Population Mean	168
7.5.2	Interval Estimation for the Proportion of Population	172
7.5.3	Estimated Error	173
7.5.4	Interval Estimation for Poisson Rate Parameter	175
7.6	Parameter Differences for Two Populations	176

7.6.1	Difference in Means 176
7.6.1.1	Case 1: σ_x^2 and σ_y^2 Known 177
7.6.1.2	Case 2: σ_x^2 and σ_y^2 Unknown, but Equal $(=\sigma^2)$ 178
7.6.1.3	Case 3: σ_x^2 and σ_y^2 Unknown and Unequal 180
7.6.2	Difference in Proportions 181
7.7	Interval Estimation for a Variance 183
7.8	Estimating the Ratio of Two Variances 184
7.9	Maximum Likelihood Estimation 185
7.10	Method of Moments 189
	Problems 194

8	**Propagation of Error** 199
8.1	Introduction 199
8.2	Error Propagation 199
8.3	Error Propagation Formulas 202
8.3.1	Sums and Differences 202
8.3.2	Products and Powers 202
8.3.3	Exponentials 203
8.3.4	Variance of the Mean 203
8.4	A Comparison of Linear and Exact Treatments 207
8.5	Delta Theorem 210
	Problems 210

9	**Measuring Radioactivity** 215
9.1	Introduction 215
9.2	Normal Approximation to the Poisson Distribution 216
9.3	Assessment of Sample Activity by Counting 216
9.4	Assessment of Uncertainty in Activity 217
9.5	Optimum Partitioning of Counting Times 222
9.6	Short-Lived Radionuclides 223
	Problems 226

10	**Statistical Performance Measures** 231
10.1	Statistical Decisions 231
10.2	Screening Samples for Radioactivity 231
10.3	Minimum Significant Measured Activity 233
10.4	Minimum Detectable True Activity 235
10.5	Hypothesis Testing 240
10.6	Criteria for Radiobioassay, HPS N13.30-1996 248
10.7	Thermoluminescence Dosimetry 255
10.8	Neyman–Pearson Lemma 262
10.9	Treating Outliers – Chauvenet's Criterion 263
	Problems 266

11	**Instrument Response** *271*
11.1	Introduction *271*
11.2	Energy Resolution *271*
11.3	Resolution and Average Energy Expended per Charge Carrier *275*
11.4	Scintillation Spectrometers *276*
11.5	Gas Proportional Counters *279*
11.6	Semiconductors *280*
11.7	Chi-Squared Test of Counter Operation *281*
11.8	Dead Time Corrections for Count Rate Measurements *284*
	Problems *290*

12	**Monte Carlo Methods and Applications in Dosimetry** *293*
12.1	Introduction *293*
12.2	Random Numbers and Random Number Generators *294*
12.3	Examples of Numerical Solutions by Monte Carlo Techniques *296*
12.3.1	Evaluation of $\pi = 3.14159265\ldots$ *296*
12.3.2	Particle in a Box *297*
12.4	Calculation of Uniform, Isotropic Chord Length Distribution in a Sphere *300*
12.5	Some Special Monte Carlo Features *306*
12.5.1	Smoothing Techniques *306*
12.5.2	Monitoring Statistical Error *306*
12.5.3	Stratified Sampling *308*
12.5.4	Importance Sampling *309*
12.6	Analytical Calculation of Isotropic Chord Length Distribution in a Sphere *309*
12.7	Generation of a Statistical Sample from a Known Frequency Distribution *312*
12.8	Decay Time Sampling from Exponential Distribution *315*
12.9	Photon Transport *317*
12.10	Dose Calculations *323*
12.11	Neutron Transport and Dose Computation *327*
	Problems *330*

13	**Dose–Response Relationships and Biological Modeling** *337*
13.1	Deterministic and Stochastic Effects of Radiation *337*
13.2	Dose–Response Relationships for Stochastic Effects *338*
13.3	Modeling Cell Survival to Radiation *341*
13.4	Single-Target, Single-Hit Model *342*
13.5	Multi-Target, Single-Hit Model *345*
13.6	The Linear–Quadratic Model *347*
	Problems *348*

14	**Regression Analysis** 353
14.1	Introduction 353
14.2	Estimation of Parameters β_0 and β_1 354
14.3	Some Properties of the Regression Estimators 358
14.4	Inferences for the Regression Model 361
14.5	Goodness of the Regression Equation 366
14.6	Bias, Pure Error, and Lack of Fit 369
14.7	Regression through the Origin 375
14.8	Inverse Regression 377
14.9	Correlation 379
	Problems 382

15	**Introduction to Bayesian Analysis** 387
15.1	Methods of Statistical Inference 387
15.2	Classical Analysis of a Problem 388
15.3	Bayesian Analysis of the Problem 390
15.4	Choice of a Prior Distribution 393
15.5	Conjugate Priors 396
15.6	Non-Informative Priors 397
15.7	Other Prior Distributions 401
15.8	Hyperparameters 402
15.9	Bayesian Inference 403
15.10	Binomial Probability 407
15.11	Poisson Rate Parameter 409
15.12	Normal Mean Parameter 414
	Problems 419

Appendix 423

Table A.1 Cumulative Binomial Distribution 423
Table A.2 Cumulative Poisson Distribution 424
Table A.3 Cumulative Normal Distribution 426
Table A.4 Quantiles $\chi^2_{\nu,\alpha}$ for the chi-squared Distribution with ν Degrees of Freedom 429
Table A.5 Quantiles $t_{\nu,\alpha}$ That Cut off Area α to the Right for Student's t-distribution with ν Degrees of Freedom 431
Table A.6 Quantiles $f_{0.95}(\nu_1, \nu_2)$ for the F Distribution 432
Table A.7 Quantiles $f_{0.99}(\nu_1, \nu_2)$ for the F Distribution 435

References 441

Index 445

Preface

Statistical Methods in Radiation Physics began as an effort to help clarify, for our students and colleagues, implications of the probabilistic nature of radioactive decay for measuring its observable consequences. Every radiological control technician knows that the uncertainty in the number of counts detected from a long-lived radioisotope is taken to be the square root of that number. But why is that so? And how is the corresponding uncertainty estimated for counts from a short-lived species, for which the count rate dies away even as the measurement is made? One of us (JET) had already been presented with these types of questions while teaching courses in the Oak Ridge Resident Graduate Program of the University of Tennessee's Evening School. A movement began in the late 1980s in the United States to codify occupational radiation protection and monitoring program requirements into Federal Regulations, and to include performance testing of programs and laboratories that provide the supporting external dosimetry and radiobioassay services. The authors' initial effort at a textbook consequently addressed statistics associated with radioactive decay and measurement, and also statistics used in the development of performance criteria and reporting of monitoring results.

What began as a short textbook grew eventually to 15 chapters, corresponding with the authors' growing realization that there did not appear to be a comparable text available. The book's scope consequently broadened from a textbook for health physicists to one useful to a wide variety of radiation scientists.

This is a statistics textbook, but the radiological focus is immediately emphasized in the first two chapters and continues throughout the book. Chapter 1 traces the evolution of deterministic classical physics at the end of the nineteenth century into the modern understanding of the wave–particle duality of nature, statistical limitations on precision of observables, and the development of quantum mechanics and its probabilistic view of nature. Chapter 2 begins with the familiar (to radiological physicists) exponential decay equation, a continuous, differentiable equation describing the behavior of large numbers of radioactive atoms, and concludes with the application of the binomial distribution to describe observations of small, discrete numbers of radioactive atoms. With the reader now on somewhat familiar ground, the next six chapters introduce probability, probability distributions, parameter and interval estimations, and error (uncertainty) propagation in derived quantities. These statistical tools are then applied in the remaining chapters to practical problems of measuring radioactivity, establishing performance measures for laboratories, instrument response, Monte Carlo modeling, dose response, and regression analysis. The final chapter introduces Bayesian analysis, which has seen increasing application in health physics in the past decade. The book is written at the senior or beginning graduate level as a text for a 1-year course in a curriculum of physics, health physics, nuclear engineering, environmental engineering, or an allied discipline. A large number of examples are worked in the text, with additional problems at the end of each chapter. SI units are emphasized, although traditional units are also used in some examples. SI abbreviations are used throughout. *Statistical Methods in Radiation Physics* is also intended as a reference for professionals in various fields of radiation physics and contains supporting tables, figures, appendices, and numerous equations.

We are indebted to our students and colleagues who first stimulated our interest in beginning such a textbook, and then who later contributed in many ways to its evolution and kept encouraging us to finish the manuscript. Some individual and institutional contributions are acknowledged in figure captions. We would like to thank Daniel Strom, in particular, for his encouragement and assistance in adding a chapter introducing Bayesian analysis.

The professional staff at Wiley-VCH has been most supportive and patient, for which we are extremely thankful. It has been a pleasure to work with Anja Tshcoertner, in particular, who regularly encouraged us to complete the manuscript. We also owe a debt of gratitude to Maike Peterson and the technical staff for their help in typesetting many equations.

We must acknowledge with great sorrow that James E. (Jim) Turner died on December 29, 2008, and did not see the publication of *Statistical Methods in Radiation Physics*. Jim conceived the idea that a statistics book applied to problems of radiological measurements would be useful, and provided the inspiration for this textbook. He was instrumental in choosing the topic areas and helped develop a large portion of the material. It was our privilege to have worked with Jim on this book, and we dedicate it to the memory of this man who professionally and personally enriched our lives and the lives of so many of our colleagues.

1
The Statistical Nature of Radiation, Emission, and Interaction

1.1
Introduction and Scope

This book is about statistics, with emphasis on its role in radiation physics, measurements, and radiation protection. That this subject is essential for understanding in these areas stems directly from the statistical nature of the submicroscopic, atomic world, as we briefly discuss in the next section. The principal aspects of atomic physics with which we shall be concerned are radioactive decay, radiation transport, and radiation interaction. Knowledge of these phenomena is necessary for success in many practical applications, which include dose assessment, shielding design, and the interpretation of instrument readings. Statistical topics will be further developed for establishing criteria to measure and characterize radioactive decay, assigning confidence limits for measured quantities, and formulating statistical measures of performance and compliance with regulations. An introduction to biological dose–response relations and to modeling the biological effects of radiation will also be included.

1.2
Classical and Modern Physics – Determinism and Probabilities

A principal objective of physical science is to discover laws and regularities that provide a quantitative description of nature as verified by observation. A desirable and useful outcome to be derived from such laws is the ability to make valid predictions of future conditions from a knowledge of the present state of a system. Newton's classical laws of motion, for example, determine completely the future motion of a system of objects if their positions and velocities at some instant of time and the forces acting between them are known. On the scale of the very large, the motion of the planets and moons can thus be calculated forward (and backward) in time, so that eclipses and other astronomical phenomena can be predicted with great accuracy. On the scale of everyday common life, Newton's laws describe all manner of diverse experience involving motion and statics. However, in the early twentieth century, the seemingly inviolate tenets of traditional physics were found to fail on the small scale

Statistical Methods in Radiation Physics, First Edition. James E. Turner, Darryl J. Downing, and James S. Bogard.
© 2012 Wiley-VCH Verlag GmbH & Co. KGaA. Published 2012 by Wiley-VCH Verlag GmbH & Co. KGaA.

of atoms. In place of a deterministic world of classical physics, it was discovered that atoms and radiation are governed by definite, but statistical, laws of quantum physics. Given the present state of an atomic system, one can predict its future, but only in statistical terms. What is the probability that a radioactive sample will undergo a certain number of disintegrations in the next minute? What is the probability that a given 100-keV gamma photon will penetrate a 0.5-cm layer of soft tissue? According to modern quantum theory, these questions can be answered as fully as possible only by giving the complete set of *probabilities* for obtaining any possible result of a measurement or observation.

By the close of the nineteenth century, the classical laws of mechanics, electromagnetism, thermodynamics, and gravitation were firmly established in physics. There were, however, some outstanding problems – evidence that all was not quite right. Two examples illustrate the growing difficulties. First, in the so-called "ultraviolet catastrophe," classical physics incorrectly predicted the distribution of wavelengths in the spectrum of electromagnetic radiation emitted from hot bodies, such as the sun. Second, sensitive measurements of the relative speed of light in different directions on earth – expected to reveal the magnitude of the velocity of the earth through space – gave a null result (no difference!). Planck found that the first problem could be resolved by proposing a nonclassical, quantum hypothesis related to the emission and absorption of radiation by matter. The now famous quantum of action, $h = 6.6261 \times 10^{-34}$ J s, was thus introduced into physics. The second dilemma was resolved by Einstein in 1905 with the revolutionary special theory of relativity. He postulated that the speed of light has the same numerical value for all observers in uniform translational motion with respect to one another, a situation wholly in conflict with velocity addition in Newtonian mechanics. Special relativity further predicts that energy and mass are equivalent and that the speed of light in a vacuum is the upper limit for the speed that any object can have. The classical concepts of absolute space and absolute time, which had been accepted as axiomatic tenets for Newton's laws of motion, were found to be untenable experimentally.

■ *Example*
In a certain experiment, 1000 monoenergetic photons are normally incident on a shield. Exactly 276 photons are observed to interact in the shield, while 724 photons pass through without interacting.

a) What is the probability that the next incident photon, under the same conditions, will not interact in the shield?
b) What is the probability that the next photon *will* interact?

Solution
a) Based on the given data, we estimate that the probability for a given photon to traverse the shield with no interaction is equal to the observed fraction that did not interact. Thus, the "best value" for the probability Pr(no) that the next photon will pass through without interacting is

$$\Pr(\text{no}) = \frac{724}{1000} = 0.724. \quad (1.1)$$

b) By the same token, the estimated probability Pr(yes) that the next photon will interact is

$$\text{Pr(yes)} = \frac{276}{1000} = 0.276, \tag{1.2}$$

based on the observation that 276 out of 1000 interacted.

This example suggests several aspects of statistical theory that we shall see often throughout this book. The sum of the probabilities for all possible outcomes considered in an experiment must add up to unity. Since only two possible alternatives were regarded in the example – either a photon interacted in the shield or it did not – we had Pr(no) + Pr(yes) = 1. We might have considered further whether an interaction was photoelectric absorption, Compton scattering, or pair production. We could assign separate probabilities for these processes and then ask, for example, what the probability is for the next interacting photon to be Compton scattered in the shield. In general, whatever number and variety of possible outcomes we wish to consider, the sum of their probabilities must be unity. This condition thus requires that there be *some* outcome for each incident photon.

It is evident, too, that a larger data sample will generally enable more reliable statistical predictions to be made. Knowing the fate of 1000 photons in the example gives more confidence in assigning values to the probabilities Pr(no) and Pr(yes) than would knowing the fate of, say, only 10 photons. Having data for 10^8 photons would be even more informative.

Indeed, the general question arises, "How can one ever know the actual, true numerical values for many of the statistical quantities that we must deal with?" Using appropriate samples and protocols that we shall develop later, one can often obtain rather precise values, but always within well-defined statistical limitations. A typical result expresses a "best" numerical value that lies within a given range with a specified degree of confidence. For instance, from the data given in the example above, we can express the "measured" probability of no interaction as

$$\text{Pr(no)} = 0.724 \pm 0.053 \quad (95\% \text{ confidence level}). \tag{1.3}$$

(The stated uncertainty, ± 0.053, is ± 1.96 standard deviations from an estimated mean of 0.724, based on the single experiment, as we shall discuss later in connection with the normal distribution.) Given the result (1.3), there is still no guarantee that the "true" value is actually in the stated range. Many such probabilities can also be accurately calculated from first principles by using quantum mechanics. In all known instances, the theoretical results are in agreement with measurements. Confirmation by observation is, of course, the final criterion for establishing the validity of the properties we ascribe to nature.

1.3
Semiclassical Atomic Theory

Following the unexpected discovery of X-rays by Roentgen in 1895, a whole series of new findings ushered in the rapidly developing field of atomic and radiation physics. Over the span of the next two decades, it became increasingly clear that classical science did not give a correct picture of the world as new physics unfolded. Becquerel discovered radioactivity in 1896, and Thomson measured the charge-to-mass ratio of the electron in 1897. Millikan succeeded in precisely measuring the charge of the electron in 1909. By 1910, a number of radioactive materials had been investigated, and the existence of isotopes and the transmutation of elements by radioactive decay were recognized. In 1911, Rutherford discovered the atomic nucleus – a small, massive dot at the center of the atom, containing all of the positive charge of the neutral atom and virtually all of its mass. The interpretation of his experiments on alpha-particle scattering from thin layers of gold pointed to a planetary structure for an atom, akin to a miniature solar system. The atom was pictured as consisting of a number of negatively charged electrons traversing most of its volume in rapid orbital motion about a tiny, massive, positively charged nucleus.

The advance made with the discovery of the nuclear atom posed another quandary for classical physics. The same successful classical theory (Maxwell's equations) that predicted many phenomena, including the existence of electromagnetic radiation, required the emission of energy by an accelerated electric charge. An electron in orbit about a nucleus should thus radiate energy and quickly spiral into the nucleus. The nuclear atom could not be stable. To circumvent this dilemma, Bohr in 1913 proposed a new, semiclassical nuclear model for the hydrogen atom. The single electron in this system moved in classical fashion about the nucleus (a proton). However, in nonclassical fashion Bohr postulated that the electron could occupy only certain circular orbits in which its angular momentum about the nucleus was quantized. (The quantum condition specified that the angular momentum was an integral multiple of Planck's constant divided by 2π.) In place of the continuum of unstable orbits allowed by classical mechanics, the possible orbits for the electron in Bohr's model were discrete. Bohr further postulated that the electron emitted radiation only when it went from one orbit to another of lower energy, closer to the nucleus. The radiation was then emitted in the form of a photon, having an energy equal to the difference in the energy the electron had in the two orbits. The atom could absorb a photon of the same energy when the electron made the reverse transition between orbits. These criteria for the emission and absorption of atomic radiation replaced the classical ideas. They also implied the recognized fact that the chemical elements emit and absorb radiation at the same wavelengths and that different elements would have their own individual, discrete, characteristic spectra. Bohr's theory for the hydrogen atom accounted in essential detail for the observed optical spectrum of this element. When applied to other atomic systems, however, the extension of Bohr's ideas often led to incorrect results.

An intensive period of semiclassical physics then followed into the 1920s. The structure and motion of atomic systems was first described by the equations of motion of classical physics, and then quantum conditions were superimposed, as Bohr had done for hydrogen. The quantized character of many variables, such as energy and angular momentum, previously assumed to be continuous, became increasingly evident experimentally.

Furthermore, nature showed a puzzling wave–particle duality in its fundamental makeup. Electromagnetic radiation, originally conceived as a purely wave phenomenon, exhibited properties of both waves and particles. The diffraction and interference of X-rays was demonstrated experimentally by von Laue in 1912, establishing their wave character. Einstein's explanation of the photoelectric effect in 1905 described electromagnetic radiation of frequency ν as consisting of packets, or photons, having energy $E = h\nu$. The massless photon carries an amount of momentum that is given by the relation

$$p = \frac{E}{c} = \frac{h\nu}{c}, \quad (1.4)$$

where $c = 2.9979 \times 10^8 \, \mathrm{m\,s^{-1}}$ is the speed of light in a vacuum. This particle-like property of momentum is exhibited experimentally, for example, by the Compton scattering of photons from electrons (1922). The wavelength λ of the radiation is given by $\lambda = c/\nu$. It follows from Eq. (1.4) that the relationship between the wavelength and momentum of a photon is given by

$$\lambda = \frac{h}{p}. \quad (1.5)$$

In 1924, de Broglie postulated that this relationship applies not only to photons, but also to other fundamental atomic particles. Electron diffraction was demonstrated experimentally by Davisson and Germer in 1927, with the electron wavelength being correctly given by Eq. (1.5). (Electron microscopes have much shorter wavelengths and hence much greater resolving power than their optical counterparts.)

There was no classical analogue to these revolutionary quantization rules and the wave–particle duality thus introduced into physics. Yet they appeared to work. The semiclassical procedures had some notable successes, but they also led to some unequivocally wrong predictions for other systems. There seemed to be elements of truth in quantizing atomic properties, but nature's secrets remained hidden in the early 1920s.

1.4
Quantum Mechanics and the Uncertainty Principle

Heisenberg reasoned that the root of the difficulties might lie in the use of nonobservable quantities to describe atomic constituents – attributes that the constituents might not even possess. Only those properties should be ascribed to an object that have an operational definition through an experiment that can be carried out to observe or measure them. What does it mean, for example, to ask whether an electron is blue or red, or even to ask whether an electron has a color? Such questions must be capable of being answered by experiment, at least in principle, or else they have no meaning in physics. Using only observable atomic quantities, such as those associated with the frequencies of the radiation emitted by an atom, Heisenberg in 1924 developed a new, matrix theory of quantum mechanics. At almost the same time, Schrödinger formulated his wave equation from an entirely different standpoint. He soon was able to show that his formulation and Heisenberg's were completely equivalent. The new quantum mechanics was born.

In the Newtonian mechanics employed by Bohr and others in the semiclassical theories, it was assumed that an atomic electron possesses a definite position and velocity at every instant of time. Heisenberg's reasoning required that, in order to have any meaning or validity, the very concept of the "position and velocity of the electron" should be defined operationally by means of an experiment that would determine it. He showed that the act of measuring the position of an electron ever more precisely would, in fact, make the simultaneous determination of its momentum (and hence velocity) more and more uncertain. In principle, the position of an electron could be observed experimentally by scattering a photon from it. The measured position would then be localized to within a distance comparable to the wavelength of the photon used, which limits its spatial resolving power. The scattered photon would, in turn, impart momentum to the electron being observed. Because of the finite aperture of any apparatus used to detect the scattered photon, its direction of scatter and hence its effect on the struck electron's momentum would not be known exactly. To measure the position of the electron precisely, one would need to use photons of very short wavelength. These, however, would have large energy and momentum, and the act of scattering would be coupled with large uncertainty in the simultaneous knowledge of the electron's momentum. Heisenberg showed that the product of the uncertainties in the position Δx in any direction in space and the component of momentum Δp_x in that direction must be at least as large as Planck's constant divided by 2π ($\hbar = h/2\pi = 1.0546 \times 10^{-34}$ J s):

$$\Delta x \Delta p_x \geq \hbar. \tag{1.6}$$

It is thus impossible to assign both position and momentum simultaneously with unlimited precision. (The equality applies only under optimum conditions.) The inequality (1.6) expresses one form of Heisenberg's uncertainty principle. A similar relationship exists between certain other pairs of variables, such as energy E and time t:

$$\Delta E \Delta t \geq \hbar. \tag{1.7}$$

The energy of a system cannot be determined with unlimited precision within a short interval of time.

These limits imposed by the uncertainty principle are not due to any shortcomings in our measurement techniques. They simply reflect the way in which the act of observation itself limits simultaneous knowledge of certain pairs of variables. To speculate whether an electron "really does have" an exact position and velocity at every instant of time, although we cannot know them together, apparently has no operational meaning. As we shall see in an example below, the limits have no practical effect on massive objects, such as those experienced in everyday life. In contrast, however, on the atomic scale the limits reflect an essential need to define carefully and operationally the concepts that are to have meaning and validity.

The subsequent development of quantum mechanics has provided an enormously successful quantitative description of many phenomena: atomic and nuclear structure, radioactive decay, lasers, semiconductors, antimatter, electron diffraction, superconductivity, elementary particles, radiation emission and absorption, the covalent chemical bond, and many others. It has revealed the dual wave–particle nature of the constituents of matter. Photons, electrons, neutrons, protons, and other particles have characteristics of both particles and waves. Instead of having a definite position and velocity, they can be thought of as being "smeared out" in space, as reflected by the uncertainty principle. They can be described in quantum mechanics by wave packets related to a probability density for observing them in different locations. They have both momentum p and wavelength λ, which are connected by the de Broglie relation (1.5). Endowed with such characteristics, the particles exhibit diffraction and interference effects under proper experimental conditions. Many quantum-mechanical properties, essential for understanding atomic and radiation physics, simply have no classical analogue in the experience of everyday life.

■ *Example*
The electron in the hydrogen atom is localized to within about 1 Å, which is the size of the atom. Use the equality in the uncertainty relation to estimate the uncertainty in its momentum. Estimate the order of magnitude of the kinetic energy that the electron (mass $= m = 9.1094 \times 10^{-31}$ kg) would have in keeping with this amount of localization in its position.

Solution
With $\Delta x = 1$ Å $= 10^{-10}$ m in Eq. (1.6), we estimate that the uncertainty in the electron's momentum is[1]

$$\Delta p \cong \frac{\hbar}{\Delta x} = \frac{1.05 \times 10^{-34}\ \text{J s}}{10^{-10}\ \text{m}} \cong 10^{-24}\ \text{kg m s}^{-1}. \quad (1.8)$$

We assume that the electron's momentum is about the same order of magnitude as this uncertainty. Denoting the electron mass by m, we estimate for its kinetic energy

$$T = \frac{(\Delta p)^2}{2m} \cong \frac{(10^{-24}\ \text{kg m s}^{-1})^2}{2 \times 9.11 \times 10^{-31}\ \text{kg}} \cong 5 \times 10^{-19}\ \text{J} \cong 3\ \text{eV}, \quad (1.9)$$

since $1\ \text{eV} = 1.60 \times 10^{-19}$ J. An electron confined to the dimensions of the hydrogen atom would be expected to have a kinetic energy in the eV range. The mean kinetic energy of the electron in the ground state of the hydrogen atom is 13.6 eV.

The uncertainty principle requires that observing the position of a particle with increased precision entails increased uncertainty in the simultaneous knowledge of its

---

1) Energy, which has the dimensions of force × distance, has units $1\ \text{J} = 1\ \text{N m}$. The newton of force has the same units as mass × acceleration: $1\ \text{N} = 1\ \text{kg m s}^{-2}$. Therefore, $1\ \text{J s m}^{-1} = 1\ \text{kg m s}^{-1}$, which are the units of momentum (mass × velocity).

momentum, or energy. Greater localization of a particle, therefore, is accompanied by greater likelihood that measurement of its energy will yield a large value. Conversely, if the energy is known with precision, then the particle must be "spread out" in space. Particles and photons can be described mathematically by quantum-mechanical wave packets, which, in place of classical determinism, provide probability distributions for the possible results of any measurement. These essential features of atomic physics are not manifested on the scale of familiar, everyday objects.

> ■ **Example**
> How would the answers to the last example be affected if
>
> a) the electron were localized to nuclear dimensions ($\Delta x \sim 10^{-15}$ m) or
> b) the electron mass were 100 g?
>
> **Solution**
> a) With $\Delta x \sim 10^{-15}$ m, the equality in the uncertainty principle (1.6) gives, in place of Eq (1.8), $\Delta p \cong 10^{-19}$ kg m s^{-1}, five orders of magnitude larger than before. The corresponding electron energy would be relativistic. Calculation shows that the energy of an electron localized to within 10^{-15} m would be about 200 MeV. (The numerical solution is given in Section 2.5 in Turner (2007), listed in the Bibliography at the end of this book.)
> b) Since Δx is the same as before (10^{-10} m), Δp in Eq. (1.8) is unchanged. With $m = 100$ g $= 0.1$ kg, the energy in place of Eq. (1.9) is now smaller by a factor of the mass ratio $(9.11 \times 10^{-31})/0.1 \cong 10^{-29}$. For all practical purposes, with the resultant extremely small value of T, the uncertainty in the velocity is negligible. Whereas the actual electron, localized to such small dimensions, has a large uncertainty in its momentum, the "100-g electron" would appear to be stationary.

Quantum-mechanical effects are generally expressed to a lesser extent with relatively massive objects, as this example shows. By atomic standards, objects in the macroscopic world are massive and have very large momenta. Their de Broglie wavelengths, expressed by Eq. (1.5), are vanishingly small. Quantum mechanics becomes important on the actual physical scale because of the small magnitude of Planck's constant.

1.5
Quantum Mechanics and Radioactive Decay

Before the discovery of the neutron by Chadwick in 1932, it was speculated that the atomic nucleus must be made up of the then known elementary subatomic particles: protons and electrons. However, according to quantum mechanics, this assumption leads to the wrong prediction of the angular momentum for certain nuclei. The nucleus of ^{6}Li, for example, would consist of six protons and three electrons, representing nine particles of half-integral spin. By quantum rules for addition of the spins of an odd number of such particles, the resulting nuclear angular momentum for ^{6}Li would also have to be a half-integral multiple of Planck's constant, $\hbar$. The measured value, however, is just $\hbar$. A similar circumstance occurs for ^{14}N. These two nuclei contain *even* numbers (6 and 14) of spin-1/2 particles (protons and neutrons), and hence must have integral angular momentum, as observed.

The existence of electrons in the nucleus would also have to be reconciled with the uncertainty principle. In part (a) of the last example, we saw that an electron confined to nuclear dimensions would have an expected kinetic energy in the range of 200 MeV. There is no experimental evidence for such large electron energies associated with beta decay or other nuclear phenomena.

If the electron is not there initially, how is its ejection from the nucleus in beta decay to be accounted for? Quantum mechanics explains the emission of the beta particle through its creation, along with an antineutrino, at the moment of the decay. Both particles are then ejected from the nucleus, causing it to recoil (slightly, because the momentum of the ejected mass is small). The beta particle, the antineutrino, and the recoil nucleus share the released energy, which is equivalent to the loss of mass ($E = mc^2$) that accompanies the radioactive transformation. Since the three participants can share this energy in a continuum of ways, beta particles emitted in radioactive decay have a continuous energy spectrum, which extends out to the total energy released. Similarly, gamma-ray or characteristic X-ray photons are not "present" in the nucleus or atom before emission. They are created when the quantum transition takes place. An alpha particle, on the other hand, is a tightly bound and stable structure of two protons and two neutrons within the nucleus. Alpha decay is treated quantum mechanically as the tunneling of the alpha particle through the confining nuclear barrier, a process that is energetically forbidden in classical mechanics. The emitted alpha particle and recoil nucleus, which separate back to back, share the released energy uniquely in inverse proportion to their masses. The resultant alpha-particle energy spectra are therefore discrete, in contrast to the continuous beta-particle spectra. The phenomenon of tunneling, which is utilized in a number of modern electronic and other applications, is purely quantum mechanical. It has no classical analogue (see Figure 1.1).

The radioactive decay of atoms and the accompanying emission of radiation are thus described in detail by quantum mechanics. As far as is known, radioactive decay occurs spontaneously and randomly, without influence from external factors. The energy thus released derives from the conversion of mass into energy, in accordance with Einstein's celebrated equation, $E = mc^2$.

■ *Example*

Each of 10 identical radioactive atoms is placed in a line of 10 separate counters, having 100% counting efficiency. The question is posed, "Which atom will decay first?" How can the question be answered, and how can the answer be verified?

Figure 1.1 An early scanning tunneling microscope (left) is used to image the electron clouds of individual carbon atoms on the surface of a highly oriented pyrolytic graphite sample. As a whisker tip just above the surface scans it horizontally in two dimensions, electrons tunnel through a classically forbidden barrier to produce a current through the tip. This current is extremely sensitive to the separation between the tip and the surface. As the separation tends to change according to the surface contours during the scan, negative feedback keeps it constant by moving a micrometer vertically up or down. These actions are translated by computer into the surface picture shown on the right. (Courtesy of R.J. Warmack.)

Solution
Since the atoms are identical and decay is spontaneous, the most one can say is that it is equally likely for *any* of the atoms, 1 through 10, to decay first. The validity of this answer, like any other, is to be decided on an objective basis by suitable experiments or observations. To perform such an experiment, in principle a large number of identical sources of 10 atoms could be prepared and then observed to see how many times the first atom to decay in a source is atom 1, atom 2, and so on. One would find a distribution, giving the relative frequencies for each of the 10 atoms that decays first. Because the atoms are identical, the distribution would be expected to show random fluctuations and become relatively flatter with an increasing number of observations.

■ *Example*
A source consists of 20 identical radioactive atoms. Each has a 90% chance of decaying within the next 24 h.

a) What is the probability that all 20 will decay in 24 h?
b) What is the probability that none will decay in 24 h?

Solution
a) The probability that atom 1 will decay in 24 h is 0.90. The probability that atoms 1 and 2 will both decay in 24 h is $0.90 \times 0.90 = (0.90)^2 = 0.81$. That is,

if the experiment is repeated many times, atom 2 is expected to decay in 90% of the cases (also 90%) in which atom 1 also decays. By extension, the probability for all atoms to decay in 24 h is

$$(0.90)^{20} = 0.12. \tag{1.10}$$

b) Since a given atom must either decay or not decay, the probability for not decaying in 24 h is $1 - 0.90 = 0.10$. The probability that none of the 20 atoms will decay is

$$(0.10)^{20} = 1.0 \times 10^{-20}. \tag{1.11}$$

As these examples illustrate, quantum mechanics does not generally predict a single, definite result for a single observation. It predicts, instead, a probability for each of all possible outcomes. Quantum mechanics thus brings into physics the idea of the essential randomness of nature. While it is the prevalent conceptual foundation in modern theory, as espoused by Bohr and others, this fundamental role of chance in our universe has not been universally acceptable to all scientists. Which atom will decay first? The contrasting viewpoint was held by Einstein, for example, summed up in the words, "God does not play dice."

Problems

1.1 The dimensions of angular momentum are those of momentum times distance. Show that Planck's constant, $h = 6.63 \times 10^{-34}$ J s, has the units of angular momentum.
1.2 Einstein's famous equation, $E = mc^2$, where c is the speed of light in a vacuum, gives the energy equivalence E of mass m. If m is expressed in kg and c in m s^{-1}, show that E is given in J.
1.3 According to classical theory, how are electromagnetic waves generated?
1.4 Why would the Bohr model of the atom be unstable, according to classical physics?
1.5 Calculate the wavelength of a 2.50-eV photon of visible light.
1.6 Calculate the wavelength of an electron having an energy of 250 eV.
1.7 What is the wavelength of a 1-MeV gamma ray?
1.8 If a neutron and an alpha particle have the same speed, how do their wavelengths compare?
1.9 If a neutron and an alpha particle have the same wavelength, how do their energies compare?
1.10 If a proton and an electron have the same wavelength, how do their momenta compare?
1.11 An electron moves freely along the X-axis. According to Eq. (1.6), if the uncertainty in its position in this direction is reduced by a factor of 2, how is the minimum uncertainty in its momentum in this direction affected?

1 The Statistical Nature of Radiation, Emission, and Interaction

1.12 Why is the uncertainty principle, so essential for understanding atomic physics, of no practical consequence for hitting a baseball?

1.13 Decay of the nuclide ^{226}Ra to the ground state of ^{222}Rn by emission of an alpha particle releases 4.88 MeV of energy.

   a) What fraction of the total mass available is thus converted into energy? (1 atomic mass unit = 931.49 MeV.)
   b) What is the initial energy of the ejected alpha particle?

1.14 Two conservation laws must be satisfied whenever a radioactive atom decays. As a result of these two conditions, the energies of the alpha particle and the recoil nucleus are uniquely determined in the two-body disintegration by alpha-particle emission. These two laws are also satisfied in beta decay, but do not suffice to determine uniquely the energy of any of the three decay products. What are these two laws, which thus require alpha-particle energies to be discrete and beta-particle energies to be continuous?

1.15 The fission of ^{235}U following capture of a thermal neutron releases an average of 195 MeV. What fraction of the total mass available (neutron plus uranium atom) is thus converted into energy? (1 atomic mass unit = 931.49 MeV.)

1.16 Five gamma rays are incident on a concrete slab. Each has a 95% chance of penetrating the slab without experiencing an interaction.

   a) What is the probability that the first three photons pass through the slab without interacting?
   b) What is the probability that all five get through without interacting?

1.17 a) In the last problem, what is the probability that photons 1, 2, and 3 penetrate the slab without interacting, while photons 4 and 5 do not?
   b) What is the probability that any three of the photons penetrate without interaction, while the other two do not?

1.18 Each photon in the last two problems has a binary fate – it either interacts in the slab or else goes through without interaction. A more detailed fate can be considered: 2/3 of the photons that interact do so by photoelectric absorption and 1/3 that interact do so by Compton scattering.

   a) What is the probability that an incident photon undergoes Compton scattering in the slab?
   b) What is the probability that it undergoes photoelectric absorption?
   c) What is the probability that an incident photon is not photoelectrically absorbed in the slab?

1.19 An atom of ^{42}K (half-life = 12.4 h) has a probability of 0.894 of surviving 2 h. For a source that consists of five atoms,

   a) what is the probability that all five will decay in 2 h and
   b) what is the probability that none of the five atoms will decay in 2 h?

1.20 What are the answers to (a) and (b) of the last problem for a source of 100 atoms?

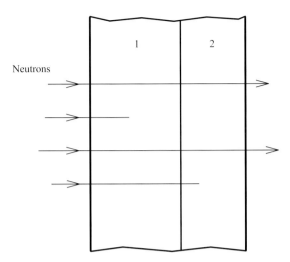

Figure 1.2 Neutrons normally incident on a pair of slabs, 1 and 2. See Problems 1.21–1.24.

1.21 Monoenergetic neutrons are normally incident on a pair of slabs, arranged back to back, as shown in Figure 1.2. A neutron either is absorbed in a slab or else goes through without interacting. The probability that a neutron gets through slab 1 is 1/3. If a neutron penetrates slab 1, then the probability that it gets through slab 2 is 1/4. What is the probability that a neutron, incident on the pair of slabs, will

a) traverse both slabs?
b) be absorbed in slab 1?
c) not be absorbed in slab 2?

1.22 If, in Figure 1.2, a neutron is normally incident from the right on slab 2, then what is the probability that it will

a) be absorbed in slab 1?
b) not be absorbed in slab 2?

1.23 For the conditions of Problem 1.21, calculate the probability that a neutron, normally incident from the left, will

a) not traverse both slabs,
b) not be absorbed in slab 1, and
c) be absorbed in slab 2.

1.24 What is the relationship among the three answers to the last problem and the corresponding answers to Problem 1.21?

2
Radioactive Decay

2.1
Scope of Chapter

This chapter deals with the random nature of radioactive decay. We begin by considering the following experiment. One prepares a source of a pure radionuclide and measures the number of disintegrations that occur during a fixed length of time t immediately thereafter. The procedure is then repeated over and over, exactly as before, with a large number of sources that are initially identical. The number of disintegrations that occur in the same fixed time t from the different sources will show a distribution of values, reflecting the random nature of the decay process. The objective of the experiment is to measure the statistical distribution of this number.

Poisson and normal statistics are often used to describe the distribution. However, as we shall see, this description is only an approximation, though often a very good one. The actual number of decays is described rigorously by another distribution, called the binomial.[1] In many applications in health physics, the binomial, Poisson, and normal statistics yield virtually indistinguishable results. Since the last two are usually more convenient to deal with mathematically, it is often a great advantage to employ one of them in place of the exact binomial formalism. This cannot always be done without large error, however, and one must then resort to the rigorous, but usually more cumbersome, binomial distribution. In Chapters 5 and 6, we shall address the conditions under which the use of one or another of the three distributions is justified.

In this chapter, we discuss radioactive disintegration from the familiar standpoint of the exponential decay of a pure radionuclide source, characterized by its half-life or, equivalently, its decay constant. We examine the relationship between activity and the number of atoms present and treat radioactive disintegration from the standpoint of

---

[1] Both the Poisson and normal distributions predict a nonzero probability for the decay of an arbitrarily large number of atoms from a source in any time t. In particular, the probability is not zero for the decay of more atoms than are in the source originally. The normal distribution, in addition, approximates the number of disintegrations as a continuous, rather than discrete, random variable.

the survival or decay probability in a specified time for each atom in a source. We shall arrive at the binomial distribution, which is confirmed by experiments like that described in the last paragraph.

2.2
Radioactive Disintegration – Exponential Decay

Exponential decay of a pure radionuclide source is a familiar concept. It is often discussed in the following way. In a short time dt, the change dN in a large number of atoms N in the source is proportional to N and to dt. The constant of proportionality is called the decay constant, λ, and one writes

$$dN = -\lambda N\, dt. \tag{2.1}$$

The negative sign indicates that N decreases as t increases. Integration gives

$$N = N_o\, e^{-\lambda t}, \tag{2.2}$$

where the constant of integration N_o represents the original number of atoms in the source at time $t=0$. Rewriting Eq. (2.1) as $\lambda = -(dN/N)/dt$, we see that the decay constant gives, at any moment, the fraction of atoms, dN/N, that decay per unit time. It thus represents the probability per unit time that a given atom will decay. The decay constant has the dimensions of reciprocal time (e.g., s^{-1}, h^{-1}).

The decay rate, or activity, of the source is

$$A = -\frac{dN}{dt} = \lambda N. \tag{2.3}$$

It follows from Eq. (2.2) that the activity as a function of time is given by

$$A = A_o\, e^{-\lambda t}, \tag{2.4}$$

where $A_o = \lambda N_o$ is the initial activity. The unit of activity is the becquerel (Bq), defined as the rate of one complete transformation of an atom per second: $1\,\text{Bq} = 1\,\text{s}^{-1}$. The older unit, curie (Ci), is now defined in terms of the becquerel: $1\,\text{Ci} \equiv 3.7 \times 10^{10}\,\text{Bq}$, exactly. It is the amount of activity associated with $1\,\text{g}$ of ^{226}Ra, as shown in an example in the next section.

In addition to its decay constant, a given radionuclide can also be characterized by its half-life. Figure 2.1 shows a plot of the relative activity $A/A_o = e^{-\lambda t}$ of a source, as follows from Eq. (2.4). The half-life $T_{1/2}$ of the nuclide is defined as the time needed for the activity (or the number of atoms) to decrease to one-half its value. Setting $A/A_o = e^{-\lambda T_{1/2}} = 1/2$ in Eq. (2.4) implies that

$$-\lambda T_{1/2} = \ln(1/2) = -\ln 2, \tag{2.5}$$

or

$$T_{1/2} = \frac{\ln 2}{\lambda} = \frac{0.693}{\lambda}. \tag{2.6}$$

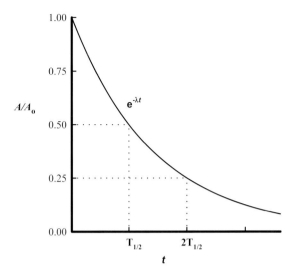

Figure 2.1 Plot of relative activity, A/A_o (=relative number of atoms, N/N_o), in a pure radionuclide source as a function of time t. The half-life $T_{1/2}$ is shown.

We thus obtain the relationship between the half-life and the decay constant. As can be seen from Figure 2.1, *starting at any time*, the activity decreases by a constant factor of 2 over successive half-lives.

The mean, or average, lifetime of an atom provides another way to characterize a radionuclide. The number of disintegrations that occur in a source during a certain time is equal to the product of the activity at the beginning of that time and the average lifetime of the atoms that decay during the time. When a source has decayed completely away, the total number of disintegrations will be equal to the number of atoms N_o originally present. This total number can also be regarded as the product of the original decay rate, A_o, and the average lifetime τ that an atom had in the source: $N_o = A_o \tau$. Since, as stated after Eq. (2.4), $A_o = \lambda N_o$, it follows that the mean life is equal to the reciprocal of the decay constant. Combining this result with Eq. (2.6), we write

$$\tau = \frac{N_o}{A_o} = \frac{1}{\lambda} = \frac{T_{1/2}}{\ln 2} = \frac{T_{1/2}}{0.693}. \tag{2.7}$$

The mean life is treated rigorously as a statistical average in Section 4.2.

■ *Example*
The radionuclide ^{32}P has a half-life of 14.3 d.

a) What is the decay constant?
b) What will be the activity of a 7.6-MBq source of ^{32}P after 1 y? (The nuclide decays by emission of a beta particle into stable ^{32}S.)
c) What is the mean life of a ^{32}P atom?

Solution

a) With $T_{1/2} = 14.3$ d, Eq. (2.6) gives

$$\lambda = \frac{0.693}{14.3 \text{ d}} = 0.0485 \text{ d}^{-1}. \tag{2.8}$$

b) From Eq. (2.4), with $A_o = 7.6 \times 10^6$ Bq, we find for the activity at time $t = 1 \text{ y} = 365$ d,

$$A = 7.6 \times 10^6 \, e^{-0.0485 \times 365} = 0.16 \text{ Bq}. \tag{2.9}$$

Because the exponent has to be a dimensionless number, λ and t in this equation must involve the same time unit. Here we expressed λ in d^{-1} and t in d.

c) The mean life is given by Eq. (2.7):

$$\tau = \frac{1}{\lambda} = \frac{1}{0.0485 \text{ d}^{-1}} = 20.6 \text{ d}. \tag{2.10}$$

As a check, we see that $T_{1/2}/\tau = 14.3/20.6 = 0.694$. To within roundoff, this ratio is equal to $\ln 2$, as required by Eq. (2.7).

2.3
Activity and Number of Atoms

The activity associated with a given radionuclide source depends on the number of atoms present and the decay constant, as related by Eq. (2.3). The relative strengths of different sources can be expressed in terms of their specific activity, defined as the disintegration rate per unit mass of the nuclide. Examples of units for specific activity are Bq kg^{-1} and Ci g^{-1}.

The specific activity of a radionuclide can be calculated from its gram atomic weight M and decay constant λ. Since M grams of the nuclide contain Avogadro's number of atoms, $N_A = 6.0221 \times 10^{23}$, the number of atoms in 1 g is

$$N = \frac{N_A}{M} = \frac{6.02 \times 10^{23}}{M}. \tag{2.11}$$

It follows that the specific activity of the radionuclide can be written as

$$S = \lambda N = \frac{\lambda N_A}{M} = \frac{6.02 \times 10^{23} \lambda}{M}. \tag{2.12}$$

Since M is in g, if λ is in s^{-1}, then this expression gives S in Bq g^{-1}. Alternatively, using Eq. (2.6) to replace λ by the half-life $T_{1/2}$, we write

$$S = \frac{N_A \ln 2}{MT_{1/2}} = \frac{4.17 \times 10^{23}}{MT_{1/2}}. \qquad (2.13)$$

With $T_{1/2}$ in s and M in g, the specific activity in Eq. (2.13) is in Bq g^{-1}.

■ *Example*

a) How many atoms are there in 1 mg of ^{226}Ra? The nuclide has a half-life of 1600 y and a gram atomic weight of 226 g.
b) Calculate the specific activity of ^{226}Ra in Bq g^{-1}.

Solution

a) With $M = 226$ g, Eq. (2.11) gives for the number of atoms in 1 mg

$$N = 10^{-3} \times \frac{6.02 \times 10^{23}}{266} = 2.66 \times 10^{18}. \qquad (2.14)$$

b) To obtain S in Bq g^{-1}, we use either Eq. (2.12) with λ in s^{-1} or Eq. (2.13) with $T_{1/2}$ in s. Choosing the latter and writing $T_{1/2} = 1600$ y $\times$ (365 d y^{-1}) $\times$ (86 400 s d^{-1}) $= 5.05 \times 10^{10}$ s, we find that

$$S = \frac{4.17 \times 10^{23}}{(226 \text{ g}) \times (5.05 \times 10^{10} \text{ s})} = 3.7 \times 10^{10} \text{ Bq g}^{-1}$$
$$= 1 \text{ Ci g}^{-1}. \qquad (2.15)$$

As mentioned after Eq. (2.4), 1 Ci $= 3.7 \times 10^{10}$ Bq exactly, by definition. The curie was originally defined as the activity of 1 g of ^{226}Ra. This fact leads to a simple formula for calculating the specific activity of other radionuclides in these units. As seen from Eq. (2.13), specific activity is inversely proportional to the half-life $T_{1/2}$ and the gram atomic weight M of a radionuclide. Comparing with ^{226}Ra, one can compute the specific activity of a nuclide by writing

$$S = \frac{1600}{T_{1/2}} \times \frac{226}{M} \text{ Ci g}^{-1}, \qquad (2.16)$$

where $T_{1/2}$ is its half-life in years.

■ *Example*

a) Calculate the specific activity of ^{60}Co (half-life 5.27 y) exactly from Eq. (2.13).
b) Calculate the approximate value from Eq. (2.16) and compare the answer with that from (a).
c) How many atoms of ^{60}Co are there in a 1-Ci source?

Solution

a) The gram atomic weight is $M = 60$ g. Expressing the half-life in s, $T_{1/2} = 5.27$ y $\times$ 365 d y^{-1} $\times$ 86 400 s d^{-1} = 1.66×10^8 s, we find from (2.13)

$$S = \frac{4.17 \times 10^{23}}{MT_{1/2}} = \frac{4.17 \times 10^{23}}{60 \text{ g} \times 1.66 \times 10^8 \text{ s}}$$

$$= 4.19 \times 10^{13} \text{ Bq g}^{-1}, \quad (2.17)$$

where the replacement, 1 Bq = 1 s^{-1}, has been made.

b) From Eq. (2.16), which requires expressing $T_{1/2}$ in y, we obtain

$$S \cong \frac{1600}{5.27} \times \frac{226}{60} = 1.14 \times 10^3 \text{ Ci g}^{-1}. \quad (2.18)$$

Converting to the same units as in part (a), we find

$$S \cong 1.14 \times 10^3 \text{ Ci g}^{-1} \times 3.7 \times 10^{10} \text{ Bq Ci}^{-1}$$
$$= 4.23 \times 10^{13} \text{ Bq g}^{-1}. \quad (2.19)$$

Comparison with Eq. (2.17) shows that the approximate formula gives the correct result to within about 4 parts in 400, or 1%.

c) Using Eqs. (2.3) and (2.6), we have for the number of atoms in a 1-Ci source ($A = 1$ Ci $= 3.7 \times 10^{10}$ s^{-1})

$$N = \frac{A}{\lambda} = \frac{AT_{1/2}}{0.693} = \frac{(3.7 \times 10^{10} \text{ s}^{-1})(1.66 \times 10^8 \text{ s})}{0.693}$$

$$= 8.86 \times 10^{18}. \quad (2.20)$$

2.4
Survival and Decay Probabilities of Atoms

In the experiment proposed at the beginning of this chapter, one can ask what the probability is for a given atom in a pure radionuclide source to survive or decay in the time t. When the number of atoms in the source is large, Eq. (2.2) gives for the fraction of undecayed atoms at time t, $N/N_o = e^{-\lambda t}$. Therefore, the probability q that a given atom will not decay in time t is just equal to this fraction:

$$q = e^{-\lambda t} \quad \text{(survival probability).} \quad (2.21)$$

The probability that a given atom will decay sometime during t is, then,

$$p = 1 - q = 1 - e^{-\lambda t} \quad \text{(decay probability).} \quad (2.22)$$

2.4 Survival and Decay Probabilities of Atoms

The fact that a given atom either decays or does not decay during time t is expressed by the requirement that $p + q = 1$. (Probability will be formally defined in Section 3.4.)

■ **Example**
What is the probability that an atom of ^{60}Co will decay in 100 d?

Solution
From the value of the half-life given in the last example, we find for the decay constant, $\lambda = 0.693/T_{1/2} = 0.693/(5.27 \text{ y} \times 365 \text{ d y}^{-1}) = 3.60 \times 10^{-4} \text{ d}^{-1}$. The probability for decay in a time $t = 100$ d is, therefore,

$$p = 1 - e^{-\lambda t} = 1 - e^{-3.60 \times 10^{-4} \times 100} = 0.0354. \tag{2.23}$$

■ **Example**
The radionuclide ^{222}Rn has a half-life of 3.82 d.

a) What is the probability that a given ^{222}Rn atom in a source will not decay in 1 wk?
b) If a ^{222}Rn atom survives for 1 wk, what is the probability that it will survive a second week?
c) What is the probability that a given ^{222}Rn atom will survive for 2 wk?
d) How are the probabilities in (a)–(c) related?
e) What is the probability that a given ^{222}Rn atom will survive the first week and then decay during the second week?

Solution

a) The survival probability is given by Eq. (2.21). The decay constant is, from Eq. (2.6),

$$\lambda = \frac{0.693}{T_{1/2}} = \frac{0.693}{3.82 \text{ d}} = 0.181 \text{ d}^{-1}. \tag{2.24}$$

It follows that the survival probability $q(7)$ for a time $t = 1$ wk $= 7$ d is

$$q(7) = e^{-\lambda t} = e^{-0.181 \times 7} = 0.282. \tag{2.25}$$

b) As far as is known, all ^{222}Rn atoms are identical and the decay is completely spontaneous and random. It is assumed that the survival probability into the future for a given atom at any moment is independent of how long that atom might have already existed. Given that the atom has survived the first week, the probability that it will survive the second week is the same as that for the first week: $q(7) = 0.282$. This example illustrates conditional probability, which is discussed in Section 3.5.

c) Like (a), the probability that a given atom will not decay in a time $t = 2$ wk $= 14$ d is

$$q(14) = e^{-\lambda t} = e^{-0.181 \times 14} = 0.0793. \tag{2.26}$$

d) An alternative way of answering the question asked in part (c) is the following. The probability for a given atom to survive 2 wk is equal to the probability that it survives for 1 wk times the probability that it survives again for 1 wk. From (a) and (b),

$$q(14) = q(7)q(7) = [q(7)]^2 = (0.282)^2 = 0.0795, \qquad (2.27)$$

which is the same result as in Eq. (2.26) to within roundoff. The probability in part (c) is thus equal to the product of the probabilities from parts (a) and (b). The equality results from the independence of the events, as we shall see in Section 3.5.

e) The probability for an atom of ^{222}Rn to decay in a week's time is (Eq. (2.22)) $p(7) = 1 - q(7) = 0.718$. The probability that a given atom will decay during week 2 is equal to the probability that it survives 1 wk times the probability that it then decays in the next week:

$$q(7)p(7) = 0.282 \times 0.718 = 0.202. \qquad (2.28)$$

This last example illustrates how probabilities can be assigned to various events, or possible alternative outcomes, for a set of observations.

2.5
Number of Disintegrations – The Binomial Distribution

We return once more to the experiment introduced at the beginning of this chapter. A large number of identical sources of a pure radionuclide, each containing exactly N atoms initially, are prepared. The number of disintegrations that occur in a time t is observed for each source, starting at time $t = 0$. One thus obtains a distribution for the number k of atoms that decay, with the possible values $k = 0, 1, 2, \ldots, N$. We can think of each source as undergoing a process in which, during the time t, each individual atom represents a single one of N trials to decay or not decay. Decay of an atom can be called a "success" and non-decay, a "failure." The outcome of each trial is binary. Since all of the atoms are identical and independent, each trial has the same probability of success or failure, and its outcome is independent of the other trials.

In statistics, the process just described for radioactive decay is called a Bernoulli process. The resulting number of disintegrations from source to source is described by the binomial distribution, which will be discussed more completely in Chapter 5. For now, we examine several examples to further illustrate the statistical nature of radioactive decay.

■ *Example*
A source that consists initially of 15 atoms of ^{11}C (half-life = 20.4 min) is observed for 5 min.

a) What is the probability that atoms 1, 2, and 7 in the source will decay in this time?

2.5 Number of Disintegrations – The Binomial Distribution

b) What is the probability that *only* these three atoms decay in this time?
c) What is the probability that exactly three atoms (any three) decay in 5 min?

Solution

a) The survival and decay probabilities for each atom in the source are given by Eqs. (2.21) and (2.22), respectively. The decay constant is $\lambda = 0.693/T_{1/2} = 0.693/(20.4 \text{ min}) = 0.0340 \text{ min}^{-1}$. The probability that a given atom will not decay in the allotted time, $t = 5$ min, is

$$q = e^{-\lambda t} = e^{-0.0340 \times 5} = 0.844. \tag{2.29}$$

The decay probability for each atom is then

$$p = 1 - q = 0.156. \tag{2.30}$$

The probability that the particular atoms, 1, 2, and 7, will decay in 5 min is, therefore,

$$p^3 = (0.156)^3 = 3.80 \times 10^{-3}. \tag{2.31}$$

Since the decay of each atom is an independent event, their probabilities multiply (Section 3.5).

b) Part (a) says nothing about the other atoms. If *only* the specified three decay, then the other 12 survive for the time $t = 5$ min, the probability being q^{12}. Therefore, the probability that *only* atoms 1, 2, and 7 decay is

$$p^3 q^{12} = (0.156)^3 (0.844)^{12} = 4.96 \times 10^{-4}. \tag{2.32}$$

c) In this part, we are asked for the probability that exactly three atoms decay, and they can be any three. The answer will be $p^3 q^{12}$ times the number of ways that the three atoms can be chosen from among the 15, without regard for the *order* in which the three are selected. The decay of atoms 1, 2, and 7 in that order, for example, is not to be distinguished from their decay in the order 2, 1, and 7. Both ways are registered as three disintegrations. To select *any* three, there are 15 independent choices for the first atom to decay, 14 for the second, and 13 for the third atom. The total number of ways in which three atoms can be chosen from 15 is, therefore, $15 \times 14 \times 13 = 2730$. Among this number, those that differ only in the order of their selection are redundant and are not to be counted as different outcomes. Therefore, we must divide the total number of choices by the number of ways (permutations) that the three can be arranged, namely, $3 \times 2 \times 1 \equiv 3! = 6$. With the help of Eq. (2.32), we find for the probability that exactly three disintegrations will occur in 5 min

$$\Pr(3) = \frac{15 \times 14 \times 13}{3!} p^3 q^{12} = \frac{2730}{6} (4.96 \times 10^{-4}) = 0.226. \tag{2.33}$$

We can generalize the results just found to calculate the probability for k disintegrations in a time t from a source consisting initially of N identical radioactive atoms. For the decay of exactly k atoms we write, in analogy with Eq. (2.33),

$$\Pr(k) = \frac{N \times (N-1) \times \cdots \times (N-k+1)}{k!} p^k q^{N-k}, \qquad (2.34)$$

where p and q are given by Eqs. (2.22) and (2.21), respectively. The coefficient in Eq. (2.34) represents the number of permutations of N distinct objects taken k at a time. In more common notation, we write it in the form (read "N choose k")

$$\binom{N}{k} = \frac{N(N-1)\cdots(N-k+1)}{k!}. \qquad (2.35)$$

Multiplying the numerator and denominator by $(N-k)!$, we can write the alternative expression

$$\binom{N}{k} = \frac{N(N-1)\cdots(N-k+1)}{k!} \times \frac{(N-k)!}{(N-k)!} = \frac{N!}{k!(N-k)!}. \qquad (2.36)$$

The probability (2.34) for exactly k disintegrations then has the compact form

$$\Pr(k) = \binom{N}{k} p^k q^{N-k}, \qquad (2.37)$$

where $k = 0, 1, 2, \ldots, N$. This result is the binomial distribution, which takes its name from the series expansion for a binomial. With $p + q = 1$,

$$(p+q)^N = \sum_{k=0}^{N} \binom{N}{k} p^k q^{N-k} = \sum_{k=0}^{N} \Pr(k) = 1, \qquad (2.38)$$

showing that the distribution is normalized. This distribution and its properties will be discussed in Section 5.4.

The binomial distribution from the last example, with $N = 15$ and $p = 0.156$, is represented in Figure 2.2. The bars show the probability that $k = 0, 1, 2, \ldots, 15$ of the original ^{11}C atoms in the source will decay in 5 min. The answer $\Pr(3) = 0.226$ (Eq. (2.33)) can be seen as the fourth bar. The function $\Pr(k) = 0$ when $k < 0$ and $k > N$. The sum of the probabilities, represented by the total height of all the bars, is unity.

■ **Example**

a) In the last example, compute the probability that none of the original 15 ^{11}C atoms will decay in 5 min.
b) What is the probability that no atoms will decay in 5 min if the source consists of 100 atoms?
c) 1000 atoms?

2.5 Number of Disintegrations – The Binomial Distribution

Solution

a) With the help of Eqs. (2.37) and (2.36), we find that

$$\Pr(0) = \frac{15!}{0!15!} p^0 q^{15} = q^{15} = (0.844)^{15} = 0.0785. \qquad (2.39)$$

With 15 atoms initially present, there is thus a 7.85% chance that none will decay in 5 min, a time equal to about one-fourth the half-life. This result can also be seen in Figure 2.2.

b) With $N = 100$,

$$\Pr(0) = q^N = (0.844)^{100} = 4.31 \times 10^{-8}, \qquad (2.40)$$

or about one chance in 23 million.

c) With $N = 1000$, the probability that none will decay in 5 min is

$$\Pr(0) = (0.844)^{1000} = 2.20 \times 10^{-74}. \qquad (2.41)$$

Note that, with 10 times the number of atoms in part (c) compared with (b), the probability (2.41) is the 10th power of that in Eq. (2.40), that is, $q^{1000} = (q^{100})^{10}$.

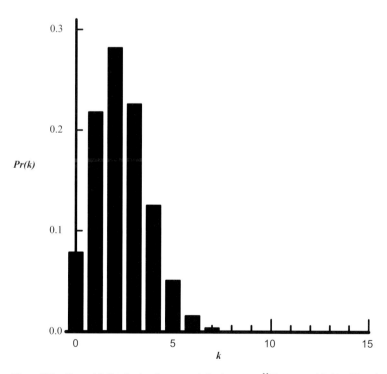

Figure 2.2 Binomial distribution for example in the text, a ^{11}C source with $N = 15$ and $p = 0.156$.

The next example illustrates how calculations with the binomial distribution rapidly get cumbersome with increasing numbers of atoms.

■ *Example*

A source consists of 2800 atoms of ^{24}Na, which has a decay constant $\lambda = 0.0462 \, \text{h}^{-1}$. What is the probability that exactly 60 atoms will disintegrate in 30 min?

Solution

The solution is given by Eq. (2.37) with $N = 2800$ and $k = 60$. For time $t = 30 \, \text{min} = 0.5 \, \text{h}$, we have $\lambda t = (0.0462 \, \text{h}^{-1})(0.5 \, \text{h}) = 0.0231$. The survival and decay probabilities are $q = e^{-\lambda t} = e^{-0.0231} = 0.9772$ and $p = 0.0228$. The factor involving these probabilities in Eq. (2.37) is

$$p^k q^{N-k} = (0.0228)^{60} (0.9772)^{2740}$$
$$= (2.99 \times 10^{-99})(3.59 \times 10^{-28}) = 1.07 \times 10^{-126}. \quad (2.42)$$

The binomial coefficient (2.36) involves the enormous number $N! = 2800!$, which is out of the range of most hand calculators. However, we can use Eq. (2.35), which has smaller numbers. There are 60 factors approximately equal to 2800 in the numerator. Thus, we obtain, approximately, from Eq. (2.35)

$$\binom{N}{k} = \binom{2800}{60} = \frac{(2800)(2799)(2798) \cdots (2741)}{60!}$$
$$\cong \frac{(2800)^{60}}{60!} = \frac{6.75 \times 10^{206}}{8.32 \times 10^{81}} = 8.12 \times 10^{124}. \quad (2.43)$$

Substituting the last two factors into Eq. (2.37), we obtain

$$\text{Pr}(60) = 8.12 \times 10^{124} \times 1.07 \times 10^{-126} = 0.0869. \quad (2.44)$$

Computations with the binomial distribution are not feasible for many or even most common sources dealt with in health physics. A 1-Ci source of ^{60}Co, for instance, contains $N = 8.86 \times 10^{18}$ atoms of the radionuclide (Eq. (2.20)). Fortunately, convenient and very good approximations to the binomial distribution exist in the form of the Poisson and the normal distributions for making many routine calculations. These will be discussed in Chapters 5 and 6.

2.6
Critique

The description of the exponential decay of a radionuclide presented in this chapter often provides an accurate and useful model for radioactive decay. However, it cannot be strictly valid. In carrying out the derivations for the number of atoms and the activity for a source as functions of time in Section 2.2, the discrete number of atoms N was treated as a continuous, differentiable function of the time. The analysis thus tacitly requires that N be very large, so that its behavior can be approximated by that of a continuous variable. However, the number of atoms in a source and the number that decay during any given time are discrete, rather than continuous. Furthermore, the decay always shows fluctuations, in contrast to what is implied by Eq. (2.3). As we did in arriving at the binomial distribution, Eq. (2.37), a rigorous description must treat the number of disintegrations as a discrete random variable.

Problems

2.1 ^{222}Rn has a half-life of 3.82 d.
   a) What is the value of the decay constant?
   b) What is the mean life?
   c) How much ^{222}Rn activity will remain from a 1.48×10^8-Bq source after 30 d?

2.2 The half-life of ^{32}P is 14.3 d. How long does it take for the activity of a ^{32}P source to decay to 0.1% of its initial value?

2.3 a) Calculate the number of ^{60}Co atoms in a 30-mCi source (half-life = 5.27 y).
   b) Calculate the mass of ^{60}Co in this source (atomic weight of ^{60}Co = 59.934).
   c) What is the specific activity of ^{60}Co in mCi g^{-1}?
   d) What is the specific activity in Bq kg^{-1}?

2.4 a) What is the decay constant of ^{238}U (half-life = 4.47×10^9 y)?
   b) What is its specific activity in Bq kg^{-1}?
   c) In Ci g^{-1}?

2.5 Calculate the specific activity of ^{3}H (half-life = 12.3 y) in (a) Bq kg^{-1} and (b) Ci g^{-1}.

2.6 Modify the formula (2.13) to give S in units of
   a) Ci kg^{-1};
   b) Bq g^{-1} when $T_{1/2}$ is expressed in years.

2.7 What mass of ^{238}U (half-life = 4.47×10^9 y) has the same activity as 1 g of ^{3}H?

2.8 A source is to be prepared with a radioisotope, having a mean life of 5.00 h. The expected value of the number of disintegrations in 3.00 h is to be as close as possible to 20.4.
   a) How many atoms should there be initially in the source?
   b) If such a source has 100 atoms initially, what is the probability that there would be exactly one undecayed atom left after five half-lives?

2.9 How many disintegrations occur in 24 h with a source of ^{24}Na (half-life = 15.0 h), having an initial activity of 4.79 mCi?

2.10 The half-life of ^{32}P is 14.3 d.
   a) What is the probability that an atom of ^{32}P will not decay within 3 wk?
   b) What is the probability that an atom of ^{32}P will decay during the fourth week?

c) If an atom has not decayed in the first 3 wk, what is the probability that it will decay during the fourth week?

2.11 A source consists of 15 atoms of ^{85}Y, having a half-life of 5.0 h.
   a) What is the probability that no atoms will decay in 4 h?
   b) What is the probability that 10 atoms will decay in 4 h?

2.12 A source consists of 10 atoms of ^{32}P, having a decay constant of 0.0485 d^{-1}.
   a) What is the probability that exactly 2 atoms will decay in 12 d?
   b) If the source consists originally of 50 atoms, what is the probability that exactly 10 atoms will decay in 12 d?
   c) Why are the answers to (a) and (b) different, even though they are the probabilities for the decay of 20% of the original atoms?

2.13 a) What is the probability that exactly 3 atoms of ^{11}C (half-life = 20.4 min) will decay in 4 min from a source that has initially 1128 atoms?
   b) What is the probability that no more that 3 atoms will decay in 4 min?
   c) How is the probability in (b) related to the probability that at least 4 of the 1128 atoms will decay in 4 min?

2.14 A source consists of 12 atoms of ^{24}Na (half-life = 15.0 h) and 10 atoms of ^{42}K (half-life = 12.4 h).
   a) What is the probability that exactly 2 atoms of ^{24}Na and exactly 2 atoms of ^{42}K will decay in the first 5 h?
   b) If exactly 6 atoms of ^{24}Na decay in 5 h, what is the probability that exactly 2 atoms of ^{42}K will decay during this time?

2.15 a) In the last problem, what is the probability that exactly three disintegrations will occur in the source in the first 5 h?
   b) What is the probability that only one atom in the source remains undecayed after 100 h?

2.16 For Figure 2.2, calculate Pr(k) for (a) k = 5, (b) k = 10, and (c) k = 15.

2.17 The half-life of ^{11}C is 0.0142 d, and the decay constant is $\lambda = 48.8$ d^{-1}. Since λ represents the probability per unit time that an atom of ^{11}C will decay, how can its numerical value exceed unity.

2.18 For the last example in the text (Eqs.), write as a sum over k an exact formula that, when evaluated, would give the probability that $50 \leq k \leq 150$ atoms will disintegrate in 30 min.

3
Sample Space, Events, and Probability

The previous chapter illustrated how the random nature of radioactive decay can be treated mathematically by means of the binomial distribution. With the present chapter we begin the development of a number of formal concepts needed as a foundation for the statistical treatments of the subjects in this book.

3.1
Sample Space

The word "experiment" is used by statisticians to describe any process that generates a set of data. An example is the experiment introduced at the beginning of Chapter 2. The raw data from the experiment are the numbers of disintegrations from a series of identical radioactive sources in the specified time t. Each observation gives a nonnegative integer, 0, 1, 2, ... The data are generated randomly in the sense that, although we know all the possible outcomes, the result of any given observation is governed by chance and cannot be predicted with certainty ahead of time. We define the set of all possible outcomes of an experiment as the *sample space*, which we denote by the symbol S.

Definition 3.1
The *sample space* S consists of all possible outcomes of an experiment.

Each of the possible outcomes is called an *individual element* of the sample space, which is defined by the experiment. In the example just considered, the sample space consists of all nonnegative integers up through the fixed number of atoms initially in a source. The individual elements are the integers themselves.

To establish the notation that describes sample spaces, we first consider a single radioactive atom that we observe for some stated time period to see if it decays. The sample space S for this experiment thus consists of two individual elements, which we denote by d or n, where d is short for decay and n is short for no decay. We describe the sample space for the experiment by writing

$$S = \{d, n\}. \tag{3.1}$$

Statistical Methods in Radiation Physics, First Edition. James E. Turner, Darryl J. Downing, and James S. Bogard.
© 2012 Wiley-VCH Verlag GmbH & Co. KGaA. Published 2012 by Wiley-VCH Verlag GmbH & Co. KGaA.

We next consider two atoms present, each having the same two possible outcomes – either d or n in the stated time – and our experiment is to observe the fate of both of them. The sample space now consists of four possible outcomes, or individual elements, which we denote in parentheses by writing for the sample space

$$S = \{(d,d), (d,n), (n,d), (n,n)\}. \tag{3.2}$$

For each individual element, enclosed in parentheses, the two symbols refer to the fate of the first and second atoms, respectively. Thus, (d, n), for instance, indicates the decay of the first atom and survival of the second. For a system consisting of an arbitrary number N of atoms, we let a_i denote the fate of the ith atom (d or n). We generalize Eq. (3.2) and write for the sample space,

$$S = \{(a_1, a_2, \ldots, a_N) | a_i = d \text{ or } n, \text{ for } i = 1, 2, \ldots, N\}. \tag{3.3}$$

The vertical bar stands for "such that" and this statement is read "S is the set consisting of N variables $(a_1, a_2, \ldots, a_N)$ such that each a_i is either d or n." This sample space applies to the experiment in which the fate of each atom in the stated time is described. Since each a_i represents one of two alternatives, there are 2^N individual elements in the sample space of this experiment.

Performing a different experiment will generally change the sample space, even for the same system under study. Returning to the system of two atoms, we can observe the number of atoms that decay in the stated time, rather than the fate of each. The sample space S then consists of three integers,

$$S = \{0, 1, 2\}, \tag{3.4}$$

in which the individual elements 0, 1, 2 describe all possible outcomes for the number of atoms that can decay. Sample spaces can thus be different for experiments that may be similar, but with outcomes recorded in different ways. We note, also, that each individual element in the sample space (3.2) is associated uniquely with one of the individual elements in Eq. (3.4), but the reverse is not true. Whereas (d, n) in Eq. (3.2) corresponds to the element 1 in Eq. (3.4), the latter in Eq. (3.4) corresponds to both (d, n) and (n, d) in Eq. (3.2). Thus, for the same system under observation, some sample spaces can evidently contain more information than others, depending on what experiment or observation is being made. Some examples of different sample spaces follow.

■ *Example*

An experiment consists of flipping a coin and recording the face that lands up, and then tossing a die and recording the number of dots on the up face. Write an expression for the sample space. How many individual elements are there in the sample space?

Solution

We can represent any outcome of the experiment by writing a pair of symbols (a_1, a_2), where a_1 denotes the result of the coin toss and a_2 denotes the result of the die toss. Specifically, we let a_1 be either H (heads) or T (tails) and a_2 be an integer from 1 to 6. Then we may write S as

$$S = \{(H,1),(H,2),(H,3),(H,4),(H,5),(H,6),(T,1),(T,2),(T,3),(T,4),(T,5),(T,6)\}. \tag{3.5}$$

There are thus 12 individual elements in the sample space of this experiment.

■ *Example*
Three solder connections on a circuit board are examined to see whether each is good (G) or defective (D). Describe the sample space.

Solution
We let the triple (a_1, a_2, a_3) denote the outcome for each of the three solder connections. Each a_i takes on the value G or D, and the sample space S can be written as

$$S = \{(G,G,G),(G,G,D),(G,D,G),(D,G,G),(G,D,D),(D,G,D),(D,D,G),(D,D,D)\}. \tag{3.6}$$

■ *Example*
Describe the sample space for recording the number of defective solder connections in the last example.

Solution
Although the system is the same as before, scoring the number of defective connections is a different experiment from seeing whether each is good or bad. The individual elements now are the integers 0 through 3. The sample space is

$$S = \{0, 1, 2, 3\}. \tag{3.7}$$

■ *Example*
Six slips of paper are numbered 1 through 6 and placed in a box. The experiment consists of drawing a slip of paper from the box and recording the number that appears on it. Write an expression for the sample space.

Solution
The sample space consists simply of the numbers 1 through 6:

$$S = \{1, 2, 3, 4, 5, 6\}. \tag{3.8}$$

■ *Example*
With the same setup as in the last example, the experiment now consists of drawing two slips of paper in succession and recording the numbers. The first slip is not replaced before the second is drawn. Write an expression for the sample space. How many individual elements are there?

Solution

We let the pair (a_1, a_2) represent the individual elements of the sample space, where a_1 denotes the number on the first slip drawn and a_2 denotes the number on the second slip. Then we can write for the sample space,

$$S = \{(1,2), (1,3), (1,4), (1,5), (1,6), (2,1), (2,3), (2,4), (2,5), (2,6),$$
$$(3,1), (3,2), (3,4), (3,5), (3,6), (4,1), (4,2), (4,3), (4,5), (4,6),$$
$$(5,1), (5,2), (5,3), (5,4), (5,6), (6,1), (6,2), (6,3), (6,4), (6,5)\}. \quad (3.9)$$

There are thus 30 individual elements, or possible outcomes, for this experiment. Note that the first slip of paper is not replaced before drawing the second slip, and so $a_1 \neq a_2$ for all of the possible outcomes in Eq. (3.9).

■ Example

We perform an experiment as in the previous example, except that the first slip is now returned to the box before the second slip is drawn. Describe the sample space. How many individual elements does it have?

Solution

This experiment differs from the last one, and the first slip now has a chance of being drawn again. The individual elements with $a_1 = a_2$ are now to be added to those expressed by Eq. (3.9). The new sample space is

$$S = \{(1,1), (1,2), (1,3), (1,4), (1,5), (1,6), (2,1), (2,2), (2,3), (2,4), (2,5), (2,6),$$
$$(3,1), (3,2), (3,3), (3,4), (3,5), (3,6), (4,1), (4,2), (4,3), (4,4), (4,5), (4,6),$$
$$(5,1), (5,2), (5,3), (5,4), (5,5), (5,6), (6,1), (6,2), (6,3), (6,4), (6,5), (6,6)\}.$$
$$(3.10)$$

There are now 36 individual elements.

■ Example

Describe the sample spaces (3.9) and (3.10) for the two experiments in a compact form like Eq. (3.3).

Solution

In place of Eq. (3.9), for which the first slip is not returned to the box, we write

$$S = \{(a_1, a_2) | a_i = 1, 2, \ldots, 6, \text{ for } i = 1, 2 \text{ and } a_1 \neq a_2\}. \quad (3.11)$$

In place of Eq. (3.10),

$$S = \{(a_1, a_2) | a_i = 1, 2, \ldots, 6, \text{ for } i = 1, 2\}. \quad (3.12)$$

Thus far, we have considered only discrete sample spaces, in which all of the possible events can be enumerated. Sample spaces can also be continuous. An example is the continuous sample space generated by the time at which an atom in a radionuclide source decays. As governed by Eq. (2.22), the time of decay t can have any value in the interval $0 \leq t < \infty$. This continuous sample space can be expressed by writing

$$S = \{t | t \in [0, \infty)\}. \tag{3.13}$$

This statement is read, "The sample space S consists of all times t such that t is contained in the semiclosed interval $[0, \infty)$." Like any continuous sample space, Eq. (3.13) has an infinite number of subsets. Discrete sample spaces can be finite or infinite and can also have an infinite number of subsets. For example, the energy levels E_n of the electron in the bound, negative-energy states of the hydrogen atom are quantized and have the values $E_n = -13.6/n^2$ eV, where $n = 1, 2, \ldots$ is any positive integer. This set of levels is discrete and countably infinite in number, with an infinite number of subsets. In addition, the unbound, positive-energy states of the electron have a continuum of values over the energy interval $[0, \infty)$. The sample space for all energies that the electron can have consists of (1) a countably infinite, discrete set of negative numbers and (2) the continuum of all nonnegative numbers.

We turn now to the concept of an *event* and related ideas that are used in statistics. As we shall see, probability theory is concerned with events and the notion of the likelihood of their occurrence.

3.2 Events

Definition 3.2
An *event* is any subset of a sample space.

Using this terminology, we also refer to the individual elements in a sample space as the *simple events*. For an experiment where the flip of a coin will result in heads (H) or tails (T), the sample space is $S = \{H, T\}$ and the simple events are H and T. One can observe the decay or survival of each of the two atoms considered before, there being four simple events, shown explicitly by Eq. (3.2). An example of an event that is not a simple event is the decay of either atom 1 or atom 2, but not both. This event is comprised of the two simple events (d, n) and (n, d), which is a subset of Eq. (3.2). Another event is the decay of either one or two atoms, which is comprised of the first three simple events in Eq. (3.2). Generally, a sample space consists of the union of all simple events.

We usually denote an event by a capital letter, for example, A, B, or C. An event can consist of each of the simple events in the sample space, any subset of them, or none of them. If the event consists of no elements of the sample space, it is called the *empty set* or *null set*, and is denoted by the symbol $\emptyset$.

These ideas can be further illustrated by considering an ordinary deck of playing cards. Drawing a card will produce one of a set of 52 different designations that identify each card. This set, which cannot be subdivided further, constitutes the 52 simple events in the sample space for drawing a card. An example of an event is drawing a queen. We can designate this event by writing

$$A = \{QS, QH, QD, QC\}, \tag{3.14}$$

where the symbols represent, respectively, the queen of spades, hearts, diamonds, and clubs. In this case, the event that a drawn card is a queen is comprised of four simple events.

Two events, A and B, associated with a sample space can have various relationships with one another. The *intersection* of two events, denoted by $A \cap B$, is the event consisting of all elements common to A and B. The *union* of two events, denoted by $A \cup B$, is the event that contains all of the elements that belong to A or B or to both. Two events are *mutually exclusive* if they have no elements in common, in which case their intersection is the empty set and one writes $A \cap B = \emptyset$. One can also consider an event A that contains some of the elements of the sample space S. The set A' that consists of all of the elements in S that are not in A is called the *complement* of A. The complement set is always taken with respect to the sample space S. Since, by definition, A and A' have no common elements, it follows that their intersection is the empty set: $A \cap A' = \emptyset$; and their union is the whole space: $A \cup A' = S$.

■ **Example**

A source consisting of four atoms of ^{11}C is observed for 5 min to see which, if any, of the atoms decay during this time. Let A be the event that atoms 1 and 2 decay, B be the event that only these two atoms decay, and C the event that exactly three atoms (any three) decay, all events within the 5 min.

a) Write the statement that designates the sample space.
b) Write expressions for the events A, B, and C.
c) What event is the intersection of A and B, $A \cap B$?
d) The union of A and B, $A \cup B$?
e) The complement B' of B?
f) How many events are possible in the sample space?

Solution

a) The whole sample space consists of $2^4 = 16$ simple events, corresponding to each atom and its two possible outcomes, decay (d) or not (n). We denote this space by writing

$$S = \{(a_1, a_2, a_3, a_4) | a_i = d \text{ or } n, \text{ for } i = 1, 2, 3, 4\}. \tag{3.15}$$

Alternatively, we show the 16 simple events of the sample space explicitly by writing

$$S = \{(d, d, d, d), (d, d, d, n), (d, d, n, d), (d, n, d, d), (n, d, d, d), (d, d, n, n),$$
$$(d, n, d, n), (n, d, d, n), (d, n, n, d), (n, d, n, d), (n, n, d, d),$$
$$(d, n, n, n), (n, d, n, n), (n, n, d, n), (n, n, n, d), (n, n, n, n)\}. \tag{3.16}$$

b) The event A that atoms 1 and 2 decay consists of all of the simple events in Eq. (3.15) or Eq. (3.16) for which $a_1 = a_2 = d$ and the other a_i can be either d or n:

$$A = \{(d, d, d, d), (d, d, d, n), (d, d, n, d), (d, d, n, n)\}. \tag{3.17}$$

The event that *only* atoms 1 and 2 decay implies that the other two do not, and so

$$B = \{(d, d, n, n)\}. \tag{3.18}$$

In this case, B itself is a simple event, the fates of all four atoms being specified. When exactly (any) three atoms decay, the event is

$$C = \{(d, d, d, n), (d, d, n, d), (d, n, d, d), (n, d, d, d)\}. \tag{3.19}$$

c) The event that is the intersection of A and B consists of all events that are common to both A and B. Since B is one of the elements of A, we see that the intersection is just the event B itself:

$$A \cap B = \{(d, d, n, n)\} = B. \tag{3.20}$$

d) The union of A and B consists of all of the events that are in A or B or both. Since B is already contained in A, the union is equivalent to A itself. Thus,

$$A \cup B = \{(d, d, d, d), (d, d, d, n), (d, d, n, d), (d, d, n, n)\}$$
$$= A. \tag{3.21}$$

e) The complement of an event is always taken with respect to the whole space, S. Thus, event B's complement, which is defined as the set of all simple events in S that are not in B, is the union of all the simple events in S except (d, d, n, n).

f) For a given event that is a subset of the sample space, each of the 16 simple events in S has two possibilities: it is either part of the event or not. Therefore, there are 2^{16} possible events in all.

The notion of events, intersections, and unions can be visualized graphically by the use of *Venn diagrams*. Figure 3.1 illustrates a sample space S that contains two events,

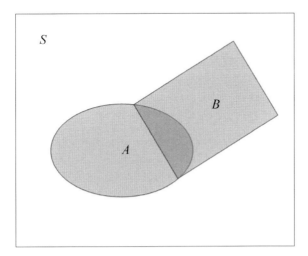

Figure 3.1 Example of Venn diagram. See the text. The dark shaded area is the intersection $A \cap B$ of two events, A and B, in a sample space S. The union of the three mutually exclusive regions $A \cap B'$, $A \cap B$, and $A' \cap B$ is the total space represented by the union of the two events, $A \cup B$.

A and B, shown as intersecting geometrical figures. The overlapping, darkly shaded area represents the intersection, $A \cap B$, of events A and B. The union, $A \cup B$, of the two events is all of the space occupied by the two figures together. The lightly shaded area inside A is $A \cap B'$ – that is, the intersection of A and the complement of B (those elements of S not in B). Likewise, the lightly shaded portion in B is the intersection, $A' \cap B$. The three pieces $A \cap B'$, $A \cap B$, and $A' \cap B$ are mutually exclusive, and their union is $A \cup B$. Venn diagrams are often useful in representing events and their subsets.

3.3
Random Variables

Definition 3.3
A *random variable* is a function that maps a sample space onto a set of real numbers.

The random variable thus associates a real number with each element in the sample space. We denote a random variable by an uppercase letter, for example, X, and the numerical values that it can have by the corresponding lowercase letter, that is, x. For example, the number of disintegrations that can occur in a given time with a radioactive source, initially containing N atoms, is a random variable, say X. The values that X can have are $x = 0, 1, \ldots, N$. These integers are the real numbers that the sample space is mapped onto by the number of disintegrations X (the random variable). We designate the probability of, say, 10 disintegrations occurring by writing $\Pr(X = 10)$. The probability for an unspecified number x of disintegrations is designated $\Pr(X = x)$. This formalism, which we adopt from now on, replaces some of our earlier notation, for example, Eq. (2.37).

A random variable is said to be *discrete* if it can take on a finite or countably infinite number of values. It is called *continuous* if it can take on the infinite number of values associated with intervals of real numbers. The number of counts from a radioactive sample in a specified time is a discrete random variable. The time of decay of an atom is a continuous random variable.

3.4
Probability of an Event

When performing an experiment that defines a discrete sample space, one is often interested in the likelihood, or *probability*, of a given outcome. Before presenting a formal definition of probability, we can see intuitively how it can be structured from the concepts we have developed up to now.

To illustrate, if a sample space consists of a set of N simple events that represent equally likely outcomes, then the probability that a random simple event will be a particular one of the set is $1/N$. Thus, the probability that the roll of an unbiased die will yield a three is $1/6$, there being six equally likely simple events. The probability for rolling either a three or a five, for example, which is not a simple event, is the sum of the probabilities for the two simple events: $1/6 + 1/6 = 1/3$. The probabilities for the different simple events in the sample space need not be the same, but their sum must be unity. If the die is biased in such a way that the likelihood of getting a three is twice that for getting any one of the other five (equally likely) numbers, then the probability for rolling a three can be represented by $2p$, where p is the probability for any of the others. Since the sum of the probabilities for the six simple events must be unity, we have $2p + 5p = 7p = 1$, so that $p = 1/7$. The probability of rolling a three now is $2p = 2/7$. The probability of getting a three or a five is $2p + p = 3p = 3/7$. The total probability for getting *some* number, one through six, is $2/7 + 5/7 = 1$. Furthermore, the probability for an event outside the sample space, such as rolling a seven with the die, is zero.

With this introduction, we now define probability for discrete sample spaces. (Continuous sample spaces will be treated in Section 4.1 in terms of probability density functions.) We consider an experiment that has an associated sample space S, comprised of n simple events, $E_1, E_2, \ldots, E_n$. We note that S is the union of all of the simple events by writing $S = \bigcup_{i=1}^{n} E_i$.

Definition 3.4

A *probability* is a numerically valued function that assigns to every event A in S a real number, Pr(A), such that the following axioms hold:

$$\Pr(A) \geq 0. \tag{3.22}$$

$$\Pr(S) = 1. \tag{3.23}$$

If A and B are mutually exclusive events in S, then $\Pr(A \cup B)$
$$= \Pr(A) + \Pr(B). \tag{3.24}$$

Axiom (3.22) states that the probability of every event must be nonnegative. Axiom (3.23) corresponds to the fact that the probability of the union of *all* simple events that make up the sample space must equal unity. Thus, the axiom is equivalent to saying that at least one event must occur, that is, that the probability for an event outside the sample space is zero: $\Pr(\emptyset) = 0$. Finally, Eq. (3.24) states that the probability of the occurrence of any two events, having no simple events in common, is the sum of their respective probabilities.

■ *Example*

A source consists of three identical radioactive atoms. What is the probability that either atom 1 or atom 3 will be the first of the three atoms to decay?

Solution

We let E_1 be the event that atom 1 decays first, E_2 the event that atom 2 decays first, and E_3 the event that atom 3 decays first. Since the atoms are identical, equal probabilities $\Pr(E_1) = \Pr(E_2) = \Pr(E_3) = 1/3$ are assigned to each of these events, their sum being unity as required by Eq. (3.23). The events E_1, E_2, and E_3 constitute the simple events in the sample space. Let A be the event that atom 1 or atom 3 is the first of the three to decay. We then write

$$A = E_1 \cup E_3. \tag{3.25}$$

Since E_1 and E_3 are mutually exclusive, we also have

$$E_1 \cap E_3 = \emptyset. \tag{3.26}$$

By axiom (3.24),

$$\Pr(A) = \Pr(E_1 \cup E_3) = \Pr(E_1) + \Pr(E_3) = \frac{1}{3} + \frac{1}{3} = \frac{2}{3}. \tag{3.27}$$

Thus, the chances are two out of three that either atom 1 or atom 3 will be the first to decay.

In the last example we see a trivial application of axiom (3.24). The important point of this example is the fact that, once we can express the event of interest as the union of simple events, the problem is solved if we can assign the probabilities to each of the simple events. We note that axiom (3.24) can be extended to the case where we have n mutually exclusive events, $A_1, A_2, \ldots, A_n$, in S. Then

$$\Pr(A_1 \cup A_2 \cup \cdots \cup A_n) = \sum_{i=1}^{n} \Pr(A_i). \tag{3.28}$$

Equation (3.28) provides a way to calculate the probability of an event that is the union of a set of simple events.

3.5
Conditional and Independent Events

Another important concept in probability theory and statistics is that of *conditional probability*. The term "conditional" is used to describe the occurrence of one event A, given that some other event B has already occurred. The notation for the conditional probability is $\Pr(A|B)$, which is read, "the probability of A given that B has occurred."

■ *Example*
A radioactive source consisted initially of 20 atoms of ^{131}I and 20 atoms of ^{32}P. The numbers of each that were observed to decay or not over a subsequent 5-day period are shown in Table 3.1. One of the 40 atoms is selected at random from the table. Let A be the event that the atom is ^{131}I and B be the event that the atom has decayed. For such a random selection, what is the probability, $\Pr(A|B)$, that the atom is ^{131}I, given that the atom decayed?

Solution
Table 3.1 shows that, of the 10 decayed atoms, 6 were ^{131}I. Hence,

$$\Pr(A|B) = \frac{6}{10}. \tag{3.29}$$

The table summarizes the complete sample space of the 40 simple events that describe the possible outcomes (decay or not) for all 40 atoms in the source. The solution (3.29) is clear when we look in the table at the reduced sample space,

Table 3.1 Decay of atoms over 5 d in a source consisting initially of 20 atoms each of ^{131}I and ^{32}P.

Status after 5 d	Number of atoms		
	^{131}I	^{32}P	Total
Decayed	6	4	10
Not decayed	14	16	30
Total	20	20	40

See example in the text.

corresponding to the event B that the atom decayed. In essence, given B, the sample space then contains only the first row of numbers in Table 3.1. Since our selection is random, each of the 10 first-row atoms has the same probability of being selected, that is, 1/10. Since six of these correspond to decayed ^{131}I atoms, the probability (3.29) is 6/10. However, one does not need to use the reduced sample space. Employing the complete space, we can write instead of Eq. (3.29),

$$\Pr(A|B) = \frac{6}{10} = \frac{6/40}{10/40} = \frac{\Pr(A \cap B)}{\Pr(B)}, \qquad (3.30)$$

where $\Pr(A \cap B)$ and $\Pr(B)$ are found from the *original* sample space. Thus, one can use either the original sample space or the subspace resulting from the conditional event to calculate conditional probabilities. Using a subspace, one always assigns to the elements probabilities that are proportional to the original probabilities and that add to unity.

The following is a formal definition of conditional probability.

Definition 3.5

The *conditional probability* of A given B, denoted by $\Pr(A|B)$, is defined as

$$\Pr(A|B) = \frac{\Pr(A \cap B)}{\Pr(B)}, \quad \text{with } \Pr(B) > 0. \qquad (3.31)$$

If $\Pr(B) = 0$, then $\Pr(A|B)$ is undefined. Since the intersection $A \cap B = B \cap A$ is the same, it also follows from the definition (3.31) that the conditional probability of B given A is $\Pr(B|A) = \Pr(A \cap B)/\Pr(A)$, with $\Pr(A) > 0$.

■ *Example*
The possible relationship between smoking and lung cancer is under investigation. In one study, 500 people were examined for lung cancer, and the results are reported in Table 3.2. Let *Sm* denote the event that a person, randomly selected in the study, is a smoker and let *C* denote the event that the individual selected has lung cancer. What is the probability that this person has lung cancer, given that he or she smokes?

Solution

Using the definition (3.31) of conditional probability, we write

$$\Pr(C|Sm) = \frac{\Pr(C \cap Sm)}{\Pr(Sm)} = \frac{25/500}{100/500} = 0.25. \qquad (3.32)$$

Thus, the probability that a randomly selected person in the study has lung cancer, given that the person smokes, is 0.25

■ Example

A source consists of three identical radioactive atoms at time zero. At the end of 1 d, it is found that a single atom has decayed. What is the probability that either atom 1 or atom 3 decayed?

Solution

The system here is the same as that in the last example of the last section. However, in this setting we are interested in the fate of each atom and not the first to decay. The statement "a single atom has decayed" describes a conditioning event. Thus, we need to investigate the conditional probability of either atom 1 or atom 3 decaying, given that a single atom decayed. As before, we represent the sample space S for the decay or not of the three atoms by writing for the eight simple events

$$S = \{(n,n,n), (d,n,n), (n,d,n), (n,n,d), (d,d,n), (d,n,d), (n,d,d), (d,d,d)\}. \qquad (3.33)$$

Letting A be the event that only atom 1 or only atom 3 decayed, we write

$$A = \{(d,n,n), (n,n,d)\}. \qquad (3.34)$$

The event B that a single atom decayed can be written as

$$B = \{(d,n,n), (n,d,n), (n,n,d)\}. \qquad (3.35)$$

The probability of A, given B, is obtained by applying Eq. (3.31). The intersection of events A and B in the numerator of Eq. (3.31) is seen from Eqs. (3.34) and (3.35) to be the same as the union of the two simple events

Table 3.2 Conditional probability for the effect of smoking on lung cancer incidence.

	Number of persons		
	Lung cancer	No lung cancer	Total
Smokers	25	75	100
Nonsmokers	5	395	400
Total	30	470	500

See example in the text.

(n, n, d) and (d, n, n). Event B in the denominator of Eq. (3.31) is the union of the three simple events in Eq. (3.33) that represent a single decay. Thus,

$$\Pr(A|B) = \frac{\Pr(A \cap B)}{\Pr(B)} = \frac{\Pr[(d, n, n) \cup (n, n, d)]}{\Pr[(d, n, n) \cup (n, d, n) \cup (n, n, d)]}. \qquad (3.36)$$

Applying axiom (3.24) for the probabilities of mutually exclusive events (either atom 1 or atom 3 decays, but not both), we have

$$\Pr(A|B) = \frac{\Pr[(d, n, n)] + \Pr[(n, n, d)]}{\Pr[(d, n, n)] + \Pr[(n, d, n)] + \Pr[(n, n, d)]}. \qquad (3.37)$$

Although we do not know the numerical value, the individual probabilities here for exactly one atom to decay are assumed to be equal, because the three atoms are identical. Therefore, the ratio (3.37) gives $\Pr(A|B) = 2/3$. It is interesting that this method gives the same result found earlier when we asked for the probability that either atom 1 or atom 3 would be the first to decay. The reduced sample space caused by the conditional event of one atom decaying is equivalent to the sample space generated by the experiment to observe the first atom to decay.

Conditional probability allows one to adjust the probability of the event under consideration in the light of other information. In the last example, we had no knowledge of what the actual probabilities were for the eight simple events in the sample space (3.33). We did not know, for instance, the probability $\Pr[(d, n, n)]$ that only atom 1 would decay. However, expression of the conditional probability effectively selected a set of simple events from Eq. (3.33) for which the individual probabilities, though unknown, were assumed to be equal. We thus were able to obtain the numerical answer. The other simple events in Eq. (3.33) generally have probabilities different from those in Eq. (3.37). In the preceding example on smoking and lung cancer (Table 3.2), one would calculate $\Pr(C) = 30/500 = 0.060$. However, given the additional information that the person was a smoker, the probability is adjusted to $25/100 = 0.25$, a considerable change.

The idea of conditional probability can be extended to the case where we may have several subevents that can occur. In the above example, the sample space was split between smokers and nonsmokers. There are situations where we may have several splits of the sample space, that is, events $A_1, A_2, \ldots, A_k$ that partition the whole sample space. These events are mutually exclusive and exhaustive, by which we mean that

$$S = \bigcup_{i=1}^{k} A_i \quad \text{and} \quad A_i \cap A_j = \emptyset \quad \text{for all } i \neq j. \qquad (3.38)$$

Since S can be partitioned completely into k disjoint sets, the following theorem holds. It is often called the *theorem of total probability*, but also goes by the name of the *rule of elimination*.

Theorem of Total Probability, or Rule of Elimination

If the events $A_1, A_2, \ldots, A_k$ constitute a mutually exclusive and exhaustive partition of the sample space S, then, for any event B in S, we have

$$\Pr(B) = \Pr(B \cap S) = \sum_{i=1}^{k} \Pr(B \cap A_i) = \sum_{i=1}^{k} \Pr(A_i)\Pr(B|A_i). \tag{3.39}$$

The proof of this theorem lies in the fact that B can be seen to be the union of k mutually exclusive events $B \cap A_1, B \cap A_2, \ldots, B \cap A_k$. By Eq. (3.28),

$$P(B) = \Pr(B \cap A_1) + \Pr(B \cap A_2) + \cdots + \Pr(B \cap A_k). \tag{3.40}$$

Next we simply apply Eq. (3.31) to each term, using it in the form of a product rather than a ratio. We obtain

$$P(B) = \Pr(A_1)\Pr(B|A_1) + \Pr(A_2)\Pr(B|A_2) + \cdots + \Pr(A_k)\Pr(B|A_k), \tag{3.41}$$

thus completing the proof.

The following example is somewhat contrived, but clearly shows the usefulness of the theorem.

> ■ **Example**
> An urn contains 5 red balls, 6 black balls, and 10 white balls. A ball is selected at random and set aside without noting its color. If a second ball is now selected, what is the probability that it is red?
>
> **Solution**
> Without knowing the color of the first ball selected, we consider three mutually exclusive and exhaustive events for the result of the first draw. These events are
>
> $A_1 =$ a red ball is selected on the first draw.
> $A_2 =$ a black ball is selected on the first draw.
> $A_3 =$ a white ball is selected on the first draw.
>
> We let B be the event that a red ball is selected on the second draw, and so we are asked to find $\Pr(B)$. Using the theorem of total probability, Eq. (3.39), we obtain
>
> $$P(B) = \Pr(A_1)\Pr(B|A_1) + \Pr(A_2)\Pr(B|A_2) + \Pr(A_3)\Pr(B|A_3) \tag{3.42}$$
>
> $$= \frac{5}{21}\frac{4}{20} + \frac{6}{21}\frac{5}{20} + \frac{10}{21}\frac{5}{20} = \frac{100}{420} = 0.238. \tag{3.43}$$

This example also suggests other questions. What is the probability that a red ball was the result of the first draw, given that red was found on the second draw? Is the probability of getting red on the first draw related to the particular event of the second draw? Is there cause and effect? (The earlier example concerning lung cancer and

3.5 Conditional and Independent Events

smoking is more appropriate in this setting, where we might ask what is the probability that a person is a smoker given that he or she has lung cancer.) Questions like these can be answered by the following rule, which is called *Bayes' theorem*.

Bayes' Theorem

If the events $A_1, A_2, \ldots, A_k$ form a mutually exclusive and exhaustive partition of the sample space S, and if B is any non-null event in S, then

$$\Pr(A_j|B) = \frac{\Pr(A_j \cap B)}{\Pr(B)} = \frac{\Pr(A_j)\Pr(B|A_j)}{\sum_{i=1}^{k} \Pr(A_i)\Pr(B|A_i)}. \tag{3.44}$$

The first equality is just the definition (3.31), which also leads to the substitution $\Pr(A_j|B) = \Pr(A_j)\Pr(B|A_j)$ in the numerator of the second equality. In the denominator, $\Pr(B)$ is, by the theorem of total probability (Eq. (3.39)), the same as the denominator in the last equality of Eq. (3.44), thus proving Bayes' theorem for discrete events. The theorem will be extended to continuous random variables in Section 4.8.

■ *Example*

An urn contains five black and seven red balls. One ball is drawn out at random. It is then put back into the urn along with three additional balls of the same color. A second ball is randomly drawn, and it is red. What is the probability that the first ball drawn was black?

Solution
As always, it is important to specify events clearly. We denote the two possible events that could occur on the first draw and their probabilities as follows:

A_1 = black ball drawn and $\Pr(A_1) = 5/12$.
A_2 = red ball drawn and $\Pr(A_2) = 7/12$.

We note that these events are mutually exclusive and exhaustive. We let B represent the event that the second ball drawn was red. We are asked to find $\Pr(A_1|B)$, the probability that the first ball was black (A_1) given that the second was red (B). The two conditional probabilities in Eq. (3.44) with $k = 2$ still need to be assigned. The probability $\Pr(B|A_1)$ that the second ball was red given that the first was black is just $7/15$, which would be the fraction of red balls in the urn after three more black balls were added. Similarly, $\Pr(B|A_2) = 10/15$ after drawing a red first and making the additions. Thus, from Eq. (3.44) it follows that

$$\Pr(A_1|B) = \frac{\Pr(A_1)\Pr(B|A_1)}{\Pr(A_1)\Pr(B|A_1) + \Pr(A_2)\Pr(B|A_2)} \tag{3.45}$$

$$= \frac{(5/12)(7/15)}{(5/12)(7/15) + (7/12)(10/15)} = \frac{1}{1+2} = \frac{1}{3}. \tag{3.46}$$

In some cases, given additional information will cause no change in the probability of the event occurring. Then, in symbols, $\Pr(A|B) = \Pr(A)$, and so the occurrence of B has no effect on the probability of the occurrence of A. We then say that event A is *independent* of event B. The lung cancer and smoking example is one in which there may be a dependence between the two events. In contrast, the events cancer and the height of a person would logically be considered to be independent. The concept of independence in statistics is defined as follows.

Definition 3.6
Two events A and B are independent if and only if

$$\Pr(A|B) = \Pr(A) \tag{3.47}$$

and

$$\Pr(B|A) = \Pr(B). \tag{3.48}$$

Otherwise A and B are dependent.

The definitions of independence and conditional probability can be used together to derive the *multiplicative rule*.

The General Multiplicative Rule
If A and B are two events in a sample space, then

$$\Pr(A \cap B) = \Pr(A)\Pr(B|A) = \Pr(B)\Pr(A|B). \tag{3.49}$$

This rule is a direct consequence of the definition (3.31) of conditional probability (Problem 3.19). The next theorem is a result of applying the definition of independence to the conditional probability statements in Eq. (3.49).

Independence Theorem
Two events, A and B, in a sample space are independent if and only if

$$\Pr(A \cap B) = \Pr(A)\Pr(B). \tag{3.50}$$

This result can be seen by noting that, for independent events, $\Pr(B|A) = \Pr(B)$ and $\Pr(A|B) = \Pr(A)$ and then applying Eq. (3.49). In general, for any set of n independent events, A_i, $i = 1, 2, \ldots, n$, it follows that (see Problem 3.20)

$$\Pr(A_1 \cap A_2 \cap \cdots \cap A_n) = \Pr(A_1)\Pr(A_2) \cdots \Pr(A_n). \tag{3.51}$$

■ *Example*

Two photons of a given energy are normally incident on a metal foil. The probability that a given photon will have an interaction in the foil is 0.2. Otherwise, it passes through without interacting. What are the probabilities that neither photon, only one photon, or both photons will interact in the foil?

Solution

The number of photons that interact in the foil is a random variable X, which can take on the possible values 0, 1, or 2. Similar to Eq. (3.2) for the decay or not of the two atoms in Section 3.1, there are four simple events for the sample space for the two photons: (n, n), (n, y), (y, n), (y, y). Here y means "yes, there is an interaction" and n means "no, there is not," the pair of symbols in parentheses denoting the respective fates of the two photons. The probability of interaction for each photon is given as 0.2, and so the probability for its having no interaction is 0.8. We are asked to find the probabilities for the three possible values of X. For the probability that neither photon interacts, we write

$$\Pr(X = 0) = \Pr[(n, n)]. \tag{3.52}$$

We can regard n for photon 1 and n for photon 2 as independent events, each having a probability of 0.8. By Eq. (3.50), the probability that neither photon interacts is, therefore,

$$\Pr(X = 0) = 0.8 \times 0.8 = 0.64. \tag{3.53}$$

The probability that exactly one photon interacts is

$$\Pr(X = 1) = \Pr[(y, n) \cup (n, y)] = \Pr[(y, n)] + \Pr[(n, y)], \tag{3.54}$$

in which the last equality makes use of Eq. (3.24). Since the probability of "yes" for a photon is 0.2 and that for "no" is 0.8, we find

$$\Pr(X = 1) = 0.2 \times 0.8 + 0.8 \times 0.2 = 0.32. \tag{3.55}$$

The probability that both photons interact is

$$\Pr(X = 2) = \Pr[(y, y)] = 0.2 \times 0.2 = 0.04. \tag{3.56}$$

This example shows how the random variable X maps the sample space of simple events onto a set of real numbers, with a probability attached to each. It demonstrates how important the idea of independence is in being able to work problems. We see, in addition, that axiom (3.23) in the definition of probability is satisfied, that is, $\Pr(X=0) + \Pr(X=1) + \Pr(X=2) = 1$.

Problems

3.1 Does the sample space (3.5) apply to an experiment in which the coin and/or die is biased?

3.2 The experiment leading to Eq. (3.9) is modified so that only the identities of the two numbers drawn are scored, without regard to the order in which they are drawn. Thus, for example, (2, 5) and (5, 2) are regarded as the same individual element, or simple event.

a) Write an expression for the new sample space.
b) How many individual elements are there now?

3.3 An experiment is to be performed in which a 12-sided, unbiased die is thrown and the number of dots on the up face is observed.
a) Write an expression that describes the sample space.
b) How is the sample space affected if the die is biased?

3.4 Using the sample space of the last problem, describe the following events:
a) A, that the up face has an even number of dots;
b) B, that the up face has an odd number of dots; and
c) C, that the number of dots on the up face is a multiple of three.

3.5 Using events A, B, and C from the last problem, find
a) $A \cap C$;
b) $B \cap C$;
c) $A \cap B$;
d) $B \cup C$;
e) $A \cup B$;
f) B'.

3.6 An experiment consists of recording the times t at which identical radioactive atoms in a source decay.
a) Describe the sample space.
b) Does the sample space change if the atoms are not identical?

3.7 In the last problem, let A be the event that the time to decay is at least 10 min and B be the event that the disintegration time is between 5 and 20 min.
a) Write the events A and B in symbolic notation.
b) Are A and B mutually exclusive events? Why or why not?
c) Describe the event $A \cap B$.

3.8 Identify the shaded areas for the Venn diagrams in Figure 3.2.

3.9 Show by using Venn diagrams that the following rules hold:
a) $(E \cup F)' = E' \cap F'$.
b) $(E \cap F)' = E' \cup F'$.
(Note: The rules can be generalized to any number of events. They are sometimes referred to as De Morgan's laws.)

3.10 For an event A in a sample space S, verify the following statements by means of Venn diagrams:
a) $A' \cap A = \emptyset$;
b) $A' \cup A = S$;
c) $S' = \emptyset$;

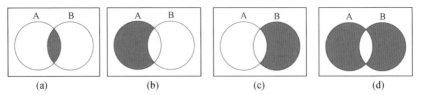

(a) (b) (c) (d)

Figure 3.2 See Problem 3.8. (Courtesy of Steven E. Smith.)

d) $A \cup \emptyset = A$;
e) $A \cap \emptyset = \emptyset$.

3.11 Draw a Venn diagram and shade in the appropriate region for each of the following events:
   a) $A \cup B$;
   b) $(A \cup B)'$;
   c) $A' \cap B$;
   d) $A \cap B'$;
   e) $(A' \cap B) \cup (A' \cap B')$;
   f) $A' \cap B'$;
   g) $(A \cap B)'$;
   h) $A' \cup B'$.

3.12 a) For two mutually exclusive events, A and B in a sample space S, represent Eq. (3.24) by means of a Venn diagram.
   b) If A and B are *any* two events in S, draw a Venn diagram to show the additivity rule for their union,
   $$\Pr(A \cup B) = \Pr(A) + \Pr(B) - \Pr(A \cap B).$$

3.13 a) Calculate Pr(A) in place of Eq. (3.27) if atom 3 is four times as likely to be the first to decay as atom 1 or 2, the latter pair having equal probabilities, as before.
   b) Same as (a), except that atom 2 has four times the likelihood of being first, compared with atoms 1 and 3, having equal probabilities.

3.14 Four identical radioactive atoms are observed to see which ones decay over a 10-min time period. The probability that a given atom will decay in this time is 0.25. For the four atoms, let (x_1, x_2, x_3, x_4) represent a point in the sample space for the experiment, where $x_i = n$ or d for $i = 1, 2, 3, 4$ and n = no decay and d = decay.
   a) Write an expression that describes the sample space.
   b) How many simple events are there in the sample space?
   c) List the items in the sample space that represent the decay of a single atom.
   d) Determine the probability of the event (n, d, d, d) by using the concept of independence and the given probability of decay.
   e) Determine the probability of each simple event in the sample space.
   f) Show that the sum of the probabilities for the sample space is unity. Which of the axioms (3.22)–(3.24) support this result?
   g) Calculate the probabilities for the events that 0, 1, 2, 3, or 4 atoms decay.
   h) Graph the results from (g), letting the ordinate denote the probability and the abscissa the number of atoms that decay.

3.15 An experiment consists of drawing a card randomly from a deck of 52 playing cards.
   a) What is the probability of drawing a black card?
   b) What assumptions are necessary to determine this probability?
   c) Given that the drawn card is red, what is the probability that it is a diamond?

d) When two cards are drawn from the full deck without replacing the first, what is the probability that the second card is black?
(*Hint*: The first card drawn must be either black or red, B or R. Thus, $\Pr(B$ on 2nd draw$) = \Pr(B$ on 2nd$|R$ on 1st$) \times \Pr(R$ on 1st$) + \Pr(B$ on 2nd$|B$ on 1st$) \times \Pr(B$ on 1st$)$.)

3.16 Refer to the example immediately following Bayes' theorem (Eq. (3.44)). Show that, if the urn originally contains b black and r red balls, and if c balls of the proper color are added after the first draw, then the probability is $b/(b + r + c)$.

3.17 A ball is drawn at random from an urn that contains three white and four black balls. The drawn ball is then placed into a second urn, which contains five white and three black balls. A ball is then randomly selected from the second urn.
a) What is the probability that the ball drawn from the second urn is white, given that the ball taken from the first urn is white?
b) What is the probability that the ball drawn from the second urn is white, given that the ball taken from the first urn is black?
c) What is the probability that the ball drawn from the second urn is white?
d) How are the answers to (a), (b), and (c) related?

3.18 A porch is illuminated with two identical, independent lights. If each light has a failure probability of 0.004 on any given evening, then
a) what is the probability that both lights fail?
b) what is the probability that neither light fails?
c) what is the probability that only one light fails?
d) what is the sum of these probabilities?

3.19 Prove Eq. (3.49), the general multiplicative rule.

3.20 Equation (3.50) states that, if two events A and B are independent, then the probability of their joint occurrence (their intersection) is equal to the product of their individual probabilities: $\Pr(A \cap B) = \Pr(A)\Pr(B)$. For three independent events A, B, and C, prove that

$$\Pr(A \cap B \cap C) = \Pr(A)\Pr(B)\Pr(C).$$

(*Hint*: Consider $B \cap C = D$ and apply Eq. (3.50) twice.)

3.21 With an unbiased pair of dice, what is the probability of rolling (a) 7, (b) 11, (c) 7 or 11, (d) 2, and (e) 2, 11, or 12?

3.22 Write an expression that describes the sample space in the last problem.

3.23 What is the probability of drawing from a well-shuffled, regular deck of 52 playing cards (a) a black card; (b) a red ace; (c) a face card; (d) a jack; (e) a black king or a red ace; (f) a 5, 6, or 7?

3.24 For a certain airline flight, experience shows that the probability for all passengers to be at the departure gate on time is 0.95, the probability for the flight to arrive on time is 0.93, and the probability for the flight to arrive on time

given that all passengers are at the departure gate on time is 0.97. Find the probability that
a) The flight will arrive on time and all passengers will be at the gate on time.
b) All passengers were at the gate on time given that the flight arrived on time.
c) Write an expression for the sample space.

3.25 A gamma-ray spectrometer is used to screen a series of samples for the presence of a certain radioisotope. The instrument will detect the isotope 99% of the time when present. It will also give a "false positive" result 1% of the time when the isotope is not there.
a) If 0.5% of the samples being screened contain the radioisotope, what is the probability that a given sample contains the isotope, given that the spectrometer indicates that it does?

(*Hint*: Let R denote the event that the isotope is present in the sample and E the event that the instrument indicates that it is. Then what is asked for is $Pr(R|E)$.)
b) Describe the sample space.

3.26 A fair coin is tossed three times.
a) What is assumed about the outcome of each toss?
b) What is the probability of obtaining three heads?
c) What is the probability of at most two heads?
d) What is the relationship between the events in parts (b) and (c)?
e) If the first two tosses result in heads, what is the probability that the third toss will give heads? Why?

3.27 One box contains three white and two black marbles, and a second box contains two white and four black marbles. One of the two boxes is selected at random, and a marble is randomly withdrawn from it.
a) What is the probability that the withdrawn marble is white?
b) Given that the withdrawn marble is white, what is the probability that box 1 was chosen?

4
Probability Distributions and Transformations

4.1
Probability Distributions

In the previous chapter, we saw how probabilities can be associated with values that a random variable takes on in a discrete sample space. The random variable X in the example at the end of the previous chapter had the three possible values, $x = 0, 1$, and 2. Corresponding to each, a probability $Pr(X = x)$ was assigned (Eqs. (3.53), (3.55), and (3.56)). The set of ordered pairs, $(x, Pr(X = x))$, is an example of a probability distribution. It associates a probability with each value of X. To simplify notation, we shall write $f(x)$ in place of $Pr(X = x)$.

Definition 4.1
The set of ordered pairs $(x, f(x))$ is called the *probability distribution* for the discrete random variable X if, for each possible outcome x,

$$Pr(X = x) = f(x), \qquad (4.1)$$

$$f(x) \geq 0, \qquad (4.2)$$

and

$$\sum_{\text{all } x} f(x) = 1. \qquad (4.3)$$

Equation (4.1) defines the shortened notation for probability, and Eq. (4.2) requires that probabilities be nonnegative, consistent with Eq. (3.22). The last relationship, Eq. (4.3), states that the sum of the probabilities over all the possible events must be unity, consistent with Eq. (3.23). Implicit in the definition is the fact that $f(x) = 0$ for all values of x that are not possible values of the random variable X.

Statistical Methods in Radiation Physics, First Edition. James E. Turner, Darryl J. Downing, and James S. Bogard
© 2012 Wiley-VCH Verlag GmbH & Co. KGaA. Published 2012 by Wiley-VCH Verlag GmbH & Co. KGaA.

■ *Example*

a) Write the ordered pairs $(x, f(x))$, giving the probability distribution for the number of photons that interact in a foil when two are incident, as considered in the last example of Chapter 3.
b) Show that condition (4.3) in the definition of a probability distribution is satisfied.

Solution

a) The probability distribution obtained from Eqs. (3.53), (3.55), and (3.56) for the number X of photons that interact can be represented by writing the ordered pairs $(x, f(x))$ as given in Table 4.1. The distribution is also shown by the plot in Figure 4.1.
b) From Table 4.1, we have

$$\sum_{\text{all } x} f(x) = 0.64 + 0.32 + 0.04 = 1.00, \tag{4.4}$$

showing that Eq. (4.3) is satisfied, as required of any discrete probability distribution.

In Chapter 1, we discussed how quantum mechanics provides a probabilistic, rather than deterministic, description of atomic and radiative phenomena. We can relate that discussion to the last example, in which two photons are incident on a foil. The basic physical factor used to interpret the observations is that the probability is $p = 0.2$ that a given photon will interact in the foil (and hence the probability is 0.8 that it will not). The numerical value of p can be determined with good precision by an experiment in which a pencil beam of photons of a given energy is directed normally at the foil. Under "good geometry" conditions,[1] the measured fraction of photons that pass through without interaction gives the value of p. Also, depending on the photon energy and the particular material of the foil, the probability p can often be

Table 4.1 Probability distribution for the number of photons that interact in a foil, each with probability 0.2, when two photons are incident.

x	f(x)
0	0.64
1	0.32
2	0.04

See example in the text.

---

1) See, for example, Section 8.7 of Turner (2007) in the Bibliography.

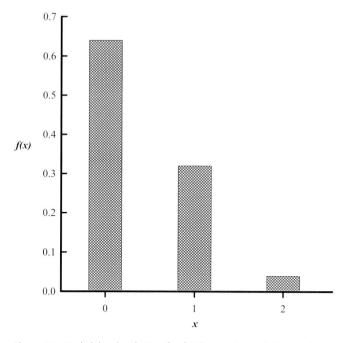

Figure 4.1 Probability distribution $f(x)$ for the number x of photons that interact in a foil when two are incident. See example in the text.

calculated from quantum mechanics. As discussed in Chapter 1, however, knowing the numerical value of p does not tell us what a given photon will do.

The two-photon problem illustrates how predictions are made in terms of probability distributions. From the basic description of $p=0.2$, the function $f(x)$ that we determined above gives the probabilities for $x=0$, 1, and 2. The predictions can be checked experimentally by bombarding the foil with a large number of photons in pairs and observing the fraction of pairs for which $x=0$, 1, or 2.

One is often interested not only in a probability distribution function $f(x)$, but also in the probability that the random variable X has a value less than or equal to some real number x. Such a probability is described by a cumulative distribution function. We denote the cumulative distribution function $F(x)$ for $f(x)$ by using a capital letter and writing $F(x) = \Pr(X \leq x)$ for the probability that the random variable X has a value less than or equal to x.

Definition 4.2

The *cumulative distribution function* $F(x)$ of a discrete random variable X with a probability distribution $f(x)$ is given by

$$F(x) = \Pr(X \leq x) = \sum_{t \leq x} f(t), \quad \text{for } -\infty < x < \infty. \tag{4.5}$$

It is implicit in this definition that $F(x)$ is defined over the whole real line ($-\infty < x < \infty$). We see from Eq. (4.3) that the cumulative distribution function increases monotonically from zero to unity as x increases. Furthermore, Eq. (4.3) implies that the probability that X has a value greater than x is given by

$$\Pr(X > x) = 1 - F(x). \tag{4.6}$$

It also follows from Eq. (4.5) that the individual probabilities $f(x)$ are the differences in the values of $F(x)$ for successive values of x:

$$\Pr(X = x_k) = f(x_k) = F(x_k) - F(x_{k-1}). \tag{4.7}$$

■ *Example*
Find the cumulative probability distribution for two photons incident on the foil in the last example.

Solution
The cumulative distribution function for the two photons is given in Table 4.2. The values of $F(x)$ are obtained by continually summing the values $f(x)$ from the probability distribution in Table 4.1, as specified in the definition (4.5). Figure 4.2 shows the function $F(x)$. Notice that the cumulative probability distribution is defined for *all* real x ($-\infty < x < \infty$), and not just for the three values (0, 1, 2) assumed by the discrete random variable. Figure 4.2 shows that the probability for fewer than zero photons to interact is zero. The probability that less than one photon (i.e., neither photon) will interact when two are incident is 0.64. The probability that fewer than two (i.e., zero or one) will interact when two are incident is 0.96. When two photons are incident, the probability for interaction is unity for fewer than any number equal to or greater than two (e.g., the probability that fewer than five photons interact when two are incident is unity).

Thus far, we have dealt only with probability distributions for discrete random variables. Analogous definitions apply to continuous random variables. When X is

Table 4.2 Cumulative distribution function $F(x)$ for the number of photons x that interact in a foil, each with probability 0.2, when two photons are incident.

x	F(x)
<0	0
$0 \leq x < 1$	0.64
$1 \leq x < 2$	0.96
≥ 2	1.00

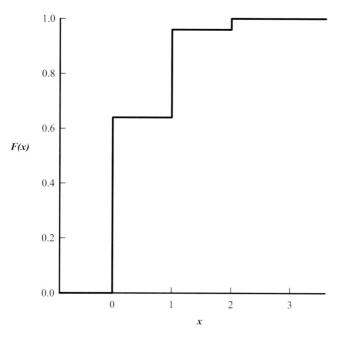

Figure 4.2 Cumulative probability distribution $F(x)$ for the number of photons x that interact when two are incident. See example in the text.

continuous, x has an uncountably infinite number of values. Consequently, the probability of occurrence for a single value $X = x$, exactly, is zero. For a continuous random variable X, we define a *probability density function*, denoted by $f(x)$. The integral of $f(x)$ over any interval of x then gives the probability that the random variable X has a value in that interval.

Definition 4.3
The function $f(x)$ is a *probability density function* for the continuous random variable X, defined over the set of real numbers R, if [2]

$$\Pr(a \leq X < b) = \int_a^b f(x) \, dx, \tag{4.8}$$

$$f(x) \geq 0 \quad \text{for all} \quad x \in R, \tag{4.9}$$

---

[2] With regard to notation, we generally assign the probability over the *semiclosed* interval, $a \leq X < b$, which includes the lower, but not the upper, boundary. Whether one includes a boundary point in the definition makes no difference mathematically, because the probability is zero for a continuous random variable at a single point, as mentioned in the last paragraph.

and

$$\int_{-\infty}^{\infty} f(x)dx = 1. \tag{4.10}$$

Equations (4.8)–(4.10) for a continuous random variable are analogous to Eqs. (4.1)–(4.3) for a discrete random variable, and can be given the same interpretation.

The corresponding cumulative distribution gives the probability that the continuous random variable has a value less than a specified value.

Definition 4.4

The *cumulative distribution* $F(x)$ of a continuous random variable X with density function $f(x)$ is given by

$$F(x) = \Pr(X < x) = \int_{-\infty}^{x} f(t)dt, \quad \text{for } -\infty < x < \infty. \tag{4.11}$$

It follows from Eqs. (4.8), (4.10), and (4.11) that (Problem 4.5)

$$\Pr(X \geq x) = 1 - \Pr(X < x) = \int_{x}^{\infty} f(t)dt. \tag{4.12}$$

Also, comparison of the definition (4.11) with Eq. (4.8) shows that the probability over any interval is given by the difference in the cumulative distribution at the end and at the beginning of the interval. Thus (Problem 4.6),

$$\Pr(a \leq X < b) = F(b) - F(a). \tag{4.13}$$

With the definition (4.11), the fundamental theorem of integral calculus implies that

$$f(x) = \frac{dF(x)}{dx}, \tag{4.14}$$

provided the derivative of $F(x)$ exists everywhere except possibly at a finite number of points. For our needs, $F(x)$ will be a continuous function of x, and hence the probability density function can be obtained from the cumulative distribution by taking its first derivative.

■ *Example*

In quantum mechanics, the probability density for the position of a particle confined to a box in one dimension with sides at $x = \pm a/2$ is

$$f(x) = \frac{2}{a}\cos^2\frac{\pi x}{a}, \quad \text{for } -\frac{a}{2} \leq x < \frac{a}{2} \tag{4.15}$$

and

$$f(x) = 0, \quad \text{elsewhere.} \tag{4.16}$$

This probability density function, which is shown in Figure 4.3, applies to the ground state, or state of lowest energy, of the particle.

a) Calculate the cumulative distribution function for the particle's position.
b) Show that the cumulative distribution function equals unity when $x = a/2$.
c) What is the probability of finding the particle between $x = 0$ and $x = a/4$?

Solution

a) The cumulative distribution is defined by Eq. (4.11). Since $f(x) = 0$ when $x < -a/2$, the cumulative distribution function is also zero in the interval on the left, outside the box in Figure 4.3:

$$F(x) = 0, \quad \text{for } x < -\frac{a}{2}. \tag{4.17}$$

For the interval inside the box, we can write

$$F(x) = \int_{-\infty}^{x} f(t)dt = \int_{-a/2}^{x} f(t)dt$$

$$= \frac{2}{a} \int_{-a/2}^{x} \cos^2\left(\frac{\pi t}{a}\right) dt, \quad \text{for } -\frac{a}{2} \leq x < \frac{a}{2}. \tag{4.18}$$

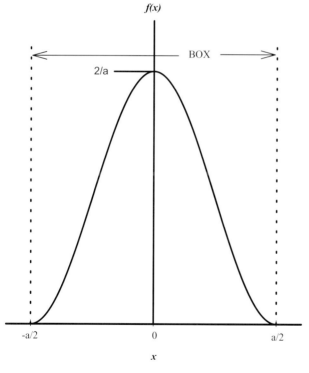

Figure 4.3 Probability density function $f(x)$ for the position X of a quantum-mechanical particle with lowest energy confined in a one-dimensional box with sides at $x = \pm a/2$. See example in the text.

Using the identity $\cos^2 \alpha = (1 + \cos 2\alpha)/2$, we find that

$$F(x) = \frac{1}{a} \int_{-a/2}^{x} \left[1 + \cos\left(\frac{2\pi t}{a}\right)\right] dt = \frac{1}{a}\left[t + \frac{a}{2\pi}\sin\frac{2\pi t}{a}\right]_{-a/2}^{x} \quad (4.19)$$

$$= \frac{x}{a} + \frac{1}{2} + \frac{1}{2\pi}\sin\frac{2\pi x}{a}, \quad \text{for } -\frac{a}{2} \leq x < \frac{a}{2}, \quad (4.20)$$

where $\sin(-\pi) = 0$ has been used in the last equality. Finally, the particle is always left of the right edge of the box. Therefore, the probability that the value of X is equal to any number greater than or equal to $a/2$ is unity:

$$F(x) = 1, \quad \text{for } x \geq \frac{a}{2}. \quad (4.21)$$

Equations (4.17), (4.20), and (4.21) constitute the desired cumulative distribution function, defined for all x.

b) Since the particle cannot be outside the box, the cumulative distribution must reach unity when x reaches or exceeds $a/2$. From Eq. (4.20), we find that, indeed,

$$F\left(\frac{a}{2}\right) = \frac{1}{2} + \frac{1}{2} + \frac{1}{2\pi}\sin\pi = 1, \quad (4.22)$$

as required.

c) From Eq. (4.8), the probability of finding the particle in the specified portion of the box is the integral of the probability density $f(x)$ from $x = 0$ to $x = a/4$:

$$\Pr\left(0 \leq X < \frac{a}{4}\right) = \int_0^{a/4} f(x)dx, \quad (4.23)$$

which can be evaluated directly. We have for this integral (see Eq. (4.19))

$$\Pr\left(0 \leq X < \frac{a}{4}\right) = \frac{1}{a}\left[t + \frac{a}{2\pi}\sin\frac{2\pi t}{a}\right]_0^{a/4} = \frac{1}{4} + \frac{1}{2\pi}\sin\frac{\pi}{2}$$
$$= 0.409. \quad (4.24)$$

Alternatively, we can use Eq. (4.13) and the cumulative function (4.20):

$$\Pr\left(0 \leq X < \frac{a}{4}\right) = F\left(\frac{a}{4}\right) - F(0) \quad (4.25)$$

$$= \frac{1}{4} + \frac{1}{2} + \frac{1}{2\pi}\sin\frac{\pi}{2} - \frac{1}{2} = 0.409, \quad (4.26)$$

in agreement with Eq. (4.24).

4.2 Expected Value

Various characteristics of discrete and continuous random variables are of interest. The mean, or average value, of the random variable is an important parameter. Equation (2.7), for example, shows how the mean life of a radionuclide is related to its decay constant, which represents the probability per unit time that a given atom will disintegrate. Another name for the mean of a random variable is its *expected value*.

Definition 4.5
The *expected value* $E(X)$ of the random variable X is defined as

$$E(X) = \sum_{\text{all } x} xf(x), \quad \text{if } X \text{ is discrete,} \tag{4.27}$$

or

$$E(X) = \int_{-\infty}^{\infty} xf(x)dx, \quad \text{if } X \text{ is continuous.} \tag{4.28}$$

In Eq. (4.27), $f(x)$ denotes the probability distribution, and in Eq. (4.28), the probability density function. The expected value is customarily denoted by the symbol μ.

From this definition, it follows that, if $X = c$ is a constant, then its expected value is the constant itself, $E(X) = E(c) = c$. Also, if $X = X_1 + X_2 + \cdots$ is the sum of two or more random variables, then its expected value is equal to the sum of the expected values, $E(X_1) + E(X_2) + \cdots$.

This definition can be generalized and put in the form of a theorem. The theorem expresses the interesting and useful result that the expected value of a function $g(X)$ of a random variable X can be obtained by taking the expected value of the function with respect to the original probability distribution function on X. For example, if $E(X^2)$ is desired, it is not necessary (though still correct) to find the probability distribution associated with $Y = X^2$ and then calculate $E(Y)$. One can simply use the above definition, replacing X with the function $g(X)$. We state the following theorem without proof.

Theorem 4.1
Let X be a random variable with probability distribution $f(x)$. The mean, or expected value, of the random variable $g(X)$ is

$$\mu_{g(X)} = E[g(X)] = \sum_{\text{all } x} g(x)f(x), \tag{4.29}$$

if X is discrete, and

$$\mu_{g(X)} = E[g(X)] = \int_{-\infty}^{\infty} g(x)f(x)dx, \tag{4.30}$$

if X is continuous.

4 Probability Distributions and Transformations

Equations (4.29) and (4.30) can be understood by regarding the distribution $f(x)$ as the weighting factor for x in determining the mean of $g(X)$.

■ **Example**

a) For the two-photon problem in the last section, what is the expected value of the number of photons that interact in the foil when two are incident?
b) What is the expected value of the number of photons that traverse the foil without interacting when two are incident?

Solution

a) The probability distribution function, $f(x)$, for the number of photons that interact is given in Table 4.1 and shown in Figure 4.1. Using the definition (4.27) for the discrete random variable, we obtain for the mean, or expected value,

$$\mu = E(X) = 0 \times 0.64 + 1 \times 0.32 + 2 \times 0.04 = 0.40. \tag{4.31}$$

Thus, an average of 0.40 photons are expected to interact in the foil when two are incident. As the example shows, the mean of a probability distribution need not equal one of the values that the random variable can take on. Such predictions can be tested against experiment. For example, if we repeated this experiment with 10 pairs and added the number of photons that interacted in each pairwise trial, the expected value for the total number of photons interacting in the foil would be four.

b) Since X represents the number of photons in a pair that interact in the foil, $Y = 2 - X$ is the number in a pair that traverse the foil without interacting. Taking the expected value of Y and using Eq. (4.29) yields

$$E(Y) = E(2 - X) = E(2) - E(X) = 2 - 0.40 = 1.60. \tag{4.32}$$

Thus, an average of 1.60 photons per pair are expected to traverse the foil without interacting. The expectation operator E in the last equation was applied to each of the terms in the difference $2 - X$. This operation is permitted through the distributive law for integration or summation of terms in Eqs. (4.29) and (4.30). Note, also, that the expected value of a constant is the constant itself – in this instance, $E(2) = 2$.

In Section 2.2, we determined that the mean, or average, life of a radionuclide is given by the reciprocal of the decay constant: $\tau = 1/\lambda$ (Eq. (2.7)). However, the argument presented there was more heuristic than rigorous. As mentioned earlier, the decay time T of a radionuclide is an example of a continuous random variable. Its mean can be obtained analytically from the above definition, Eq. (4.28), provided one knows the probability density function $f(t)$ for the decay times T. We carry out this calculation next.

The probability that a given atom present at time $t=0$ in a radioactive source will decay during a short time between t and $t + dt$ is equal to the product of (1) the probability that the atom has survived until time t and (2) the probability that it subsequently decays during dt. The former probability is $e^{-\lambda t}$, as given by Eq. (2.21). For very small dt, the latter probability is proportional to dt and can be written as $C\,dt$, where C is the constant of proportionality. Thus, the probability for an atom to decay between t and $t + dt$ is $Ce^{-\lambda t}\,dt$. The probability that the decay will occur between arbitrary times a and b is then given by the integral

$$\Pr(a \leq T < b) = C \int_a^b e^{-\lambda t}\,dt. \tag{4.33}$$

Comparison with Eq. (4.8) shows that the probability density function for the random decay time of an atom is

$$f(t) = Ce^{-\lambda t}. \tag{4.34}$$

The constant C is determined by the requirement (4.10) that this function be normalized (i.e., have unit area). Thus,

$$C \int_0^\infty e^{-\lambda t}\,dt = 1. \tag{4.35}$$

(Since we start at time $t=0$, the probability density for decay during the time $t<0$ is zero.) Integration of Eq. (4.35) gives $C=\lambda$, and so the probability density function for the random decay time T of a radionuclide is

$$f(t) = \lambda e^{-\lambda t}, \quad t \geq 0, \tag{4.36}$$

and

$$f(t) = 0, \quad t < 0. \tag{4.37}$$

The average decay time, or mean life for radioactive decay, is usually denoted by the special symbol τ. We find that

$$\tau = E(T) = \int_0^\infty tf(t)\,dt = \lambda \int_0^\infty te^{-\lambda t}\,dt = \frac{1}{\lambda}, \tag{4.38}$$

where the integration has been performed by parts (Problem 4.10). This result confirms that expressed in Eq. (2.7).

▉ *Example*

a) Use the probability density function $f(t)$ for the decay time of a radionuclide to construct the cumulative distribution $F(t)$.
b) Use $F(t)$ to find the relationship between the decay constant λ and the half-life $T_{1/2}$ of a radionuclide.

Solution

a) We first find the function $F(t)$ and then, for part (b), we set $F(t) = 1/2$ with $t = T_{1/2}$. Combining Eqs. (4.36) and (4.37) with the definition (4.11) of the cumulative distribution for a continuous random variable T, one has

$$F(t) = 0, \quad \text{for } t < 0, \qquad (4.39)$$

and

$$F(t) = \int_{-\infty}^{t} f(t')dt' = \lambda \int_{0}^{t} e^{-\lambda t'} dt' = 1 - e^{-\lambda t}, \quad \text{for } t \geq 0. \qquad (4.40)$$

b) When t equals the half-life, the cumulative distribution has the value $F(T_{1/2}) = 1/2$. That is, at time $T_{1/2}$, the probability that the random decay time has a value less than $T_{1/2}$ is $1/2$, reflecting the condition that one-half of the original atoms are still present. We obtain from Eq. (4.40)

$$\frac{1}{2} = 1 - e^{-\lambda T_{1/2}}, \qquad (4.41)$$

giving $T_{1/2} = (\ln 2)/\lambda$, in agreement with Eq. (2.6).

In addition to the expected value, or mean, another important characteristic of a distribution is the *median*. For a continuous distribution, the median is defined as that value m_e such that the probability $\Pr(X \leq m_e) = 1/2$, exactly. We write

$$\int_{-\infty}^{m_e} f(x)dx = \frac{1}{2}. \qquad (4.42)$$

The median divides the cumulative probability function into two equal portions. Comparison with the cumulative function, Eq. (4.11), shows that it is equally likely that the value of the continuous random variable X will occur on either side of the median. If the random variable is discrete, then the median is defined in terms of the cumulative function $F(x)$ as the number m_e such that

$$\lim_{x \to m_e^-} F(x) = F(m_e - 0) \leq \frac{1}{2} \leq F(m_e + 0) = \lim_{x \to m_e^+} F(x). \qquad (4.43)$$

Here the symbol $x \to m_e^-$ implies that x approaches m_e from below, and $x \to m_e^+$ implies that the approach is from above. Since the variable is discrete, there may be an interval of points that satisfy this equation. When this is the case, one uses the midpoint of the interval as the median. The mean of the absolute value of the deviation of a random variable about a given value in a distribution is a minimum at the median of the distribution (Problem 4.12).

4.3 Variance

Another important property of a random variable is its *variance*, which measures the variability, or dispersion (spread), of the probability distribution.

Definition 4.6
The *variance* is defined as the expected value of the square of the difference between the random variable and its mean:

$$\text{Variance}(X) = \text{Var}(X) = E[(X - \mu)^2]. \tag{4.44}$$

For a discrete random variable one has

$$\text{Var}(X) = \sum_{\text{all } x} (x - \mu)^2 f(x); \tag{4.45}$$

and for a continuous random variable,

$$\text{Var}(X) = \int_{-\infty}^{\infty} (x - \mu)^2 f(x) dx. \tag{4.46}$$

The variance of sums or differences of independent random variables is the sum of their variances.

The variance, which is in squared units, is usually denoted by the symbol σ^2. The positive square root of the variance is called the *standard deviation*, σ. Taking the square root converts the measure of variability to the original units of the random variable. The standard deviation is very important in describing a distribution, not only measuring its spread, but also serving as a yardstick to gauge the probability that an observation is some distance from the mean. The mean determines a center for the distribution, and the standard deviation tells us how far most observations range from that center. As we shall see in future applications, for most probability distributions it would be rare to observe outcomes that exceed three standard deviations from the mean in either direction.

To calculate the variance, one can use the definition (4.44) or the following expression:

$$\text{Var}(X) = E[(X - \mu)^2] = E(X^2 - 2X\mu + \mu^2) = E(X^2) - 2\mu E(X) + \mu^2. \tag{4.47}$$

Since $E(X) = \mu$, one can combine the last two terms to obtain

$$\text{Var}(X) = E(X^2) - \mu^2. \tag{4.48}$$

The expression (4.48) is usually more convenient to use for computing variances than the definition (4.44), although either may be used.

Example
Calculate the standard deviation for the decay time of a radionuclide.

Solution
We employ Eq. (4.48). The distribution for the decay time T is given by Eqs. (4.36) and (4.37). The second term on the right-hand side of Eq. (4.48) is the square of the mean life, given by Eq. (4.38): $\mu = \tau = 1/\lambda$. The first term on the right-hand side in Eq. (4.48) is

$$E(T^2) = \int_{-\infty}^{\infty} t^2 f(t) dt = \lambda \int_0^{\infty} t^2 e^{-\lambda t} dt. \tag{4.49}$$

Integration by parts (Problem 4.14) gives

$$E(T^2) = -\frac{2}{\lambda^2} e^{-\lambda t} \Big|_0^{\infty} = \frac{2}{\lambda^2}. \tag{4.50}$$

Hence, the variance of the distribution of decay times is, from Eq. (4.48),

$$\sigma^2 = E(T^2) - \frac{1}{\lambda^2} = \frac{2}{\lambda^2} - \frac{1}{\lambda^2} = \frac{1}{\lambda^2}. \tag{4.51}$$

The standard deviation is

$$\sigma = \frac{1}{\lambda}. \tag{4.52}$$

Thus, the distribution of decay times has a mean and standard deviation both equal to $1/\lambda$.

As already mentioned, one can use the standard deviation as a measure of distance in the space of a probability distribution, as the next example illustrates.

Example
What is the probability of observing a nuclide decay time T that is at least two standard deviations later than the mean?

Solution
We seek the probability

$$\Pr(T - \tau \geq 2\sigma) = \Pr(T \geq \tau + 2\sigma). \tag{4.53}$$

Substituting $\tau = \sigma = 1/\lambda$ for the mean and standard deviation, we write

$$\Pr(T - \tau \geq 2\sigma) = \Pr\left[T \geq \frac{1}{\lambda} + 2\left(\frac{1}{\lambda}\right)\right] = \Pr\left(T \geq \frac{3}{\lambda}\right). \tag{4.54}$$

Applying Eq. (4.12) for the cumulative probability and using the probability density function, Eqs. (4.36) and (4.37), for the decay time T, we obtain

$$\Pr\left(T \geq \frac{3}{\lambda}\right) = \lambda \int_{3/\lambda}^{\infty} e^{-\lambda t}\, dt = -e^{-\lambda t}\big|_{3/\lambda}^{\infty} = e^{-3} = 0.0498. \qquad (4.55)$$

Thus, only about 5 times in 100 would we expect an atom in the original source to decay at a time later than two standard deviations beyond the mean. Comparison of the last equality in Eq. (4.55) with q in Eq. (2.21) shows that the probability calculated here is, as it should be, just the same as that for an atom in the original source to survive for a time equal to at least three mean lives.

4.4
Joint Distributions

We consider next the variation of several random variables at once and introduce the idea of joint distributions. Just as we define probability functions for discrete and continuous random variables, we do the same for situations in which we observe two or more random variables simultaneously. Joint distributions can occur when we describe outcomes by giving the values of several random variables. For example, we might measure the weight and hardness of materials; the color, pH, and temperature for certain chemical reactions; or the height, weight, and fat content of different individuals. If $x_1, x_2, \ldots, x_k$ are the values of k random variables, we shall refer to a function, f, with values $f(x_1, x_2, \ldots, x_k)$ as the joint probability density function of these variables. For simplicity, we shall usually deal with the bivariate case, $k=2$. Extensions to larger k are straightforward. As with the univariate case, described by Eqs. (4.8)–(4.10), only certain functions can qualify as joint probability functions.

Definition 4.7
A function $f(x_1, x_2)$ is a *joint probability density function* for the continuous random variables X_1 and X_2, defined over the set of real numbers R, if

$$\Pr(a_1 \leq X_1 < b_1, a_2 \leq X_2 < b_2) = \int_{a_2}^{b_2}\int_{a_1}^{b_1} f(x_1, x_2)\, dx_1\, dx_2, \qquad (4.56)$$

$$f(x_1, x_2) \geq 0 \quad \text{for all } x_1, x_2 \in R, \qquad (4.57)$$

and

$$\int_{-\infty}^{\infty}\int_{-\infty}^{\infty} f(x_1, x_2)\, dx_1\, dx_2 = 1. \qquad (4.58)$$

4 Probability Distributions and Transformations

One can similarly define *joint probability functions* for discrete random variables, replacing the integrals in this definition with appropriate sums.

■ **Example**

Let X_1 and X_2 have the joint probability density function

$$f(x_1, x_2) = \begin{cases} e^{-x_1-x_2}, & \text{where } 0 < x_1 < \infty \text{ and } 0 < x_2 < \infty \\ 0, & \text{otherwise.} \end{cases} \quad (4.59)$$

a) Show that this probability function satisfies requirements (4.57) and (4.58) in the above definition.
b) Determine $\Pr(X_1 + X_2 < 5)$.

Solution

a) First, inspection of Eq. (4.59) shows that $f(x_1, x_2) \geq 0$ for all values of x_1 and x_2; hence, condition (4.57) is satisfied. Second, we show that this function integrates to unity:

$$\int_{-\infty}^{\infty}\int_{-\infty}^{\infty} f(x_1, x_2) dx_1\, dx_2 = \int_{0}^{\infty}\int_{0}^{\infty} e^{-x_1-x_2}\, dx_1\, dx_2$$

$$= \int_{0}^{\infty} e^{-x_1}\, dx_1 \int_{0}^{\infty} e^{-x_2}\, dx_2 = (1)(1) = 1. \quad (4.60)$$

b) To determine $\Pr(X_1 + X_2 < 5)$, we consider the region in the x_1, x_2 plane that satisfies the relation $x_1 + x_2 < 5$, as shown in Figure 4.4. We can integrate over x_1 from 0 to $5 - x_2$, and then over x_2 from 0 to 5. Thus,

$$\Pr(X_1 + X_2 < 5) = \int_{0}^{5}\int_{0}^{5-x_2} e^{-x_1-x_2}\, dx_1\, dx_2 = \int_{0}^{5} e^{-x_2}[-e^{-x_1}]_0^{5-x_2}\, dx_2 \quad (4.61)$$

$$= \int_{0}^{5} e^{-x_2}(1 - e^{-5+x_2})dx_2 = \int_{0}^{5} (e^{-x_2} - e^{-5})dx_2 \quad (4.62)$$

$$= [-e^{-x_2} - x_2 e^{-5}]_0^5 = 1 - 6e^{-5} = 0.960. \quad (4.63)$$

For random variables $X_1, X_2, \ldots, X_k$, we can define the *joint cumulative distribution function* (or, simply, joint distribution function), $F(x_1, x_2, \ldots, x_k)$. This function gives the probability that $X_1 < x_1, X_2 < x_2, \ldots, X_k < x_k$.

An important concept in the discussion of joint distributions is that of *independence* among the random variables. Independence is defined in terms of the *marginal densities* of all of the individual variables. The marginal densities are obtained when a continuous joint probability density is integrated over all but a single random

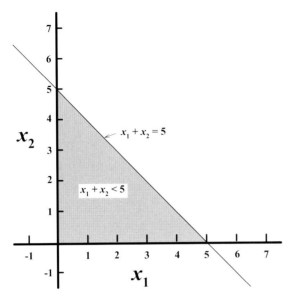

Figure 4.4 Shaded area shows the region of nonnegative probability for $x_1 + x_2 < 5$, calculated for example in the text (Eq. (4.63)).

variable, or a discrete *joint probability distribution* is summed over all but a single variable. The following definition and examples address these ideas.

Definition 4.8
$X_1, X_2, \ldots, X_k$ are independent random variables if and only if

$$f(x_1, x_2, \ldots, x_k) = f_1(x_1) f_2(x_2) \cdots f_k(x)_k, \qquad (4.64)$$

where $f_1(x_1), f_2(x_2), \ldots, f_k(x_k)$ are, respectively, the k marginal densities obtained by integrating out the $(k-1)$ other variables in the joint probability distribution $f(x_1, x_2, \ldots, x_k)$.

■ *Example*
The discrete random variables X_1 and X_2 have the following joint probability function:

$$f(x_1, x_2) = \begin{cases} \dfrac{x_1 + x_2}{32}, & \text{when } x_1 = 1, 2 \text{ and } x_2 = 1, 2, 3, 4 \\ 0, & \text{otherwise.} \end{cases} \qquad (4.65)$$

a) Find the marginal distribution on X_1.
b) Find the marginal distribution on X_2.
c) Are X_1 and X_2 independent?
d) Obtain $\Pr(X_1 + X_2 = 5)$.

Solution

a) We sum over X_2 to obtain the marginal distribution, $f_1(x_1)$, on X_1:

$$f_1(x_1) = \begin{cases} \sum_{x_2=1}^{4} \dfrac{x_1 + x_2}{32} = \dfrac{1}{32}[(x_1 + 1) + (x_1 + 2) + (x_1 + 3) + (x_1 + 4)] \\ \dfrac{4x_1 + 10}{32} = \dfrac{2x_1 + 5}{16}, \quad \text{when } x_1 = 1, 2 \\ 0, \quad \text{otherwise.} \end{cases} \quad (4.66)$$

One should check to see whether the marginal distribution $f_1(x_1)$ is a true probability function. It is, because it is nonnegative and

$$\sum_{x_1=1}^{2} f_1(x_1) = \frac{7}{16} + \frac{9}{16} = 1. \qquad (4.67)$$

b) The marginal distribution on X_2 is

$$f_2(x_2) = \begin{cases} \sum_{x_1=1}^{2} \dfrac{x_1 + x_2}{32} = \dfrac{2x_2 + 3}{32}, \quad x_2 = 1, 2, 3, 4 \\ 0, \quad \text{elsewhere.} \end{cases} \quad (4.68)$$

It is straightforward to verify that $f_2(x_2)$ is a true probability function (Problem 4.15).

c) The definition of independence states that X_1 and X_2 are independent if and only if $f(x_1, x_2) = f_1(x_1) f_2(x_2)$. From Eqs. (4.66) and (4.68), we see that this is not true, and so X_1 and X_2 are dependent.

d) To obtain $\Pr(X_1 + X_2 = 5)$ we need to find the values of X_1 and X_2 whose sums equal 5. Those pairs are $(x_1, x_2) = (1, 4)$ and $(x_1, x_2) = (2, 3)$. Therefore,

$$\Pr(X_1 + X_2 = 5) = f(1, 4) + f(2, 3) = \frac{1+4}{32} + \frac{2+3}{32} = \frac{10}{32} = \frac{5}{16}. \quad (4.69)$$

■ **Example**
If X_1 and X_2 have the joint probability density function

$$f(x_1, x_2) = \begin{cases} 30e^{-5x_1 - 6x_2}, & x_1 > 0, x_2 > 0 \\ 0, & \text{elsewhere,} \end{cases} \quad (4.70)$$

show that they are independent random variables.

4.4 Joint Distributions

Solution

We calculate the marginal densities for the two variables and see whether their product is equal to the joint probability density given by Eq. (4.70). We have

$$f_1(x_1) = \begin{cases} \int_0^\infty 30 e^{-5x_1 - 6x_2}\, dx_2 = 30 e^{-5x_1} \left[\dfrac{-e^{-6x_2}}{6}\right]_0^\infty = 5 e^{-5x_1}, & x_1 > 0 \\ 0, & \text{elsewhere.} \end{cases} \quad (4.71)$$

Similarly (Problem 4.16, or by inspection of (4.71)),

$$f_2(x_2) = \begin{cases} 6 e^{-6x_2}, & x_2 > 0 \\ 0, & \text{elsewhere.} \end{cases} \quad (4.72)$$

We see from the last three equations that $f(x_1)f(x_2) = f(x_1, x_2)$, and hence X_1 and X_2 are independent random variables.

We next show how expected values are calculated for joint distributions. For continuous distributions, integrations are carried out over all variables. The expected value of a single random variable X_i in a joint distribution, for example, is given by

$$E(X_i) = \int_{-\infty}^{\infty} \int_{-\infty}^{\infty} \cdots \int_{-\infty}^{\infty} x_i\, f(x_1, x_2, \ldots, x_k)\, dx_1\, dx_2 \cdots dx_k. \quad (4.73)$$

For the product of two random variables,

$$E(X_i X_j) = \int_{-\infty}^{\infty} \int_{-\infty}^{\infty} \cdots \int_{-\infty}^{\infty} x_i x_j f(x_1, x_2, \ldots, x_k)\, dx_1\, dx_2 \cdots dx_k. \quad (4.74)$$

Higher order products are obtained by extending this procedure. Similar equations apply for discrete random variables, with the integrals replaced by appropriate sums. In addition to products of random variables, other functions can be treated in similar fashion.

Conditional probability is another important aspect of joint distributions. In Chapter 3 (Eq. (3.31)), we defined the conditional probability of one event A, given knowledge that another event B had occurred, as

$$\Pr(A|B) = \dfrac{\Pr(A \cap B)}{\Pr(B)}, \quad \text{with } P(B) > 0. \quad (4.75)$$

For two discrete random variables, if A and B represent events given by $X = x$ and $Y = y$, respectively, then we may write

$$\Pr(X = x | Y = y) = \dfrac{\Pr(X = x, Y = y)}{\Pr(Y = y)} = \dfrac{f(x, y)}{f_2(y)}, \quad (4.76)$$

which defines functions $f(x, y)$ and $f_2(y)$, with $f_2(y) > 0$. The quantity $f(x, y)$ is thus a function of x alone when y is fixed. One can show that it satisfies all of the properties of a probability distribution. The same conditions hold also when X and Y are continuous random variables. The joint probability density function is then $f(x, y)$, and the marginal probability density on Y is $f_2(y)$.

Definition 4.9

Let X and Y be two (discrete or continuous) random variables with joint probability density function $f(x, y)$. The *conditional distribution* of X, given that $Y = y$, is denoted by the symbol $f(x|y)$ and defined as

$$f(x|y) = \frac{f(x, y)}{f_2(y)}, \quad f_2(y) > 0. \tag{4.77}$$

Similarly, the symbol $f(y|x)$ denotes the conditional distribution of Y, given that $X = x$,

$$f(y|x) = \frac{f(x, y)}{f_1(x)}, \quad f_1(x) > 0. \tag{4.78}$$

■ **Example**

The joint density of X and Y is given by

$$f(x, y) = \begin{cases} 6xy(2 - x - y), & 0 < x < 1, \ 0 < y < 1 \\ 0, & \text{elsewhere.} \end{cases} \tag{4.79}$$

Obtain the conditional probability of X, given that $Y = y = 1/2$.

Solution
From the definition (4.77), we write

$$f\left(x|y = \frac{1}{2}\right) = \frac{f(x, 1/2)}{\int_0^1 f(x, 1/2) dx}. \tag{4.80}$$

Substitution from Eq. (4.79) then gives

$$f\left(x|y = \frac{1}{2}\right) = \frac{3x(3/2 - x)}{\int_0^1 3x(3/2 - x) dx} = \frac{12x}{5}\left(\frac{3}{2} - x\right). \tag{4.81}$$

4.5 Covariance

When dealing with the joint distribution function of two random variables, X_1 and X_2, it is useful to recognize the existence of any association that might exist between the two. For instance, large values of X_1 might tend to occur when X_2 is large. Unless the random variables are independent, values of X_1 and X_2 will be correlated in some way with one another. A quantity that reflects such an association is the *covariance* of two random variables.

Definition 4.10
For any pair of random variables X_1 and X_2 with means μ_1 and μ_2, the *covariance*, denoted by σ_{12}, is defined as

$$\text{Cov}(X_1, X_2) = \sigma_{12} = E[(X_1 - \mu_1)(X_2 - \mu_2)]. \tag{4.82}$$

The covariance expresses how the random variables vary jointly about their means. Thus, if both X_1 and X_2 tend to be relatively large or relatively small together, then the product $(X_1 - \mu_1)(X_2 - \mu_2)$ will tend to be positive. On the other hand, if large X_1 and small X_2 are apt to occur together and vice versa, then $(X_1 - \mu_1)(X_2 - \mu_2)$ will tend to be negative. Therefore, the *sign* of the covariance indicates whether there is a positive or negative relationship between two random variables. (If X_1 and X_2 are independent, it can be shown that the covariance is zero. The converse, however, is not true.) In place of the definition (4.82), it is often convenient to use the equivalent expression,

$$\sigma_{12} = E(X_1 X_2) - \mu_1 \mu_2, \tag{4.83}$$

for the covariance (Problem 4.23).

Although the covariance provides information on the nature of the relationship between two random variables, it is not a quantitative measure of the strength of that relationship. Its numerical value depends on the units chosen for X_1 and X_2. To overcome this limitation, one defines the following dimensionless *coefficient of correlation*.

Definition 4.11
The *coefficient of correlation*, $\text{Corr}(X_1, X_2)$, denoted by ρ, between any pair of random variables X_1 and X_2, having variances σ_1^2 and σ_2^2, is

$$\text{Corr}(X_1, X_2) = \rho = \frac{\text{Cov}(X_1, X_2)}{\sqrt{\text{Var}(X_1)\text{Var}(X_2)}} = \frac{\sigma_{12}}{\sqrt{\sigma_1^2 \sigma_2^2}}. \tag{4.84}$$

If the relationship between X_1 and X_2 is not linear, then the correlation coefficient is a poor estimate of the dependence of the two variables on one another.

■ *Example*
Using the discrete probability function, Eq. (4.65), and the results obtained in that example, find

a) the expected values, μ_1 and μ_2, of X_1 and X_2;
b) the variances, σ_1^2 and σ_2^2;
c) the covariance, σ_{12}; and
d) the coefficient of correlation ρ between X_1 and X_2.

Solution

a) The expected values of the single variables can be obtained from the joint distribution by applying the prescription analogous to Eq. (4.73) for discrete random variables. However, it is simpler to use the marginal distributions straightaway, which we already determined through Eqs. (4.66) and (4.68). We have

$$\mu_1 = E(X_1) = \sum_{x_1=1}^{2} \frac{x_1(2x_1+5)}{16} = \frac{1(2+5)+2(4+5)}{16} = \frac{25}{16} \quad (4.85)$$

and

$$\mu_2 = E(X_2) = \sum_{x_2=1}^{4} \frac{x_2(2x_2+3)}{32}$$

$$= \frac{1(2+3)+2(4+3)+3(6+3)+4(8+3)}{32} = \frac{45}{16}. \quad (4.86)$$

b) For the variances, with $i=1, 2$, we calculate $E(X_i^2)$ and use Eq. (4.48): $\sigma_i^2 = E(X_i^2) - [E(X_i)]^2$. Thus,

$$E(X_1^2) = \sum_{x_1=1}^{2} \frac{x_1^2(2x_1+5)}{16} = \frac{1(2+5)+4(4+5)}{16} = \frac{43}{16}. \quad (4.87)$$

Using this result and that of Eq. (4.85) in Eq. (4.48), we obtain for the variance of X_1,

$$\sigma_1^2 = \frac{43}{16} - \left(\frac{25}{16}\right)^2 = \frac{63}{256}. \quad (4.88)$$

In similar fashion, one obtains (Problem 4.24) $\sigma_2^2 = 295/256$.

c) To determine the covariance, σ_{12}, we use Eq. (4.83). The first term on the right-hand side is

$$E(X_1 X_2) = \sum_{x_2=1}^{4}\sum_{x_1=1}^{4} x_1 x_2 \frac{x_1+x_2}{32} = \sum_{x_2=1}^{4} x_2 \sum_{x_1=1}^{2} \frac{x_1(x_1+x_2)}{32} \qquad (4.89)$$

$$= \sum_{x_2=1}^{4} x_2 \frac{1+x_2+4+2x_2}{32} = \sum_{x_2=1}^{4} \frac{5x_2+3x_2^2}{32} = \frac{140}{32} = \frac{35}{8}. \qquad (4.90)$$

Combining this result with Eqs. (4.85) and (4.86) for μ_1 and μ_2, we find that Eq. (4.83) gives for the covariance

$$\sigma_{12} = \frac{35}{8} - \left(\frac{25}{16}\right)\left(\frac{45}{16}\right) = -\frac{5}{256}. \qquad (4.91)$$

d) It follows from the definition (4.84) of the correlation coefficient and the results obtained thus far that

$$\rho = \frac{\sigma_{12}}{\sqrt{\sigma_1^2 \sigma_2^2}} = \frac{-5/256}{\sqrt{(63/256)(295/256)}} = -0.0367. \qquad (4.92)$$

Thus, there is a very weak, negative linear relationship between X_1 and X_2.

The following two theorems apply to the coefficient of correlation and indicate its usefulness in a quantitative way.

Theorem 4.2
If X_1 and X_2 are independent random variables, then the coefficient of correlation is zero. That is,

$$\text{if } X_1 \text{ and } X_2 \text{ are independent, then } \rho = 0. \qquad (4.93)$$

The proof of this theorem is left as Problem 4.25 at the end of the chapter. (One can use the fact that $E(X_1 X_2) = E(X_1)E(X_2)$.)

Theorem 4.3
The value of the correlation coefficient ρ lies in the closed interval $[-1, 1]$; that is,

$$-1 \leq \rho \leq 1. \qquad (4.94)$$

To prove this theorem, we form the following nonnegative function of an arbitrary variable t:

$$H(t) = E\{[(X_1 - \mu_1)t + (X_2 - \mu_2)]^2\} \qquad (4.95)$$

$$= E[(X_1 - \mu_1)^2]t^2 + 2E[(X_1 - \mu_1)(X_2 - \mu_2)]t + E[(X_2 - \mu_2)^2]. \qquad (4.96)$$

Because $H(t) \geq 0$ for all values of t, the discriminant associated with this quadratic function of t cannot be positive,[3] and so we may write

$$\{2E[(X_1 - \mu_1)(X_2 - \mu_2)]\}^2 - 4E[(X_1 - \mu_1)^2]E[(X_2 - \mu_2)^2] \leq 0. \qquad (4.97)$$

The expected values are the squares of the covariance and the variances. Thus (dividing out the common factor of four),

$$\sigma_{12}^2 - \sigma_1^2 \sigma_2^2 \leq 0; \qquad (4.98)$$

and so

$$\rho^2 = \frac{\sigma_{12}^2}{\sigma_1^2 \sigma_2^2} \leq 1, \qquad (4.99)$$

which is equivalent to Eq. (4.94).

The two theorems, (4.93) and (4.94), provide some interesting quantitative information regarding the joint relationship between two random variables. If the variables are independent, then the correlation coefficient will be zero. (As noted above, however, the converse is not necessarily true.) Also, the largest magnitude that the correlation coefficient can have is unity. The value of unity implies that the variables have a perfect linear relationship; that is, $X_1 = a + bX_2$, where a and b are constants.

Finally, we state without proof an additional theorem, which is important for later applications, especially when we investigate the properties of estimators.

Theorem 4.4
For a collection of n random variables X_i, $i = 1, 2, \ldots, n$, having means μ_i, variances σ_i^2, and covariances σ_{ij} ($i \neq j$), the variance of a linear combination of the n variables is

$$\begin{aligned} \mathrm{Var}\left(\sum_{i=1}^n a_i X_i\right) &= \sum_{i=1}^n a_i^2 \, \mathrm{Var}(X_i) + 2 \sum_{i=1}^{n-1} \sum_{j=i+1}^n a_i a_j \, \mathrm{Cov}(X_i, X_j) \\ &= \sum_{i=1}^n a_i^2 \sigma_i^2 + 2 \sum_{i=1}^{n-1} \sum_{j=i+1}^n a_i a_j \sigma_{ij}, \end{aligned} \qquad (4.100)$$

---

[3] The quadratic function (4.96) of t is associated with the quadratic equation of the form $at^2 + bt + c = 0$. The two roots are $t = (-b \pm \sqrt{b^2 - 4ac})/(2a)$. If $H(t)$ does not change sign over the infinite domain of t, then the quadratic equation does not have two distinct real roots. Thus, the discriminant, $b^2 - 4ac$, cannot be positive, and so $b^2 - 4ac \leq 0$.

where the a_i and a_j are constants. Note that, for independent random variables, this equation simplifies to (Problem 4.26)

$$\text{Var}\left(\sum_{i=1}^{n} a_i X_i\right) = \sum_{i=1}^{n} a_i^2 \text{Var}(X_i) = \sum_{i=1}^{n} a_i^2 \sigma_i^2. \tag{4.101}$$

■ *Example*
Let X_1 and X_2 have the joint probability density function

$$f(x_1, x_2) = \begin{cases} x_1 + x_2, & 0 < x_1 < 1 \text{ and } 0 < x_2 < 1, \\ 0, & \text{elsewhere}. \end{cases} \tag{4.102}$$

a) Calculate μ_1, μ_2, σ_1^2, σ_2^2, and σ_{12}.
b) Determine the mean and variance of the average, $M = (X_1 + X_2)/2$, of X_1 and X_2.

Solution

a) The mean of X_1 is

$$\mu_1 = E(X_1) = \int_0^1 \int_0^1 x_1(x_1 + x_2) dx_1\, dx_2$$

$$= \int_0^1 \left[\frac{x_1^3}{3} + \frac{x_1 x_2}{2}\right]_{x_1=0}^1 dx_2 \tag{4.103}$$

$$= \int_0^1 \left(\frac{1}{3} + \frac{x_2}{2}\right) dx_2 = \left[\frac{x_2}{3} + \frac{x_2^2}{4}\right]_0^1 = \frac{7}{12}. \tag{4.104}$$

Also, the mean of X_2 is, by symmetry,

$$\mu_2 = E(X_2) = \int_0^1 \int_0^1 x_2(x_1 + x_2) dx_1\, dx_2 = \frac{7}{12}. \tag{4.105}$$

To calculate the variance, we employ Eq. (4.48). Since

$$E(X_1^2) = \int_0^1 \int_0^1 x_1^2(x_1 + x_2) dx_1\, dx_2 = \int_0^1 \left[\frac{x_1^4}{4} + \frac{x_1^3 x_2}{3}\right]_{x_1=0}^1 dx_2 \tag{4.106}$$

$$= \int_0^1 \left[\frac{1}{4} + \frac{x_2}{3}\right] dx_2 = \left[\frac{x_2}{4} + \frac{x_2^2}{6}\right]_0^1 = \frac{5}{12}, \tag{4.107}$$

we obtain

$$\sigma_1^2 = \frac{5}{12} - \left(\frac{7}{12}\right)^2 = \frac{11}{144}. \tag{4.108}$$

Again, by symmetry, one finds (Problem 4.27) that $\sigma_2^2 = \sigma_1^2 = 11/144$. For the covariance, we use Eq. (4.83). Thus,

$$E(X_1 X_2) = \int_0^1 \int_0^1 x_1 x_2 (x_1 + x_2) dx_1 \, dx_2 + \int_0^1 \left[\frac{x_1^3 x_2}{3} + \frac{x_1^2 x_2^2}{2}\right]_{x_1=0}^1 dx_2 \tag{4.109}$$

$$= \int_0^1 \left(\frac{x_2}{3} + \frac{x_2^2}{2}\right) dx_2 = \left[\frac{x_2^2}{6} + \frac{x_2^3}{6}\right]_0^1 = \frac{1}{3}; \tag{4.110}$$

and so

$$\sigma_{12} = \frac{1}{3} - \left(\frac{7}{12}\right)\left(\frac{7}{12}\right) = -\frac{1}{144}. \tag{4.111}$$

b) The mean of M is

$$E(M) = \frac{1}{2} E(X_1 + X_2) = \frac{1}{2}(\mu_1 + \mu_2) = \frac{7}{12}. \tag{4.112}$$

The variance is, with the help of Eq. (4.100),

$$\text{Var}(M) = \text{Var}\left(\frac{X_1}{2} + \frac{X_2}{2}\right)$$

$$= \frac{\text{Var}(X_1)}{4} + \frac{\text{Var}(X_2)}{4} + 2\left(\frac{1}{2}\right)\left(\frac{1}{2}\right) \text{Cov}(X_1, X_2) \tag{4.113}$$

$$= \frac{1}{4}\left(\frac{11}{144}\right) + \frac{1}{4}\left(\frac{11}{144}\right) + \frac{1}{2}\left(\frac{-1}{144}\right) = \frac{5}{144}. \tag{4.114}$$

4.6
Chebyshev's Inequality

The rarity of observations that exceed several standard deviations from the mean is confirmed by *Chebyshev's inequality*, which we present without proof.

Chebyshev's Inequality
Let the random variable X have a probability distribution with finite mean μ and variance σ^2. Then, for every $k > 0$,

$$\Pr(|X - \mu| \geq k\sigma) \leq \frac{1}{k^2}. \tag{4.115}$$

Accordingly, the probability that a random variable has a value at least as far away as $k\sigma$ from the mean, either to the right or left, is at most $1/k^2$. This relationship provides a rigorous, but often very loose, upper bound to the actual probability. When $k=2$, for example, Chebyshev's inequality states that the probability of an observation being at least two standard deviations away from the mean does not exceed 0.25. In the last example in Section 4.3, we found that the probability for the random decay time to be at least two standard deviations beyond the mean was 0.0498 (Eq. (4.55)). Thus, we see how the upper bound of 0.25 from Chebyshev's inequality is satisfied in this instance, although it is far from the actual value of the true probability. Nevertheless, the general applicability of Chebyshev's inequality to *any* distribution with finite mean and variance often makes it very useful. Chebyshev's inequality is valid for both continuous and discrete random variables, as long as they possess finite means and variances.

4.7
Transformations of Random Variables

Change of variables is common practice in analysis. Given the kinetic energy of a beta particle, for instance, one can transform this quantity into the particle's velocity or momentum for various purposes. Such a change of variable is straightforward. Nonrelativistically, for instance, if the kinetic energy is E, then the velocity is $V = \sqrt{2E/m}$, where m is the electron mass, and the momentum is $P = \sqrt{2mE}$. Less straightforward are transformations of a *function* of a random variable. Given the *spectrum* (i.e., the probability density function) of energies E for a source of beta particles, one might want to describe the probability density function for the velocity or the momentum of the emitted electrons. In this section, we show how to transform a probability distribution, given as a function of one random variable, into a distribution in terms of another, related variable. We treat first discrete, and then continuous, random variables.

We let X be a discrete random variable and Y be a single-valued function of X, which we denote by writing $Y=u(X)$. Given a value of X, the function $u(X)$ then provides a unique value of the random variable Y. We restrict the inverse transformation, $X=w(Y)$, of Y into X to be single-valued also, so that there is a one-to-one relation between the values of X and Y. (For example, $Y=X^2$ implies that $X=\pm\sqrt{Y}$, which is not single-valued. To make the transformation single-valued, we can select either $X=+\sqrt{Y}$ or $X=-\sqrt{Y}$, depending on the application at hand.) Given the probability distribution on X, we wish to find the distribution on Y. The transformation is accomplished by writing, for all y,

$$\Pr(Y = y) = \Pr[u(X) = y] = \Pr[X = w(y)]. \tag{4.116}$$

An example will illustrate the performance of these operations.

■ *Example*
A random variable X has the binomial distribution shown for $x=k$ in Eq. (2.37), with parameters N and p.

a) Find the distribution $\Pr(Y=y)$ on the random variable $Y=X^2$.
b) With $N=50$ and $p=0.70$, find $\Pr(Y=961)$.

Solution

a) Since X consists of nonnegative integers, we choose $X=+\sqrt{Y}$ (rather than $-\sqrt{Y}$) to be the single-valued inverse $w(Y)$ that makes the transformation one-to-one. This choice also requires X to be nonnegative, as desired. Using Eq. (4.116), we find that the probability distribution on Y is given by

$$\Pr(Y=y) = \Pr(X^2=y) = \Pr(X=\sqrt{y}). \tag{4.117}$$

Substituting $k=x=\sqrt{y}$ and $q=1-p$ in Eq. (2.37) and applying the last equality in (4.117) gives

$$\Pr(Y=y) = \binom{N}{\sqrt{y}} p^{\sqrt{y}} (1-p)^{N-\sqrt{y}}, \tag{4.118}$$

in which $y=0, 1, 4, \ldots, N^2$.

b) When $Y=961$, $\sqrt{Y}=31$. For $N=50$, Eq. (4.118) yields

$$\Pr(Y=961) = \binom{50}{31}(0.70)^{31}(0.30)^{50-31} = 0.0558. \tag{4.119}$$

As a check, we see that Eq. (2.37) with $k=x$ gives the same numerical expression as Eq. (4.119) for the probability $\Pr(X=31)$.

We next consider two continuous random variables, X and Y, with $Y=u(X)$ and $X=w(Y)$ representing their functional relationship. Given the probability density $f(x)$ on X, we want to determine the density $g(y)$ on Y. We shall assume that the transformation between X and Y is either an increasing or decreasing monotonic function over the entire domain of X, thus assuring a one-to-one relationship between the two random variables. To be specific, we first select an increasing function. Then an event $c \leq Y < d$, where c and d are two arbitrary values of Y, must be equivalent to the event $w(c) \leq X < w(d)$. (If the relationship is monotonically decreasing, then the latter condition is $w(d) < X \leq w(c)$.) Thus,

$$\Pr(c \leq Y < d) = \Pr[w(c) \leq X < w(d)]. \tag{4.120}$$

It follows from Eq. (4.8) that

$$\int_c^d g(y)\,dy = \int_{w(c)}^{w(d)} f(x)\,dx. \qquad (4.121)$$

Changing variables in the integrand on the right-hand side to equivalent Y quantities, one has

$$f(x) = f[w(y)] \quad \text{and} \quad dx = \frac{dw(y)}{dy}\,dy. \qquad (4.122)$$

Equation (4.121) can thus be written as

$$\int_c^d g(y)\,dy = \int_{w(c)}^{w(d)} f[w(y)]\frac{dw(y)}{dy}\,dy, \qquad (4.123)$$

from which it follows that

$$g(y) = f[w(y)]\frac{dw(y)}{dy}. \qquad (4.124)$$

This expression enables one to transform a probability density function $f(x)$ on X to the corresponding probability density function $g(y)$ on Y when the derivative $dw(y)/dy$ is positive. If the transformation between X and Y is monotonically decreasing, then to maintain nonnegative probabilities, we use the absolute value and write

$$g(y) = f[w(y)]\left|\frac{dw(y)}{dy}\right|. \qquad (4.125)$$

Since Eq. (4.125) is valid for both increasing and decreasing monotonic transformations, we shall henceforth always use it for transformations.

> ■ *Example*
> Let X be a random variable with probability density given by
>
> $$f(x) = \begin{cases} \dfrac{x^3}{4}, & 0 \le x < 2 \\ 0, & \text{elsewhere}. \end{cases} \qquad (4.126)$$
>
> a) Find the probability density function $g(y)$ for $Y = X^2$.
> b) Show that $g(y)$ integrates to unity, as required of a probability density function (Eq. (4.10)).
>
> *Solution*
>
> a) Choosing the positive square root, $X = w(Y) = +\sqrt{Y}$, gives an increasing monotonic function in which Y goes from 0 to 4 over the domain of X between 0 and 2. Using Eq. (4.125), we write

$$g(y) = f(\sqrt{y})\frac{d}{dy}(y^{1/2}) = \frac{y^{3/2}}{4} \times \frac{y^{-1/2}}{2}$$

$$= \begin{cases} \dfrac{y}{8}, & 0 \le y < 4 \\ 0, & \text{elsewhere.} \end{cases} \tag{4.127}$$

b) Integration of the probability density function (4.127) yields

$$\int_{-\infty}^{\infty} g(y)\,dy = \int_{0}^{4} g(y)\,dy = \frac{1}{8}\int_{0}^{4} y\,dy = \left.\frac{y^2}{16}\right|_{0}^{4} = 1. \tag{4.128}$$

The region where the density function is nonzero is called the *support* of a continuous random variable. In the last example, the support of Y is $0 \le y < 4$ over the domain of X.

■ *Example*
The random variable X has the probability density function

$$f(x) = \begin{cases} 2x, & 0 \le x < 1 \\ 0, & \text{elsewhere.} \end{cases} \tag{4.129}$$

Find the probability density function for $Y = 1 - 2X$.

Solution
The inverse function is $x = (1-y)/2$, and the support of y is $-1 < y \le 1$ over the domain of X. This example thus involves a monotonic decreasing transformation. From Eq. (4.125),

$$g(y) = f\!\left(\frac{1-y}{2}\right)\left|\frac{d}{dy}\frac{1-y}{2}\right| = (1-y)\left|-\frac{1}{2}\right| = \frac{1}{2}(1-y). \tag{4.130}$$

Hence, the probability density function on Y is

$$g(y) = \begin{cases} \dfrac{1-y}{2}, & -1 < y \le 1 \\ 0, & \text{elsewhere.} \end{cases} \tag{4.131}$$

This function is normalized (Problem 4.34).

■ *Example*

The normalized energy-loss spectrum for fast neutrons of energy E_o, scattered elastically from protons initially at rest, is shown in Figure 4.5a. For the flat spectrum, the probability $f(q)dq$ that the neutron loses an amount of energy between q and $q + dq$ is simply (Turner, 2007, Section 9.6)

$$f(q)dq = \begin{cases} \dfrac{1}{E_o} dq, & 0 \leq q < E_o \\ 0, & \text{elsewhere.} \end{cases} \qquad (4.132)$$

a) Determine the probability density function for the recoil velocity of the proton.
b) What are the MKS units of the velocity density function?
c) Show that the velocity density function is normalized.
d) What is the probability that a neutron of energy E_o loses between 1/4 and 1/2 its energy in a collision?
e) What is the probability that a struck proton recoils with a velocity between 1/4 and 1/2 the original velocity of the neutron?

Solution

a) We can express the proton recoil energy Q by writing $Q = MV^2/2$, where M and V are the proton mass and velocity. Since the scattering is elastic, Q is also equal to the energy lost by the neutron in the collision. Therefore, the spectrum shown in Figure 4.5a also represents the probability density for the proton recoil energy. We are asked to transform $f(q)$ into the probability density function, which we shall call $g(v)$, for the recoil velocity V of the proton. For the inverse transformation between V and Q we choose the positive square root, $V = \sqrt{2Q/M}$, giving a monotonically increasing relationship between the proton recoil energy and velocity. The neutron mass is assumed to be the same as that of the proton, M. Therefore, the proton can acquire any amount of the neutron energy E_o and thus recoil with any velocity up to a maximum of $v_o = \sqrt{2E_o/M}$ in a head-on, billiard-ball-like collision. Applying the formalism of Eq. (4.125) with $X = Q$, $Y = V$, $f = 1/E_o$, and $w(y) = w(v) = Mv^2/2$, we find that

$$g(v) = \begin{cases} \dfrac{1}{E_o} \dfrac{d}{dv}\left(\dfrac{Mv^2}{2}\right) = \dfrac{Mv}{E_o}, & 0 \leq v < v_o \\ 0, & \text{elsewhere.} \end{cases} \qquad (4.133)$$

This function is shown in Figure 4.5b. Whereas all proton recoil *energies* up to E_o are equally probable, this result shows that the probability density for the recoil *velocity* of the proton increases linearly with v up to a maximum given by $v_o = \sqrt{2E_o/M}$, equal to the velocity of the incident neutron.

b) Since $g(v)dv$ is dimensionless (a probability) and dv has the dimensions of velocity, it follows that the MKS units of $g(v)$ must be the reciprocal velocity units, $(m\,s^{-1})^{-1} = m^{-1}\,s$. To show this result explicitly, we see from Eq. (4.133) that the dimensions of g are those of $E_o^{-1}Mv$. The energy E_o is in joules, where $1\,J = 1\,kg\,m^2\,s^{-2}$ (i.e., work = force × distance = mass × acceleration × distance). Thus, we write

$$\text{Units } [E_o^{-1}Mv] = (kg\,m^2\,s^{-2})^{-1}kg\,m\,s^{-1} = m^{-1}\,s. \tag{4.134}$$

c) From Eq. (4.133) it follows that

$$\int_{-\infty}^{\infty} g(v)dv = \int_0^{v_o} g(v)dv = \frac{M}{E_o}\int_0^{v_o} v\,dv = \frac{M}{E_o} \times \frac{v_o^2}{2} = 1. \tag{4.135}$$

The probability density function for the velocity must be normalized.

d) The probability that a neutron loses between $1/4$ and $1/2$ its energy E_o in a collision can be found by integrating the probability density function in Eq. (4.132) between these two limits. However, by inspection of the flat spectrum in Figure 4.5a one sees that this probability must be $1/4$.

e) The probability that the struck proton acquires a velocity between $1/4$ and $1/2$ that of the original neutron, v_o, can be found by integrating the probability density function (4.133):

$$\Pr\left(\frac{v_o}{4} \leq V < \frac{v_o}{2}\right) = \frac{M}{E_o}\int_{v_o/4}^{v_o/2} v\,dv = \frac{M}{2E_o}\left(\frac{v_o^2}{4} - \frac{v_o^2}{16}\right) = \frac{3Mv_o^2}{32E_o}$$

$$= \frac{3}{16}. \tag{4.136}$$

Transformations do not alter the general conditions on probability density functions. Two quick checks are often useful to see whether calculated transformations meet these conditions. The transformed probability density must be nonnegative everywhere, and the new density function must integrate to unity. These conditions are readily seen in the last example.

Transformations for more than one variable can also be performed. The interested reader can find multivariable transformations discussed in the books by Hogg and Craig (1978) and by Taylor (1974) listed in the Bibliography.

4.8
Bayes' Theorem

We next extend Bayes' theorem for discrete events to continuous random variables. In place of the discrete event B and the partitioning A_j of the sample space in Eq. (3.44), we consider a continuous random variable X, having a probability density function $f(x)$ that also depends on another variable Θ. We denote the conditional

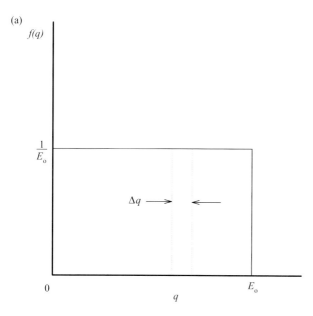

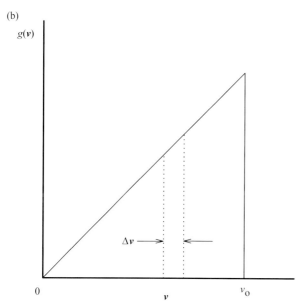

Figure 4.5 Normalized probability density functions (a) $f(q)$ for the energy loss Q by a neutron and (b) $g(v)$ for the recoil velocity V of the struck proton in the collision of a fast neutron with a proton.

probability density function of X given Θ as $f(x|\theta)$ and the probability density function of Θ as $g(\theta)$. In analogy with Eq. (3.44), we can write Bayes' theorem for continuous random variables as follows. For every x such that $f(x) > 0$ exists, the probability density function of Θ given $X=x$ is

$$f(\theta|x) = \frac{f(x,\theta)}{f(x)} = \frac{f(x|\theta)g(\theta)}{f(x)} = \frac{f(x|\theta)g(\theta)}{\int f(x|\theta)g(\theta)d\theta}. \qquad (4.137)$$

The denominator is the marginal probability density function of X (Section 4.4). The statement (4.137) for continuous random variables can be compared directly with Eq. (3.44) for discrete events.

Chapter 15 on Bayesian analysis will show applications of Bayes' theorem in the form Eq. (4.137). In that context, $g(\theta)$ is referred to as the *prior probability density function* of Θ, and $g(\theta|x)$ is called the *posterior probability density function* of Θ. In contrast to classical methods, Bayesian analysis offers an alternative – and conceptually different – basis for making statistical inferences.

Problems

4.1 An experiment has exactly four possible outcomes: E_1, E_2, E_3, and E_4. Check whether the following assignments of probabilities are possible and state why or why not:
  a) $\Pr(E_1)=0.26$, $\Pr(E_2)=0.24$, $\Pr(E_3)=0.17$, $\Pr(E_4)=0.33$.
  b) $\Pr(E_1)=0.27$, $\Pr(E_2)=0.31$, $\Pr(E_3)=0.26$, $\Pr(E_4)=0.17$.
  c) $\Pr(E_1)=0.08$, $\Pr(E_2)=0.57$, $\Pr(E_3)=-0.08$, $\Pr(E_4)=0.43$.
  d) $\Pr(E_1)=1.01$, $\Pr(E_2)=0.01$, $\Pr(E_3)=0.04$, $\Pr(E_4)=0.31$.

4.2 Show that the function

$$f(x) = \begin{cases} \dfrac{2x}{n(n+1)}, & x = 1, 2, \ldots, n \\ 0, & \text{elsewhere,} \end{cases}$$

satisfies Eqs. (4.2) and (4.3) and hence can be called a probability distribution.

4.3 Using the probability function in the last problem, calculate its (a) mean and (b) variance.

4.4 Use the probability function from Problem 4.2.
  a) Determine the cumulative probability function for arbitrary n.
  b) Plot the cumulative function for $n=5$.

4.5 Verify Eq. (4.12).

4.6 Verify Eq. (4.13).

4.7 A continuous random variable X has the density function

$$f(x) = \begin{cases} 6x(1-x), & 0 < x < 1, \\ 0, & \text{elsewhere.} \end{cases}$$

  a) Determine the cumulative distribution function.
  b) Plot both the density function and the cumulative distribution function.
  c) Determine $\Pr(0.2 < X < 0.6)$ by straightforward integration.

d) Determine Pr(0.2 < X < 0.6) by using the cumulative distribution function and Eq. (4.13).

4.8 A continuous random variable X has the following probability density function:

$$f(x) = \begin{cases} 4x\,e^{-2x}, & 0 < x < \infty, \\ 0, & \text{elsewhere.} \end{cases}$$

a) Plot $f(x)$.
b) Determine the cumulative distribution function, $F(x)$, and plot it.
c) Obtain the expected value of X and locate it on the plots in parts (a) and (b).

4.9 In a sample of five urine specimens, let X denote the numbers that are found to contain reportable activity. If we assume that the specimens are independent and that the probability of being found radioactive is 0.05 for each, then X follows the binomial distribution.
a) Show that the probability distribution for X is

$$\Pr(X = x) = \begin{cases} \binom{5}{x}(0.05)^x(0.95)^{5-x}, & \text{for } x = 0, 1, 2, \ldots, 5; \\ 0, & \text{elsewhere.} \end{cases}$$

Graph the probability distribution.
b) What is the probability of observing at most two radioactive specimens?
c) What is the expected value of X? Explain what this value means. Can X have this value?
d) Plot the cumulative distribution function for X.

4.10 Verify Eq. (4.38).

4.11 The single-parameter ($\lambda > 0$) exponential distribution is often used to model the lifetime X (or time to failure) of an item:

$$f(x) = \begin{cases} \lambda\,e^{-\lambda x}, & \text{for } x > 0 \\ 0, & \text{elsewhere.} \end{cases}$$

The expected value of the lifetime is the reciprocal of the parameter λ: $E(X) = 1/\lambda$ (cf. Eq. (4.38)).
a) A manufacturer claims that his light bulbs have a expected life of 2000 h. Assuming that the exponential distribution is an adequate representation of the length-of-life distribution for the bulbs, determine the probability that a given bulb will last more than 2400 h.
b) Determine the lifetime that only 5% of the bulbs will exceed.
c) What is the probability that a light bulb will last at least 1800 h, but not more than 4000 h?
d) Find the cumulative distribution function for the model.
e) Check your answer to part (c) by using the cumulative distribution.

4.12 For a continuous random variable X, show that $E(|X-v|)$ has its minimum value when v is the median of the distribution.
(Hint: Write

$$u(v) = E(|X-v|) = \int_{-\infty}^{\infty} |x-v| f(x) dx$$

$$= \int_{-\infty}^{v} (v-x) f(x) dx + \int_{v}^{\infty} (x-v) f(x) dx.$$

Solve $du/dv=0$ for v, taking into account the fact that the variable v appears in the limits of integration as well as in the integrands.)

4.13 In place of Eq. (4.46), one can consider the mean squared deviation from an arbitrary value v of a continuous random variable X. Show that $E[(X-v)^2]$ has its minimum value when v is the mean of the distribution.

4.14 Verify Eq. (4.50).

4.15 Show that the marginal distribution (4.68) is a true probability function.

4.16 Verify that the marginal density function on X_2 from Eq. (4.70) is given by Eq. (4.72).

4.17 The random variables X_1 and X_2 have the following joint probability distribution:

$$\begin{matrix} x_1 & = & 1 & 2 & 3 & 1 & 2 & 3 & 1 & 2 & 3 \\ x_2 & = & 1 & 1 & 1 & 2 & 2 & 2 & 3 & 3 & 3 \\ f(x_1, x_2) & = & 0 & \dfrac{1}{6} & \dfrac{1}{12} & 0 & \dfrac{1}{5} & \dfrac{1}{9} & \dfrac{2}{15} & \dfrac{1}{6} & \dfrac{5}{36} \end{matrix}$$

a) Find the marginal distribution of X_1.
b) Find the marginal distribution of X_2.
c) Are X_1 and X_2 independent random variables? Why or why not?
d) Find $\Pr(X_2 = 3 | X_1 = 2)$.

4.18 Let X_1 and X_2 be the larger and smaller, respectively, of the two numbers showing when a pair of fair dice is thrown.
a) Find the joint distribution of X_1 and X_2.
b) Obtain the marginal distributions on X_1 and on X_2.
(Hint: List all possibilities in a table and find the ones that are possible for X_1 and X_2, with $X_1 \leq X_2$.)

4.19 The random variables X_1 and X_2 have the following joint probability density:

$$f(x_1, x_2) = \begin{cases} \dfrac{2}{3}(x_1 + 2x_2), & 0 \leq x_1 \leq 1, \ 0 \leq x_2 \leq 1 \\ 0, & \text{elsewhere}. \end{cases}$$

a) Find the marginal density on X_1.
b) Find the marginal density on X_2.
c) Are X_1 and X_2 independent? Why or why not?
d) Calculate $\Pr(X_1 < X_2)$.

4.20 The random variables X_1 and X_2 denote the lifetimes in years of two electronic components, having the following joint density:

$$f(x_1, x_2) = \begin{cases} 2e^{-x_1 - 2x_2}, & x_1 > 0,\ x_2 > 0, \\ 0, & \text{elsewhere.} \end{cases}$$

a) Determine the marginal density on X_1.
b) Determine the marginal density on X_2.
c) Show that X_1 and X_2 are independent.
d) Use the results from (a), (b), and (c) to calculate $\Pr(X_1 > 1,\ X_2 < 2)$.

4.21 A company produces blended oils, each blend containing various proportions of type A and type B oils, plus others. The proportions X and Y of types A and B in a blend are random variables, having the joint density function

$$f(x, y) = \begin{cases} 24xy, & 0 < x \le 1,\ 0 < y \le 1,\ x+y \le 1 \\ 0, & \text{elsewhere.} \end{cases}$$

a) Obtain the marginal probability densities on X and Y.
b) For a given can of oil, find the probability that type A accounts for over one-half of the blend.
c) If a given can of oil contains 3/4 of type B oil, find the probability that the proportion of type A will be less than 1/8.

4.22 Given the joint density function

$$f(x, y) = \begin{cases} \dfrac{1}{27}(7 - x - y), & 0 < x < 3,\ 1 < y < 4 \\ 0, & \text{elsewhere.} \end{cases}$$

a) Find the marginal densities on X and Y.
b) Obtain the conditional probability density of Y given X.
c) Using the result from part (b), compute $\Pr(1 \le Y \le 3 | X = 2)$.

4.23 Show that Eq. (4.83) follows from the definition (4.82) of covariance.
4.24 After Eq. (4.88), show that $\sigma_2^2 = 295/256$.
4.25 Prove the theorem (4.93).
4.26 For independent random variables, show that Eq. (4.101) follows from Eq. (4.100).
4.27 Following Eq. (4.108), show that $\sigma_2^2 = 11/144$.
4.28 If it is known that X has a mean of 25 and a variance of 16, use Chebyshev's inequality to obtain
  a) a lower bound for $\Pr(17 < X < 33)$

(Hint: Chebyshev's inequality, Eq. (4.115), states that $\Pr(|X-\mu| \geq k\sigma) \leq 1/k^2$; and so the complementary event yields $\Pr(|X-\mu| < k\sigma) \geq 1 - 1/k^2$.)
and
b) an upper bound for $\Pr(|X-25| \geq 14)$.

4.29 If $E(X) = 10$ and $E(X^2) = 136$, use Chebyshev's inequality to obtain
a) a lower bound for $\Pr(-2 < X < 22)$ and
b) an upper bound for $\Pr(|X-10| \geq 18)$.

4.30 Let X be uniformly distributed over the open interval (0, 1).
a) Determine the distribution of $Y = X^3$.
b) Calculate $E(Y)$.
c) Use Eq. (4.30) to obtain $E(Y) = E(X^3)$ directly from the density function on X.

4.31 A continuous random variable X has a probability density function $f_X(x)$ and cumulative distribution function $F_X(x)$. Consider the transformation $Y = X^2$. For $y < 0$, the set $\{x: x^2 \leq y\}$ is the empty set of real numbers. Consequently, $F_Y(y) = 0$ for $y < 0$. For $y \geq 0$,

$$F_Y(y) = \Pr(Y \leq y) = \Pr(X^2 \leq y) = \Pr(-\sqrt{y} \leq X \leq \sqrt{y})$$
$$= F_X(\sqrt{y}) - F_X(-\sqrt{y}).$$

Show that the probability density for Y is given by

$$f_Y(y) = \begin{cases} \dfrac{1}{2\sqrt{y}}[f_X(\sqrt{y}) + f_X(-\sqrt{y})], & \text{for } y \geq 0 \\ 0, & \text{for } y < 0. \end{cases}$$

4.32 A random variable X has the following density function:

$$f_X(x) = \frac{1}{\sqrt{2\pi}} e^{-x^2/2}, \quad -\infty < x < \infty.$$

Obtain the probability density for the random variable $Y = X^2$ by using the result of the last problem. (Note: The random variable X has the density function associated with a normal random variable with zero mean and unit variance. The distribution on Y is that of a chi-squared variable with degrees of freedom equal to unity. The chi-squared distribution is discussed in Section 6.8.)

4.33 The distribution of temperatures X in degrees Fahrenheit of normal, healthy persons has a density function approximated by

$$f(x) = \frac{1}{2\sqrt{\pi}} e^{-(1/4)(x-98.6)^2}, \quad -\infty < x < \infty.$$

Find the density function for the temperature $Y = 5(X - 32.0)/9$, measured in degrees centigrade.

4.34 Show that the probability density function (4.131) integrates to unity.

4.35 For the example involving the probability density function (4.132), calculate, as functions of E_o,
   a) the expected value of the proton recoil energy and
   b) the variance of the proton recoil energy.

4.36 Repeat the last problem for the proton recoil velocity, rather than the energy.

4.37 With reference to the last two problems, show that the ratio of (1) the expected value of the proton recoil velocity and (2) the velocity corresponding to the expected value of the proton recoil energy is $2\sqrt{2}/3 = 0.943$. Why are the two velocities not equal?

4.38 What is the median proton recoil velocity for the distribution given by Eq. (4.133)?

4.39 A point source emits radiation isotropically, that is, uniformly in all directions over the solid angle of 4π steradians. In spherical polar coordinates (θ, φ), where θ and φ are, respectively, the polar and azimuthal angles, the probability for isotropic emission into the solid angle $d\theta\, d\varphi$ is

$$p(\theta, \varphi)d\varphi\, d\theta = \begin{cases} \dfrac{1}{4\pi}\sin\theta\, d\theta\, d\varphi, & 0 \leq \theta < \pi,\ 0 \leq \varphi < 2\pi \\ 0, & \text{elsewhere}, \end{cases}$$

where $p(\theta, \varphi)$ is the probability density function.
   a) Show that the density function for emission at a polar angle $0 \leq \theta < \pi$ is $(1/2)\sin\theta$ and zero elsewhere.
   b) How is the answer to (a) related to the marginal density function on θ?
   c) Find the cumulative density function for emission at a polar angle not exceeding θ.
   d) Derive the expression given above for the probability density function $p(\theta, \varphi)$ and show that it is normalized.

4.40 If Θ is uniformly distributed on $(-\pi/2, \pi/2)$, show that $Y = \tan\Theta$ has the density function

$$f(y) = \dfrac{1}{\pi(1+y^2)}, \quad -\infty < y < \infty.$$

The random variable Y satisfies this Cauchy distribution, and has the property that its mean as well as all higher moments are infinite. Graphically, it is similar to the normal distribution, except that its tails are much thicker, leading to its infinite mean. The Student's distribution with one degree of freedom has the Cauchy distribution (Section 6.9).

5
Discrete Distributions

5.1
Introduction

Having laid the foundation of probability theory, essential to the application of statistics, we now focus on specific probability distributions that are capable of describing many practical applications. The exponential function and the binomial distribution were employed in Chapter 2 for computations of radioactive disintegration. We shall discuss other models that provide useful approximations to observed data. As we shall see, a variety of probability distributions, characterized by a few parameters, describe many phenomena that occur naturally. We treat discrete probability distributions in this chapter and continuous distributions in Chapter 6. Additional discrete distributions are discussed in Johnson et al. (2005).

5.2
Discrete Uniform Distribution

The *discrete uniform distribution* describes a case in which the probability is the *same* for all values that a discrete random variable can take on. The gambler's roulette wheel provides an example. It has 38 identical holes, designated 1–36, 0, and 00. If the wheel is balanced and the ball is rolled in a fair way, then all 38 holes have the same chance of being the one in which the ball will land. The discrete uniform distribution is then a good model for describing the probability of landing on any given number 1–36, 0, or 00. When the random variable X takes on the k values $x_1, x_2, \ldots, x_k$ with equal likelihood, the discrete uniform distribution is given by writing

$$\Pr(X = x) = \frac{1}{k}, \quad \text{for } x = x_1, x_2, \ldots, x_k. \tag{5.1}$$

Other examples of the discrete uniform distribution occur in tossing a fair die, randomly drawing a card from a deck, and randomly drawing a number from a hat. The discrete uniform distribution is often ascribed to a random sample drawn from a

---

Statistical Methods in Radiation Physics, First Edition. James E. Turner, Darryl J. Downing, and James S. Bogard
© 2012 Wiley-VCH Verlag GmbH & Co. KGaA. Published 2012 by Wiley-VCH Verlag GmbH & Co. KGaA.

population, when the assumption is made that each sample point is chosen with equal probability.

Similar ideas apply to continuous uniform random variables, as described in Section 6.2. In principle, a computer random number generator produces a uniform distribution of numbers $0 < r \leq 1$. However, because of their finite precision, digital computers generate a discrete distribution of pseudorandom numbers, ideally with a uniform distribution.

▪ **Example**
A hat contains 50 identical pieces of paper, each marked with a different integer from 1 to 50. Let X denote the number on a piece of paper drawn randomly from the hat. Determine the expected value and variance of X.

Solution
The appropriate probability model for this experiment is the discrete uniform distribution, for which X takes on the values x from 1 to 50 with probability distribution given by

$$\Pr(X = x) = \frac{1}{50}, \quad \text{for } x = 1, 2, \ldots, 50. \tag{5.2}$$

The expected value for X is, from Eq. (4.27),

$$E(X) = \sum_{i=1}^{50} x_i \Pr(x_i) = \sum_{i=1}^{50} i \frac{1}{50} = \frac{1}{50} \sum_{i=1}^{50} i = \frac{1}{50} \times \frac{1+50}{2} \times 50$$
$$= \frac{51}{2}. \tag{5.3}$$

The variance can be obtained by using Eq. (4.48). The sum of the squares of the first k integers is $k(k+1)(2k+1)/6$, and so $E(X^2) = (k+1)(2k+1)/6$. Thus, from Eq. (4.48), we find

$$\sigma_X^2 = \frac{(51)(101)}{6} - \left(\frac{51}{2}\right)^2 = 858.50 - 650.25 = 208.25. \tag{5.4}$$

5.3
Bernoulli Distribution

A statistical experiment can consist of repeated, identical, independent trials, each trial resulting in either one or another of only two possible outcomes, which can be labeled "success" or "failure." When independent trials have two possible outcomes and the probability for success or failure remains the same from trial to trial, the experiment is termed a *Bernoulli process*. Each trial is called a *Bernoulli trial*.

One can formally describe the probability distribution for Bernoulli trials. We let X denote a Bernoulli random variable and arbitrarily assign the value 1 to success and the value 0 to failure. Then, if p is the probability of success, we write for the *Bernoulli distribution*,

$$\Pr(X = x) = p^x(1-p)^{1-x}, \quad \text{for } x = 0, 1. \tag{5.5}$$

We see that the probability for success in a given trial is then $\Pr(X=1)=p$, and that for failure is $\Pr(X=0)=1-p$.

A familiar example of Bernoulli trials is the repeated tossing of a coin. The coin toss is considered fair if the probability of either side landing face up is exactly 1/2. In this case, $p=1/2$, irrespective of whether heads or tails is regarded as success. It follows from Eq. (5.5) that $\Pr(X=0)=\Pr(X=1)=1/2$. As long as each toss is independent of the others and the probability of a given side landing face up is constant (not necessarily 1/2), then repeated tosses of the coin constitute a series of independent Bernoulli trials.

■ *Example*
A biased coin, which turns up heads (H) with probability 0.7, is tossed five times in succession. Determine the probability of obtaining the sequence H, H, T, H, T in five tosses.

Solution
The Bernoulli probability model is appropriate for this experiment. Letting X_i ($i = 1, 2, \ldots, 5$) denote the outcomes of each toss, where $X_i = 0$ if heads appears on the ith toss, then we desire

$$\Pr(X_1 = 0, X_2 = 0, X_3 = 1, X_4 = 0, X_5 = 1)$$
$$= (0.7)(0.7)(0.3)(0.7)(0.3) = (0.7)^3(0.3)^2 = 0.0309. \tag{5.6}$$

5.4
Binomial Distribution

The binomial distribution was introduced in Section 2.5 to describe the number of atoms that decay in a given period of time from a radioactive source with a fixed initial number of identical atoms. This distribution is a model for statistical experiments in which there are only two possible outcomes. In Chapter 2, the alternatives for each atom in the source were to "decay" or "not decay," which we described there as "success" or "failure," respectively. The following set of statements formalizes the specific conditions that lead to the binomial distribution.

A *binomial experiment* consists of a number n of repeated Bernoulli trials made under the following conditions:

1) The outcome of each trial is either one of two possibilities, success or failure.
2) The probability of either outcome is constant from trial to trial.
3) The number of trials n is fixed.
4) The repeated trials are independent.

A binomial experiment of n trials can be repeated over and over. The number of successes in a given experiment is thus itself a random variable, taking on the discrete, integral values 0, 1, 2, ..., n. If p represents the probability of success in an individual trial and $1 - p$ the probability of failure, then the number of successes X has the binomial distribution with parameters n and p. We write

$$\Pr(X = x) = b(x; n, p) = \binom{n}{x} p^x (1-p)^{n-x}, \quad \text{for } x = 0, 1, \ldots, n, \tag{5.7}$$

where

$$\binom{n}{x} = \frac{n!}{x!(n-x)!}. \tag{5.8}$$

With slightly different notation, Eq. (5.7) is the same as Eq. (2.37) in Chapter 2. It represents the number of combinations of n things taken x at a time. It can also be thought of as the number of ways we can partition n objects into two groups, one containing x successes and the other $(n - x)$ failures. In the above notation, $b(x; n, p)$ denotes the probability that the binomial random variable takes on the value x (i.e., one observes x successes), when there are n trials and the probability of success on any trial is p.

We shall use the notation $B(r; n, p)$ to denote the *cumulative binomial distribution*, that is, the probability $\Pr(X \leq r)$ of observing r or fewer successes in n trials, where the probability of success on any trial is p. We write

$$B(r; n, p) = \Pr(X \leq r) = \sum_{x=0}^{r} b(x; n, p), \tag{5.9}$$

where $b(x; n, p)$ is given by Eq. (5.7). The cumulative binomial distribution is tabulated in Table A.1 in the Appendix for selected values of r, n, and p. One is frequently interested in the probability $\Pr(X > x)$ that the binomial random variable has a value *greater* than x. From Eq. (4.6) it follows that $\Pr(X > r) = 1 - B(r; n, p)$. Also, since the cumulative distribution gives the values for $\Pr(X \leq r)$, individual probabilities $b(x; n, p)$ can be obtained by subtracting values of $B(r; n, p)$ for successive values of r. Thus, from Eq. (4.7),

$$\Pr(X = r) = B(r; n, p) - B(r-1; n, p). \tag{5.10}$$

These uses of Table A.1 are illustrated in the examples that follow and in the problems at the end of the chapter.

Comparison of Eqs. (5.7) and (5.5) shows that the binomial model becomes the same as the Bernoulli model when the number of trials is $n = 1$ (Problem 5.4). This fact can be used to calculate the mean and variance of the binomial distribution, as we show next. (The mean and variance can also be found directly by using the probability distribution (5.7) (Problem 5.10).) Since each trial is a Bernoulli trial, we can regard the binomial distribution as the result of summing the outcomes of n individual Bernoulli trials. We let $X_1, X_2, \ldots, X_n$ denote the consecutive outcomes of each Bernoulli trial, each X_i having the value 0 or 1, according to whether a failure or success occurred. Then, if X denotes the number of successes in n trials, we have

$$X = \sum_{i=1}^{n} X_i. \tag{5.11}$$

Since the X_i are independent random variables, we may write for the mean

$$E(X) = E\left(\sum_{i=1}^{n} X_i\right) = \sum_{i=1}^{n} E(X_i). \tag{5.12}$$

Since each X_i takes on the value 0 with probability $(1-p)$, and the value 1 with probability p, its expected value is $E(X_i) = p$ (Problem 5.3). Therefore, we obtain for the mean of the binomial distribution,

$$E(X) = \sum_{i=1}^{n} E(X_i) = np. \tag{5.13}$$

For the variance of each X_i, we have, with the help of Eqs. (4.47) and (4.48),

$$\operatorname{Var}(X_i) = E[(X_i - p)^2] = E(X_i^2) - p^2 = (0)^2(1-p) + (1)^2 p - p^2$$
$$= p(1-p). \tag{5.14}$$

Thus, the variance of the binomial distribution is given by

$$\operatorname{Var}(X) = \operatorname{Var}\left(\sum_{i=1}^{n} X_i\right) = \sum_{i=1}^{n} \operatorname{Var}(X_i) = np(1-p). \tag{5.15}$$

We see from Eq. (5.13) that the mean of the binomial distribution is simply the proportion p of the total number of n trials that are successes. On the other hand, the expression (5.15) for the variance, which is a measure of the spread of the distribution, is not intuitively apparent.

■ **Example**
To monitor for possible intakes of radioactive material, the urine of radiation workers is sometimes tested by counting for high radiation content. At a certain installation, the probability that the test procedure will falsely declare a high content for an individual's specimen when it should not is 0.10. In one survey, specimens from a group of 20 workers are counted, and 5 are found to score in the high range. Could this finding be a random occurrence, or is there reason to be concerned that some individuals in the group have elevated intakes?

Solution
The observed result of 5 high out of 20 tested appears to be rather large, if some individuals in the group did not, in fact, experience elevated intakes. One way to gauge this frequency is to look at how far the observed number is from the mean, as measured by the standard deviation. The testing procedure meets the conditions of a binomial experiment, in which $n = 20$ and $p = 0.10$

is the probability that an individual outcome is declared "high" when it should not be, the alternative being "not high." From Eq. (5.13), the expected value of the number of tests declared high, when they should not be so declared, is 20 (0.10) = 2. We are asked whether the observed result, 5 − 2 = 3 more than the mean, is too large a deviation to expect. From Eq. (5.15), the variance is 20(0.10) (0.90) = 1.80, and so the standard deviation is $\sqrt{1.80} = 1.34$. The observed result is thus $3/1.34 = 2.24$ standard deviations beyond the mean. Chebyshev's inequality, Eq. (4.115), provides a rigorous upper bound to the probability for the occurrence of this result. It follows from Eq. (4.115) that the probability is no larger than $1/k^2 = 1/(2.24)^2 = 0.199$. Since Chebyshev's inequality is usually very imprecise (it applies to *any* distribution), the actual probability is presumably considerably less than 20%. Based on this evidence, we should suspect that finding 5 false high specimens is not just a random occurrence among 20 unexposed individuals. While the application of Chebyshev's inequality is of some use, it is not decisive in this example. We can, however, find the actual probability from the binomial distribution, as we do next.

We let X be the number of specimens that the procedure declares to have high radiation content when they should not. Then Eq. (5.7) gives the probability $\Pr(X = x) = b(x; n, p)$ of observing exactly x high results. Our concern is whether any result as large as 5 should occur randomly in the absence of elevated intakes. Therefore, we evaluate the probability for $X \geq 5$. The evaluation can be conveniently carried out with the help of the cumulative binomial distribution function. As discussed in the paragraph following Eqs. (5.7) and (5.8), we can use Table A.1 to write

$$\Pr(X \geq 5) = \sum_{x=5}^{20} b(x; 20, 0.10) = 1 - B(4; 20, 0.10) = 1 - 0.957$$
$$= 0.043. \qquad (5.16)$$

Thus, it would be rare to find five or more high results in the group, if no individuals had experienced elevated intakes. The evidence strongly suggests that elevated intakes might have occurred and thus been detected. Additional measurements are called for. Also, the analytical procedures could be checked for possible errors.

We note in this example that the probability of finding *exactly* five high results is, of itself, not as significant as finding *any* number as large as five. Therefore, our judgment is based on the value of $\Pr(X \geq 5)$ rather than $\Pr(X = 5)$.

■ *Example*

In the last example with 20 specimens,

a) What is the probability that exactly five specimens falsely show high radiation content?
b) What is the probability that at most two specimens will falsely read high?

c) What is the probability that between two and five specimens will falsely read high?

Solution

a) The first question asks for

$$\Pr(X = 5) = b(5; 20, 0.10) = B(5; 20, 0.10) - B(4; 20, 0.10)$$
$$= 0.989 - 0.957 = 0.032. \qquad (5.17)$$

(This quantity can also be computed directly from Eq. (5.7).) Comparison with Eq. (5.16) shows that $\Pr(X \geq 5)$ is about half again as large as the probability of finding *exactly* five high readings.

b) The probability for at most two false high readings is

$$\Pr(X \leq 2) = B(2; 20, 0.10) = 0.667, \qquad (5.18)$$

where, again, we have employed Table A.1.

c) The probability that between two and five specimens are falsely declared as high is conveniently found as the difference in the values from Table A.1:

$$\sum_{x=3}^{4} b(x, 20, 0.10) = B(4; 20, 0.10) - B(2; 20, 0.10) = 0.957 - 0.677 = 0.280.$$
$$(5.19)$$

Answers like (a)–(c) would be likely to occur in the absence of elevated intakes.

■ *Example*
Use the cumulative distribution to calculate the probability that exactly five persons will be falsely declared high in the last examples when the probability of that happening for an individual is 0.50 and the group size is 20.

Solution
The solution we desire is $b(5; 20, 0.5)$, which we are asked to evaluate from the cumulative distribution, rather than directly from Eq. (5.7). To do this, we need to subtract $B(4; 20, 0.5)$ from $B(5; 20, 0.5)$. Using Table A.1, we obtain

$$b(5; 20, 0.50) = B(5; 20, 0.50) - B(4; 20, 0.50) = 0.021 - 0.006$$
$$= 0.015. \qquad (5.20)$$

(Alternatively, Eq. (5.7) yields the result 0.014785....)

Similarly, to calculate the probability that the random variable lies in any interval, we simply subtract the two table values for that interval.

Example

Determine the probability that anywhere from 4 to 10 false high results would be observed in a group of 20, when the probability is 0.10 for a single specimen.

Solution

In this case, from Table A.1,

$$\Pr(4 \leq X \leq 10) = B(10; 20, 0.10) - B(3; 20, 0.10)$$
$$= 1.000 - 0.867 = 0.133. \tag{5.21}$$

5.5 Poisson Distribution

In his lifetime, Poisson (Figure 5.1) made a number of fundamental contributions to mathematics and physics. One of the most remarkable statistical distributions in these fields bears his name. The *Poisson distribution* describes *all* random processes that occur with a probability that is both small and constant. As such, it finds an

Figure 5.1 Siméon-Denis Poisson (1781–1840), French mathematician and teacher, did pioneering work on definite integrals, electromagnetic theory, mechanics, and probability theory. (Courtesy of Académie des sciences Institut de France.)

enormous variety of applications. It describes the statistical fluctuations in such phenomena as radioactive decay for long-lived nuclides, the number of traffic accidents per week in a city, the number of skunks per acre in a rural county on a summer day, and the number of calvary men killed annually in an army by mule kicks. These and many other observations have in common the counting of random events that occur in a specified time or place at a constant average rate. The Poisson distribution was originally worked out as an approximation to the binomial under certain circumstances. We shall first present this derivation of the Poisson distribution and, later in this section, derive it from first principles.

As with the binomial experiment in the last section, a particular set of conditions constitutes a Poisson process.

A *Poisson process* has the following properties:

1) The number of successes in one interval of time or space is independent of the number that occur in any other disjoint interval. (In other words, the Poisson process has no memory.)
2) The probability of success in a very small interval is proportional to the size of the interval.
3) The probability that more than one success will occur in such a small interval is negligible.

There are similarities as well as differences between the binomial and Poisson experiments.

We next show that the binomial distribution (5.7), in a particular limit of large n, yields a distribution for which the conditions of the Poisson process are satisfied. Starting with Eqs. (5.7) and (5.8), we write

$$b(x; n, p) = \frac{n(n-1) \cdots (n-x+1)}{x!} p^x (1-p)^{n-x}. \tag{5.22}$$

We consider an experiment in which events take place at a constant mean rate, λ, over a time interval, t, of any size. An example of such an experiment is the recording of counts from a long-lived radionuclide, in which case $\mu = \lambda t$ is the expected value of the number of counts in time t. We divide t into a large number n of equal subintervals of length t/n. By making n very large, we assume that the probability of recording two or more counts in an interval can be made negligible (property 3, above). Furthermore, whether an event happens in one subinterval is independent of whether an event has happened in any other subinterval (property 1). The probability of registering an event in a given subinterval is then given by $p = \lambda(t/n) = \mu/n$ (property 2). Substituting $p = \mu/n$ into Eq. (5.22) gives

$$b(x; n, p) = \frac{n(n-1) \cdots (n-x+1)}{x!} \left(\frac{\mu}{n}\right)^x \left(1 - \frac{\mu}{n}\right)^{n-x} \tag{5.23}$$

$$= \left(1 - \frac{1}{n}\right)\left(1 - \frac{2}{n}\right) \cdots \left(1 - \frac{x-1}{n}\right) \frac{\mu^x}{x!} \left(1 - \frac{\mu}{n}\right)^{n-x}. \tag{5.24}$$

To preserve the conditions applicable to the Poisson experiment, we evaluate Eq. (5.24) in the limit where $n \to \infty$, $p \to 0$, and $\mu = np$ remains constant. All of the $(x-1)$ multiplicative factors in front on the right-hand side approach unity as $n \to \infty$. We obtain a new distribution $p(x; \mu)$, which depends only upon x and μ:

$$p(x; \mu) = \lim_{n \to \infty} b(x; n, p) = \frac{\mu^x}{x!} \lim_{n \to \infty} \left(1 - \frac{\mu}{n}\right)^n \left(1 - \frac{\mu}{n}\right)^{-x}. \qquad (5.25)$$

The limit of the last factor here is also unity. The middle factor is just the definition of e, the base of the natural logarithms, raised to the power $-\mu$:

$$\lim_{n \to \infty} \left(1 - \frac{\mu}{n}\right)^n \equiv e^{-\mu}. \qquad (5.26)$$

Therefore, Eq. (5.25) becomes

$$p(x; \mu) = \lim_{n \to \infty} b(x; n, p) = \frac{\mu^x e^{-\mu}}{x!}, \qquad (5.27)$$

which is the Poisson distribution with parameter $\mu = \lambda t$. It describes the probability $\Pr(X = x)$ for the number of successes x that occur over an interval t at a constant mean rate λ per unit interval. As we show explicitly later in this section, $E(X) = \mu$, as we have already anticipated in deriving Eq. (5.27). Table A.2 in the Appendix gives values of the cumulative Poisson distribution sums $P(r; \mu) = \Pr(X \leq r)$ for selected values of r and μ. Equations exactly analogous to Eqs. (5.9) and (5.10) for the binomial functions, $b(x; n, p)$ and $B(r; n, p)$, apply to the Poisson functions, $p(x; \mu)$ and $P(r; \mu)$.

As we showed in Chapter 2, the decay of a radionuclide is rigorously described by the binomial distribution. If we deal with a long-lived source, then the average rate of decay over our observation time can be considered constant. Since only a small fraction of the total number of atoms present decay over an observation period, and since all of the atoms are identical and independent, the conditions of a Poisson process are satisfied. Thus, we may use Eq. (5.27) in place of Eq. (5.7), if desired, to describe the decay of long-lived radionuclides to a very good approximation. The Poisson distribution is often much more convenient than the binomial for numerical computations when n is large and p is small.

Figure 5.2 shows a comparison of the binomial and Poisson distributions, having the same mean $\mu = pn = 10$, in each panel, but with various values of p and n. The steady merging of the binomial into the Poisson can be seen as one progresses through larger values of n and smaller values of p. They are indistinguishable, for all practical purposes, in the last panel.

Other comparisons in Figure 5.2 are instructive. Since the mean of the distributions stays fixed, the Poisson distribution itself is the same in each panel, while the binomial changes. (Compare Eqs. (5.27) and (5.7).) Where the distributions are clearly different, as in the first panel, the Poisson probabilities are smaller than the binomial around the mean and larger in the wings of the distributions. This is understandable, because the two distributions are normalized, and the Poisson is positive for all nonnegative integers, whereas the binomial probabilities

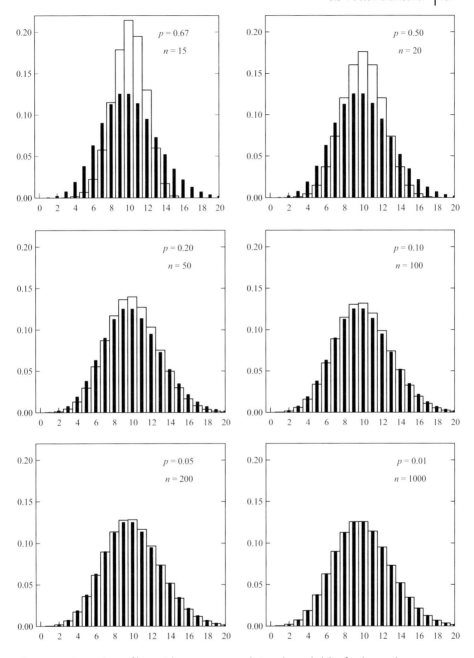

Figure 5.2 Comparison of binomial (histogram) and Poisson (solid bars) distributions, having the same mean μ, but different values of the sample size n and probability of success p. The ordinate in each panel gives the probability for the number successes shown on the abscissa. Because μ is the same in each panel, the Poisson distribution is the same throughout.

span only the interval $0 \leq x \leq n$. (One cannot have more atoms decay than are present!)

In anticipation of the use of the normal distribution for radioactive decay, we note that the two distributions in Figure 5.2 merge into a bell-shaped form. This fact is a result of n becoming relatively large (1000) in the last panel. In contrast, in Figure 5.3, $n = 100$ throughout, with p changing from 0.90 to 0.01. The mean thus shifts to smaller values as one progresses through the figure. The two distributions virtually coincide in the last panel, but not into a bell-shaped form. The first panel also illustrates clearly how the nonzero probabilities for the Poisson distribution can extend well beyond the limit ($n = 100$ in this instance) after which the binomial probabilities are identically zero.

■ **Example**

A long-lived radionuclide being monitored shows a steady average decay rate of 3 min^{-1}.

a) What is the probability of observing exactly five decays in any given minute?
b) What is the probability of observing 13 or more decays in a 5-min period?

Solution

a) Since we deal with a long-lived source, we may employ the Poisson formula (5.27). (In fact, we cannot use the binomial formula (5.7), because we are not given values for p and n. Only the mean decay rate, $\lambda = 3$ min^{-1}, is specified.) In an interval of $t = 1$ min, the mean number of counts would be $\mu = \lambda t = 3$ min$^{-1} \times 1$ min $= 3$. For $X = 5$, Eq. (5.27) then gives

$$\Pr(X = 5) = p(5; 3) = \frac{3^5 e^{-3}}{5!} = 0.101. \tag{5.28}$$

b) For the 5-min interval, $\mu = 3$ min$^{-1} \times 5$ min $= 15$. With the help of the cumulative Poisson distribution in Table A.2, we find

$$\Pr(X \geq 13) = \sum_{x=13}^{\infty} \frac{15^x e^{-15}}{x!} = 1 - P(12; 15) = 1 - 0.268$$
$$= 0.732. \tag{5.29}$$

This example illustrates an important aspect of the Poisson process. The 5-min interval in part (b) can be considered as the sum of five 1-min intervals as in part (a). The sum of identically distributed Poisson random variables is also Poisson distributed with parameter equal to the sum of the parameters for each of the individual distributions.

■ **Example**

A certain particle counter, which locks when more than four particles enter it in a millisecond, is used to monitor a steady radiation field. If the average number of particles entering the counter is 1.00 ms^{-1}, what is the probability that the counter will become locked in any given millisecond?

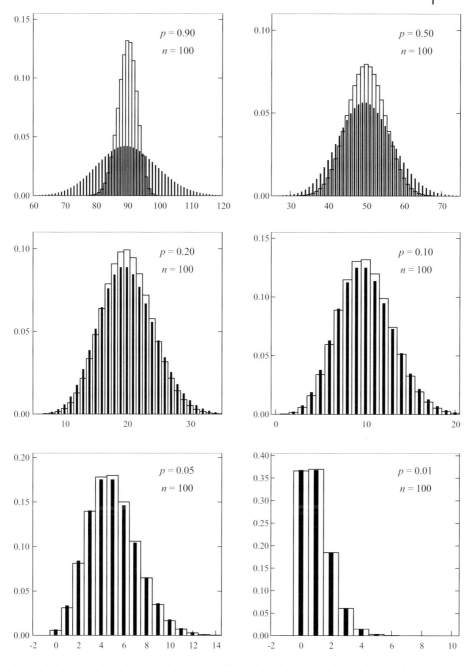

Figure 5.3 Comparison of binomial (histogram) and Poisson (solid bars) distributions for fixed sample size, $n = 100$, and different values of the probability of success p. The ordinate in each panel gives the probability for the number successes shown on the abscissa. The mean of the two distributions in a given panel is the same.

Solution

We let X denote the number of particles entering the counter in a millisecond. For the rate of incidence $\lambda = 1\,\text{ms}^{-1}$ and time $t = 1\,\text{ms}$, X has the Poisson distribution with parameter $\mu = \lambda t = 1\,\text{ms}^{-1} \times 1\,\text{ms} = 1$. The counter locks when $X > 4$ in this interval. With the help of Table A.2, we find

$$\Pr(X > 4) = \sum_{k=5}^{\infty} \frac{1^k e^{-1}}{k!} = 1 - P(4; 1) = 1 - 0.996 = 0.004. \tag{5.30}$$

We next derive the Poisson distribution function from first principles, based on the Poisson process, rather than a limiting case of the binomial distribution. If λ is the (constant) average rate that events occur, then the average number that happen during a time t is $\mu = \lambda t$. We let $p_x(t)$ denote the probability that exactly $X = x$ events happen in t. During a very short subsequent time Δt between t and $t + \Delta t$, the probability that an event occurs is $\lambda \Delta t$, and the probability that no event occurs is $1 - \lambda \Delta t$. (The probability for more than one event to happen in Δt is, by the conditions of the Poisson process, negligible.) The probability that $X = x$ events happen over the entire time $t + \Delta t$ is the sum of the probabilities (1) for having $(x - 1)$ events in t and one event in Δt and (2) for having x events in t and none in Δt. This is expressed by writing

$$p_x(t + \Delta t) = p_{x-1}(t)\lambda \Delta t + p_x(t)(1 - \lambda \Delta t), \tag{5.31}$$

or

$$\frac{p_x(t + \Delta t) - p_x(t)}{\Delta t} = \lambda[p_{x-1}(t) - p_x(t)]. \tag{5.32}$$

In the limit as $\Delta t \to 0$, the left-hand side of this equation is the derivative, $dp_x(t)/dt$, and so

$$\frac{dp_x(t)}{dt} = \lambda[p_{x-1}(t) - p_x(t)]. \tag{5.33}$$

Solution of a recurrence-type differential equation can often be facilitated by use of an exponential integrating factor. Substituting a solution of the form

$$p_x(t) = q_x(t)e^{-\lambda t} \tag{5.34}$$

reduces Eq. (5.33) to

$$\frac{dq_x(t)}{dt} = \lambda q_{x-1}(t). \tag{5.35}$$

The solution is

$$q_x(t) = \frac{(\lambda t)^x}{x!}, \tag{5.36}$$

as can be seen by direct substitution:

$$\frac{dq_x(t)}{dt} = \frac{\lambda^x x t^{x-1}}{x!} = \lambda \frac{(\lambda t)^{x-1}}{(x-1)!} = \lambda q_{x-1}(t). \tag{5.37}$$

Combining Eqs. (5.34) and (5.36) gives

$$p_x(t) = \frac{(\lambda t)^x e^{-\lambda t}}{x!}, \tag{5.38}$$

which is the Poisson distribution with parameter $\mu = \lambda t$, as derived earlier (Eq. (5.27)).

We next show that the Poisson distribution function is normalized and then obtain the expected value and variance. For the normalization, we return to Eq. (5.27) and sum over all nonnegative integers. Thus,

$$\sum_{x=0}^{\infty} p(x;\mu) = e^{-\mu} \sum_{x=0}^{\infty} \frac{\mu^x}{x!} = e^{-\mu} e^{\mu} = 1, \tag{5.39}$$

in which the sum is identically the power-series expansion of e^μ.

The mean of the distribution is given by

$$E(X) = \sum_{x=0}^{\infty} x p(x;\mu) = \sum_{x=0}^{\infty} x \frac{\mu^x e^{-\mu}}{x!} = e^{-\mu} \sum_{x=1}^{\infty} \frac{x\mu^x}{x!} = e^{-\mu} \sum_{x=1}^{\infty} \frac{\mu^x}{(x-1)!}, \tag{5.40}$$

where the zero contribution from $x=0$ has been omitted in writing the third summation. The quantity x is a dummy integer in these summations. Letting $x = y + 1$, we can transform the last summation in Eq. (5.40) in the following way without changing the value of $E(X)$:

$$E(X) = \mu e^{-\mu} \sum_{x=1}^{\infty} \frac{\mu^{x-1}}{(x-1)!} = \mu e^{-\mu} \sum_{y=0}^{\infty} \frac{\mu^y}{y!}. \tag{5.41}$$

Again, as in Eq. (5.39), the summation is just e^μ, leaving the important result that, for the Poisson distribution,

$$E(X) = \mu. \tag{5.42}$$

For the variance, Eq. (4.48), we need to evaluate $E(X^2)$:

$$E(X^2) = \sum_{x=0}^{\infty} x^2 p(x;\mu) = e^{-\mu} \sum_{x=0}^{\infty} \frac{x^2 \mu^x}{x!} = e^{-\mu} \sum_{x=1}^{\infty} \frac{x^2 \mu^x}{x!}$$

$$= \mu e^{-\mu} \sum_{x=1}^{\infty} \frac{x \mu^{x-1}}{(x-1)!}. \tag{5.43}$$

Letting $x = y + 1$ again, we write (Problem 5.21)

$$E(X^2) = \mu e^{-\mu} \sum_{y=0}^{\infty} \frac{(y+1)\mu^y}{y!} = \mu e^{-\mu} \sum_{y=0}^{\infty} \left(\frac{y\mu^y}{y!} + \frac{\mu^y}{y!} \right) = \mu(\mu+1). \tag{5.44}$$

The last equality can be seen from the sums in Eqs. (5.40) and (5.41). Equation (4.48) then gives

$$\text{Var}(X) = E(X^2) - \mu^2 = \mu(\mu+1) - \mu^2 = \mu. \tag{5.45}$$

The variance of the Poisson distribution is thus identical with its mean.

■ *Example*

A 1.00-nCi tritium source is placed in a counter having an efficiency of 39.0%. Tritium is a pure beta emitter with a half-life of 12.3 y. If X is the number of counts recorded in 5-s time intervals, find

a) $E(X)$;
b) the distribution of X;
c) the standard deviation of X.

Solution

a) The observation time is negligibly short compared with the half-life, and so we may use Poisson statistics. The mean count rate μ_c is the product of the mean disintegration rate λ and the counter efficiency ε:

$$\mu_c = \lambda\varepsilon = (1 \times 10^{-9} \text{ Ci}) \times (3.70 \times 10^{10} \text{ s}^{-1} \text{ Ci}^{-1}) \times 0.390$$
$$= 14.4 \text{ s}^{-1}. \tag{5.46}$$

In 5-s intervals, the mean number of counts is

$$E(X) = \mu = \mu_c t = 14.4 \text{ s}^{-1} \times 5 \text{ s} = 72.2. \tag{5.47}$$

b) The distribution of the number of counts in 5-s intervals is

$$\Pr(X = x) = \frac{72.2^x e^{-72.2}}{x!}. \tag{5.48}$$

c) From Eq. (5.45), $\text{Var}(X) = 72.2$, and so the standard deviation is $\sigma_c = \sqrt{72.2} = 8.49$.

In Chapter 7, we shall discuss the relative error, or ratio of the standard deviation and the mean, in counting measurements. One can see from this example how the relative error with a long-lived radionuclide decreases as a longer time interval is used for taking a single count as an estimate of the mean. The ratio σ_c/μ varies as $\sqrt{\lambda t}/(\lambda t) = 1/\sqrt{\lambda t}$. Thus, when counting for a longer time, the relative error decreases as the square root of the time.

5.6
Hypergeometric Distribution

We have seen that the binomial distribution results when one counts the number of successes from n independent trials, having only two possible outcomes, in which the probability of success p (or failure, $1 - p$) is constant each time. If a trial consists of randomly selecting an object from a population of finite size, then the binomial distribution results only when one returns the selected object to the population pool each time before making the next selection. This method of repeated *sampling with replacement* from the finite population ensures that p is constant for each independent trial.

Sampling from a finite population *without replacement* is a different, but related, procedure. The probability of success on any given trial after the first then depends on the particular outcomes of the previous trials. The resulting frequency distribution of the number of successes is called the *hypergeometric distribution*. An example is afforded by randomly drawing successive balls without replacement from an initial collection of three white and seven black balls. We designate drawing a white ball as success. The probability that the second draw will be a success depends on the result of the first draw. If the first draw was a success, then the probability of success on the second draw will be 2/9; if the first ball was black, then 3/9 is the chance of success on the second draw.

To describe the hypergeometric distribution formally, we let k denote the number of elements to be regarded as successes in a population of size N. The number of elements that are failures is then $N - k$. If we randomly sample n items from this population without replacement, then the number of successes X has the hypergeometric distribution with probability function

$$\Pr(X = x) = \frac{\binom{k}{x}\binom{N-k}{n-x}}{\binom{N}{n}}, \quad x = 0, 1, 2, \ldots, \min(n, k), \tag{5.49}$$

where $\min(n, k)$ denotes the smaller of the values, n and k. (The interested reader is referred to pp. 81–82 of Scheaffer and McClane (1982), listed in the Bibliography, for a derivation of the hypergeometric distribution.) It can be shown (Problem 5.22) that the mean and variance are given by

$$E(X) = \frac{nk}{N} \tag{5.50}$$

and

$$\mathrm{Var}(X) = n\left(\frac{k}{N}\right)\left(\frac{N-k}{N}\right)\left(\frac{N-n}{N-1}\right). \tag{5.51}$$

The hypergeometric distribution is applied in quality control, acceptance testing, and finite population sampling.

The relationship between the binomial and hypergeometric distributions can be seen as follows. If we consider $p = k/N$ as the proportion of successes, then Eq. (5.50) is equivalent to the binomial mean, Eq. (5.13). The variances (5.51) and (5.15) are equivalent except for the factor $(N - n)/(N - 1)$, which is sometimes called the *finite population correction factor*. If the population size N is large compared with the sample size n, then we can approximate the hypergeometric distribution with the binomial by using n and $p = k/N$ as the binomial parameters.

■ *Example*
A large, outdoor site is to be assessed for possible pockets of radioactive contamination. In lieu of surveying the entire site, it is divided uniformly into a grid of 100 squares, which can be individually monitored. The survey team randomly selects 20 of the squares for a thorough search, which will detect contamination, if present, with certainty. Unknown to the team, 10 of the squares, also located randomly on the site, have contamination; the other 90 are clean.

a) What is the probability that the surveyors will find exactly three contaminated squares?
b) What is the probability that none of the 20 squares they have chosen will have contamination?

Solution
a) The conditions can be modeled by the hypergeometric distribution, with $N=100$, $k=10$, and $n=20$. If X represents the number of contaminated grid squares found in the survey, then from Eq. (5.49) we find

$$\Pr(X=3) = \frac{\binom{10}{3}\binom{100-10}{20-3}}{\binom{100}{20}} = \frac{10!}{3!7!}\frac{90!}{17!73!}\frac{20!80!}{100!}$$

$$= 0.209. \qquad (5.52)$$

b) With $k=0$, we obtain from Eq. (5.49)

$$\Pr(X=0) = \frac{\binom{10}{0}\binom{100-10}{20-0}}{\binom{100}{20}} = 0.0951. \qquad (5.53)$$

Thus, the chances are about 1 in 10 that the survey protocol would not detect any of the 10 contaminated squares that are on the site.

■ *Example*
Use the binomial approximation to the hypergeometric distribution to obtain $\Pr(X=3)$ in the last example. Compare the approximate and exact solutions.

Solution
We let $p=k/N=10/100=0.1$ and $n=20$. Then from Eq. (5.7) we write

$$\Pr(X=3) \cong b(3;20,0.1) = \binom{20}{3}(0.1)^3(0.9)^{17} = 0.190. \qquad (5.54)$$

The relative error in the binomial approximation is

$$\frac{0.209 - 0.190}{0.209} = 0.0909, \tag{5.55}$$

and so the approximation is about 9% too low. This inaccuracy is not surprising, however, because the finite population factor,

$$\frac{N-n}{N-1} = \frac{100-20}{100-1} = 0.808, \tag{5.56}$$

differs significantly from unity.

■ *Example*

In the last two examples, suppose the site consists of 10 000 uniform squares and that the proportion of contaminated squares is the same as before, that is, 10%, randomly distributed on the site. What is the probability of finding three contaminated squares if, again, 20 randomly selected squares are surveyed? How does the binomial approximation compare with the exact answer in this case?

Solution

The hypergeometric distribution now has parameters $k = 1000$, $N = 10\,000$, and $n = 20$. Thus, from Eq. (5.49),

$$\Pr(X = 3) = \frac{\binom{1000}{3} \binom{10\,000 - 1000}{20 - 3}}{\binom{10\,000}{20}} = 0.190. \tag{5.57}$$

The binomial approximation to the hypergeometric result is, as before, $b(20, 3, 0.1) = 0.190$ (Eq. (5.54)). Thus, the approximation is excellent in this case, since the finite population correction factor is nearly unity (Problem 5.29).

The hypergeometric distribution can be extended to the case where we have m categories, rather than just two. The population N is partitioned into these m categories, so that there are k_1 elements of category 1, k_2 of category 2, ..., and k_m of category m. In this situation, we take a random sample of size n and consider the probability that x_1 members of the sample are from category 1, x_2 are from category 2, ..., and x_m are from category m. We can represent this *multivariate hypergeometric* probability distribution by writing

$$\Pr(X_1 = x_1, X_2 = x_2, \ldots, X_m = x_m) = \frac{\binom{k_1}{x_1} \binom{k_2}{x_2} \cdots \binom{k_m}{x_m}}{\binom{N}{n}}, \tag{5.58}$$

where $\sum_{i=1}^{m} k_i = N$ and $\sum_{i=1}^{m} x_i = n$.

Example

A City Beautification Committee of 4 is to be randomly selected from a group of 10 politicians, consisting of 3 Republicans, 5 Democrats, and 2 Independents. What is the probability that this random sample will have two Republicans, one Democrat, and one Independent?

Solution

Using the above notation, we have $k_1 = 3$, $k_2 = 5$, and $k_3 = 2$, respectively, for the three named parties. The sample yields $x_1 = 2$, $x_2 = 1$, and $x_3 = 1$. By Eq. (5.58), we have

$$\Pr(X_1 = 2, X_2 = 1, X_3 = 1) = \frac{\binom{3}{2}\binom{5}{1}\binom{2}{1}}{\binom{10}{4}} = \frac{1}{7} = 0.143. \qquad (5.59)$$

5.7 Geometric Distribution

We have considered the frequency distribution of the number of successes in repeated Bernoulli trials, in which the probability of success p and failure $q = 1 - p$ is constant and the trials are independent. Another aspect of this procedure is to consider the frequency distribution of the number X of the trial in which the *first success* occurs. Since the first success comes immediately after $X - 1$ failures, it follows that

$$\Pr(X = x) = (1-p)^{x-1} p, \quad x = 1, 2, \ldots. \qquad (5.60)$$

This function is called the *geometric distribution*. To show that it is normalized, we substitute $q = 1 - p$ and write

$$\sum_{x=1}^{\infty} (1-p)^{x-1} p = p \sum_{x=1}^{\infty} q^{x-1} = p(1 + q + q^2 + \cdots) = p\left(\frac{1}{1-q}\right) = 1. \qquad (5.61)$$

The geometric progression (hence the name) is the binomial expansion of $1/(1-q) = 1/p$, and so the normalization condition follows in Eq. (5.61). The mean and variance are

$$E(X) = \frac{1}{p} \qquad (5.62)$$

and

$$\mathrm{Var}(X) = \frac{1-p}{p}. \qquad (5.63)$$

Example

A pulse height analyzer is used with a high discriminator setting to monitor by the hour random background events that deposit a relatively large amount of energy. It is found that one recordable event occurs on average every 114 min.

a) What is the probability that the first recordable event happens during the first hour?
b) During the fourth hour?
c) On the average, during which hour does the first event occur?
d) What is the probability that one must wait at least 3 h before observing the first event?

Solution

We divide the observation time into successive 1-h intervals, starting at time $t=0$. The average event rate is $1/(114\,\text{min}) = 1/(1.9\,\text{h}) = 0.526\,\text{h}^{-1}$. We let $p = 0.526$ be the probability that an event occurs (success) in any of the given 1-h periods and X represent the number of the interval in which the first event happens. Equation (5.60) then gives for the probability distribution of the number of that interval

$$\Pr(X = x) = (1-0.526)^{x-1}(0.526) = 0.526(0.474)^{x-1}. \tag{5.64}$$

a) The probability that the first event occurs in the first hour ($x=1$) is

$$\Pr(X = 1) = 0.526(0.474)^0 = 0.526. \tag{5.65}$$

b) Similarly,

$$\Pr(X = 4) = 0.526(0.474)^3 = 0.0560. \tag{5.66}$$

c) The average number of the interval in which the first event occurs is, from Eq. (5.62),

$$\mu = \frac{1}{p} = \frac{1}{0.526} = 1.90. \tag{5.67}$$

Thus, on the average, the first event occurs during the second 1-h time interval.

d) If one has to wait at least 3 h for the first event to happen, then we are asked to find $\Pr(X \geq 4)$, the first event thus coming in the fourth or later interval. This probability is equal to unity minus the probability that the first event happens in any of the first three intervals. From Eq. (5.60),

$$\Pr(X \geq 4) = 1 - \Pr(X \leq 3) = 1 - [\Pr(X = 1) + \Pr(X = 2) + \Pr(X = 3)] \tag{5.68}$$

$$= 1 - [(0.474)^0(0.526) + (0.474)^1(0.526) + (0.474)^2(0.526)]$$
$$= 1 - 0.894 = 0.106.$$

$$\tag{5.69}$$

5.8
Negative Binomial Distribution

With the geometric distribution we were concerned with the probability distribution of the trial at which the first success occurred. We consider next the probability distribution for the trial number at which the rth success occurs. The distribution that describes this behavior is called the *negative binomial distribution*.

We let X denote the number of the trial at which the rth success happens, where, as before, each trial is an independent Bernoulli trial with constant probability p of success. The probability function for X is then given by

$$\Pr(X = x) = \binom{x-1}{r-1} p^r (1-p)^{x-r}, \quad x = r, r+1, \ldots \quad (5.70)$$

Since the rth success must occur on the xth trial, the preceding $(r-1)$ successes must happen in the previous $(x-1)$ trials. The number of different ways in which this can occur is $\binom{x-1}{r-1}$. Each trial is independent; r of them result in successes, giving the term p^r in Eq. (5.70). The remaining $(x-r)$ trials result in failure, giving the term $(1-p)^{x-r}$. The product of these three factors gives the negative binomial probability distribution (5.70) on X.

■ *Example*
The probability that a person will purchase high-test gasoline at a certain service station is 0.20.

a) What is the probability that the first high-test purchase will be made by the fifth customer of the day?
b) What is the probability that the third high-test purchase will be made by the 10th customer of the day?

Solution
a) This part is solved by the geometric distribution. If purchasing high-test gasoline is considered a success, then we use $p = 0.20$ and $x = 5$ in Eq. (5.60) to obtain

$$\Pr(X = 5) = (1-0.20)^{5-1}(0.20) = (0.80)^4(0.20) = 0.0819. \quad (5.71)$$

b) The negative binomial distribution applies. Using $p = 0.20$, as before, $r = 3$ (purchase number), and $x = 10$ (customer number) in Eq. (5.70), we find

$$\Pr(X = 10) = \binom{10-1}{3-1}(0.20)^3(1-0.20)^{10-3}$$

$$= \binom{9}{2}(0.20)^3(0.80)^7 = 0.0604. \quad (5.72)$$

The negative binomial distribution can result as the sum of r geometric random variables in the following way. The geometric distribution describes the trial X at which the first success occurs, and the negative binomial describes the trial Y at which the rth success occurs. Assume that the probability of success, p, is the same for both of these experiments. Also, let X_1 be the number of trials until the first success, X_2 be the number of trials from the first until the second success occurs, and so on with X_r being the number of trials from the $(r-1)$th success until the rth success. Then

$$Y = X_1 + X_2 + \cdots + X_r, \tag{5.73}$$

since each X_i has the geometric distribution with parameter p and they are independent. This relationship allows for simplified calculations of $E(Y)$ and $\mathrm{Var}(Y)$. Since the X_i are independent and identically distributed geometric random variables, it follows that

$$E(Y) = E(X_1 + X_2 + \cdots + X_r) = r\frac{1}{p} = \frac{r}{p} \tag{5.74}$$

and

$$\mathrm{Var}(Y) = \sum_{i=1}^{r} \mathrm{Var}(X_i) = r\frac{q}{p^2} = \frac{rq}{p^2}. \tag{5.75}$$

An alternative form of the negative binomial distribution is provided by the distribution of the number of failures Y that precede the rth success:

$$\mathrm{Pr}(Y = y) = \binom{y + r - 1}{y} p^r (1-p)^y, \quad y = 1, 2, \ldots \tag{5.76}$$

In this case, the factors p^r and $(1-p)^y$ account, respectively, for r successes and y failures among the $r + y$ independent trials. The last trial consists of the rth success. The binomial coefficient in Eq. (5.76) then gives the number of different ways in which y failures and the other $(r-1)$ successes can occur in the remaining $(y + r - 1)$ trials.

Problems

5.1 Show that the probability distribution (5.1) is normalized.

5.2 Ten identical paper slips numbered 5, 10, ..., 50 are placed in a box. A slip is randomly drawn and its number, X, is recorded. It is then placed back in the box.
   a) Describe a probability distribution that may be associated with X, stating any assumptions necessary.
   b) Using the distribution, obtain the mean value of X.
   c) Using the distribution, obtain the variance of X.

5.3 Let X be a Bernoulli random variable such that $X=0$ with probability $(1-p)$ and $X=1$ with probability p. Find
   a) $E(X)$;
   b) $Var(X)$.

5.4 Show that the binomial model (5.7) is the same as the Bernoulli model with $n=1$.

5.5 A neutron either penetrates a target with a probability of 0.20 or else is deflected with a probability of 0.80. If five such neutrons are incident in succession on the target, what is the probability that
   a) the first three pass through and the last two are deflected?
   b) all five pass through undeflected?
   c) none pass through undeflected?
   d) at least one passes through?
   e) What is the relationship between the last two answers?

5.6 Tests show that 0.20% of TLD chips produced by a new process are defective. Find the probability that a given batch of 1000 chips will contain the following numbers of defective units:
   a) zero;
   b) one;
   c) three;
   d) five.

5.7 In the last problem, what is the probability that a batch of 1000 chips will have no more than three defective units?

5.8 For a binomial random variable with $n=20$ and $p=0.30$, use either Eq. (5.7) or Table A.1 to find
   a) $\Pr(X < 4)$;
   b) $\Pr(2 \leq X \leq 4)$;
   c) $\Pr(X > 9)$;
   d) $\Pr(X=6)$.

5.9 a) For a fixed number of trials, how does the spread of the binomial distribution vary with the probability p of success of each Bernoulli trial?
   b) For what value of p does the variance have an extremum?
   c) Is the extremum a maximum or a minimum?

5.10 Use Eq. (5.7) to derive the mean, Eq. (5.13), and variance, Eq. (5.15), of the binomial distribution.

5.11 a) Calculate $b(6; 10, 0.4)$ from Eq. (5.7).
   b) Use Table A.1 to obtain $b(6; 10, 0.4)$.

5.12 a) Use Table A.1 to evaluate $\Pr(X \leq 12)$ with $p=0.70$ and $n=20$.
   b) What is the value of $\Pr(X \geq 12)$?

5.13 Calculate the value of the first term ($x=5$) in the summation in Eq. (5.16) by using
   a) the binomial distribution (5.7);
   b) the Poisson distribution (5.27).

5.14 a) For the binomial distribution, prove that
$$\frac{b(x+1; n, p)}{b(x; n, p)} = \frac{p(n-x)}{(1-p)(x+1)}$$
for $x = 0, 1, 2, \ldots, n-1$.
b) Use this recursion formula to calculate the binomial distribution for $n = 5$ and $p = 0.30$.

5.15 A group of 1000 persons selected at random is sampled for a rare disease, which occurs in 0.10% of the population. What is the probability of finding in the group
a) two persons with the disease?
b) at least two persons with the disease?

5.16 A long-lived radioisotope is being counted at a mean rate of $20\,\text{min}^{-1}$. Find
a) the distribution of the number of particles X recorded in half-minute intervals;
b) $E(X)$;
c) $\text{Var}(X)$;
d) $\Pr(X \geq 12)$.

5.17 An alpha-particle monitor has a steady background rate of 8.0 counts per hour. What is the probability of observing
a) five counts in 30 min?
b) zero counts in 5 min?
c) zero counts in 60 min?

5.18 a) In the last problem, if a 5-min interval passes with zero counts, what is the probability that there will be no counts in the next 5 min?
b) What is the probability for having no counts in a 10-min interval?
c) What is the relationship between the two probabilities in (a) and (b)?

5.19 Show by direct substitution that the Poisson distribution function (5.38) satisfies Eq. (5.33).

5.20 For the Poisson distribution, show that the probability of observing one fewer than the mean is equal to the probability of observing the mean.

5.21 Verify Eqs. (5.43) and (5.44).

5.22 If X has the hypergeometric distribution (5.49), show that the mean and variance of X are given by Eqs. (5.50) and (5.51), respectively.

5.23 Six squares in a uniform grid of 24 have hot spots. An inspector randomly selects four squares for a survey. What is the probability that he or she selects
a) no squares having hot spots?
b) exactly two squares with hot spots?
c) at least two squares with hot spots?

5.24 A certain process makes parts with a defective rate of 1.0%. If 100 parts are randomly selected from the process, find $\Pr(f \leq 0.03)$, where $f = X/n$ is the sample fraction defective, defined as the ratio of the number of defective parts X and the sample size n.
a) Use the binomial distribution.
b) Use the Poisson approximation.

5.25 A process makes parts with a defective rate of 1.0%. A manager is alerted if the sample fraction defective in a random sample of 50 parts is greater than or equal to 0.060.
  a) If the process is in control (i.e., the defective rate is within 1.0%), what is the probability that the manager will be alerted?
  b) If the process worsens and the defective rate increases to 5.0%, what is the probability that the manager will not be alerted, given this higher rate?

5.26 Solve the last problem by means of Poisson statistics.

5.27 Let X have the hypergeometric distribution (5.49). Show that the distribution on X converges to a binomial $b(x; n, p)$, with $p = k/N$ remaining fixed while $N \to \infty$. (As a rule of thumb, the binomial distribution may be used as an approximation to the hypergeometric when $n \leq N/10$.)
(Hint: Begin by showing that

$$\Pr(X = m) = \frac{\binom{k}{m}\binom{N-k}{n-m}}{\binom{N}{n}} = \binom{n}{m}\frac{(k)_m(N-k)_{n-m}}{(N)_n},$$

for $m = 0, 1, \ldots, n$,

where $(r)_j = (r-1)(r-2) \cdots (r-j+1)$. Then rewrite the equation in terms of $p = k/N$.)

5.28 A crate contains $N = 10\,000$ bolts. If the manufacturing process is in control, then no more than $p = 0.050$ of the bolts will be defective. Assume that the process is in control and that a sample of $n = 100$ bolts is randomly selected. Using the binomial approximation, find the probability that exactly three defective bolts are found. (Exact treatment with the hypergeometric distribution gives the result 0.1394.)

5.29 Compute the finite population correction factor for the example involving Eq. (5.57).

5.30 A potential customer enters a store every hour on the average. A clerk has probability $p = 0.25$ of making a sale. If the clerk is determined to work until making three sales, what is the probability that the clerk will have to work exactly 8 h?

5.31 Skiers are lifted to the top of a mountain. The probability of any skier making it to the bottom of the run without falling is 0.050. What is the probability that the first skier to make the run without falling is
  a) the 15th skier of the day?
  b) the 20th of the day?
  c) either the 15th, the 16th, or the 17th?

5.32 In a match against one another, two tennis players both have an 85% chance of successfully making their shot (including service). Any given point ends the first time either player fails to keep the ball in play.

a) What is the probability that a given point will take exactly eight strokes, ending on the missed eighth shot?
b) What is the probability that a point will last longer than three strokes, including the last shot?
c) If the first player serves and has a success probability of 0.85 and the second player has a success probability of only 0.70, repeat (a).
d) Which player loses the point in (c)?
e) Repeat (b) under the conditions given in (c).

5.33 A ^{60}Co atom decays (99 + % of the time) by nuclear emission of a β^- particle, followed immediately by two γ photons (1.17 and 1.33 MeV) in succession. Assume that these radiations are emitted isotropically from the nucleus (although the two photon directions are, in fact, correlated). A small detector is placed near a point source of ^{60}Co.
a) What is the probability that the first five entities to strike the detector are γ photons and that the sixth is a β^- particle?
b) What is the probability that the first five are β^- particles and the sixth is a photon?
c) What is the probability that the first β^- particle to strike the target is either the fifth, sixth, or seventh entity that hits it?

5.34 In the original statement (5.70) of the negative binomial distribution, X is the number of the trial at which the rth success occurs. In Eq. (5.76), Y refers to the number of failures that occur until the rth success. Thus, $Y = X - r$. By using this transformation in Eq. (5.70), show that Eq. (5.76) results.

6
Continuous Distributions

6.1
Introduction

Conditions and constraints associated with discrete random variables usually lead to a probability model that can be chosen with little doubt. In contrast, for continuous data there might be several probability distributions that fit a given set of observations well. The true distribution associated with a continuous random variable is hardly ever known. There are goodness-of-fit techniques (Chapter 14) that allow one to check assumptions about distributions, but these are applied after data collection. A number of continuous distributions are available to provide alternatives for modeling empirical data. There are many situations where the physical model of the process under study leads to a good choice of the probability distribution, as we shall see with the exponential function in Section 6.7. In this chapter, we look at several of the most prominent continuous distributions used in statistical analysis of data. Additional continuous distributions are discussed in Johnson et al. (1994).

A number of useful tools, including statistical calculators, tables, and demonstrations of statistical principles, are available on the World Wide Web. Values of the probability densities and cumulative functions can be obtained for many distributions.

6.2
Continuous Uniform Distribution

Here we derive the probability density function for the *continuous uniform distribution*, corresponding to that for discrete variables in Section 5.2. We consider a random variable X that has *support*, or positive probability, only over a finite closed interval $[a, b]$. If the probability that X lies in any subspace $A = [x_1, x_2]$ of this interval is proportional to the ratio of $(x_2 - x_1)$ and $(b - a)$, then we may write

$$\Pr(X \in A) = c \frac{x_2 - x_1}{b - a}, \tag{6.1}$$

Statistical Methods in Radiation Physics, First Edition. James E. Turner, Darryl J. Downing, and James S. Bogard.
© 2012 Wiley-VCH Verlag GmbH & Co. KGaA. Published 2012 by Wiley-VCH Verlag GmbH & Co. KGaA.

where c is the constant of proportionality. Since the probability that X lies in $[a, b]$ must be unity, it follows that $c = 1$. The cumulative probability function is

$$F(x) = \Pr(X \in [a, x]) = \begin{cases} 0, & x < a, \\ \dfrac{x-a}{b-a}, & a \leq x < b, \\ 1, & x \geq b. \end{cases} \tag{6.2}$$

We find by differentiation (see Eq. (4.14)) that the probability density function for the random variable X is

$$f(x) = \begin{cases} \dfrac{1}{b-a}, & a \leq x < b, \\ 0, & \text{elsewhere.} \end{cases} \tag{6.3}$$

The term *uniform* is attached to a random variable whose probability density function is constant over its support; thus, we say here that X has the uniform distribution.

■ *Example*
Let X have the uniform distribution over the interval $[a, b]$.

a) Obtain the mean of X as a function of a and b.
b) Obtain the variance of X.
c) If $a = 0$ and $b = 1$, determine the numerical values of the mean and variance.

Solution

a) Using Eq. (6.3), we find for the mean,

$$\mu = \int_{-\infty}^{\infty} x f(x) dx = \int_a^b \frac{x}{b-a} dx = \frac{b^2 - a^2}{2(b-a)} = \frac{a+b}{2}. \tag{6.4}$$

b) To find the variance, we use Eq. (4.48), for which we need

$$E(X^2) = \int_a^b \frac{x^2}{b-a} dx = \frac{b^3 - a^3}{3(b-a)} = \frac{b^2 + ab + a^2}{3}. \tag{6.5}$$

From Eq. (4.48), then

$$\text{Var}(X) = \frac{b^2 + ab + a^2}{3} - \frac{(a+b)^2}{4} = \frac{(b-a)^2}{12}. \tag{6.6}$$

c) Direct substitution into Eqs. (6.4) and (6.6) gives

$$\mu = \frac{0+1}{2} = \frac{1}{2} \tag{6.7}$$

and
$$\text{Var}(X) = \frac{(1-0)^2}{12} = \frac{1}{12}. \tag{6.8}$$

Note that the value of μ coincides with the midpoint of the support over X.

The continuous uniform distribution plays an important role in the generation of a random sample from the probability distribution function $f(x)$ for a given continuous random variable X. If we make the transformation $Y = F(X)$, where $F(x)$ is the cumulative distribution for X, then $0 \leq Y \leq 1$ over the entire range of values of X. With the help of Eq. (4.125), we can write the transformed probability density function $g(y)$ on Y, where $0 \leq y \leq 1$. Substituting the inverse function $x = F^{-1}(y) = w(y)$ and $y = F(x)$ into Eq. (4.125), we obtain

$$g(y) = f[w(y)] \left| \frac{dw(y)}{dy} \right| = f(x) \left| \frac{dx}{dy} \right| = f(x) \frac{1}{|dF(x)/dx|}. \tag{6.9}$$

Since, by Eq. (4.14), the derivative of the cumulative distribution function is equal to the probability density function, it follows that

$$g(y) = 1, \quad 0 \leq y \leq 1. \tag{6.10}$$

We thus have the important result that, if X is a continuous random variable having a probability density function $f(x)$ and cumulative distribution function $F(x)$, then the random variable $Y = F(X)$ has a uniform distribution $g(y)$ on [0, 1]. A random sample drawn from the uniform distribution can then be used to generate a corresponding random sample from any continuous distribution by use of the inverse function $x = F^{-1}(y)$.

■ *Example*
Given the exponential probability density function

$$f(x) = \begin{cases} 5e^{-5x}, & x > 0, \\ 0, & \text{elsewhere}, \end{cases} \tag{6.11}$$

show how a random sample of values of X can be generated.

Solution
We know that the random variable $Y = F(X)$ has a uniform distribution. Thus, observing Y is equivalent to observing X from the distribution $F(x)$, such that $y = F(x)$. To obtain the distribution in X, we need to find the inverse function, $x = F^{-1}(y)$. From (6.11), we obtain for the cumulative distribution function,

$$F(x) = \int_0^x 5e^{-5t}\, dt = -e^{-5t}\big|_0^x = 1 - e^{-5x}, \quad 0 < x < \infty, \tag{6.12}$$

and $F(x) = 0$ elsewhere. Then, letting $y = F(x)$ and solving for x yields

$$x = -\frac{\ln(1-y)}{5}, \tag{6.13}$$

which is defined over the range $0 \leq y < 1$. Random numbers Y selected uniformly over the interval $[0, 1)$ in Eq. (6.13) will generate values of X having the probability density function $f(x)$ given by Eq. (6.11).

This example illustrates how a uniform random number generator allows one to generate random samples from any continuous distribution. The solution was simple here, because the cumulative distribution function has an analytical form. When the cumulative function is not known analytically, one can still use the result implied by Eq. (6.10). However, one then has to use algorithms to compute the values of x that satisfy the relationship $y = F(x)$. The use of random number generators and such numerical, rather than analytical, calculations are readily handled with digital computers and with many hand-held calculators. In Chapter 12, we shall see examples of the generation of random samples from given probability distribution functions – samples needed in order to solve problems by Monte Carlo techniques. Such computational methods are extensively used in radiation protection for shielding and dosimetry.

■ **Example**

Individual photons from a point source in Figure 6.1 can be directed at an adjustable angle θ to travel in the plane of the page toward a flat screen (shown edgewise in Figure 6.1), placed a unit distance away. A photon thus strikes the screen somewhere along a vertical line at a position y relative to an arbitrary reference point O, as indicated. The source itself is located opposite the point $y = \tau$. If photons are emitted randomly with a uniform distribution of angles, $-\pi/2 < \Theta < +\pi/2$, find the probability density function for their arrival coordinates Y on the screen.

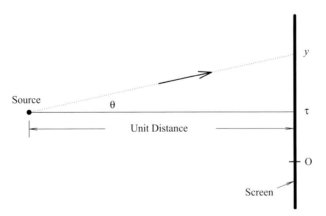

Figure 6.1 Screen, seen edge-on, located at unit distance from an isotropic point source of photons. See example in the text.

Solution
The uniform probability density function $f(\theta)$ on Θ is given by

$$f(\theta) = \begin{cases} \dfrac{1}{\pi}, & -\dfrac{\pi}{2} \le \theta \le \dfrac{\pi}{2}, \\ 0, & \text{elsewhere.} \end{cases} \qquad (6.14)$$

The density on Y can be found from the transformation procedure, Eq. (4.125). The relationship between the random variables Θ and Y is given by $\tan\theta = y - \tau$, the source being unit distance away from the screen. The inverse relationship is $\theta = \tan^{-1}(y - \tau)$. Differentiating, one has

$$\frac{d\theta}{dy} = \frac{1}{1 + (y - \tau)^2}. \qquad (6.15)$$

Substitution into Eq. (4.125), with $\theta = w(y) = \tan^{-1}(y - \tau)$, gives

$$g(y) = f\left[\tan^{-1}(y - \tau)\right] \frac{1}{1 + (y - \tau)^2} = \frac{1}{\pi} \frac{1}{1 + (y - \tau)^2}, \qquad (6.16)$$

with $-\infty < y < \infty$. A random variable Y having this probability density function, shown in Figure 6.2, is called a *Cauchy random variable*. As discussed below in Section 6.9, this distribution is the same as the *Student's t-distribution* with one degree of freedom. A unique characteristic of the distribution is the fact that it has an infinite mean.

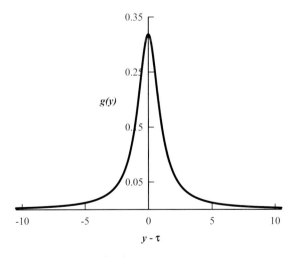

Figure 6.2 Cauchy distribution, Eq. (6.16).

6.3
Normal Distribution

The *normal*, or *Gaussian*, distribution is the most widely used probability distribution in statistics. Its importance is due to its general applicability in a large number of areas and due to the fact that many sample statistics tend to follow the normal distribution. This latter circumstance is described by the *central limit theorem*, which is discussed in the next section. In addition, observations often entail a combination of errors from several sources, thus increasing the tendency toward a Gaussian distribution. For example, a temperature measurement might be affected by pressure, vibration, and other factors, which can cause slight changes in the measuring equipment, leading to variability in the result. It can be shown that sums of random variables often tend to have a normal distribution. Thus, a measurement, which includes the sum of errors induced by different factors, frequently tends to be normally distributed.

The *normal density function* for a random variable X is given by

$$f(x) = \frac{1}{\sqrt{2\pi}\sigma} e^{-\frac{1}{2}\left(\frac{x-\mu}{\sigma}\right)^2}, \quad -\infty < x < \infty. \tag{6.17}$$

The distribution is characterized by two independent parameters, μ and σ, the mean and standard deviation, with $-\infty < \mu < \infty$ and $\sigma > 0$. The density is a symmetric, bell-shaped curve with inflection points at $\mu \pm \sigma$ (Problem 6.11). We introduce the symbol "$\sim$" to be read "is distributed as." The fact that a random variable X is normally distributed with mean μ and standard deviation σ can then be conveniently indicated by using the shorthand notation, $X \sim N(\mu, \sigma)$.

One can employ the so-called *standard normal distribution* in place of Eq. (6.17) to deal with many different problems having a wide range of possible values for μ and σ. The standard normal is, in essence, a dimensionless form of the normal distribution, having zero mean and unit standard deviation. We introduce the (dimensionless) standardized variable,

$$Z = \frac{X - \mu}{\sigma}. \tag{6.18}$$

The quantity Z expresses the displacement of X from the mean in multiples of the standard deviation. To transform the distribution (6.17) from X into Z, we employ Eq. (4.125). The inverse function for doing so is, from Eq. (6.18), $x = w(z) = \sigma z + \mu$, and so $dw/dz = \sigma$. When we substitute into Eq. (4.125), the factor σ from the derivative cancels the factor σ in the denominator of Eq. (6.17). The transformed function, therefore, does not depend explicitly on either σ or μ. It thus represents the normal distribution (6.17) with zero mean and unit standard deviation ($\mu = 0$ and $\sigma = 1$), and so we write it as

$$f(z) = \frac{1}{\sqrt{2\pi}} e^{-(1/2)z^2}, \quad -\infty < z < \infty. \tag{6.19}$$

The probability density function $f(z)$ is called the standard normal distribution. Since it is normally distributed, we may also describe it by writing $Z \sim N(0, 1)$. The cumulative distribution is given by

$$F(z) = \Pr(Z \le z) = \frac{1}{\sqrt{2\pi}} \int_{-\infty}^{z} e^{-(1/2)t^2} \, dt. \tag{6.20}$$

The two distributions, (6.19) and (6.20), are shown in Figure 6.3. Also, numerical values for the cumulative standard normal distribution (6.20) are given in Table A.3. Often one is interested in the probability that a measurement lies within a given number of standard deviations from the mean. The shaded area in Figure 6.4a, for instance, gives the probability that Z falls within the interval one standard deviation on either side of the mean. Numerical integration of Eq. (6.19) gives the result

$$\Pr(-1 \le Z < 1) = \Pr(|Z| < 1) = 0.6826. \tag{6.21}$$

The probability that Z falls outside this interval is

$$\Pr(|Z| \ge 1) = 1 - 0.6826 = 0.3174, \tag{6.22}$$

shown by the shaded regions in Figure 6.4b. This latter value is commonly referred to as a "two-tail" area, giving the probability that Z has a value somewhere in either tail of the distribution, *outside* the symmetric shaded area in Figure 6.4a. In counting statistics, the probability for exceeding a certain number of counts is often used for control purposes. Such a probability can be represented by a "one-tail" portion of the standard normal curve to the right of the mean. Since the distribution is symmetric, the one-tail probability is one-half the two-tail value. From Eq. (6.22) we have

$$\Pr(Z \ge 1) = 0.1587, \tag{6.23}$$

as indicated by the one-tail area in Figure 6.4c.

■ *Example*
Use Table A.3 to verify the results given by Eqs. (6.21)–(6.23) for Figure 6.4.

Solution
Table A.3 gives numerical values of the cumulative standard normal distribution $F(Z)$ defined by Eq. (6.20). The shaded area in Figure 6.4a is equal to the difference in the cumulative function at the two boundaries (see Eq. (4.13)). From the table,

$$\Pr(-1 \le z < 1) = F(1) - F(-1) = 0.8413 - 0.1587$$
$$= 0.6826, \tag{6.24}$$

as given by Eq. (6.21), above. Using Table A.3 to evaluate the probability outside the shaded area, we write

$$\Pr(|Z| \ge 1) = F(-1) + [1 - F(1)] = 0.1587 + [1 - 0.8413]$$
$$= 0.3174, \tag{6.25}$$

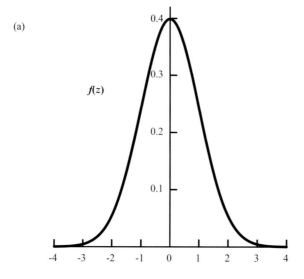

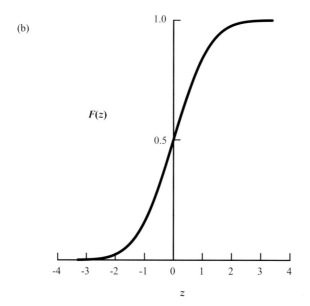

Figure 6.3 (a) Standard normal distribution, Eq. (6.19). (b) Cumulative distribution, Eq. (6.20).

as obtained before in Eq. (6.22). Finally,

$$\Pr(Z \geq 1) = 1 - F(1) = 1 - 0.8413 = 0.1587. \quad (6.26)$$

We see that the areas of the two tails are equal, and that the one-tail area in this example is given numerically by $F(-1)$, whether to the right or to the left of the shaded area.

(a)
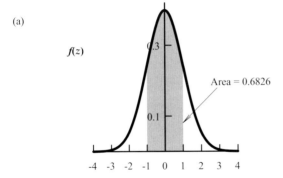

(b)
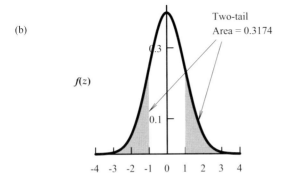

(c)
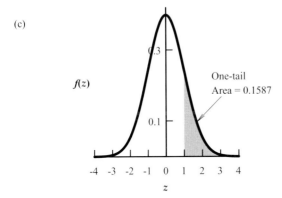

Figure 6.4 Areas under several different portions of the standard normal distribution determined by one standard deviation from the mean, $\mu = 0$. (a) Area of the shaded region gives the probability, 0.6826, that the random variable has a value within one standard deviation ($z = \pm 1$) of the mean. (b) Two-tail shaded area outside one standard deviation is 0.3174. (c) One-tail area outside one standard deviation is 0.1587.

Table 6.1 One-tail areas under the standard normal distribution from z to ∞.

z	Area
0.000	0.5000
0.675	0.2500
1.000	0.1587
1.282	0.1000
1.645	0.0500
1.960	0.0250
2.000	0.0228
2.236	0.0100
2.576	0.0050
3.000	0.0013
3.500	2.3×10^{-4}
4.000	3.2×10^{-5}
4.753	1.0×10^{-6}
5.000	2.9×10^{-7}
6.000	1.0×10^{-9}
7.000	1.3×10^{-12}

Table 6.1 lists some one-tail areas for the standard normal distribution to the right of the mean. For instance, one finds the value given by Eq. (6.26) for $z = 1$. This result implies that, for any normally distributed random variable, the probability is 0.1587 that its value exceeds the mean μ by one standard deviation σ or more. One sees, also, that there is a 5% chance that a randomly chosen value from a normal distribution will lie beyond 1.645σ. The probability quickly becomes very small for an observation to be outside several standard deviations. The chances are one in a million that it will be more than 4.753σ beyond the mean and only one in a billion that it will be beyond 6.000σ. These areas, given by $1 - F(z)$ $(=F(-z))$, are complementary to values found in Table A.3 for the cumulative normal distribution. It is interesting to compare the behavior shown in Table 6.1 with Chebyshev's inequality, Eq. (4.115) (see Problem 6.19).

■ *Example*

Find the following probabilities for the standard normal distribution:

a) $\Pr(0.34 < Z < 1.31)$;
b) $\Pr(-0.56 < Z < 1.10)$;
c) $\Pr(-1.20 < Z < -0.60)$.

Solution

We use Table A.3 throughout:

$$\text{(a)} \quad \Pr(0.34 < Z < 1.31) = F(1.31) - F(0.34)$$
$$= 0.9049 - 0.6331 = 0.2718. \tag{6.27}$$

(b) $\Pr(-0.56 < Z < 1.10) = F(1.10) - F(-0.56)$
$= 0.8643 - 0.2877 = 0.5766.$ (6.28)

(c) $\Pr(-1.20 < Z < -0.60) = F(-0.60) - F(-1.20)$
$= 0.2743 - 0.1151 = 0.1592.$ (6.29)

These probabilities are shown by the shaded areas in Figure 6.5.

For typical problems in practice, the mean will not be zero and the standard deviation will not be unity. One can still use the standard normal table for numerical work by transforming the problem into the standardized variable Z, defined by Eq. (6.18). If X is normally distributed with mean μ and standard deviation σ (indicated by the notation $X \sim N(\mu, \sigma)$), then

$$\Pr(a < X < b) = \Pr\left(\frac{a-\mu}{\sigma} < \frac{X-\mu}{\sigma} < \frac{b-\mu}{\sigma}\right)$$
$$= \Pr\left(\frac{a-\mu}{\sigma} < Z < \frac{b-\mu}{\sigma}\right). \quad (6.30)$$

Here we have subtracted μ from each of the three quantities a, X, b and then divided by σ, operations that do not change the equalities in (6.30). The middle term thus becomes the standard normal variable Z. The probability that X lies in the originally specified interval (a, b) is thus identical with the probability that the transformed variable Z lies in the corresponding interval shown in this equation. Both the X and the Z distributions are normal. The respective intervals in the second equality of (6.30) are the same, expressed by the number of standard deviations to the left and to the right of the means of both distributions. Since $Z \sim N(0, 1)$, one can employ the standard normal Table A.3 to find probabilities for X. The next example illustrates this procedure.

■ *Example*
Given $X \sim N(100, 10)$, find $\Pr(80 < X < 120)$.

Solution
We are given the normally distributed random variable X with mean $\mu = 100$ and standard deviation $\sigma = 10$. As shown by Eq. (6.30), we transform the required probability statement into one for the standardized variable Z in order to use Table A.3. To this end, we subtract $\mu = 100$ from each part of the given probability statement and then divide by $\sigma = 10$ in each part. Thus, with the help of Table A.3, we find that

$$\Pr(80 < X < 120) = \Pr\left(\frac{80-100}{10} < \frac{X-100}{10} < \frac{120-100}{10}\right) \quad (6.31)$$

$$= \Pr(-2 < Z < 2) = F(2) - F(-2) = 0.9772 - 0.0228$$
$$= 0.9544. \quad (6.32)$$

6 Continuous Distributions

(a)
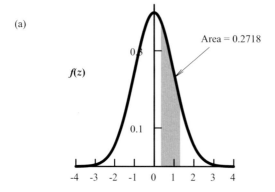

(b)
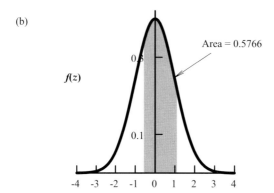

(c)
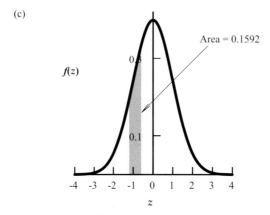

Figure 6.5 See example in the text, Eqs. (6.27)–(6.29).

This probability is equal to the area of the shaded region under the given normal curve in Figure 6.6a, which, in turn, is equal to the corresponding shaded area under the standard normal curve in Figure 6.6b. Note the differences in the scales of the axes in Figure 6.6a and b. Compared with

(a)

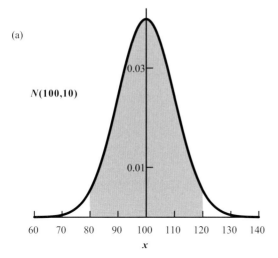

(b)

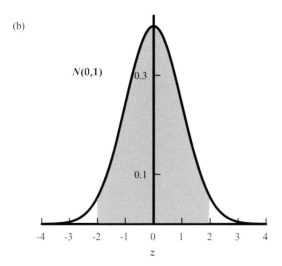

Figure 6.6 In (a), $X \sim N(100, 10)$, and in (b), $Z \sim N(0, 1)$. The shaded area in (a), included in the interval $\mu \pm 2\sigma$ about the mean, is equal to the shaded area in (b) under the standard normal distribution in the interval $(-2 < z < 2)$.

the random variable X, the scale of the abscissa for the standard normal Z is 10 times smaller and its ordinate is 10 times larger. The areas in Figure 6.6a and b are identical in size, namely, 0.9544.

■ *Example*
The chromium content of soil samples in a particular region is observed to follow a normal distribution with a mean of 5 ppm and a standard deviation of

1.2 ppm. What is the probability of a particular soil sample yielding a reading of 7.4 ppm or greater?

Solution
If X denotes the amount of chromium in a soil sample, then we are given $X \sim N(5, 1.2)$. We are asked to make the following evaluation:

$$\Pr(X \geq 7.4) = \Pr\left(\frac{X - \mu}{\sigma} \geq \frac{7.4 - 5}{1.2}\right) \tag{6.33}$$

$$= \Pr(Z \geq 2.0) = 1 - \Pr(Z < 2.0) = 0.0228. \tag{6.34}$$

This probability is the same as that represented by the unshaded area to the right under the standard normal curve in Figure 6.6b.

The normal distribution is also called the *Gaussian distribution* or, simply, "the Gaussian," in honor of the German mathematician, Carl Friedrich Gauss (1777–1855). It was, however, actually discovered in 1733 by the French mathematician, Abraham de Moivre (1667–1754) who, among other things, was particularly skilled in the application of probability theory to gambling. The numerous contributions of Gauss and his place in scientific history were also recognized in his native country's currency. The former German 10-Mark bill, shown in Figure 6.7, displays his picture and the famous formula and its plot. This note was legal tender until the beginning of 2002, when the euro replaced the traditional currencies of Germany and most other member nations of the European Union.

6.4
Central Limit Theorem

In the last section, we mentioned the special importance of the normal distribution, because, among other things, it describes the behavior of the *sampling distribution* of many statistics, such as sample mean, median, and others.

To illustrate, we consider the distribution of average values, $\overline{X} = (1/n) \sum_{i=1}^{n} X_i$, obtained when random samples of size n are drawn from a population having finite

Figure 6.7 German 10-Mark bank note honoring Gauss and the normal distribution.

mean μ and standard deviation σ. The mean $\mu_{\overline{X}}$ of this distribution, which is called the sampling distribution of the means, is the same as the mean of the population from which the drawing is made: $\mu_{\overline{X}} = \mu$. We state without proof that the standard deviation $\sigma_{\overline{X}}$ of the sampling distribution is equal to $\sigma/\sqrt{n}$. (A proof is given later, Eq. (8.24).) Thus, the standard deviation of the sampling distribution of the means is smaller than that of the sampled distribution by the factor $\sqrt{n}$. Although it is unlikely that a given value of $\overline{X}$ will exactly equal μ, values of $\overline{X}$ might be expected to be closely distributed about μ, particularly if the sample size n is large. These relationships between the sampling distribution of $\overline{X}$ and the original population, characterized by μ and σ, are summarized by writing

$$\mu_{\overline{X}} = \mu \quad \text{and} \quad \sigma_{\overline{X}} = \frac{\sigma}{\sqrt{n}}. \tag{6.35}$$

The quantity $\sigma_{\overline{X}}$, which is called the *standard error of the mean*, becomes vanishingly small as n increases without limit, and the mean $\overline{X}$ approaches the true mean μ. Whereas σ reflects the uncertainty in a single measurement, $\sigma/\sqrt{n}$ reflects the uncertainty in the sample mean from a sample of size n.

The sampled population itself might have a normal distribution. If so, then the X_i in a sample are also normally distributed, and the sampling distribution of means will have a normal distribution, exactly, with parameters given by Eq. (6.35). For other population distributions, the sampling distribution of the mean will be approximately normal when n is sufficiently large, generally when $n \geq 30$. The approximation to the normal improves as the sample size increases, approaching the normal in the limit as $n \to \infty$.

These remarkable results are formalized in the following statement, which casts the random variable $\overline{X}$ in terms of the standard normal variable.

Central Limit Theorem

If $\overline{X}$ is the mean of a random sample of size n drawn from a population with finite mean μ and standard deviation σ, then in the limit as $n \to \infty$ the distribution on $\overline{X}$ converges to the normal with mean μ and standard deviation $\sigma/\sqrt{n}$. Using the symbol "$\dot{\sim}$" to be read "is approximately distributed as," we represent the central limit theorem for means in a compact form by writing

$$\overline{X} \dot{\sim} N\left(\mu, \frac{\sigma}{\sqrt{n}}\right). \tag{6.36}$$

The symbol "$\dot{\sim}$" is meant to imply that the distribution converges to the normal in the limit as n increases without bound. Alternatively, we can express the central limit theorem for the mean in terms of the standard normal variable,

$$Z = \frac{\overline{X} - \mu}{\sigma/\sqrt{n}}, \tag{6.37}$$

by writing

$$Z \dot{\sim} N(0, 1). \tag{6.38}$$

Proof of the central limit theorem can be found in advanced texts (e.g., Walpole and Myers (1989), p. 216).

The power of the central limit theorem lies in the fact that it holds for *any* distribution, discrete or continuous, that has a finite mean and finite standard deviation. Furthermore, if the sampled population is normal, then the sampling distribution will always be normal exactly, without the above-mentioned restriction, $n \geq 30$. In place of Eq. (6.38) one can then write $Z \sim (0, 1)$, where the symbol "$\sim$" without the dots denotes an exact relationship, as defined following Eq. (6.17).

The central limit theorem can also be applied to the sum $Y_n = \sum_{i=1}^{n} X_i = n\bar{X}$ of n independent and identically distributed random variables X_i with the same mean μ and standard deviation σ. If we multiply the numerator and denominator on the right-hand side of Eq. (6.37) by n, we obtain

$$Z = \frac{n\bar{X} - n\mu}{\sqrt{n}\sigma} = \frac{Y_n - n\mu}{\sqrt{n}\sigma}, \qquad (6.39)$$

which still satisfies Eq. (6.38). Comparison with Eq. (6.36) implies that

$$Y_n \dot\sim N(n\mu, \sqrt{n}\sigma). \qquad (6.40)$$

Returning to Eq. (6.35), we note that the second relationship applies strictly to an infinitely large population or to a finite population that is sampled with replacement. If a finite population of size $n_p > n$ is sampled without replacement, then the standard deviation of the sample distribution of the mean is given by (Problem 6.23)

$$\sigma_{\bar{X}} = \frac{\sigma}{\sqrt{n}} \sqrt{\frac{n_p - n}{n_p - 1}}, \qquad (6.41)$$

in place of that shown in Eq. (6.35). The additional factor, not present in Eq. (6.35), is called the finite *population correction factor*. The relationship $\mu_{\bar{X}} = \mu$ given by Eq. (6.36) remains the same.

■ Example

A total of 20 random air samples taken one afternoon in a large outdoor area were analyzed for their airborne radon concentration. The mean Rn concentration of the 20 samples was found to be 8.1 Bq m^{-3} and their standard deviation, 1.6 Bq m^{-3}. The true, underlying distribution of the airborne Rn concentration is not known.

a) What information can one offer about the population of the means of samples of size 20 taken from this site?
b) Estimate the probability of finding a mean greater than 8.4 Bq m^{-3} when 20 such independent, random samples are analyzed.

Solution
a) If we assume that each of the samples is independent and identically distributed, then we can apply the central limit theorem. Although the number of samples, $n = 20$, is smaller than the general criterion of 30

given for the theorem to assure a very good approximation, it is reasonable to assume that the distribution of the sample means would be at least approximately normal. Using the only measurements provided, we estimate that the unknown population mean is $\bar{x} = 8.1 \text{ Bq m}^{-3}$ and the standard deviation is $s = 1.6 \text{ Bq m}^{-3}$. The central limit theorem as expressed by Eq. (6.35) then gives for the distribution of the sample mean,

$$\hat{\mu}_{\bar{X}} = \hat{\mu} = \bar{x} = 8.1 \text{ Bq m}^{-3} \quad \text{and} \quad \hat{\sigma}_{\bar{X}} = \frac{\hat{\sigma}}{\sqrt{n}} = \frac{s}{\sqrt{n}} = \frac{1.6}{\sqrt{20}}$$
$$= 0.36 \text{ Bq m}^{-3}. \tag{6.42}$$

With the help of Eq. (6.36), we thus describe the distribution of the sample mean by writing

$$\bar{X} \stackrel{.}{\sim} N(8.1, 0.36), \tag{6.43}$$

with the units Bq m^{-3} implied. The central limit theorem states that the sample mean is approximately normally distributed with a mean of 8.1 Bq m^{-3} and a standard deviation, or standard error, of 0.36 Bq m^{-3}. Notice how the central limit theorem provides a factor of $\sqrt{n}$ smaller uncertainty for the mean than that for the individual measurements.

b) Using the result (6.42) and $Z \stackrel{.}{\sim} N(0,1)$, as implied by Eq. (6.38), we find from Table A.3 that

$$\Pr(\bar{X} > 8.4) = \Pr\left(\frac{\bar{X} - \mu}{\sigma/\sqrt{n}} > \frac{8.4 - 8.1}{0.36}\right) \cong \Pr(Z > 0.83)$$
$$= 0.2033. \tag{6.44}$$

We thus estimate the probability to be about 0.20.

6.5
Normal Approximation to the Binomial Distribution

In addition to describing sample statistics, the normal distribution can often be used to approximate another distribution. If the latter has a finite mean and variance, then, under certain conditions, the normal will do a good job of representing it. The binomial distribution, Eq. (5.7), with mean $\mu = np$ and variance $\sigma^2 = np(1-p)$ (Eqs. (5.13) and (5.15)), is a case in point. Its relationship to the normal distribution can be described as follows.

Theorem 6.1
For a random variable X having the binomial distribution, $X \sim b(x; n, p)$,

$$\frac{X - np}{\sqrt{np(1-p)}} \stackrel{.}{\sim} N(0, 1). \tag{6.45}$$

The theorem states that the normal distribution with parameters n and p provides an accurate representation of the binomial when n is large. Furthermore, the binomial distribution approaches the normal in the limit as $n \to \infty$. The degree to which the normal approximates the binomial distribution as represented by Eq. (6.45) depends on the values of n and p. When $p = 1/2$, the binomial distribution is, like the normal, symmetric. For a given degree of accuracy in the approximation, the restriction of large n is then less severe for p near $1/2$ than for p closer to 0 or 1. As a rule of thumb, the normal approximation to the binomial distribution is adequate for many purposes when np and $n(1-p)$ are both greater than about five.

■ Example

Find the probability of getting from 10 to 15 (inclusively) heads with 20 tosses of an unbiased coin by using

a) the exact binomial model and
b) the normal approximation to the binomial distribution.

Solution

a) The number of heads X is distributed according to the binomial probabilities (Eq. (5.7)) $b(x; n, p)$ with $n = 20$ and $p = 1/2$. Using the cumulative probabilities in Table A.1, we find

$$\Pr(10 \le X \le 15) = B(15; 20, 0.5) - B(9; 20, 0.5)$$
$$= 0.994 - 0.412 = 0.582. \tag{6.46}$$

b) We employ the normal distribution with the same mean and standard deviation as in (a), namely,

$$\mu = np = 10 \quad \text{and} \quad \sigma = \sqrt{np(1-p)} = \sqrt{5}. \tag{6.47}$$

The histogram of $b(x; 20, 0.5)$ and the superimposed normal distribution $N(10, \sqrt{5})$ are shown in Figure 6.8. The probability for each value of the random variable X is equal to the area of the histogram bar of unit width centered about that value of X on the abscissa. The exact probability calculated in (a) is indicated by the shaded area of the histogram elements between $x = 9.5$ and $x = 15.5$. Converting these boundaries into values of the standard normal variable and using Table A.3, we find for the approximate probability as included under the standard normal curve,

$$\Pr(9.5 \le X \le 15.5) \cong \Pr\left(\frac{9.5 - 10}{\sqrt{5}} \le Z \le \frac{15.5 - 10}{\sqrt{5}}\right) \tag{6.48}$$

$$= F(2.46) - F(-0.22) = 0.9931 - 0.4129 = 0.580. \tag{6.49}$$

Comparison of Eqs. (6.49) with (6.46) shows that the normal approximation gives a result that differs from the exact answer by about 0.3%. In this example,

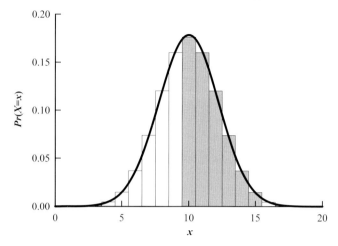

Figure 6.8 Exact binomial (histogram) and normal approximation (solid curve) for example in the text.

$np = n(1 - p) = 10$, which satisfies the rule of thumb value (≥ 5) given above for the general validity of the approximation. In addition, the approximation is best when $p = 1/2$, as is the case here. Figure 6.9 illustrates how the normal approximation to the binomial distribution appears under two altered conditions from this example. In Figure 6.9a, n has been changed from 20 to 40; in Figure 6.9b, n is again 20, while p has been changed from 1/2 to 1/5. Compared with Figure 6.8, the change to larger n in Figure 6.9a improves the approximation. The change of p away from 1/2 while keeping $n = 20$ in Figure 6.9b results in a poorer approximation.

In the last example, the binomial probabilities were represented by histogram bars with heights $\Pr(X = x) = b(x; n, p)$ and unit widths, centered on integral values of x along the abscissa. The areas of the bars are thus equal to the binomial probabilities. As seen from Figure 6.8, when the histogram is approximated by using a continuous probability distribution, such as the normal, some portions of the rectangular bars are wrongly excluded from under the curve, while other portions outside the bars are wrongly included. When calculating continuous approximations to the discrete probabilities, one should apply the additive continuity correction factors of $\pm 1/2$ to the continuous random variable. The particular factors to be applied depend on the information sought. The following indicates some of the rules, in which the exact binomial probabilities are approximated by using the cumulative probability, $F(z)$, of the continuous variable:

$$\text{for } \Pr(X \leq a) \text{ use } F\left(\frac{a + 1/2 - \mu}{\sigma}\right), \tag{6.50}$$

$$\text{for } \Pr(X \geq a) \text{ use } 1 - F\left(\frac{a - 1/2 - \mu}{\sigma}\right), \tag{6.51}$$

(a)

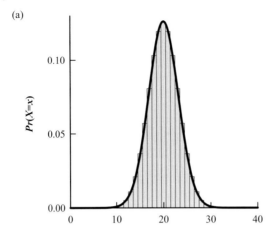

(b)

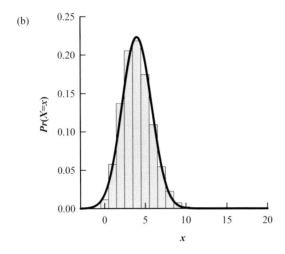

Figure 6.9 (a) Same conditions as in Figure 6.8, except that n has been changed from 20 to 40, improving the approximation. (b) Same as Figure 6.8, except that p has been changed from 1/2 to 1/5, showing that the approximation is not as good as when $p = 1/2$.

$$\text{for } \Pr(a \leq X \leq b) \text{ use } F\left(\frac{b+1/2-\mu}{\sigma}\right) - F\left(\frac{a-1/2-\mu}{\sigma}\right). \tag{6.52}$$

Other rules can be similarly formulated (Problem 6.24).

■ *Example*
If $X \sim b(x;\ 100,\ 0.05)$, use the normal approximation to determine the following probabilities, with and without the continuity correction factors:

a) $\Pr(X \geq 3)$;
b) $\Pr(10 \leq X \leq 20)$.

Solution

a) Ignoring the continuity correction factors, we simply use $X = 3$ in Eq. (6.45). The mean of the binomial distribution is $np = 100 \times 0.05 = 5$, and the standard deviation is

$$\sqrt{np(1-p)} = \sqrt{100 \times 0.05(1-0.05)} = 2.18. \tag{6.53}$$

Therefore, from Eq. (6.45),

$$\Pr(X \geq 3) \cong \Pr\left(Z \geq \frac{3-5}{2.18}\right) = \Pr(Z \geq -0.92) \tag{6.54}$$

$$= 1 - F(-0.92) = 1 - 0.1788 = 0.8212. \tag{6.55}$$

With the continuity correction factor, in this case Eq. (6.51),

$$\Pr(X \geq 3) \cong \Pr\left(Z \geq \frac{2.5-5}{2.18}\right) = \Pr(Z \geq -1.15) \tag{6.56}$$

$$= 1 - F(-1.15) = 1 - 0.1251 = 0.8749. \tag{6.57}$$

This value is considerably closer to the exact answer, 0.8817, than the uncorrected result (6.55).

b) With no continuity correction factor, Eq. (6.45) implies that

$$\Pr(10 \leq X \leq 20) \cong \Pr\left(\frac{10-5}{2.18} \leq Z \leq \frac{20-5}{2.18}\right) = \Pr(2.29 \leq Z \leq 6.88) \tag{6.58}$$

$$= F(6.88) - F(2.29) = 1.000 - 0.9890 = 0.0110. \tag{6.59}$$

With the correction factor (6.52), one finds

$$\Pr(10 \leq X \leq 20) \cong \Pr\left(\frac{9.5-5}{2.18} \leq Z \leq \frac{20.5-5}{2.18}\right) = \Pr(2.06 \leq Z \leq 7.11) \tag{6.60}$$

$$= F(7.11) - F(2.06) = 1.000 - 0.9803 = 0.0197. \tag{6.61}$$

The exact answer is 0.0282. The corrected result (6.61) is closer than (6.59), but still has a *relatively* big error. The parameter n is large here, but p is rather far removed from the optimum value of $1/2$ for use of the approximation. Whereas the normal approximation is always symmetric, the binomial distribution is very skewed in this example. The nonzero binomial probabilities, though small, extend out to $x = 20$ in Figure 6.10. The normal distribution extends to infinity in both directions.

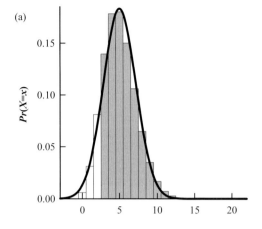
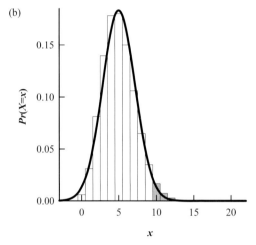

Figure 6.10 Binomial distribution, $b(x; n, p)$ (histogram), and normal approximation (solid curve) for example in the text, parts (a) and (b).

The normal distribution is very convenient to use as an approximating distribution because its standard cumulative distribution is readily tabulated. The next example illustrates its use in radioactive decay to solve a practical problem that would offer considerable difficulty to compute by using the exact binomial formulation.

■ *Example*
A 37-Bq (1-nCi) source of pure ^{42}K contains 2.39×10^6 atoms. The half-life is 12.36 h. Consider the probability that from 27 to 57 atoms (inclusive) will decay in 1 s.

a) Solve for the probability by using the normal approximation to the binomial distribution.
b) Use the exact binomial model for radioactive decay to set up the solution.

6.5 Normal Approximation to the Binomial Distribution

Solution

a) We let p be the probability that a given atom will decay in 1 s. Since $\mu = 37$ is the mean number of atoms that decay in this time from among the $n = 2.39 \times 10^6$ present,

$$p = \frac{\mu}{n} = \frac{37}{2.39 \times 10^6} = 0.0000155. \qquad (6.62)$$

(Note that this probability for decay in 1 s is numerically equal to the decay constant of the nuclide in s^{-1}.) The standard deviation is

$$\sqrt{np(1-p)} = \sqrt{2.39 \times 10^6 (0.0000155)(1 - 0.0000155)} = 6.08. \qquad (6.63)$$

Applying Eq. (6.52) with appropriate continuity correction factors, we find for the probability that the number of disintegrations X in 1 s will be in the declared interval,

$$\Pr(27 \leq X \leq 57) \cong \Pr\left(\frac{26.5 - 37}{6.08} \leq Z \leq \frac{57.5 - 37}{6.08}\right)$$

$$= \Pr(-1.73 \leq Z \leq 3.37) \qquad (6.64)$$

$$= F(3.37) - F(-1.73) = 0.9996 - 0.0418 = 0.9578. \qquad (6.65)$$

The problem is thus readily and accurately solved by using the normal approximation.

b) We are asked to set up, but not solve, the problem with the exact binomial model. One has (Eqs. (5.7) and (5.8))

$$\Pr(X = x) = b(x; n, p) = \frac{n!}{x!(n-x)!} p^x (1-p)^{n-x}. \qquad (6.66)$$

Substituting the numerical values of n and p, we have for the exact answer,

$$\Pr(27 \leq X \leq 57)$$

$$= \sum_{x=27}^{57} \frac{(2.39 \times 10^6)!}{x!(2.39 \times 10^6 - x)!} (0.0000155)^x (0.9999845)^{2.39 \times 10^6 - x}. \qquad (6.67)$$

This expression involves the sum of products of some extremely large and small numbers, necessitating the use of practical approximations. In so doing, one is led naturally to the more readily handled Poisson approximation to the binomial, which we described in Chapter 5 (Eq. (5.27)). (This example is an extension of one given in Turner (2007), pp. 310–311. Also, see the last example in Section 2.5 of the present book.)

6.6
Gamma Distribution

The *gamma distribution* is useful in the analysis of reliability and queuing, for example, times to the failure of a system or times between the arrivals of certain events. Two special cases of the gamma distribution find widespread use in health physics. These are the exponential and the chi-square distributions, discussed in the next two sections.

The gamma distribution for a continuous random variable X has two parameters, $k > 0$ and $\lambda > 0$. Its density function is[1]

$$f(x; k, \lambda) = \begin{cases} \dfrac{\lambda^k x^{k-1} e^{-\lambda x}}{\Gamma(k)}, & \text{for } x > 0, \\ 0, & \text{elsewhere.} \end{cases} \tag{6.68}$$

The mean and variance are (Problem 6.30)

$$\mu = \frac{k}{\lambda} \quad \text{and} \quad \sigma^2 = \frac{k}{\lambda^2}. \tag{6.69}$$

By way of review, the gamma function is defined for $k > 0$ as

$$\Gamma(k) = \int_0^\infty x^{k-1} e^{-x} \, dx. \tag{6.70}$$

It satisfies the recursion relation

$$\Gamma(k+1) = k\Gamma(k), \tag{6.71}$$

as can be shown by integration of the right-hand side of Eq. (6.70) by parts (Problem 6.31). When k is a positive integer, then repeated application of Eq. (6.71) gives

$$\Gamma(k+1) = k!, \tag{6.72}$$

the factors terminating with $\Gamma(1) = 0! = 1$.

Gamma distributions can assume a variety of shapes, depending on the values of the parameters k and λ. Figure 6.11 shows the density functions for $\lambda = 1$ and several values of k. We turn now to the density function for $k = 1$.

6.7
Exponential Distribution

When $k = 1$, the gamma distribution (6.68) becomes the *exponential probability density function*

$$f(x; 1, \lambda) = \begin{cases} \lambda e^{-\lambda x}, & x > 0, \\ 0, & \text{elsewhere.} \end{cases} \tag{6.73}$$

1) Some texts use $\beta = 1/\lambda$ in place of λ in the definition (6.68).

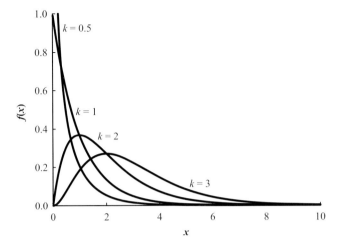

Figure 6.11 Gamma distributions for $\lambda = 1$ and different values of k from Eq. (6.68).

As we have seen, this important continuous random variable can be used to describe the discrete process of radioactive decay. We next discuss its relationship to the Poisson process (Section 5.5).

We showed by Eq. (5.38) that the Poisson distribution describes the probability $p_x(t)$ for the number of events X that occur in time t when the mean number of events per unit time is λ. We now consider the random variable T that describes the time taken for the first event to happen. The probability that no events ($X = 0$) take place during the time span from zero up to $T = t$ is, from Eq. (5.38),

$$p_0(t) = \frac{(\lambda t)^0 e^{-\lambda t}}{0!} = e^{-\lambda t}. \tag{6.74}$$

Letting the random variable T be the time to the first Poisson event, we have for the probability that no event occurs in the time interval $(0, t)$,

$$\Pr(T \geq t) = e^{-\lambda t}. \tag{6.75}$$

Therefore,

$$\Pr(T < t) = 1 - e^{-\lambda t}, \tag{6.76}$$

thus providing the cumulative distribution for the first-event times. (These same functions describe the survival and decay probabilities, Eqs. (2.21) and (2.22), for radioactive decay of an atom.) Differentiation of the right-hand side of Eq. (6.76) with respect to t (see Eq. (4.14)) gives the probability density function for the *arrival times* of the first Poisson event:

$$f(t) = \begin{cases} \lambda e^{-\lambda t}, & t > 0, \\ 0, & \text{otherwise}, \end{cases} \tag{6.77}$$

in agreement with Eq. (6.73). We derived this density function in an earlier chapter from another standpoint (Eqs. (4.36) and (4.37)). We showed there that its mean and standard deviation are both equal to $1/\lambda$ (Eqs. (4.38) and (4.52)), which agrees with Eq. (6.69) when $k = 1$.

We see from Eq. (6.77) that the time to first arrival for a Poisson process has an exponential distribution. If failures happen randomly according to a Poisson process, then the resulting times to occurrence have an exponential distribution. In this way, the exponential as well as the more general gamma distributions prove useful for describing reliability and the time to failure of industrial products and components. Applied to radioactive decay, Eq. (6.77) describes the distribution of the decay times of a large number of identical radionuclides. (See Eq. (5.27) and the discussion following it.) The parameter λ is the decay constant, and the mean life is $1/\lambda$.

■ Example

An alpha-particle counter has a steady average background rate of 30 counts per hour.

a) What fraction of the intervals between successive counts will be longer than 5 min?
b) What fraction will be shorter than 30 s?
c) What is the probability that, between two successive counts, a time interval will occur whose length is within two standard deviations of the mean length of the intervals?

Solution

a) The number of alpha particles counted per unit time is Poisson distributed. The probability density function for the decay events is described by Eq. (6.77) with parameter $\lambda = 30\,\text{h}^{-1}$ and t in hours:

$$f(t) = \begin{cases} 30e^{-30t}, & t > 0, \\ 0, & \text{otherwise.} \end{cases} \tag{6.78}$$

The fraction of successive intervals that are longer than 5 min = $1/12$ h is given by

$$\Pr\left(T > \frac{1}{12}\right) = 30\int_{1/12}^{\infty} e^{-30t}\,dt = e^{-30/12} = 0.0821. \tag{6.79}$$

b) The fraction of successive intervals shorter than $T = 30\,\text{s} = 1/120\,\text{h}$ is

$$\Pr\left(T < \frac{1}{120}\right) = 30\int_{0}^{1/120} e^{-30t}\,dt = 0.221. \tag{6.80}$$

c) The count rate of $30\,\text{h}^{-1}$ corresponds to a mean interval length of 2 min, which is also the standard deviation for the exponential distribution. The relevant time interval for this example, therefore, goes from

$t = 2 - 4 = -2$ min to $t = 2 + 4 = 6$ min, or from $-1/30$ to $1/10$ h. Using the density function (6.78), we find ($f = 0$ when $T \leq 0$)

$$\Pr\left(T < \frac{1}{10}\right) = 30 \int_0^{1/10} e^{-30t}\, dt = 0.950. \tag{6.81}$$

Note from Table 6.1 how close this answer is to the one-tail area under the standard normal curve within two standard deviations of the mean.

An interesting aspect of the exponential density is its "no memory" feature. If a radionuclide has been observed not to decay for a time t, what is the chance that it will not decay during an additional time s? In other words, what is the probability for a random decay time $T > t + s$, given that $T > t$ has been observed? If we let A denote the event $T > t + s$ and B the event $T > t$, then we ask for the conditional probability $\Pr(A|B)$, discussed in Section 3.5. According to Eq. (3.31),

$$\Pr(A|B) = \frac{\Pr(A \cap B)}{\Pr(B)}. \tag{6.82}$$

The intersection of A and B is just A itself:

$$A \cap B = \{T > t + s\} \cap \{T > t\} = \{T > t + s\} = A. \tag{6.83}$$

Using Eq. (6.77) to describe radioactive decay, we obtain from the last two equations,

$$\Pr(A|B) = \frac{\Pr(A)}{\Pr(B)} = \frac{\Pr(T > t + s)}{\Pr(T > t)} = \frac{e^{-\lambda(t+s)}}{e^{-\lambda t}} = e^{-\lambda s} = \Pr(T > s). \tag{6.84}$$

The last term is the probability that T will exceed s, irrespective of the fact that we have observed T for the time t already. In other words, an "old" radioactive atom that has already lived a time t has the same probability of living any additional time s as a newly formed identical atom. That the older atom might have been "at risk for decay" for some time is irrelevant. Thus, one can characterize the exponential distribution as "having no memory."

It can be shown that, if $X_1, X_2, \ldots, X_k$ is a random sample from the exponential distribution with parameter λ, then $Y = \sum_{i=1}^{k} X_i$ has the gamma distribution with parameters k and λ. Since each X_i has mean $1/\lambda$ and variance $1/\lambda^2$, it follows that Y has mean k/λ and variance k/λ^2, as given by Eq. (6.69).

6.8 Chi-Square Distribution

Another important continuous distribution with many applications in statistics is the *chi-square distribution*. As we shall see in later chapters, "chi-square testing" is often used to help judge whether a given hypothesis can reasonably account for observed data. Another application is estimation of the population variance when sampling from a normal population. The density function for the chi-square distribution is a special case of the gamma distribution with parameters $\lambda = 1/2$ and $k = v/2$. From Eq. (6.68), we write it in the form

$$f\left(\chi^2; \frac{v}{2}, \frac{1}{2}\right) = \begin{cases} \dfrac{(\chi^2)^{(v/2)-1} e^{-\chi^2/2}}{2^{v/2} \Gamma(v/2)}, & \chi^2 \geq 0, \\ 0, & \text{elsewhere.} \end{cases} \quad (6.85)$$

The density depends on the single parameter v, which is called the number of degrees of freedom, or simply the *degrees of freedom*. Its role will become clearer in later applications. Since the chi-square distribution is a gamma distribution, its mean and variance are given by Eq. (6.69) (Problem 6.37):

$$\mu = v \quad \text{and} \quad \sigma^2 = 2v. \quad (6.86)$$

Figure 6.12 shows some plots of the distribution for several values of v.

Letting $\chi^2_{v,\alpha}$ denote the value of χ^2 with v degrees of freedom for which the cumulative probability is α, we write

$$\alpha = \Pr(\chi^2 \leq \chi^2_{v,\alpha}) \frac{1}{2^{v/2} \Gamma(v/2)} \int_0^{\chi^2_{v,\alpha}} (\chi^2)^{(v/2)-1} e^{-\chi^2/2} \, d\chi^2. \quad (6.87)$$

Table A.4 in the Appendix gives the quantiles for $\chi^2_{v,\alpha}$ for $v = 1$ to 30 degrees of freedom and various probability values α. (A *quantile*, such as $\chi^2_{v,\alpha}$, refers to the value of the variable that marks the division of the distribution into that fraction α to the left or to the right, depending on how the quantile is defined.)

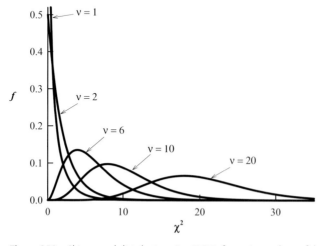

Figure 6.12 Chi-squared distribution, Eq. (6.85), for various values of the degrees of freedom v.

Example

Answer the following with the help of Table A.4.

a) What is the value of χ^2 with 6 degrees of freedom below which 30% of the distribution lies?
b) Find the value of χ^2 with 7 degrees of freedom that cuts off 5% of the distribution to the right.
c) Find the probability that the value of χ^2 with 12 degrees of freedom could be as large as 15.0.
d) What is the probability that χ^2 with 3 degrees of freedom will be greater than 12.2?

Solution

a) With $v=6$ and $\alpha=0.300$ in Table A.4, we find $\chi^2_{v,\alpha} = \chi^2_{6,0.30} = 3.828$.
b) In this case, we want the value of χ^2 with 7 degrees of freedom below which 95% of the area lies. Table A.4 gives $\chi^2_{v,\alpha} = \chi^2_{7,0.95} = 14.07$.
c) With reference to Eq. (6.87), we are asked to find α when $\chi^2_{12,\alpha} = 15.0$, where α is the area to the left of $\chi^2 = 15.0$. In Table A.4, we need to interpolate between the two entries $\chi^2_{12,0.75} = 14.85$ and $\chi^2_{12,0.80} = 15.81$. Linear interpolation gives $\chi^2_{12,0.758} = 15.0$, and so $\Pr(\chi^2 \leq 15.0) = \alpha = 0.758$.
d) From Table A.4, $\chi^2_{3,0.990} = 11.35$ and $\chi^2_{3,0.995} = 12.84$. Linear interpolation gives $\chi^2_{3,0.993} = 12.2$, or $\Pr(\chi^2 \leq 12.2) = \alpha = 0.993$. The area to the right of 12.2 is $\Pr(\chi^2 > 12.2) = 1 - \alpha = 0.007$.

When the degrees of freedom $v > 30$ are beyond the range given in Table A.4, one can proceed as follows. It turns out that the quantity $Z = (\sqrt{2\chi^2} - \sqrt{2v-1})$ is then very nearly normally distributed with zero mean and unit standard deviation. Letting $\chi^2_{v,\alpha}$ and z_α denote the 100αth percentiles of the chi-square and the standard normal distributions, it follows that the normal approximation to the chi-square distribution yields

$$\chi^2_{v,\alpha} = \frac{1}{2}(z_\alpha + \sqrt{2v-1})^2, \quad v > 30. \tag{6.88}$$

The cumulative standard normal Table A.3 can then be employed.

Often one is interested in the boundaries of an interval that cuts off equal areas of a chi-square distribution on both ends. Unless stated otherwise, we shall assume by convention that such an interval is always implied. As we have seen, the boundaries for the interval are symmetric for the standard normal curve. In contrast, the chi-square distribution is asymmetric, and therefore fixing the interval is somewhat more involved, as the next example illustrates.

Example

A random variable has a chi-square distribution with 21 degrees of freedom. Find the constants b_1 and b_2 such that $\Pr(b_1 \leq \chi^2 \leq b_2) = 0.90$.

Solution

The interval (b_1, b_2) is chosen to cut off 5% of the chi-square distribution on both ends (in accordance with the convention, $\Pr(\chi^2 < b_1) = \Pr(\chi^2 > b_2)$). The relevant values of α are 0.05 and 0.95 with $\nu = 21$. From Table A.4 we find that $b_1 = \chi^2_{21,0.05} = 11.591$ and that $b_2 = \chi^2_{21,0.95} = 32.67$. Thus, the interval $(b_1, b_2) = (11.591, 32.67)$ encloses the middle 90% of the distribution, as shown in Figure 6.13, and excludes the two shaded areas of 5% on either end.

The chi-square distribution has the property of *additivity*, which we state here, but do not prove. If $Y_1, Y_2, \ldots, Y_n$ are independent random variables having chi-square distributions with $\nu_1, \nu_2, \ldots, \nu_n$ degrees of freedom, respectively, then $Y = \sum_{i=1}^{n} Y_i$ is chi-square distributed with $\nu = \sum_{i=1}^{n} \nu_i$ degrees of freedom.

A very important relationship exists between the chi-square and normal distributions. If the random variable Z has the standard normal distribution, then $Y = Z^2$ has the chi-square distribution with $\nu = 1$ degree of freedom. To show this, we focus on the cumulative probability distribution function $G(y)$, writing

$$G(y) = \Pr(Y \leq y) = \Pr(Z^2 \leq y) = \Pr(-\sqrt{y} \leq Z \leq \sqrt{y}). \tag{6.89}$$

In terms of the cumulative standard normal distribution $F(z)$ given by Eq. (6.20), it follows that

$$G(y) = F(\sqrt{y}) - F(-\sqrt{y}). \tag{6.90}$$

The probability density function $g(y)$ for Y is the derivative of the cumulative function $G(y)$ with respect to y, as shown by Eq. (4.14). Using the chain rule, we write from Eq. (6.90)

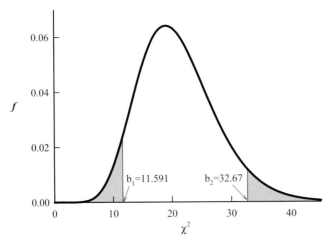

Figure 6.13 See example in the text. The values b_1 and b_2 cut off 5% of the area (shaded) on the left and right, respectively, of the chi-squared distribution with $\nu = 21$ degrees of freedom.

$$g(y) = \frac{dG}{dy} = \frac{dF(\sqrt{y})}{dz}\frac{d(\sqrt{y})}{dy} - \frac{dF(-\sqrt{y})}{dz}\frac{d(\sqrt{y})}{dy}. \quad (6.91)$$

The quantity $dF/dz = f(z)$ is the standard normal density function (6.19). Also, $d(\sqrt{y})/dy = (1/2)y^{-1/2}$. Thus, with the help of Eq. (6.19), Eq. (6.91) becomes

$$g(y) = \frac{1}{2}y^{-1/2}[f(\sqrt{y}) + f(-\sqrt{y})] = \frac{y^{-1/2}}{2\sqrt{2\pi}}(e^{-y/2} + e^{-y/2})$$

$$= \frac{1}{\sqrt{2\pi}}y^{-1/2}e^{-y/2}. \quad (6.92)$$

Comparison with Eq. (6.85) shows that $Y = Z^2$ has the chi-square distribution with $\nu = 1$ degree of freedom. The constant $\Gamma(1/2) = \sqrt{\pi}$, and the density (6.92) is zero when $y < 0$.

For a normal distribution with mean μ and standard deviation σ, we have thus shown that the square of the standard normal variable $Z = (X - \mu)/\sigma$ defined by Eq. (6.18) has a chi-square distribution with one degree of freedom. Combining this finding with the additivity property of chi-square distributions leads to a powerful result that we shall use in later sections. If we have $X_1, X_2, \ldots, X_n$ independent random variables with the same normal distribution with mean μ and standard deviation σ, then additivity implies that the random variable

$$Y = \sum_{i=1}^{n}\left(\frac{X - \mu}{\sigma}\right)^2 \sim \chi_n^2, \quad (6.93)$$

where the notation χ_n^2 means that Y has a chi-square distribution with n degrees of freedom.

6.9
Student's t-Distribution

The random variable T is defined as follows in terms of two independent random variables Z and Y that have, respectively, the standard normal distribution and the chi-square distribution with ν degrees of freedom:

$$T = \frac{Z}{\sqrt{Y/\nu}}. \quad (6.94)$$

The sampling distribution of T has Student's t-distribution with ν degrees of freedom. Its density function is given by

$$h(t) = \frac{1}{\sqrt{\pi\nu}}\frac{\Gamma((\nu+1)/2)}{\Gamma(\nu/2)}\left(1 + \frac{t^2}{\nu}\right)^{-(\nu+1)/2}, \quad -\infty < t < \infty, \quad \nu > 0. \quad (6.95)$$

Figure 6.14 shows the t-distribution for $\nu = 1$ and $\nu = 4$ degrees of freedom and the standard normal distribution. One sees that the t-distribution, which is symmetric about zero, approaches the standard normal as ν increases. The two nearly coincide

6 Continuous Distributions

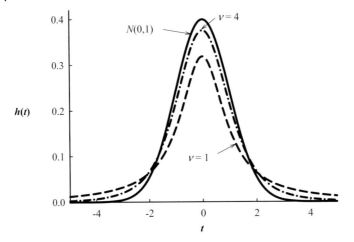

Figure 6.14 Student's t-distribution, Eq. (6.95), for $v=1$ and $v=4$ degrees of freedom and the standard normal distribution $N(0, 1)$.

when $v \geq 30$. In fact, many tables of the t-distribution simply refer to the normal distribution after 30 degrees of freedom. At the other extreme, when $v=1$, $h(t)$ coincides with the Cauchy distribution, Eq. (6.16) with $\tau = 0$ (Problem 6.46).

Table A.5 in the Appendix gives the quantiles $t_{v,\alpha}$ of the t-distribution with v degrees of freedom, such that the fraction α of the distribution lies to the *right*:

$$\alpha = \Pr(T > t_{v,\alpha}) = \int_{t_{v,\alpha}}^{\infty} h(t) dt. \tag{6.96}$$

Values of α in the table range from 0.100 to 0.005. Because of the symmetry of the t-distribution, $-t_{v,\alpha}$ is the quantile for which the fraction α of the area lies to the left.

■ *Example*

Use Table A.5 to answer the following.

a) With 10 degrees of freedom, what value of t leaves 5% of the t-distribution to the right?
b) What is the value of t with 6 degrees of freedom that cuts off 2.5% of the distribution to the left?
c) Find the value of t_1 with 21 degrees of freedom such that $\Pr(-t_1 \leq T \leq t_1) = 0.900$.
d) Find $\Pr(-1.325 < T < 2.086)$ for the t-distribution with 20 degrees of freedom.

Solution

a) The quantile asked for is given directly in Table A.5, $t_{10,0.050} = 1.812$.

b) In the notation of Table A.5, we are asked to find $t_{6,0.975}$, which leaves 2.5% of the distribution on its left. Using the table and the fact that the t-distribution is symmetric, we find that $t_{6,0.975} = -t_{6,0.025} = 2.447$.
c) The interval $(-t_1, t_1)$ leaves $\alpha = 0.050$ of the symmetric distribution outside on the left and the same fraction outside on the right. With $\nu = 21$, Table A.5 gives $t_1 = t_{21,0.050} = 1.721$.
d) With $\nu = 20$ in Table A.5, we find that $t_{20,0.100} = 1.325$ and $t_{20,0.025} = 2.026$. The value $t = -1.325$ at the lower boundary of the interval excludes 0.100 of the area to the left. The value 2.086 excludes 0.025 to the right. Since the total area outside the interval between the last two values of t is $0.100 + 0.025 = 0.125$, it follows that

$$\Pr(-1.325 < T < 2.086) = 1 - 0.125 = 0.875. \tag{6.97}$$

The discoverer of the t-distribution in the early twentieth century was W.S. Gosset, who published under the pseudonym "Student"; consequently, the designation "Student's t-distribution." The distribution is important for comparing sample means for a normal population when the population variance is unknown.

6.10
F Distribution

The *F distribution* is useful for obtaining confidence intervals or tests of hypothesis in comparing two variances. It is also used to test whether two or more mean values are equal. The *F* random variable is defined as the ratio of two independent chi-square random variables, U_1 and U_2, each divided by its respective degrees of freedom, ν_1 and ν_2:

$$F(\nu_1, \nu_2) = \frac{U_1/\nu_1}{U_2/\nu_2} = F_{\nu_1,\nu_2}. \tag{6.98}$$

By convention, when describing F the number of degrees of freedom ν_1 associated with the function U_1 in the numerator is always stated first, followed by ν_2. The probability density function for an F random variable with ν_1 and ν_2 degrees of freedom is given by

$$h(f) = \begin{cases} \dfrac{\Gamma((\nu_1+\nu_2)/2)}{\Gamma(\nu_1/2)\Gamma(\nu_2/2)} \left(\dfrac{\nu_1}{\nu_2}\right)^{\nu_1/2} \dfrac{f^{(\nu_1/2)-1}}{(1+(\nu_1 f/\nu_2))^{(\nu_1+\nu_2)/2}}, & 0 < f < \infty, \\ 0, & \text{elsewhere}. \end{cases} \tag{6.99}$$

Three examples of the density function are shown in Figure 6.15. Each depends on the two parameters ν_1 and ν_2. Quantiles for the cumulative probability,

$$\alpha = \int_0^{f_\alpha(\nu_1,\nu_2)} h(f) df, \tag{6.100}$$

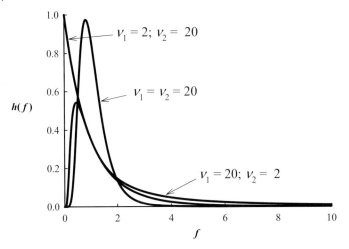

Figure 6.15 Three examples of the probability density function, Eq. (6.99), for the F distribution with degrees of freedom v_1 and v_2 shown.

are given for different values of v_1 and v_2 in Tables A.6 and A.7 in the Appendix. The quantity $f_\alpha(v_1, v_2)$ is the value of f that includes the fraction α of the area under the curve F on its left and cuts off $(1 - \alpha)$ on its right. Tables A.6 and A.7 include only values for $\alpha = 0.95$ and $\alpha = 0.99$, respectively, for various combinations of v_1 and v_2. However, these tables can also be used to obtain values for $\alpha = 0.05$ and $\alpha = 0.01$ by means of the following relation between the lower and upper quantiles of the F distribution (Problem 6.47):

$$f_{1-\alpha}(v_1, v_2) = \frac{1}{f_\alpha(v_2, v_1)}. \tag{6.101}$$

■ *Example*
Find the F value with 5 and 10 degrees of freedom that leaves an area 0.95 to the right.

Solution
We are asked to find $f_\alpha(v_1, v_2) = f_{0.05}(5, 10)$. Since Tables A.6 and A.7 provide values only for $\alpha = 0.95$ and 0.99, we use Eq. (6.101). With $\alpha = 0.95$, $v_1 = 5$, and $v_2 = 10$, substitution into Eq. (6.101) gives, from Table A.6,

$$f_{0.05}(5, 10) = \frac{1}{f_{0.95}(10, 5)} = \frac{1}{4.735} = 0.211. \tag{6.102}$$

■ *Example*
Find values b_1 and b_2 such that $\Pr(b_1 < F < b_2) = 0.90$, where F is an F random variable with 14 and 19 degrees of freedom.

Solution

We follow the convention that, unless otherwise specified, intervals are selected to cut off equal areas on both ends of the distribution. Accordingly, we need to determine b_1 and b_2 by finding the quantiles that cut off 5% on each end. The area between the two boundaries will then be 0.90, as required. The right-hand boundary b_2 is tabulated directly in Table A.6. With $\alpha = 0.95$, $v_1 = 14$, and $v_2 = 19$, we find $b_2 = f_\alpha(v_1, v_2) = f_{0.95}(14, 19) = 2.256$. The value of b_1 would be given in Table A.6 by $b_1 = f_{0.05}(14, 19)$, but this value is not tabulated. We can still determine the b_1 boundary from the table, however, with the help of the relation (6.101). Substituting $\alpha = 0.95$ into Eq. (6.101), keeping v_1 and v_2 the same as before, and referring to Table A.6, we obtain

$$f_{0.05}(14, 19) = \frac{1}{f_{0.95}(19, 14)} = \frac{1}{2.400} = 0.417. \tag{6.103}$$

Thus, $\Pr(0.417 < F < 2.256) = 0.90$.

6.11 Lognormal Distribution

Many population distributions are skewed with a long tail in one direction or the other. Distributions of this kind can arise, for example, in survival analysis, environmental measurements, and salaries earned. A frequently used model for such data is the *lognormal distribution*. In this case, it is the natural logarithm $Y = \ln X$ of a random variable X that is normally distributed, rather than the variable X itself.

We obtain the distribution for X by transforming variables. For the normally distributed Y with mean μ_y and standard deviation σ_y, we write from the definition (6.17)

$$f(y) = \frac{1}{\sqrt{2\pi}\sigma_y} e^{-(y-\mu_y)^2/2\sigma_y^2}, \quad -\infty < y < \infty. \tag{6.104}$$

The distribution on X can be inferred from Eq. (4.125). In the present notation, we substitute $y = \ln x$ into Eq. (6.104) and then multiply by the derivative $d(\ln x)/dx = 1/x$. Thus, the lognormal distribution is

$$g(x) = \begin{cases} \dfrac{1}{\sqrt{2\pi}\sigma_y x} e^{-(\ln x - \mu_y)^2/2\sigma_y^2}, & x > 0, \ -\infty < \mu_y < \infty, \ \sigma_y > 0, \\ 0, & \text{elsewhere.} \end{cases} \tag{6.105}$$

The two parameters μ_y and σ_y of the distribution are the true mean and standard deviation of the transformed random variable $Y = \ln X$. Put another way, if Y is normally distributed with mean μ_y and standard deviation σ_y, then $X = e^Y$ has the lognormal distribution with density function given by Eq. (6.105).

The mean of the original variable X is

$$\mu_x = e^{\mu_y + (1/2)\sigma_y^2} \tag{6.106}$$

and its variance is

$$\sigma_x^2 = e^{2\mu_y + 2\sigma_y^2} - e^{2\mu_y + \sigma_y^2} = e^{2\mu_y + \sigma_y^2}(e^{\sigma_y^2} - 1). \tag{6.107}$$

■ *Example*
If $Y = \ln X$ has the normal distribution with $\mu_y = 1$ and $\sigma_y = 2$, determine the mean and standard deviation for X.

Solution
From Eq. (6.106), the mean is

$$\mu_x = e^{1 + (1/2)(2)^2} = e^3 = 20.1. \tag{6.108}$$

From Eq. (6.107), the variance is

$$\sigma_x^2 = e^{2(1) + 2(2)^2} - e^{2(1) + (2)^2} = e^{10} - e^6 = e^6(e^4 - 1)$$
$$= 2.16 \times 10^4. \tag{6.109}$$

The standard deviation is

$$\sigma_x = e^3 \sqrt{e^4 - 1} = \sqrt{2.16 \times 10^4} = 147. \tag{6.110}$$

6.12
Beta Distribution

The *beta distribution* is commonly used in modeling the reliability, X, of a system. The distribution also arises naturally in the statistical area of Bayesian analysis, which will be discussed in Chapter 15. The probability density function of a beta random variable, which has two parameters $\alpha > 0$ and $\beta > 0$, is given by

$$f(x; \alpha, \beta) = \begin{cases} \dfrac{\Gamma(\alpha + \beta)}{\Gamma(\alpha)\Gamma(\beta)} x^{\alpha - 1}(1 - x)^{\beta - 1}, & 0 \le x \le 1, \\ 0, & \text{elsewhere.} \end{cases} \tag{6.111}$$

The mean and variance are (Problem 6.53)

$$\mu = \frac{\alpha}{\alpha + \beta} \quad \text{and} \quad \sigma^2 = \frac{\alpha \beta}{(\alpha + \beta + 1)(\alpha + \beta)^2}. \tag{6.112}$$

An important relationship exists between the gamma and beta distributions. We consider two independent gamma random variables, X_1 and X_2, with parameters (α, λ) and (β, λ), respectively. Since X_1 and X_2 are independent, the random variables $Y_1 = X_1 + X_2$ and $Y_2 = X_1/(X_1 + X_2)$ are also independently distributed. The variable Y_1 represents the sum of the two variables, and Y_2 represents the fraction of the sum due to X_1. It can be shown that the joint probability density function for Y_1 is a gamma distribution with parameters $(\alpha + \beta, \lambda)$ and that Y_2 has a beta distribution with parameters (α, β) (see, for example, Hogg and Tanis (1993)).

This relationship provides an interesting result. Consider, for example, two sources of the same radionuclide (decay constant $= \lambda$), source A having originally α atoms and source B, β atoms. If X_1 and X_2, respectively, denote the times for each of the two sources to decay away completely, it can be shown that X_1 and X_2 are independent gamma variables, having parameters (α, λ) and (β, λ), respectively. Then the above result states that, independently of the time needed for all $\alpha + \beta$ disintegrations to occur, the proportion of the total time that comes from source A has a beta distribution with parameters (α, β). This result is illustrated in the next example.

■ *Example*

Consider two sources of a radionuclide with decay constant $\lambda = 2.0\,\text{min}^{-1}$. Initially, source A has two atoms and source B has five atoms.

a) Describe the distribution of the total time for all atoms from both sources to decay.
b) Calculate the mean and variance of the total time for all atoms to decay.
c) Write the density function and determine the mean and variance of the proportion of time for source A to decay totally.

Solution

a) We let $\alpha = 2$ and $\beta = 5$, respectively, be the numbers of atoms initially present in the two sources. The total time in minutes for both sources to decay then has a gamma distribution with parameters $\alpha + \beta = 2 + 5 = 7$ and $\lambda = 2.0\,\text{min}^{-1}$.

b) The mean and variance for the total time are given by Eq. (6.69):

$$\mu = \frac{\alpha + \beta}{\lambda} = \frac{7}{2.0\,\text{min}^{-1}} = 3.5\,\text{min} \tag{6.113}$$

and

$$\sigma^2 = \frac{\alpha + \beta}{\lambda^2} = \frac{7}{(2.0\,\text{min}^{-1})^2} = 1.75\,\text{min}^2. \tag{6.114}$$

c) The density is the beta distribution (6.111) with parameters $\alpha = 2$ and $\beta = 5$:

$$f(x) = \begin{cases} \dfrac{\Gamma(7)}{\Gamma(2)\Gamma(5)} x(1-x)^4, & 0 < x < 1, \\ 0, & \text{elsewhere.} \end{cases} \tag{6.115}$$

The mean and variance are, from Eq. (6.112),

$$\mu = \frac{\alpha}{\alpha + \beta} = \frac{2}{7} = 0.2857 \tag{6.116}$$

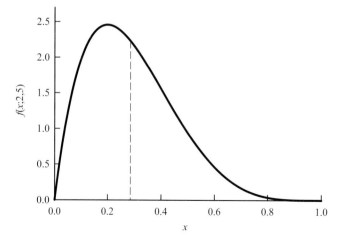

Figure 6.16 Beta distribution, Eq. (6.115), with parameters $\alpha=2$ and $\beta=5$. Vertical dashed line marks position of the mean, $\mu=0.2857$, for proportion of source A to decay completely. See example in the text.

and

$$\sigma^2 = \frac{(2)(5)}{(2+5+1)(2+5)^2} = 0.02551. \tag{6.117}$$

Figure 6.16 shows a plot of the beta distribution (6.115) with $\alpha=2$ and $\beta=5$ for the portion of the time for source 1 to decay completely. The dashed line marks the location of the mean (6.116).

Problems

6.1 The random variable X has a uniform distribution on the interval [0, 1]. Find
 a) $\Pr(X>0.5)$;
 b) $\Pr(0.3<X<0.7)$;
 c) $\Pr(X<0.3)$.
6.2 If X has the uniform distribution on the interval [0, 100], determine
 a) $E(X)$;
 b) $\mathrm{Var}(X)$.
6.3 Use the defining Eq. (4.44) for the variance in place of Eq. (4.48) to compute the result given by Eq. (6.8).
6.4 Verify the probability density function (6.14) and show that it is normalized.
6.5 The spectrum of energy losses q for the scattering of a fast neutron of energy E_o from hydrogen, shown in Figure 4.5a, is a uniform distribution.
 a) What is the probability that a collision of a 5.21-MeV neutron will reduce its energy to a value between 3.37 and 3.39 MeV?

b) What is the probability that the energy of the 5.21-MeV neutron after the collision will lie in any energetically possible interval of width 0.020 MeV?

c) Write the cumulative energy-loss spectrum from Eq. (4.132) for $-\infty < q < \infty$ and sketch it.

6.6 Show that the Cauchy distribution (6.16) is normalized.

6.7 Refer to Figure 6.1 and consider an unshielded source of photons that are randomly emitted isotropically (in three dimensions) from a point source and strike the screen over its surface in two dimensions. Let ϱ be the distance from the intersection of the perpendicular to the screen (at r) and the point where a photon hits the screen.

a) Find the probability density function on ϱ, where $0 \leq \varrho < \infty$.

b) Show that this probability density function is normalized.

6.8 Show that Eq. (6.19) follows from Eqs. (6.17) and (6.18).

6.9 Verify that the standard normal distribution (6.19) is normalized.

6.10 For the normal distribution, Eq. (6.17), show that $E(X) = \mu$ and $\text{Var}(X) \equiv E[(X-\mu)^2] = \sigma^2$. (*Hint:* Starting with Eq. (6.17), make the change of variables (6.18) and then integrate by parts, remembering that the function (6.19) is normalized.)

6.11 Show that the normal distribution (6.17) has inflection points at $x = \mu \pm \sigma$.

6.12 For the random variable Z with the standard normal distribution, determine the following:

a) $\Pr(Z > 1.96)$;
b) $\Pr(-1.96 < Z < 1.96)$;
c) $\Pr(Z < 1.28)$;
d) $\Pr(-1.28 < Z < 1.28)$.

6.13 The amount of coffee, X, dispensed per cup by an automatic machine has a normal distribution with a mean of 6.00 ounces and a standard deviation of 0.25 ounce. What is the probability that the amount of coffee dispensed in a cup will be

a) less than 5.50 ounces?
b) between 5.50 and 6.50 ounces?
c) more than 6.25 ounces?

6.14 The average neutron fluence rate from a sealed ^{252}Cf source, determined from many measurements, is normally distributed with mean 2.58×10^{10} cm^{-2} s^{-1} and standard deviation 0.11×10^{10} cm^{-2} s^{-1}. (The change in source strength over the measurement period is negligible.) What is the probability that a subsequent measurement will indicate a neutron fluence rate of

a) less than 2.36×10^{10} cm^{-2} s^{-1}?
b) between 2.36×10^{10} and 2.79×10^{10} cm^{-2} s^{-1}?
c) greater than 2.69×10^{10} cm^{-2} s^{-1}?

6.15 The sealed ^{252}Cf source in the last problem is stored in a pool of water, which provides a biological shield when the source is not in use. Water samples are periodically taken from the pool and analyzed for total alpha-particle activity to verify that the source encapsulation has not failed and that radioactive material is not leaking. Measurements of the net alpha-particle activity concentration in the water have a mean of $0.0253\,\text{Bq}\,\text{l}^{-1}$ and a standard deviation of $0.0128\,\text{Bq}\,\text{l}^{-1}$. (The net activity concentration is the difference between the sample measurement and that of an otherwise identical sample known to have no added activity.) For control purposes, net results exceeding $0.0509\,\text{Bq}\,\text{l}^{-1}$ are taken to indicate an abnormal condition that requires additional testing. Assume that the source is not leaking and that measurement results are from the net background alpha-particle activity alone.

  a) What is the probability that a measurement will yield a net total alpha-particle activity concentration exceeding $0.0278\,\text{Bq}\,\text{l}^{-1}$?
  b) What proportion of measurements will yield net total alpha-particle activity concentration between 0.0003 and $0.0509\,\text{Bq}\,\text{l}^{-1}$?
  c) What proportion of measurements will exceed the level above which additional testing is required?

6.16 A manufacturing process produces bolts that have a length that is normally distributed with mean $1.0000\,\text{cm}$ and standard deviation $0.0100\,\text{cm}$.
  a) What is the probability that a bolt's length will exceed $1.0196\,\text{cm}$?
  b) What proportion of bolts will have lengths between 0.9800 and $1.0200\,\text{cm}$?

6.17 The amount of rainfall in a year for a certain city is normally distributed with a mean of $89\,\text{cm}$ and standard deviation of $5\,\text{cm}$.
  a) Determine the annual rainfall amount that will be exceeded only 5% of the time.
  b) Determine what percentage of annual rainfall amounts will qualify as a drought year of $74\,\text{cm}$ or less.
  c) What is the probability that rainfall is between 79 and $99\,\text{cm}$ in a given year?

6.18 The time it takes a health physicist to travel daily from work to home is normally distributed with mean $43.0\,\text{min}$ and standard deviation $4.2\,\text{min}$.
  a) What is the probability that it will take longer than $50\,\text{min}$ to get home from work?
  b) What travel time will be exceeded in 99% of the trips?

6.19 Chebyshev's inequality as expressed by the relation (4.115) gives a rigorous upper bound for the two-tail probability for any random variable X.
  a) Determine this upper bound when X lies outside the interval $\mu \pm 3.2\sigma$.
  b) From Table 6.1, what is the exact two-tail probability for a normally distributed random variable?

6.20 Passengers flying on a certain airline have baggage weights that are normally distributed with mean weight $15.0\,\text{kg}$ and standard deviation $3.0\,\text{kg}$. If 30 passengers board a flight,
  a) What is the probability that the average baggage weight lies between 14.3 and $15.7\,\text{kg}$?
  b) What is the probability that the total baggage weight exceeds $482.2\,\text{kg}$?

6.21 Let X be the random number of spots that show when an unbiased die is rolled.
  a) Calculate the mean and standard deviation of X.
  b) Determine the mean and standard deviation of the sampling distribution for the mean number of spots $\bar{X}$ that show when the die is rolled twice.

c) Make a plot of Pr(X).
d) Make a plot of Pr($\overline{X}$).

6.22 The die in the last problem is rolled 20 times.
   a) What are the mean and standard deviation of the sampling distribution?
   b) Make a rough sketch of Pr($\overline{X}$).

6.23 Verify Eq. (6.41).

6.24 Write appropriate rules, similar to Eqs. (6.50)–(6.52), for
   a) Pr($X < a$);
   b) Pr($X > a$);
   c) Pr($a \leq X < b$).

6.25 For $X \sim b(x; 20, 0.4)$, use the normal approximation to determine the following, with and without continuity correction factors:
   a) Pr($X \leq 5$);
   b) Pr($6 \leq X \leq 13$);
   c) Pr($6 < X < 13$).

6.26 Compare the answers to the last problem with the exact answers found from Table A.1.

6.27 The example involving Eq. (6.48) in the text used continuity correction factors.
   a) Repeat the calculations without applying these factors.
   b) What are the percentages of error made with and without the factors?

6.28 In order to see whether her new proposal will be favored by the public, a politician performs a survey on 100 randomly selected voters.
   a) Use the normal approximation to the binomial distribution to find the probability that 60 or more of the persons sampled would say that they favored the proposal, if the true proportion in favor were 0.50.
   b) Express the exact probability by assuming that the binomial model is correct, but do not calculate the value.

6.29 A new manufacturing process for special thermoluminescent dosimeter (TLD) chips is said to be in control if no more than 1% of its product is defective. A random sample of 100 specimens from the process is examined.
   a) If the process is in control, what is the probability of finding at most three defective chips?
   b) Suppose that the process has slipped, and now 5% of the product is defective. What is then the probability of finding at most three defective chips from a random sample of 100?

6.30 Show that the mean and variance of the gamma distribution are given by Eq. (6.69).

6.31 a) Show that the recursion relation (6.71) follows from the definition (6.70) of the gamma function.
   b) Show that $0! = 1$.
   c) Show that $\Gamma(1/2) = \sqrt{\pi}$.

6.32 Show that the exponential probability density function (6.73) is normalized.

6.33 For a random variable X that has the exponential distribution (6.73), show that
   a) $E(X) = 1/\lambda$;
   b) $\text{Var}(X) = 1/\lambda^2$.

6.34 The number of persons entering a store is Poisson distributed with parameter $\lambda = 20$ customers per hour.
   a) What is the mean time in minutes between successive customers?
   b) What is the probability that the time between two successive customers is from 1 to 5 min?

6.35 The length of time that a patron waits in queue to buy popcorn at a certain movie theater is a random variable, having an exponential distribution with mean $\mu = 4$ min.
   a) What is the probability that a patron will be served within 3 min?
   b) What is the probability that exactly three of the next five persons will be served in less than 3 min?
   c) What is the probability that at least three of the next five persons will be served in less than 3 min?

6.36 Show that Eq. (6.84) can be written as
$$\Pr(T > s) = \Pr(T > t + s | X > t).$$

6.37 Verify Eq. (6.86) for the mean and variance of the chi-square distribution with v degrees of freedom.

6.38 a) With 17 degrees of freedom, what value of χ^2 cuts off 5% of the area of the distribution to the right?
   b) Find $\Pr(\chi^2 \leq 20.00)$.
   c) Find $\Pr(\chi^2 > 31.00)$.
   d) Calculate $\Pr(20.00 \leq \chi^2 \leq 31.00)$.

6.39 a) What range of values includes 99% of the chi-square distribution with 10 degrees of freedom?
   b) What is the mean value of χ^2?
   c) What is the standard deviation?

6.40 Use Eq. (6.88) to find the upper 5% point for the chi-square distribution with 40 degrees of freedom.

6.41 a) The text states that $\sqrt{2\chi^2} - \sqrt{2v - 1}$ is approximately distributed as standard normal when v is large. Use this fact to approximate the upper 2.5 percentile for the chi-square distribution with $v = 30$.
   b) In addition, the central limit theorem implies that $(\chi^2 - v)/\sqrt{2v}$ is also approximately standard normal for large v. Work part (a) by using this approximation.
   c) Compare with the exact value in Table A.5.

6.42 Let $X_1, X_2, \ldots, X_{10}$ represent a random sample from a normal population with zero mean and unit standard deviation. For the variable $Y = \sum_{i=1}^{10} X_i^2$, determine
   a) $\Pr(Y \leq 7.267)$;
   b) $\Pr(3.940 < Y < 18.31)$.

6.43 The random variable X has a uniform distribution on the interval $[0, 1]$.
   a) Determine the probability density function for $Y = -2 \ln X$.
   b) Show that Y has a chi-square distribution with 2 degrees of freedom.

6.44 a) For Student's t-distribution with 5 degrees of freedom, find the value of t below which 90% of the distribution lies.
b) Find the value of t with 8 degrees of freedom that cuts off 5% of the distribution to the right.

6.45 a) Find the probability that the value of t in the t-distribution with 15 degrees of freedom could be as large as 1.90. (Use linear interpolation.)
b) With 13 degrees of freedom, find the value of a such that $\Pr(-a < T < a) = 0.95$.

6.46 Show that Eq. (6.95) with one degree of freedom leads to the Cauchy distribution (6.16) with $\tau = 0$.

6.47 Prove Eq. (6.101).

6.48 For the F distribution with 4 and 12 degrees of freedom, find the value of f that
a) leaves an area of 0.05 to the right;
b) leaves an area of 0.95 to the right.

6.49 Find values b_1 and b_2 such that $\Pr(b_1 < F < b_2) = 0.90$ for an F random variable with 8 and 12 degrees of freedom.

6.50 The random variable $Y = \ln X$ has a normal distribution with mean 1 and variance 4.
a) Determine c_1 and c_2 such that $\Pr(c_1 < Y < c_2) = 0.95$.
b) Use part (a) and the fact that $Y = \ln X$ to determine a_1 and a_2 such that $\Pr(a_1 < X < a_2) = 0.95$.

6.51 Given that $Y = \ln X$ is normally distributed with mean 1.5 and variance 3.0, determine $\Pr(0.0954 < X < 28.4806)$.

6.52 If $Y = \ln X$ has a normal distribution with $\mu_Y = 2$ and $\sigma_Y = \sqrt{3}$, determine the mean and standard deviation for X.

6.53 Derive the relations (6.112) for the mean and variance of the beta distribution.

6.54 Let Y be distributed as a beta random variable with $\alpha = 1$ and $\beta = 3$.
a) Calculate the mean and variance of Y.
b) Find the value of b for which $\Pr(Y > b) = 0.05$.
c) Find the value of a for which $\Pr(Y < a) = 0.05$.
d) Using these values of a and b, show that $\Pr(a < Y < b) = 0.90$.

6.55 Consider two sources of a radionuclide with decay constant $\lambda = 2.0 \, \text{min}^{-1}$. Initially, source A has three atoms and source B has five atoms.
a) Describe the distribution of the total time for all atoms from both sources to decay.
b) Calculate the mean and variance of the total time for all atoms to decay.
c) Write an equation whose solution would determine the median of the total time for all atoms to decay.
d) Describe the distribution of the portion of the time that comes from source A.
e) Determine the mean and variance of the proportion of the time for source A to decay totally.

7
Parameter and Interval Estimation

7.1
Introduction

This chapter treats the practical problem of estimating (1) the numerical value of a parameter that characterizes a large population from which we can sample and (2) the uncertainty associated with that estimate. We discuss point and interval estimates, how they can be calculated, and their interpretation. For example, what is the "best value" to report for a population mean, estimated on the basis of a given sample of data? What is the uncertainty in the value thus reported? What statistical distribution is associated with the estimator itself?

The next chapter, on the propagation of errors, treats the uncertainty in a quantity that is derived from a *combination* of random variables, each with its own random error.

7.2
Random and Systematic Errors

How reliable is the numerical value assigned to a physical quantity that one measures or observes? A practical way to answer this question is to repeat the measurement a number of times, in exactly the same way, and examine the set of values obtained. Experimental uncertainties that can be thus revealed are called *random errors*. For example, the period of a pendulum can be measured directly by using a stopwatch to determine the time it takes the pendulum to return to its position of maximum displacement in one complete swing. Repeated measurements will generally yield a distribution of the times that are obtained for different swings. An obvious source of error in this procedure is the variability in the instant at which the watch is started or stopped relative to the precise location of the pendulum at the beginning or end of a swing. Without bias, the watch will sometimes be started a bit early or a bit late. From the distribution of the timed values for a number of observations made in the same way, one can get a reliable estimate of the random error that is thus occurring in the determination of the pendulum's period. In general, the *precision* in the

Statistical Methods in Radiation Physics, First Edition. James E. Turner, Darryl J. Downing, and James S. Bogard.
© 2012 Wiley-VCH Verlag GmbH & Co. KGaA. Published 2012 by Wiley-VCH Verlag GmbH & Co. KGaA.

determination of a quantity is reflected in the spread, or range of values, obtained when repeated, independent measurements are made in the same way. Precision can be expressed quantitatively in terms of the variance or standard deviation computed from the results of the measurements.

In contrast to random errors, *systematic* errors are not revealed through repeated measurements. If the stopwatch runs too fast or too slow, this condition will cause a systematic error in the individual measurements of the period. This kind of error will not be detected by repeated observations, and it will always affect the results in the same way. If the watch runs too fast, for instance, the effect is to make the measured value of the pendulum's period too large. Systematic errors always reduce the *accuracy* of the result obtained. Whereas random errors can usually be assessed from repeated observations, systematic errors are often hard to detect or even recognize and evaluate. However, steps can be taken to reduce them and improve accuracy. Such steps include careful calibration of equipment, comparison with measurements made at other laboratories, and review and analysis of one's measurement procedures. The accuracy of the stopwatch should be checked against a certified time standard for possible corrections to its readings. Calibration is an extremely important element of any radiation monitoring program.

In the treatment of errors and error propagation in this and the next chapter, we shall not address systematic errors in experimental data. While this subject is extremely important, it is outside the scope of this book. We shall deal only with the effects and assessment of random errors.

7.3
Terminology and Notation

A *parameter* is a value associated with a probability distribution that helps characterize or describe that distribution. For example, the binomial distribution depends on two parameters, n and p, representing the number of trials and the probability of success. If we know them both, then we can write down the distribution explicitly and calculate probabilities. Similarly, the Poisson distribution is completely determined by the single rate parameter, λ. The normal distribution has two parameters, the mean and variance, usually denoted by μ and σ^2, that identify it completely. The single parameter, called degrees of freedom, characterizes the chi-squared distribution. A parameter can also be a function of other parameters. For instance, the variance of the binomial random variable with parameters n and p is also a parameter, given by $\sigma^2 = np(1-p)$.

The dimensionless *coefficient of variation* (CV) is defined as the ratio σ/μ of the standard deviation σ and the mean μ of a distribution. It thus represents the standard deviation of the distribution in multiples of its mean. The coefficient of variation is often used to compare populations with varying means or to express the relative variation of the population as a percentage of the mean. For example, one might want to measure the variability in the responses of two ionization chambers, having sensitive volumes of different sizes, when exposed to X-rays under the same conditions. The larger chamber would tend to show the greater variation due simply to the larger amount of charge it collects. A way of standardizing the variation for comparing the two chambers is to use the relative response, or coefficient of variation.

A *statistic* is any function of the observable random variables from a random sample that does not depend upon any unknown parameters. An example of a statistic is the sample mean or median. Another important example of a statistic is the sample variance, defined as

$$S^2 = \frac{\sum_{i=1}^{n}(X_i - \bar{X})^2}{n-1}, \qquad (7.1)$$

where $\bar{X}$ is the sample mean and n is the number of observations. In contrast, the quantity

$$S^{*2} = \frac{\sum_{i=1}^{n}(X_i - \mu)^2}{n} \qquad (7.2)$$

qualifies as a statistic only when the population mean μ is known.

A *point estimate* of a population parameter is the single value of a statistic obtained for that parameter from data collected from the population. The statistic that one uses is called the *estimator*. In general notation, for the population parameter θ we let $\hat{\theta}$ denote the value, or point estimate, of the statistic $\hat{\Theta}$. For example, the value $\hat{\mu} = \bar{x}$ of the statistic $\bar{X}$ obtained from a random sample $X_1, X_2, \ldots, X_n$ is a point estimate of the population mean μ. The sample mean $\bar{X}$ is the estimator, and the sample value $\bar{x}$ is called the point estimate, or simply the estimate.

7.4
Estimator Properties

In addition to $\bar{X}$, there are other statistics that one can use to estimate the population mean μ. For instance, the sample median or any single X_i could serve as an estimator of μ. However, these estimators behave differently from $\bar{X}$. In general, we want an estimator to be unbiased, consistent, and have small variability. We discuss these properties next.

The first characteristic, *unbiased*, means that the expected value of the estimator equals the parameter being estimated. For the random sample $X_1, X_2, \ldots, X_n$, the statistic $\hat{\Theta}$ is said to be an *unbiased estimator* for the parameter θ if

$$\mu_{\hat{\Theta}} = E(\hat{\Theta}) = \theta. \qquad (7.3)$$

Thus, in repeated samples, if we obtain $\hat{\theta}_1, \hat{\theta}_2, \ldots, \hat{\theta}_n$, then the average value of these estimators should be close to the true value θ. In other words, $\hat{\Theta}$ is a statistic, having its own associated distribution; if $\hat{\Theta}$ is unbiased for θ, then the mean of the distribution on $\hat{\Theta}$ is θ.

The second property, *consistency*, implies that a larger sample size will provide a more accurate estimate of a population characteristic. In a practical sense, more sampling leads to a better estimate. Consistency is one reason why we do not use any single observation from a sample as an estimator. It is not a *consistent* estimator. For instance, we might sample an entire population completely (a process called a *census*). We would then know the population mean exactly, without error. We could elect to use any single observation to estimate the population mean. However, the variance associated with this estimate is the same, regardless of how many items in the population we sample. Using all the observations in the population and calculating the average yields μ with zero variability. A much more precise definition of consistency is found in Garthwaite, Jolliffe, and Jones (2002).

With the third property, we want to select an estimator having the *smallest variability*, restricted, however, to those that are unbiased. An unbiased estimator with a smaller variance than any other unbiased estimator is called the *minimum variance unbiased estimator* (MVUE).

Any estimator that is a function of the random sample will generally vary from sample to sample. One does not expect an estimator to yield the exact value of the parameter it is estimating – there will generally be some random error in the estimation. The variability of this error in repeated sampling will be reflected in the variability in the sampling distribution of the estimator. For two unbiased estimators $\hat{\Theta}_1$ and $\hat{\Theta}_2$ for θ, if the variance $\sigma^2_{\hat{\Theta}_1}$ of $\hat{\Theta}_1$ is smaller than the variance $\sigma^2_{\hat{\Theta}_2}$ of $\hat{\Theta}_2$, then one says that $\hat{\Theta}_1$ is *more efficient* than $\hat{\Theta}_2$.

■ *Example*

Let the random sample $X_1, X_2, \ldots, X_n$ of size n represent the number of counts in a given, fixed amount of time from a long-lived radioactive source. If the distribution of counts is Poisson with parameter λ, then it can be shown that the MVUE is

$$\hat{\lambda} = \frac{\sum_{i=1}^{n} X_i}{n}. \tag{7.4}$$

a) Show that $\hat{\lambda}$ is an unbiased estimator for λ.
b) Find the variance of $\hat{\lambda}$.
c) Is $\hat{\lambda}$ a consistent estimator?

Solution

a) To prove that $\hat{\lambda}$ is unbiased, we must show that its expected value is λ. From Eq. (7.4) we write

$$E(\hat{\lambda}) = E\left(\frac{1}{n}\sum_{i=1}^{n} X_i\right) = \frac{1}{n}\sum_{i=1}^{n} E(X_i). \tag{7.5}$$

7.4 Estimator Properties

Since each X_i is distributed as a Poisson random variable with parameter λ, it follows that $E(X_i) = \lambda$ for each i. Hence, the sum on the right-hand side of Eq. (7.5) is just $n\lambda$, and so the last equality gives $E(\hat{\lambda}) = \lambda$.

b) The variance of $\hat{\lambda}$ is given by

$$\text{Var}(\hat{\lambda}) = \text{Var}\left(\frac{1}{n}\sum_{i=1}^{n} X_i\right) = \frac{1}{n^2}\text{Var}\left(\sum_{i=1}^{n} X_i\right). \tag{7.6}$$

Since the variance of the sum of the independent random variables X_i is equal to the sum of the variances (Section 4.3), we find that

$$\text{Var}(\hat{\lambda}) = \frac{1}{n^2}\sum_{i=1}^{n} \text{Var}(X_i) = \frac{1}{n^2}\sum_{i=1}^{n} \lambda = \frac{\lambda}{n}. \tag{7.7}$$

c) The above result also shows that $\hat{\lambda}$ is a consistent estimator. That is, as n gets larger, the variance of $\hat{\lambda}$ gets smaller. All of the estimators that we discuss in this chapter are consistent.

Estimation theory is a rich and extensive field in itself, which we can only touch upon here. Table 7.1 summarizes the estimation of parameters for several common distributions, or populations, for a random sample $X_1, X_2, \ldots, X_n$ of size n. In each case, the estimator is the unique MVUE for that population. Table 7.1 lists the point estimator for the parameter of interest.

Table 7.1 Minimum variance unbiased estimator for parameters of several distributions (n = number of observations).

Population	Parameter	Estimator
Normal with mean μ and variance σ^2: $f(X;\mu,\sigma^2) = \frac{1}{\sqrt{2\pi}\sigma}e^{-(1/2\sigma^2)(X-\mu)^2}$	μ σ^2	$\hat{\mu} = \bar{X}$ $\hat{\sigma}^2 = s^2 = \frac{\sum_{i=1}^{n}(X_i - \bar{X})^2}{n-1}$
Poisson with parameter λ: $P(X=k) = \frac{\lambda^k e^{-\lambda}}{k!}, k = 0, 1, 2, \ldots$	λ	$\hat{\lambda} = \bar{X}$
Discrete uniform with values $1, 2, \ldots, \theta$: $P(X_i = k) = \frac{1}{\theta}$ for $k = 1, 2, \ldots, \theta$	θ	$\hat{\Theta} = \frac{(\max X_i)^{n+1} - (\max X_i - 1)^{n+1}}{(\max X_i)^n - (\max X_i - 1)^n}$
Exponential with parameter λ: $f(x;\lambda) = \lambda e^{-\lambda x}, x > 0$	λ	$\hat{\lambda} = \frac{n}{\sum_{i=1}^{n} X_i} \cdot \frac{n-1}{n} = \frac{n-1}{n\bar{x}}$
Binomial with parameters p and n: $\binom{n}{x}p^x(1-p)^{n-x}$	p	$\hat{p} = \frac{\sum_{i=1}^{n} X_i}{n}$, where $X_i = 1$ if success, or $X_i = 0$ if not

It can be shown that the statistic S^2 defined by Eq. (7.1) is an unbiased estimator of the population variance σ^2. Consider a random sample of size n drawn from a normal population with variance σ^2. From Eq. (7.1) we write the following and state without proof that

$$\frac{(n-1)S^2}{\sigma^2} = \sum_{i=1}^{n} \left(\frac{X_i - \bar{X}}{\sigma}\right)^2 \sim \chi^2(n-1), \qquad (7.8)$$

where the notation means that the quantities have the chi-squared distribution with $(n-1)$ degrees of freedom. This statement can be compared with Eq. (6.93), in which the population mean appears rather than $\bar{X}$. The sum (6.93) from the normal population has the chi-squared distribution with n degrees of freedom. Since $\bar{X}$ is a computed quantity for the sample, the sum (7.8) is distributed as a chi-squared random variable with $(n-1)$ degrees of freedom. When drawn from a normal population, the sample mean itself has a normal distribution. Thus,

$$\bar{X} \sim N\left(\mu, \frac{\sigma}{\sqrt{n}}\right). \qquad (7.9)$$

It can be shown that the distributions of $\bar{X}$ and S^2 given by the two important results embodied in Eqs. (7.8) and (7.9) are independent.

7.5
Interval Estimation of Parameters

While valuable, a point estimate will rarely give the true parameter value exactly. Often of more practical importance is knowledge of a numerical interval in which the true parameter value lies with a high degree of confidence. Such a determination is called an *interval estimate*, the subject of this section.

Each of the parameter estimators shown in the last column of Table 7.1 for different distributions has, itself, an associated sampling distribution. For example, we can randomly sample from a population and estimate the population mean by using the sample mean $\bar{X}$. If we sample the population again (for convenience, keeping the sample size n the same), we obtain a second value for $\bar{X}$, which is likely different from the first value. Sampling k times produces a distribution of sample means, $\bar{X}_1, \bar{X}_2, \ldots, \bar{X}_k$, which form the *sampling distribution*. The standard deviation of a sampling distribution is called the *standard error*.

7.5.1
Interval Estimation for Population Mean

We consider first the sampling distribution of the means obtained by sampling from a population that is normally distributed with standard deviation σ. Equation (7.9) tells us that the sampling distribution of $\bar{X}$ is normal, with the same mean value as the population from which it is derived and a standard deviation that is smaller than that

7.5 Interval Estimation of Parameters

of the original population by a factor $\sqrt{n}$. Thus, the standard error for the mean, when sampling from a normal population, is $\sigma/\sqrt{n}$. By using this information, it is possible to obtain an interval estimate about the mean within which we are confident that the true mean lies. We obtain this interval from Eq. (7.9) and the laws of probability theory in the following way.

From Eq. (7.9) and Eq. (6.18), it follows that, for the sampling distribution of the standardized mean when the population standard deviation σ is known,

$$Z = \frac{\bar{X} - \mu}{\sigma/\sqrt{n}} \sim N(0, 1). \tag{7.10}$$

We can write, for the standard normal distribution,

$$\Pr(-z_{\alpha/2} < Z < z_{\alpha/2}) = 1 - \alpha, \tag{7.11}$$

where $z_{\alpha/2}$ is the quantity that cuts off an area of size $\alpha/2$ to the right under the standard normal curve. By symmetry, $-z_{\alpha/2}$ cuts off an area $\alpha/2$ to the left. According to Eq. (7.11), the probability that the value of Z lies between these two limits is, therefore, $1 - 2(\alpha/2) = 1 - \alpha$. Substituting for Z from Eq. (7.10) gives, in place of Eq. (7.11),

$$\Pr\left(-z_{\alpha/2} < \frac{\bar{X} - \mu}{\sigma/\sqrt{n}} < z_{\alpha/2}\right) = 1 - \alpha. \tag{7.12}$$

Solving the inequality for μ yields (Problem 7.8)

$$\Pr\left(\bar{X} - z_{\alpha/2}\frac{\sigma}{\sqrt{n}} < \mu < \bar{X} + z_{\alpha/2}\frac{\sigma}{\sqrt{n}}\right) = 1 - \alpha. \tag{7.13}$$

Thus, the interval

$$\left(\bar{X} - z_{\alpha/2}\frac{\sigma}{\sqrt{n}}, \bar{X} + z_{\alpha/2}\frac{\sigma}{\sqrt{n}}\right) \tag{7.14}$$

forms what is called a $100(1-\alpha)\%$ confidence interval for μ. One must be careful to understand the interpretation of such an interval. When a sample is collected, $\bar{X}$ computed, and the interval (7.14) determined, there is no guarantee that the interval actually contains the true value μ. The interpretation is that $100(1-\alpha)\%$ of the intervals so constructed in repeated sampling are expected to contain μ. Thus, for a given sample, one can be $100(1-\alpha)\%$ confident that the interval contains the true value μ.

Depending on the application at hand, different confidence intervals may be employed. Perhaps the most common is the standard error. The width of the symmetric interval about the estimated mean is then two standard deviations. For the standard normal curve, the value of $z_{\alpha/2}$ in Eq. (7.11) is then exactly 1, and so $\alpha/2 = 0.1587$ (Table 6.1 or A.3). The $100(1-\alpha)\%$ confidence interval for the one standard error is thus $100(1 - 2 \times 0.1587)\% = 68.3\%$. A measured mean count rate and its one standard error for a long-lived source might be reported, for example, as 850 ± 30 cpm. This statement implies that the true mean count rate has been found to be between 820 and 880 cpm, with a probability of 0.683. (As emphasized in the last paragraph, however, there is no certainty that the true mean even lies within the stated interval.) Another quantity, the *probable error*, is defined such that the confidence interval is 50%; that is, there is a 50–50 probability that the true value lies within the range specified, which is ± 0.675 standard deviations (Problem 7.9). When a result is reported without a specific statement that defines the interval, the one standard error is usually, but not always, implied.

■ *Example*

The mean of potassium concentrations in body tissue from whole-body counts of 36 randomly selected women aged 22 y is 1685 mg kg^{-1} of body weight. State the standard error and find the 95 and 99% confidence intervals for the mean of the entire female population of this age if the population standard deviation is 60 mg kg^{-1}.

Solution

The point estimate of the population mean μ is $\bar{x} = 1685$ mg kg^{-1}, and the population standard deviation is $\sigma = 60$ mg kg^{-1}. The standard error for the sample mean with sample size $n = 36$ is $\sigma/\sqrt{n} = 60/6 = 10$. (We shall omit writing the units for now and insert them at the end.) With the standard normal function, the z value leaving an area of 0.025 to the right and, therefore, an area of 0.975 to the left is $z_{0.025} = 1.960$ (Table 6.1 or A.3). Hence, the 95% confidence interval for μ is, from (7.14),

$$\left(\bar{X} - 1.960 \frac{\sigma}{\sqrt{n}}, \bar{X} + 1.960 \frac{\sigma}{\sqrt{n}} \right). \tag{7.15}$$

This reduces to

$$(1665 \text{ mg kg}^{-1}, 1705 \text{ mg kg}^{-1}). \tag{7.16}$$

For the 99% confidence interval, the value that cuts off 0.005 of the area to the right is $z_{0.005} = 2.575$. One finds

$$\left(\bar{X} - 2.575 \frac{\sigma}{\sqrt{n}}, \bar{X} + 2.575 \frac{\sigma}{\sqrt{n}} \right), \tag{7.17}$$

or

$$(1659 \text{ mg kg}^{-1}, 1711 \text{ mg kg}^{-1}). \tag{7.18}$$

We note that a higher confidence level requires a wider interval.

Determination of the interval (7.14) depends on knowing the true population variance, σ^2. This quantity (as well as the true mean) is rarely known in practice. One can then use the sample variance S^2 to estimate σ^2 and then proceed as described above, using, however, the Student's t-distribution (Section 6.9). When the sample

7.5 Interval Estimation of Parameters

comes from a normal distribution with both mean μ and variance σ^2 unknown, we can use $\bar{X}$ and S^2 to form, like Eq. (7.10),

$$T = \frac{\bar{X} - \mu}{S/\sqrt{n}}. \tag{7.19}$$

From Eqs. (7.8), (7.9), and (6.94), one can show that the sampling distribution of T is the Student's t-distribution with $(n-1)$ degrees of freedom (Problem 7.13). Because T as defined by Eq. (7.19) contains μ, which is unknown, it is *not* a statistic. The quantity T is called, instead, a *pivotal quantity*. If μ is assumed known, the distribution of T is known, and this pivotal quantity can be used to derive confidence limits. (Some texts do not distinguish between statistics and pivotal quantities.) We can now use the quantiles of the Student's t-distribution in a manner similar to that employed for the normal distribution in the last section. That is, using the fact that the t-distribution is symmetric about the origin, we define the confidence interval by first writing the probability statement

$$\Pr(-t_{n-1,\alpha/2} < T < t_{n-1,\alpha/2}) = 1 - \alpha. \tag{7.20}$$

Substituting for T from Eq. (7.19) and arranging terms to get μ alone between the inequality signs, we find that

$$\Pr\left(\bar{X} - t_{n-1,\alpha/2} \frac{S}{\sqrt{n}} < \mu < \bar{X} + t_{n-1,\alpha/2} \frac{S}{\sqrt{n}}\right) = 1 - \alpha. \tag{7.21}$$

Thus, the $100(1-\alpha)\%$ confidence interval for μ when σ^2 is unknown is given by (Problem 7.14)

$$\left(\bar{X} - t_{n-1,\alpha/2} \frac{S}{\sqrt{n}}, \bar{X} + t_{n-1,\alpha/2} \frac{S}{\sqrt{n}}\right). \tag{7.22}$$

The interpretation of this interval is the same as that described for (7.14). With repeated sampling, the true mean is expected to lie within the interval $100(1-\alpha)\%$ of the time.

■ **Example**
Cobalt pellets are being fabricated in the shape of right circular cylinders. The diameters in cm of nine pellets drawn at random from the production line are 1.01, 0.97, 1.04, 1.02, 0.95, 0.99, 1.01, 1.03, and 1.03. Find a 95% confidence interval for the mean diameter of pellets, assuming an approximately normal distribution.

Solution
The sample mean and standard deviation are $\bar{x} = 1.01$ cm and $s = 0.03$ cm (Problem 7.15). Using Table A.5 with $\nu = 8$, we find $t_{0.025,8} = 2.306$. From Eq. (7.21) we obtain

$$1.01 - 2.306\left(\frac{0.03}{\sqrt{9}}\right) < \mu < 1.01 + 2.306\left(\frac{0.03}{\sqrt{9}}\right). \tag{7.23}$$

The 95% confidence interval for μ is, therefore, (0.987 cm, 1.033 cm).

7.5.2
Interval Estimation for the Proportion of Population

The above methods work well for determining a confidence interval for the mean of a population. Similar methods may be applied when interest centers on the *proportion* of the population that is defined by some criterion. For example, we might be interested in estimating the proportion of people in favor of a particular political candidate, or the proportion of cancer patients that respond favorably to some treatment. In such instances, we can run a binomial experiment (Section 5.4) and use the results to estimate the proportion p. For the binomial parameter p, the estimator $\hat{P}$ is simply the proportion of successes in the sample. We let $X_1, X_2, \ldots, X_n$ represent a random sample of size n from a Bernoulli population such that $X_i = 1$ with probability p and $X_i = 0$ with probability $(1 - p)$. Thus,

$$\hat{P} = \frac{1}{n}\sum_{i=1}^{n} X_i = \bar{X}. \tag{7.24}$$

By the central limit theorem (Section 6.4), for sufficiently large n, $\hat{P}$ is approximately normally distributed with mean

$$\mu_{\hat{P}} = E(\hat{P}) = E\left(\frac{1}{n}\sum_{i=1}^{n} X_i\right) = \frac{np}{n} = p \tag{7.25}$$

and variance

$$\sigma_{\hat{P}}^2 = \text{Var}(\hat{P}) = \text{Var}\left(\frac{1}{n}\sum_{i=1}^{n} X_i\right) = \frac{1}{n^2}\text{Var}\left(\sum_{i=1}^{n} X_i\right). \tag{7.26}$$

We recall from Section 5.4 (Eq. (5.11)) that the sum of n Bernoulli random variables is a binomial random variable with parameters n and p. The variance of the binomial random variable with these parameters is, by Eq. (5.15), $np(1-p)$. Replacing the term in the last equality of Eq. (7.26) by this quantity, we find

$$\sigma_{\hat{P}}^2 = \frac{p(1-p)}{n}. \tag{7.27}$$

The distribution of $\bar{X}$ is approximately normal because n is large, and so it can now be described by writing

$$\bar{X} \dot{\sim} N\left(p, \sqrt{\frac{p(1-p)}{n}}\right). \tag{7.28}$$

7.5 Interval Estimation of Parameters

We can now standardize $\bar{X}$ and use the standard normal distribution to write

$$\Pr\left(-z_{\alpha/2} < \frac{\bar{X} - p}{\sqrt{p(1-p)/n}} < z_{\alpha/2}\right) = 1 - \alpha, \tag{7.29}$$

or

$$\Pr\left(\bar{X} - z_{\alpha/2}\sqrt{\frac{p(1-p)}{n}} < p < \bar{X} + z_{\alpha/2}\sqrt{\frac{p(1-p)}{n}}\right) = 1 - \alpha. \tag{7.30}$$

This expression can be solved for p. However, when n is large, little error is introduced if we substitute $\hat{P}$ for p under the radical. An approximate $100(1 - \alpha)\%$ confidence interval for p is then (Problem 7.16)

$$\left(\hat{P} - z_{\alpha/2}\sqrt{\frac{\hat{P}(1-\hat{P})}{n}}, \hat{P} + z_{\alpha/2}\sqrt{\frac{\hat{P}(1-\hat{P})}{n}}\right). \tag{7.31}$$

■ **Example**
A random sample of blood specimens from 100 different workers in a manufacturing plant yielded 9 cases that were outside, either above or below, the normal range for a certain chemical marker. Use these data to calculate a 95% confidence interval for the proportion of the worker population whose reading will fall outside the normal range.

Solution
The sample proportion is $\hat{P} = 9/100 = 0.09$. Using Eq. (7.31) and $z_{0.025}$, we have

$$0.09 - 1.96\sqrt{\frac{0.09(1 - 0.09)}{100}} < p < 0.09 + 1.96\sqrt{\frac{0.09(1 - 0.09)}{100}}, \tag{7.32}$$

giving the 95% confidence interval, (0.034, 0.146).

7.5.3
Estimated Error

The interval estimates we have discussed thus far are of the form

$$\hat{\Theta} \pm q_{\alpha/2}\sqrt{\operatorname{Var}\hat{\Theta}}, \tag{7.33}$$

where $q_{\alpha/2}$ is the $\alpha/2$ quantile associated with the sampling distribution of $\hat{\Theta}$. The quantity $\sqrt{\operatorname{Var}\hat{\Theta}}$ is the *standard error* of the estimator $\hat{\Theta}$, that is, the standard deviation associated with the sampling distribution on $\hat{\Theta}$. In cases where the variance of $\hat{\Theta}$ contains unknown parameters, we estimate them (e.g., using S^2 for σ^2 and $\hat{P}$ for p). The square root of this estimated variance is then called the *estimated standard error*.

The *error of estimation* is the difference, $\hat{\Theta} - \Theta$, between the estimate and the true value. Using the interval estimators, we can say that we are $100(1-\alpha)\%$ confident that the error in estimation is bounded by $\pm q_{\alpha/2}\sqrt{\text{Var }\hat{\Theta}}$. This condition is expressed by writing

$$\Pr(|\hat{\Theta} - \Theta| < q_{\alpha/2}\sqrt{\text{Var }\hat{\Theta}}) = 1 - \alpha. \tag{7.34}$$

■ *Example*
In the example before last, dealing with cobalt pellets, by how much is the estimate of the mean bounded?

Solution
We previously found $s = 0.03$ cm with a sample size $n = 9$. Noting that $t_{0.025,8} = 2.306$, we computed

$$2.306\sqrt{\frac{(0.03)^2}{9}} = 0.023. \tag{7.35}$$

We may thus say that we are 95% confident that the error in estimating the mean is bounded by ± 0.023 cm.

It is often important to assure with $100(1-\alpha)\%$ confidence that an error of estimation will not exceed some fixed amount E. This can generally be accomplished by using sufficiently large samples, as we now describe. We see from Eq. (7.34) that the condition imposed is

$$E = q_{\alpha/2}\sqrt{\text{Var }\hat{\Theta}}. \tag{7.36}$$

Letting $\Theta = \mu$ and $\hat{\Theta} = \bar{X}$, and assuming we are sampling from a normal population, we have

$$E = z_{\alpha/2}\sqrt{\frac{\sigma^2}{n}}. \tag{7.37}$$

For the error not to exceed E at the stated level of confidence, it follows that the sample size needed is

$$n = \frac{z_{\alpha/2}^2 \sigma^2}{E^2}. \tag{7.38}$$

If $\Theta = p$, and the sample size is large enough to invoke the central limit theorem, then the same argument that led to Eq. (7.38) yields

$$n = \frac{z_{\alpha/2}^2 p(1-p)}{E^2}. \tag{7.39}$$

This expression involves p, which, like σ^2, is unknown. We can either (1) substitute $\hat{P}$ for p or (2) set $p = 1/2$, thus maximizing $p(1-p)$ and making n as large as possible, yielding a conservative estimate for n.

■ **Example**
In the example before last, 9 out of 100 urine samples were found to be outside normal range, for a proportion of 0.09. What sample size would be required to estimate the true proportion within an error 0.02 with 95% confidence if one

a) assumes $p = 0.09$ or
b) makes no assumption regarding the value of p?

Solution

a) Assuming $p = 0.09$ and a 95% confidence level, we find from Eq. (7.39) that a sample size of

$$n = \frac{(1.96)^2 (0.09)(1 - 0.09)}{(0.02)^2} = 787 \tag{7.40}$$

would be needed.

b) With no knowledge about the value of p, we use $p = 0.50$ in Eq. (7.39), thus maximizing n:

$$n = \frac{(1.96)^2 (0.50)(1 - 0.50)}{(0.02)^2} = 2401. \tag{7.41}$$

7.5.4
Interval Estimation for Poisson Rate Parameter

We next consider estimating the rate parameter λ for a Poisson population. In this case, $\hat{\lambda} = \bar{X}$. By the central limit theorem, $\hat{\lambda}$ for large n is approximately normally distributed with mean

$$\mu_{\hat{\lambda}} = E(\hat{\lambda}) = E\left(\frac{1}{n}\sum_{i=1}^{n} X_i\right) = \frac{n\lambda}{n} = \lambda \tag{7.42}$$

and variance

$$\sigma_{\hat{\lambda}}^2 = \text{Var}(\hat{\lambda}) = \text{Var}\left(\frac{1}{n}\sum_{i=1}^{n} X_i\right) = \frac{n\lambda}{n^2} = \frac{\lambda}{n}. \tag{7.43}$$

Using the same techniques as before (e.g., Eqs. (7.29) and (7.30)), we find that a 100(1 − α)% confidence interval for λ is

$$\left(\bar{X} - z_{\alpha/2}\sqrt{\frac{\bar{X}}{n}}, \bar{X} + z_{\alpha/2}\sqrt{\frac{\bar{X}}{n}}\right), \tag{7.44}$$

where $\sqrt{\bar{X}/n}$ is the estimated standard error.

■ *Example*

The number of disintegrations for a radioactive source was measured in 10 successive 1-min intervals, yielding the results: 26, 24, 26, 32, 26, 26, 27, 32, 17, and 22. Use this sample, assuming it came from a Poisson distribution, to obtain an approximate 95% confidence interval for the rate parameter λ in counts per minute.

Solution

We use Eq. (7.44). The mean of the numbers in the sample is $\bar{x} = 25.8$. For a 95% confidence interval, $z_{0.975} = 1.96$ (Table A.3), the required interval for the 1 min count numbers is

$$\left(25.8 - 1.96\sqrt{\frac{25.8}{10}}, 25.8 + 1.96\sqrt{\frac{25.8}{10}}\right). \tag{7.45}$$

It follows that the 95% confidence interval for λ is

$$(22.7\ \text{min}^{-1}, 28.9\ \text{min}^{-1}). \tag{7.46}$$

(Using a computer to randomly select the count numbers for this example, we sampled from a Poisson distribution with a true rate (decay constant), $\lambda = 25.00\ \text{min}^{-1}$. In this instance, the method did capture the true value in the interval found.)

7.6
Parameter Differences for Two Populations

In addition to point and interval estimates of a parameter for a single population, one is often interested in such estimates for the *difference* in parameters for two populations. An example of another type is the difference in the proportion of Democratic and Republican senators who are in favor of a certain bill before Congress. Estimating the differences between means or proportions is straightforward, but forming confidence intervals requires some assumptions.

7.6.1
Difference in Means

We let $X_1, X_2, \ldots, X_{n_1}$ denote a random sample of size n_1 from a normally distributed population with mean μ_x and variance σ_x^2. Similarly, $Y_1, Y_2, \ldots, Y_{n_2}$ denotes a sample of size n_2 from another normally distributed population with mean μ_y and variance σ_y^2. For the difference in means, $\mu_x - \mu_y$, the point estimate, which we denote by $\hat{\mu}_{x-y}$, is

$$\hat{\mu}_{x-y} = \hat{\mu}_x - \hat{\mu}_y = \bar{X} - \bar{Y}. \tag{7.47}$$

To obtain confidence intervals for this difference, we shall consider three special cases with respect to the population variances: (1) σ_x^2 and σ_y^2 both known; (2) σ_x^2 and σ_y^2 unknown, but equal; and (3) σ_x^2 and σ_y^2 unknown and not equal. Recall from the discussion following the definition (4.44) that the variance of the sum or difference of two independent random variables is the sum of their individual variances.

7.6.1.1 Case 1: σ_x^2 and σ_y^2 Known

For the two normal distributions, we have

$$\hat{\mu}_{x-y} = \hat{\mu}_x - \hat{\mu}_y = \bar{X} - \bar{Y} \sim N\left(\mu_x - \mu_y, \sqrt{\frac{\sigma_x^2}{n_1} + \frac{\sigma_y^2}{n_2}}\right). \tag{7.48}$$

Transforming to the standard normal distribution, we write for the $100(1-\alpha)\%$ confidence interval on the mean,

$$\Pr\left(-z_{\alpha/2} < \frac{\bar{X} - \bar{Y} - (\mu_x - \mu_y)}{\sqrt{(\sigma_x^2/n_1) + (\sigma_y^2/n_2)}} < z_{\alpha/2}\right) = 1 - \alpha. \tag{7.49}$$

Solving this inequality for $\mu_x - \mu_y$, we obtain

$$\bar{X} - \bar{Y} - z_{\alpha/2}\sqrt{\frac{\sigma_x^2}{n_1} + \frac{\sigma_y^2}{n_2}} < (\mu_x - \mu_y) < \bar{X} - \bar{Y} + z_{\alpha/2}\sqrt{\frac{\sigma_x^2}{n_1} + \frac{\sigma_y^2}{n_2}}. \tag{7.50}$$

■ *Example*

An experiment was performed to compare the effectiveness of two chemical compounds, A and B, in blocking the thyroid uptake of iodine. A number of mice of the same age, sex, and size were selected and considered to be identical for this experiment. A fixed amount of radioactive iodine was injected into each of $n_1 = 10$ mice (group 1) after administration of compound A and into $n_2 = 15$ other mice (group 2) after they were given compound B, all other conditions being the same. The mice were later monitored by observing the activity of the iodine in the thyroid at a particular time. The average activity for the mice in group A was $\bar{x} = 25$ kBq, and that for group B was $\bar{y} = 27$ kBq. Assume that the variances in the activities for the two groups were $\sigma_x^2 = 25$ kBq^2 and $\sigma_y^2 = 45$ kBq^2. Obtain a 95% confidence interval for the difference between the true means of the two groups. Which thyroid blocking compound is the more effective, based on this experiment?

Solution
Denoting the true means by μ_x and μ_y and using units of kBq, we write from Eq. (7.50) with $z_{\alpha/2} = z_{0.025} = 1.96$,

$$25 - 27 - 1.96\sqrt{\frac{25}{10} + \frac{45}{15}} < \mu_x - \mu_y < 25 - 27 + 1.96\sqrt{\frac{25}{10} + \frac{45}{15}}. \quad (7.51)$$

Completing the arithmetic operations gives

$$-6.6 \text{ kBq} < \mu_x - \mu_y < 2.6 \text{ kBq}. \quad (7.52)$$

Since zero is included in this 95% confidence interval, the true means are likely to be close to each other. The two compounds thus appear to have comparable effects in blocking the uptake of iodine by the thyroid.

7.6.1.2 Case 2: σ_x^2 and σ_y^2 Unknown, but Equal ($=\sigma^2$)

In place of Eq. (7.48), we write

$$\hat{\mu}_{x-y} = \bar{X} - \bar{Y} \sim N\left(\mu_x - \mu_y, \sqrt{\frac{\sigma^2}{n_1} + \frac{\sigma^2}{n_2}}\right). \quad (7.53)$$

The variance of the estimator is

$$\text{Var}(\bar{X} - \bar{Y}) = \sigma^2 \left(\frac{1}{n_1} + \frac{1}{n_2}\right). \quad (7.54)$$

In order to find a good estimator of σ^2, we note that the sample variance S_x^2 is the best estimator of $\sigma_x^2 = \sigma^2$ and, similarly, S_y^2 is the best estimator of $\sigma_y^2 = \sigma^2$. Since the variances are the same, S_x^2 and S_y^2 should be similar in value, and we should combine them in some way. We could average them, using as the estimator $\bar{S}^2 = (S_x^2 + S_y^2)/2$, but this choice does not account for possibly different sample sizes. A better selection is the *pooled estimator for the variance*.

$$S_p^2 = \frac{(n_1 - 1)S_x^2 + (n_2 - 1)S_y^2}{n_1 + n_2 - 2}. \quad (7.55)$$

We see that, if $n_1 = n_2$, then $S_p^2 = (S_x^2 + S_y^2)/2$, as first suggested. The weighting factors $(n_1 - 1)$ and $(n_2 - 1)$ reflect the fact that S_x^2 and S_y^2 have, respectively, that many independent pieces of information. Thus, using these factors and dividing by $(n_1 + n_2 - 2)$ in Eq. (7.55) apportions the weights appropriately. It can be shown theoretically that, if we are sampling from a normal population,

$$\frac{\bar{X} - \bar{Y} - (\mu_x - \mu_y)}{S_p\sqrt{(1/n_1) + (1/n_2)}} \sim t_{n_1+n_2-2}. \quad (7.56)$$

The quantity on the left-hand side has Student's t-distribution with $(n_1 + n_2 - 2)$ degrees of freedom. Thus, we may write

$$\Pr\left(-t_{n_1+n_2-2,\alpha/2} < \frac{\bar{X} - \bar{Y} - (\mu_x - \mu_y)}{S_p\sqrt{(1/n_1) + (1/n_2)}} < t_{n_1+n_2-2,\alpha/2}\right) = 1 - \alpha. \quad (7.57)$$

Solving the inequality for $\mu_x - \mu_y$, we find that, for the $100(1 - \alpha)\%$ confidence interval,

$$\bar{X} - \bar{Y} - S_p\sqrt{\frac{1}{n_1} + \frac{1}{n_2}}\, t_{n_1+n_2-2,\alpha/2} < \mu_x - \mu_y \quad (7.58)$$
$$< \bar{X} - \bar{Y} + S_p\sqrt{\frac{1}{n_1} + \frac{1}{n_2}}\, t_{n_1+n_2-2,\alpha/2}.$$

■ *Example*

The time until first decay is measured 10 times for each of two pure, long-lived radioisotope samples, A and B. The results are given in Table 7.2. Using these data, obtain a 95% confidence interval for the difference in the mean first-decay times of the two isotopes, assuming normality and equal variances. What can one conclude about the decay rates of the two isotopes?

Solution

From the data in Table 7.2, we compute the following values (with time expressed in seconds) for the means and variances of the samples for the two isotopes:

$$\bar{x}_A = 2.109, \quad s_A^2 = 3.753 \quad (7.59)$$

Table 7.2 Ten measurements of time in seconds to first decay of two isotopes.

Measurement	Time (s)	
	Isotope A	Isotope B
1	0.66	0.28
2	4.45	1.44
3	2.01	12.04
4	2.94	0.37
5	2.11	0.30
6	1.31	8.61
7	0.73	4.21
8	0.15	2.76
9	0.54	0.61
10	6.19	3.86

Data are used in several examples in the text.

and

$$\bar{x}_B = 3.448, \quad s_B^2 = 15.911. \tag{7.60}$$

Assuming that the variances are equal for the two isotopes, we pool them according to Eq. (7.55). With $n_1 = n_2 = 10$, we have

$$s_p^2 = \frac{9(3.753) + 9(15.911)}{18} = 9.832. \tag{7.61}$$

The 0.975 quantile of the Student's t-distribution with 18 degrees of freedom is (Table A.5)

$$t_{18, 0.025} = 2.101. \tag{7.62}$$

From Eq. (7.58) we write for the 95% confidence interval for the difference in the means of the decay times

$$(2.109 - 3.448) - 3.136\sqrt{\frac{1}{10} + \frac{1}{10}}(2.101) < \mu_A - \mu_B$$
$$< (2.109 - 3.448) + 3.136\sqrt{\frac{1}{10} + \frac{1}{10}}(2.101). \tag{7.63}$$

Thus, to the appropriate number of significant figures,

$$-4.29 \text{ s} < \mu_A - \mu_B < 1.61 \text{ s}. \tag{7.64}$$

Since the interval includes zero, we can conclude that the two isotopes have similar decay rates. They could even be the same radionuclide.

7.6.1.3 Case 3: σ_x^2 and σ_y^2 Unknown and Unequal

As before, Eq. (7.48) holds; however, we do not know either variance. In this case, we use the best estimators for them, namely, S_x^2 and S_y^2, and form the *approximate* $100(1 - \alpha)\%$ confidence interval for $\mu_x - \mu_y$:

$$\bar{X} - \bar{Y} - t_{v,\alpha/2}\sqrt{\frac{S_x^2}{n_1} + \frac{S_y^2}{n_2}} < \mu_x - \mu_y < \bar{X} - \bar{Y} + t_{v,\alpha/2}\sqrt{\frac{S_x^2}{n_1} + \frac{S_y^2}{n_2}}. \tag{7.65}$$

Here, $t_{v,\alpha/2}$ is the t-value with degrees of freedom v. It can be shown (Satterthwaite, 1946) that the t-distribution in (7.65) can be approximated by a Student's t-distribution with degrees of freedom given by Satterthwaite's approximation,

$$v = \frac{((S_x^2/n_1) + (S_y^2/n_2))^2}{(1/(n_1 - 1))(S_x^2/n_1)^2 + (1/(n_2 - 1))(S_y^2/n_2)^2}. \tag{7.66}$$

In most instances, v as defined by this equation will not be an integer, in which case we round down to the nearest integer to be conservative.

■ *Example*

Repeat the last example for the data in Table 7.2, assuming in place of equal variances, that $\sigma_A^2 \neq \sigma_B^2$.

Solution

We apply Eq. (7.66) with $n_A = n_B = 10$ and express times in seconds. The sample variances were previously calculated (Eqs. (7.59) and (7.60)). The approximate degrees of freedom for the *t*-value are, from Eq. (7.66),

$$\nu = \frac{((3.753/10) + (15.911/10))^2}{(1/(10-1))(3.753/10)^2 + (1/(10-1))(15.911/10)^2}$$
$$= 13.02. \tag{7.67}$$

Rounding down to $\nu = 13$, we find the 0.975 quantile of the Student's *t*-distribution to be (Table A.5) $t_{13, 0.025} = 2.160$. The 95% confidence interval for $\mu_A - \mu_B$ is then, from Eq. (7.65),

$$(2.109 - 3.448) - 2.160\sqrt{\frac{3.753}{10} + \frac{15.911}{10}} < \mu_A - \mu_B$$
$$< (2.109 - 3.448) + 2.160\sqrt{\frac{3.753}{10} + \frac{15.911}{10}}. \tag{7.68}$$

It follows that

$$-4.37 \text{ s} < \mu_A - \mu_B < 1.69 \text{ s}. \tag{7.69}$$

This interval is slightly larger than that found earlier (see Eq. (7.64)). The previous assumption of equal variances for the two samples implies more knowledge about the populations from which we are sampling than we had in the present example. The increased knowledge translates into a smaller confidence interval. That the two confidence intervals are not too different implies that the variances are not too different. In Section 7.8, we shall see how to compare two variances.

7.6.2
Difference in Proportions

Differences in proportions present similar problems. To form confidence intervals, we must rely on the central limit theorem. We let $\hat{P}_1$ represent the proportion of successes in a random sample of size n_1 from one population and $\hat{P}_2$ the proportion in a random sample of size n_2 from another population. By the central limit theorem,

$$\hat{P}_1 - \hat{P}_2 \overset{.}{\sim} N\left(p_1 - p_2, \sqrt{\frac{p_1 q_1}{n_1} + \frac{p_2 q_2}{n_2}}\right), \tag{7.70}$$

where $q_1 = 1 - p_1$ and $q_2 = 1 - p_2$. An approximate $100(1 - \alpha)\%$ confidence interval is then

$$\Pr\left(-z_{\alpha/2} < \frac{\hat{P}_1 - \hat{P}_2 - (p_1 - p_2)}{\sqrt{p_1 q_1/n_1 + p_2 q_2/n_2}} < z_{\alpha/2}\right) = 1 - \alpha. \qquad (7.71)$$

Solving for $p_1 - p_2$, we find that

$$\hat{P}_1 - \hat{P}_2 - z_{\alpha/2}\sqrt{\frac{p_1 q_1}{n_1} + \frac{p_2 q_2}{n_2}} < p_1 - p_2 < \hat{P}_1 - \hat{P}_2 + z_{\alpha/2}\sqrt{\frac{p_1 q_1}{n_1} + \frac{p_2 q_2}{n_2}}. \qquad (7.72)$$

Under the radical we see that these limits require values of p_1 and p_2, which are unknown. As before, for the confidence interval on a single proportion, we simply substitute our estimates $\hat{P}_1$ and $\hat{P}_2$ for p_1 and p_2 under the radical in Eq. (7.72).

■ *Example*
New safety signs were proposed to replace existing ones at a plant. Workers on two shifts were shown the new signs and asked whether they thought they would be an improvement. Of the 75 workers queried on the first shift, 60 responded favorably toward the change; 85 out of 95 on the second shift were also in favor. Using this information, obtain a 95% confidence interval on the difference in the proportions of the two shifts that favor the new signs. Based on this survey, are the proportions of favorable responses among workers on the two shifts different?

Solution
For use in Eq. (7.72), we compute the favorable proportions for the two shifts: $\hat{P}_1 = 60/75 = 0.80$ and $\hat{P}_2 = 85/95 = 0.89$. Recalling that $z_{0.025} = 1.96$, we write

$$(0.80 - 0.89) - 1.96\sqrt{\frac{(0.80)(0.20)}{75} + \frac{(0.89)(0.11)}{95}} < p_1 - p_2$$

$$< (0.80 - 0.89) + 1.96\sqrt{\frac{(0.80)(0.20)}{75} + \frac{(0.89)(0.11)}{95}}. \qquad (7.73)$$

The solution is

$$-0.20 < p_1 - p_2 < 0.02. \qquad (7.74)$$

We see that zero is contained in the interval. We might conclude with 95% confidence that the proportion in favor of the new signs is essentially the same on both shifts.

7.7
Interval Estimation for a Variance

In discussing interval estimation for a variance, we assume that we sample from a normal population with mean μ and variance σ^2, both unknown. The chi-squared distribution, discussed in Chapter 6, then plays a special role. According to Eq. (7.8), the quantity

$$Q = \frac{(n-1)S^2}{\sigma^2} \tag{7.75}$$

has a chi-squared distribution with $(n-1)$ degrees of freedom. Thus,

$$\Pr(\chi^2_{n-1,\alpha/2} < Q < \chi^2_{n-1,1-\alpha/2}) = 1 - \alpha. \tag{7.76}$$

We note that $\chi^2_{n-1,\alpha}$ is the value that leaves the fraction α of the area to the left under the chi-squared distribution with $(n-1)$ degrees of freedom (Table A.4). Unlike the normal and Student's t-distributions, the chi-squared distribution is not symmetric. Choosing $\chi^2_{n-1,\alpha/2}$ and $\chi^2_{n-1,1-\alpha/2}$ will provide an interval that is nearly the shortest possible, although one might do better by numerical methods. For most practical purposes, these values will result in an acceptable interval. Using Eqs. (7.75) and (7.76), we write

$$\Pr\left(\chi^2_{n-1,\alpha/2} < \frac{(n-1)S^2}{\sigma^2} < \chi^2_{n-1,1-\alpha/2}\right) = 1 - \alpha. \tag{7.77}$$

We thus obtain the following inequality for the variance:

$$\frac{(n-1)S^2}{\chi^2_{n-1,1-\alpha/2}} < \sigma^2 < \frac{(n-1)S^2}{\chi^2_{n-1,\alpha/2}}. \tag{7.78}$$

■ Example

In a previous example, we analyzed the time to the first decay of a radionuclide. Data for isotope A in Table 7.2 for a sample size $n = 10$ yielded $\bar{x}_A = 2.109$ and $s^2_A = 3.753$, with the times in seconds (Eq. (7.59)). Obtain a 95% confidence interval for the variance σ^2_A.

Solution
We need the 0.025 and 0.975 quantiles of the chi-squared distribution with 9 degrees of freedom ($n = 10$). From Table A.4, we find $\chi^2_{9,0.025} = 2.70$ and $\chi^2_{9,0.975} = 19.02$. Using Eq. (7.78), we find that

$$\frac{9(3.753)}{19.02} < \sigma^2 < \frac{9(3.753)}{2.70}, \tag{7.79}$$

or

$$1.776 < \sigma^2 < 12.51. \tag{7.80}$$

The units are s^2.

7.8
Estimating the Ratio of Two Variances

Just as important as the comparison of two population means is the comparison of their variances. Usually, one is interested in the *ratio* of the variances, which are both measures of the spread of the two distributions. A ratio near unity indicates near equality of the variances. On the other hand, either a very large or a very small ratio occurs when there is a wide difference in variation between the two populations.

We consider two normally distributed populations, having variances σ_1^2 and σ_2^2. We select from the first a random sample of size n_1 and estimate the variance using S_1^2 and, similarly, from the second a sample of size n_2 and estimate its variance using S_2^2. Recalling the chi-squared distribution from Eq. (7.75), we write

$$Q_1 = \frac{(n_1-1)S_1^2}{\sigma_1^2} \sim \chi_{n_1-1}^2 \tag{7.81}$$

and

$$Q_2 = \frac{(n_2-1)S_2^2}{\sigma_2^2} \sim \chi_{n_2-1}^2. \tag{7.82}$$

The ratio of two independent chi-squared random variables, divided by their respective degrees of freedom, has the F distribution (Section 6.10). From the last two equations and the definition (6.98), it follows, therefore, that

$$F = \frac{Q_1/(n_1-1)}{Q_2/(n_2-1)} = \frac{S_1^2 \sigma_2^2}{S_2^2 \sigma_1^2} \sim F(n_1-1, n_2-1). \tag{7.83}$$

We express a confidence interval by writing

$$\Pr(f_{\alpha/2}(n_1-1, n_2-1) < F < f_{1-\alpha/2}(n_1-1, n_2-1)) = 1 - \alpha. \tag{7.84}$$

Combining Eqs. (7.83) and (7.84) gives

$$\Pr\left(f_{\alpha/2}(n_1-1, n_2-1) < \frac{S_1^2 \sigma_2^2}{S_2^2 \sigma_1^2} < f_{1-\alpha/2}(n_1-1, n_2-1)\right) = 1 - \alpha. \tag{7.85}$$

The ratio of the variances thus satisfies the following inequality:

$$\frac{S_1^2}{S_2^2} \frac{1}{f_{1-\alpha/2}(n_1-1, n_2-1)} < \frac{\sigma_1^2}{\sigma_2^2} < \frac{S_1^2}{S_2^2} \frac{1}{f_{\alpha/2}(n_1-1, n_2-1)}. \tag{7.86}$$

Lower values of the F distribution are not usually tabulated. However, by Eq. (6.101) we have

$$f_{1-\alpha/2}(n_1-1, n_2-1) = \frac{1}{f_{\alpha/2}(n_2-1, n_1-1)}, \tag{7.87}$$

and so, in place of Eq. (7.86), we can write (Problem 7.28)

$$\frac{S_1^2}{S_2^2} \frac{1}{f_{1-\alpha/2}(n_1-1, n_2-1)} < \frac{\sigma_1^2}{\sigma_2^2} < \frac{S_1^2}{S_2^2} f_{1-\alpha/2}(n_2-1, n_1-1). \tag{7.88}$$

■ *Example*
The times for first decay of two radioisotopes, 1 and 2, are measured in an experiment. The following data summarize the results, with time in seconds.

Isotope, i	Mean time, $\bar{x}_i$	Variance, s_i^2	Sample size, n_i
1	6.95	13.75	8
2	13.75	24.39	12

Obtain a 90% confidence interval for the ratio σ_1^2/σ_2^2 of the two population variances in the time to first decay.

Solution
The interval is found from Eq. (7.88) with $\alpha = 0.10$. Using Table A.6, we find for the upper 0.95 quantile of the f distribution with $n_1 - 1 = 7$ and $n_2 - 1 = 11$ degrees of freedom, $f_{0.95}(7, 11) = 3.603$. For the lower quantile, Eq. (7.87) gives $f_{0.95}(11, 7) = 3.012$. Substitution into Eq. (7.88) yields

$$\frac{13.75}{24.39} \frac{1}{3.012} < \frac{\sigma_1^2}{\sigma_2^2} < \frac{13.75}{24.39} 3.603, \tag{7.89}$$

or

$$0.187 < \frac{\sigma_1^2}{\sigma_2^2} < 2.03. \tag{7.90}$$

Since this interval includes unity, we can be 90% confident that these population variances are approximately the same.

7.9
Maximum Likelihood Estimation

In the previous sections we looked at various estimators and their properties, largely without derivations. In this section we discuss a method that allows one to derive an estimator. We consider a random sample $X_1, X_2, \ldots, X_n$ drawn from a population having unknown parameters that we wish to estimate. For example, we might select from a large population, having a characteristic that occurs in some (unknown) proportion p of that population. Each draw results in a Bernoulli trial with outcome X_i, $i = 1, 2, \ldots, n$, and probability distribution

$$f(x_i; p) = \Pr(X_i = x_i) = p^{x_i}(1-p)^{1-x_i}, \tag{7.91}$$

with $x_i = 0, 1$ and $0 \leq p \leq 1$. We seek a function $u(X_1, X_2, \ldots, X_n)$ (the estimator) such that the sample value $u(X_1, X_2, \ldots, X_n)$ is a good point estimate of the population proportion p.

We consider the joint distribution of $X_1, X_2, \ldots, X_n$ under the assumption that this is a random sample from the same, identical distribution (so that each draw is independent and identically distributed). We then write for the observed values

$$\Pr(X_1 = x_1, X_2 = x_2, \ldots, X_n = x_n) = \prod_{i=1}^{n} p^{x_i}(1-p)^{1-x_i}$$
$$= p^{\Sigma x_i}(1-p)^{n-\Sigma x_i}, \tag{7.92}$$

in which the sums in the exponents go from $i = 1$ to n. We can treat this joint probability function as a function of p, rather than the x_i, and then find the value of p that maximizes it, this value being the one most likely to have produced the set of observations. As a function of p, the resulting joint probability function is called the *likelihood function*. In the illustration (7.91), the likelihood function is

$$L(p) = \Pr(X_1 = x_1, X_2 = x_2, \ldots, X_n = x_n; p) \tag{7.93}$$

$$= f(x_1; p) f(x_2; p) \cdots f(x_n; p) \tag{7.94}$$

$$= p^{\Sigma x_i}(1-p)^{n-\Sigma x_i}, \quad 0 \leq p \leq 1. \tag{7.95}$$

To find the value of p that maximizes $L(p)$, we first differentiate:

$$\frac{dL(p)}{dp} = \left(\sum x_i\right) p^{\Sigma x_i - 1}(1-p)^{n-\Sigma x_i} - \left(n - \sum x_i\right) p^{\Sigma x_i}(1-p)^{n-\Sigma x_i - 1}. \tag{7.96}$$

Setting the derivative equal to zero and solving for p gives the maximum likelihood estimator, which we denote by $\hat{p}$. That is,

$$\hat{p}^{\Sigma x_i}(1-\hat{p})^{n-\Sigma x_i}\left(\frac{\sum x_i}{\hat{p}} - \frac{n - \sum x_i}{1 - \hat{p}}\right) = 0. \tag{7.97}$$

The cases $\hat{p} = 0$ and $\hat{p} = 1$ are uninteresting in the sense that the characteristic is either never present or always present. Hence, we treat only $0 < \hat{p} < 1$. Then Eq. (7.97) can be satisfied only if the term in the parentheses vanishes:

$$\frac{\sum x_i}{\hat{p}} - \frac{n - \sum x_i}{1 - \hat{p}} = 0. \tag{7.98}$$

Multiplying by $\hat{p}(1-\hat{p})$ and simplifying, we find that

$$\hat{p} = \frac{\sum x_i}{n} = \bar{x}. \tag{7.99}$$

The corresponding statistic, $(\sum X_i)/n = \bar{X}$, which maximizes the likelihood function (Problem 7.33), is called the *maximum likelihood estimator*, abbreviated MLE. Our earlier estimator, Eq. (7.24), is the same as the MLE when the random variables are replaced by the sample values.

This maximization procedure can also be applied to the logarithm of the likelihood function, thus often simplifying the mathematics involved. In place of Eq. (7.95), for example, we can write

$$\ln L(p) = \sum x_i \ln p + \left(n - \sum x_i\right)\ln(1-p). \tag{7.100}$$

The first derivative is

$$\frac{d \ln L(p)}{dp} = \frac{\sum x_i}{p} - \frac{n - \sum x_i}{1-p}, \tag{7.101}$$

which, when set equal to zero, leads to Eq. (7.98).

■ *Example*
Assume that one can obtain a random sample of size n for a decay process that follows a Poisson distribution with parameter λ. Obtain the MLE for λ.

Solution
Let $X_1, X_2, \ldots, X_n$ denote the random sample from the Poisson population with decay parameter λ. The probability function is given by

$$f(x_i; \lambda) = \frac{e^{-\lambda}\lambda^{x_i}}{x_i!}, \tag{7.102}$$

with $i = 1, 2, \ldots, n$ and $0 < \lambda < \infty$. The likelihood function is

$$L(\lambda) = \prod_{i=1}^{n} \frac{e^{-\lambda}\lambda^{x_i}}{x_i!} = \frac{e^{-n\lambda}\lambda^{\sum x_i}}{\prod x_i!}. \tag{7.103}$$

Taking natural logarithms gives

$$\ln L(\lambda) = -n\lambda + \sum x_i \ln \lambda - \ln\left(\prod x_i!\right). \tag{7.104}$$

Differentiating, we find that

$$\frac{d \ln L(\lambda)}{d\lambda} = -n + \frac{\sum x_i}{\lambda}. \tag{7.105}$$

Setting the derivative equal to zero then yields

$$\hat{\lambda} = \frac{\sum x_i}{n} = \bar{x}. \tag{7.106}$$

Comparison with Eq. (7.4) shows that the MLE $= \bar{X}$ is the same as the MVUE (Section 7.4).

A population might have more than one parameter. The likelihood function is then differentiated for each parameter, yielding coupled equations to be solved simultaneously.

■ *Example*
A normal population is characterized by the two parameters (μ, σ^2). For a random sample of size n, determine the maximum likelihood estimators for μ and σ^2.

Solution
We let $X_1, X_2, \ldots, X_n$ denote the random sample from the normal population. The probability density function is given by

$$f(x_i; \mu, \sigma^2) = \frac{1}{\sqrt{2\pi}\sigma} e^{-(x_i-\mu)^2/2\sigma^2}, \tag{7.107}$$

where $-\infty < x_i < \infty$ ($i = 1, 2, \ldots, n$), $-\infty < \mu < \infty$, and $\sigma^2 > 0$. The likelihood function is

$$L(\mu, \sigma^2) = \prod_{i=1}^{n} \frac{1}{\sqrt{2\pi}\sigma} e^{-(x_i-\mu)^2/2\sigma^2} \tag{7.108}$$

$$= (2\pi)^{-n/2} (\sigma^2)^{-n/2} e^{-(1/2\sigma^2) \sum (x_i-\mu)^2}, \tag{7.109}$$

and its natural logarithm is

$$\ln L(\mu, \sigma^2) = -\frac{n}{2}\ln(2\pi) - \frac{n}{2}\ln \sigma^2 - \frac{1}{2\sigma^2}\sum (x_i - \mu)^2. \tag{7.110}$$

Taking the partial derivatives with respect to μ and σ^2 and equating them to zero, we write

$$\frac{\partial \ln L(\hat{\mu}, \hat{\sigma}^2)}{\partial \hat{\mu}} = \frac{1}{\hat{\sigma}^2}\sum (x_i - \hat{\mu}) = 0 \tag{7.111}$$

and

$$\frac{\partial \ln L(\hat{\mu}, \hat{\sigma}^2)}{\partial \hat{\sigma}^2} = -\frac{n}{2\hat{\sigma}^2} + \frac{1}{2\hat{\sigma}^4}\sum (x_i - \hat{\mu})^2 = 0. \tag{7.112}$$

The two equations yield

$$\hat{\mu} = \frac{\sum x_i}{n} = \bar{x} \tag{7.113}$$

and

$$\hat{\sigma}^2 = \frac{\sum (x_i - \hat{\mu})^2}{n}. \tag{7.114}$$

Since the estimator $\hat{\mu}$ is $\bar{x}$, we have

$$\hat{\sigma}^2 = \frac{\sum (x_i - \bar{x})^2}{n}. \tag{7.115}$$

We note that the relation $\hat{\mu} = \bar{x}$ corresponds to the estimator discussed in Section 7.3, but that $\hat{\sigma}^2$ is not equal to S^2, our usual estimator for the variance. Comparing Eqs. (7.1) and (7.115), we see that

$$\hat{\sigma}^2 = \frac{(n-1)S^2}{n}. \tag{7.116}$$

It can be shown that S^2 is an unbiased estimator; that is, $E(S^2) = \sigma^2$. This follows from Eq. (7.8) and the fact that the expected value of a chi-squared random variable is equal to its degrees of freedom (Problem 7.34). Hence,

$$E(\hat{\sigma}^2) = \frac{n-1}{n} E(S^2) = \frac{n-1}{n} \sigma^2, \tag{7.117}$$

and it follows that $\hat{\sigma}^2$ is biased. The bias diminishes with increasing sample size, but tends to underestimate the true value of the variance for small samples. This feature is one of the drawbacks of maximum likelihood estimation, although the bias can usually be removed by some multiplicative adjustment (e.g., $n/(n-1)$ in this example).

There are many nice properties of MLEs, also. Under some general conditions, their asymptotic (large-sample) distributions converge to normal distributions. If a MVUE exists, it can be shown to be a function of the MLE. For example, S^2 is the MVUE of σ^2, and the MLE of $\hat{\sigma}^2$ is $(n-1)S^2/n$. Thus, MLEs are good estimators. In the multiparameter case, there might be no analytical solution, and iterative numerical techniques must be applied. The interested reader can find out much more about MLEs in the works by Hogg and Tanis (1993), Bickel and Doksum (1977), and Edwards (1972), listed in the Bibliography.

7.10
Method of Moments

Another means of estimation is provided by the *method of moments*. Let X_1, $X_2, \ldots, X_n$ be a random sample of size n from a population with distribution function given by $f(x; \theta_1, \theta_2, \ldots, \theta_r)$, where the θ_k, $k = 1, 2, \ldots, r$, are parameters of the distribution. (For example, we might deal with the normal population $f(x; \mu, \sigma^2)$, in which case $\theta_1 = \mu$ and $\theta_2 = \sigma^2$.) The *j*th moment of the random variable X about the origin is defined as the expected value of the *j*th power,

$$m_j = E(X^j), \quad j = 1, 2, \ldots. \tag{7.118}$$

Similarly, the *j*th moment of the *sample* is defined as

$$\hat{m}_j = \frac{1}{n} \sum_{i=1}^{n} x_i^j, \quad j = 1, 2, \ldots. \tag{7.119}$$

To obtain estimates of the θ_k by the method of moments, we can use Eq. (7.118) to generate a number of moments that express the m_j in terms of the θ_k. We start with $j = 1$ and continue until there are enough equations to provide a set that can be solved for the θ_k as functions of the m_j. We then replace the m_j by the sample moments $\hat{m}_j$, thus providing the estimates for θ_k.

■ **Example**

Let $X_1, X_2, \ldots, X_n$ be a random sample from a normal population with unknown mean μ and unknown variance σ^2. Use the method of moments to obtain point estimates of μ and σ^2.

Solution
The first two moments of the normal distribution about the origin are (Problem 7.38)

$$m_1 = E(X) = \mu \tag{7.120}$$

and

$$m_2 = E(X^2) = \sigma^2 + \mu^2. \tag{7.121}$$

The latter expression follows directly from Eq. (4.48). Solving Eqs. (7.120) and (7.121) for μ and σ^2, we find

$$\mu = m_1 \tag{7.122}$$

and

$$\sigma^2 = m_2 - \mu^2 = m_2 - m_1^2. \tag{7.123}$$

The estimates of μ and σ^2 are obtained by replacing the m_j by the $\hat{m}_j$. Thus,

$$\hat{\mu} = \hat{m}_1 = \frac{1}{n} \sum_{i=1}^{n} X_i = \bar{X} \tag{7.124}$$

and

$$\hat{\sigma}^2 = \hat{m}_2 - \hat{m}_1^2 = \frac{1}{n}\sum_{i=1}^{n} X_i^2 - \bar{X}^2 = \frac{1}{n}\sum_{i=1}^{n}(X_i - \bar{X})^2. \tag{7.125}$$

In this example, the method of moments gives the same estimators as the maximum likelihood method. This will not be true in general. The example also shows that the moment estimators may be biased estimators, since $\hat{\sigma}^2$ above is biased (Eq. (7.117)). In some cases, the moment estimators may not be unique. An example is afforded by sampling from a Poisson distribution with parameter λ. If we want moment estimators of the mean and variance, we can proceed as in the last example to obtain Eqs. (7.122) and (7.123). For the Poisson distribution, $\mu = \sigma^2 = \lambda$, and so we can estimate λ by using either $\hat{\mu}$ or $\hat{\sigma}^2$. On the positive side, moment estimators are easy to obtain and they are consistent.

To use this method easily, one needs a means to calculate the moments. One way utilizes the *moment generating function*, which we now define for the random variable X. Let there be a positive number h such that, for $-h < t < h$, the mathematical expectation $E(e^{tx})$ exists. If X is continuous with density function $f(x)$, then its moment generating function is

$$M(t) = E(e^{tx}) = \int_{-\infty}^{\infty} e^{tx} f(x) dx. \tag{7.126}$$

If X is discrete with probability function $p(x)$, then its moment generating function is

$$M(t) = E(e^{tx}) = \sum_{x} e^{tx} p(x). \tag{7.127}$$

Note that the existence of $E(e^{tx})$ for $-h < t < h$ (with $h > 0$) implies that derivatives of $M(t)$ to all orders exist at $t = 0$. Thus, we may write, for instance,

$$\frac{d}{dt}M(t) = M^{(1)}(t) = \int_{-\infty}^{\infty} x\, e^{tx} f(x) dx \tag{7.128}$$

for a continuous random variable X, or

$$\frac{d}{dt}M(t) = M^{(1)}(t) = \sum_{x} x\, e^{tx} p(x) \tag{7.129}$$

if X is discrete. Setting $t = 0$ in either of the last two equations yields

$$M^{(1)}(0) = E(X) = \mu. \tag{7.130}$$

In general, if j is a positive integer and $M^{(j)}(t)$ denotes the jth derivative of $M(t)$ with respect to t, then repeated differentiation of Eq. (7.126) or Eq. (7.127) and setting

$t = 0$ gives

$$M^{(j)}(0) = E(X^j) = m_j. \qquad (7.131)$$

These derivatives of the moment generating function $E(e^{tx})$ with respect to t, evaluated at $t=0$, are thus the moments defined by Eq. (7.118).

We note also that these are moments defined about the origin. One can also investigate moments about points other than the origin. Let X denote a random variable and j a positive integer. Then the expected value $E[(X-b)^j]$ is called the jth moment of the random variable about the point b. If we replace b by μ, then we would say the jth moment about the mean. We could replace $M(t)$ in Eq. (7.126) by $R(t) = E(e^{t(x-b)})$ with the same conditions as for $M(t)$ and obtain a moment generating function about the point b. However, for our needs it is easier to work with $M(t)$.

■ **Example**
Let X be a Bernoulli random variable taking on the values 1 and 0 with probability p and $(1-p)$, respectively. Determine the moment generating function for X and use it to obtain the first two moments.

Solution
By definition,

$$M(t) = E(e^{tx}) = \sum_{x=0}^{1} e^{tx} p^x (1-p)^{1-x} = (1-p) + p e^t. \qquad (7.132)$$

Taking the first derivative of $M(t)$ with respect to t and setting $t=0$, we find for the first moment

$$M^{(1)}(0) = p. \qquad (7.133)$$

Taking the second derivative and setting $t=0$ gives

$$M^{(2)}(0) = p. \qquad (7.134)$$

Since $\sigma^2 = E(X^2) - [E(X)]^2 = M^{(2)}(0) - [M^{(1)}(0)]^2$, we see that $\sigma^2 = p - p^2 = p(1-p)$ for the Bernoulli random variable.

Not all distributions have moment generating functions (e.g., the Cauchy distribution). When such a function does exist, its representation is unique. As a consequence, for example, given the moment generating function, $1/2 + (1/2)e^t$, of a random variable X, we can say that X has a Bernoulli distribution with $p = 1/2$. (Replace p by $1/2$ in Eq. (7.132).) The uniqueness property comes from the theory of transforms in mathematics. Thus, if X_1 and X_2 have the same moment generating function, then both have the same distribution. In more advanced texts, the moment generating function is replaced by the characteristic function, $\phi(t) = E(e^{itx})$, where $i = \sqrt{-1}$. The characteristic function exists for all distributions. The interested reader can consult Parzen (1960).

Table 7.3 Moment generating functions of some common distributions.

Sampling distribution	Moment generating function
Bernoulli: $p(x) = p(1-p)^{1-x}$, $x = 0, 1$	$pe^t + 1 - p$
Binomial: $p(x) = \binom{n}{x} p^x (1-p)^{n-x}$, $x = 0, 1, \ldots, n$	$(pe^t + 1 - p)^n$
Poisson: $p(x) = \frac{e^{-\lambda} \lambda^x}{x!}$, $x = 0, 1, 2, \ldots$	$e^{\lambda(e^t - 1)}$
Geometric: $p(x) = p(1-p)^{x-1}$, $x = 1, 2, \ldots$	$\frac{p e^t}{1 - (1-p)e^t}$
Uniform over (a, b): $f(x) = \frac{1}{b-a}$, $a < x < b$	$\frac{e^{tb} - e^{ta}}{t(b-a)}$
Normal: $f(x) = \frac{1}{\sigma\sqrt{2\pi}} e^{-(1/2)[(x-\mu)/\sigma]^2}$	$e^{t\mu + (t^2 \sigma^2 / 2)}$
Exponential: $f(x) = \lambda e^{-\lambda x}$	$\frac{\lambda}{\lambda - t} = \left(1 - \frac{t}{\lambda}\right)^{-1}$
Gamma: $f(x) = \frac{\beta^\alpha x^{\alpha-1} e^{-\beta x}}{\Gamma(\alpha)}$	$\left(1 - \frac{t}{\beta}\right)^{-\alpha}$

Another interesting property of the moment generating function is the following. If $X_1, X_2, \ldots, X_n$ are n independent random variables with moment generating functions $M_1(t), M_2(t), \ldots, M_n(t)$, then the moment generating function of their sum, $Y = X_1 + X_2 + \cdots + X_n$, is the product $M_1(t) \cdot M_2(t) \cdots M_n(t)$. This property is seen by applying the definition (7.126) or (7.127):

$$M_Y = E(e^{tY}) = E(e^{t(X_1 + X_2 + \cdots + X_n)}) \tag{7.135}$$

$$= E(e^{tX_1}) \cdot E(e^{tX_2}) \cdots E(e^{tX_n}) = M_1(t) \cdot M_2(t) \cdots M_n(t). \tag{7.136}$$

Note that the expectation values factor because $X_1, X_2, \ldots, X_n$ are independent.

■ *Example*
Let X have a Poisson distribution with parameter λ.

a) Show that the moment generating function is $e^{\lambda(e^t - 1)}$.
b) Let $X_1, X_2, \ldots, X_n$ be independent, identically distributed Poisson random variables with parameter λ. Show that $Y = X_1 + X_2 + \cdots + X_n$ is also Poisson distributed, but with parameter $n\lambda$.

Solution

a) For the moment generating function (7.127), we write

$$M_X(t) = E(e^{tX}) = \sum_{x=0}^{\infty} e^{tx} \frac{e^{-\lambda} \lambda^x}{x!} \tag{7.137}$$

$$= e^{-\lambda} \sum_{x=0}^{\infty} \frac{(\lambda e^t)^x}{x!} = e^{-\lambda} e^{\lambda e^t} = e^{\lambda(e^t - 1)}. \tag{7.138}$$

(Note that we have used the series expansion, $\sum_{x=0}^{\infty} a^x / x! = e^a$).

b) With $Y = X_1 + X_2 + \cdots + X_n$, we write from Eqs. (7.136) and (7.138)

$$M_Y(t) = \prod_{i=1}^{n} M_{X_i}(t) = \prod_{i=1}^{n} e^{\lambda(e^t - 1)} = e^{n\lambda(e^t - 1)}. \qquad (7.139)$$

This result is identical with the Poisson function (7.138), but with parameter $n\lambda$. By the uniqueness property of the moment generating function, we can conclude that Y has the Poisson distribution with parameter $n\lambda$.

For reference, Table 7.3 provides a number of moment generating functions.

Problems

7.1 The heights of 50 randomly selected male military recruits were measured, and the average height was reported to be 175.2 ± 7.9 cm.
  a) Give an example of a systematic error that could affect this finding.
  b) What sources of random error contribute to the uncertainty?
  c) Distinguish between the accuracy and the precision of the reported result.

7.2 The following gives results of 10 observations in each of two samples. Obtain the sample means, standard deviations, and coefficients of variation for
  Sample (a): 6.6, 4.4, 6.1, 4.2, 4.0, 6.0, 6.2, 5.5, 5.7, 2.4.
  Sample (b): -6.4, 25.7, -1.3, 38.3, 41.6, 37.2, 28.2, 4.9, 40., -11.6.

7.3 a) What is a statistic?
  b) What is an estimator?
  c) What is an unbiased estimator?

7.4 Unknown to an investigator, a population consists simply of the first five positive integers 1, 2, 3, 4, 5. The investigator chooses to use any single observation as an estimator for the population mean.
  a) Is the estimator unbiased? Explain.
  b) Calculate the variance for this estimator.
  c) Is the estimator consistent? Why or why not?

7.5 A sample of n observations is drawn from a normal population with mean μ and variance σ^2. If we use

$$S^{*2} = \frac{1}{n} \sum_{i=1}^{n} (X_i - \bar{X})^2$$

as the estimator for σ^2, is it unbiased or biased? Why or why not? If it is biased, is there a way to make it unbiased? If so, show how it can be done.

7.6 Ten observations taken from a Poisson population with unknown parameter λ yield the following data: $X = 3, 2, 5, 5, 4, 4, 5, 6, 9, 6$. Use Eq. (7.4) to obtain the MVUE of λ.

7.7 Using Eq. (7.7) and the estimator obtained in the last problem, estimate the variance of $\hat{\lambda}$.

7.8 Show that Eq. (7.13) follows from Eqs. (7.10) and (7.11).

7.9 a) In the notation of (7.14), what values of α determine the probable error and the standard error?
   b) How many standard deviations determine the half-widths of the intervals (7.14) for the probable error and the standard error?

7.10 A random sample of 16 men yielded an average potassium concentration in body tissues of $1895\ \text{mg kg}^{-1}$ of body weight. Assume that the data are normally distributed and that the population standard deviation is $80\ \text{mg kg}^{-1}$. Obtain
   a) the standard error of the mean;
   b) a 90% confidence interval on the true mean.

7.11 In the last problem, if the population standard deviation were not known and if the sample standard deviation came out to be $94\ \text{mg kg}^{-1}$, what would be the answers to (a) and (b)?

7.12 Using the sample mean of $1895\ \text{mg kg}^{-1}$ and population standard deviation of $94\ \text{mg kg}^{-1}$ from the last problem, find the confidence interval for the mean that corresponds to the probable error.

7.13 Show that the random variable T in Eq. (7.19) has the Student's t-distribution with $(n-1)$ degrees of freedom.

7.14 Verify that (7.22) follows from Eqs. (7.19) and (7.20).

7.15 In the example leading to Eq. (7.23), show that $\bar{x} = 1.01$ cm and $s = 0.03$ cm.

7.16 Starting with Eq. (7.29), verify (7.30) and (7.31).

7.17 Swipes in a laboratory building were collected at random from 100 different areas, known to be uncontaminated, and analyzed for total activity. It was found that five of the swipes exceeded the maximum total activity used as a control for posted contamination areas.
   a) Estimate the proportion of swipes expected to exceed the control limit in similarly contaminated areas.
   b) Obtain an approximate 95% confidence interval on this value.

7.18 In the last problem we estimated the proportion p of smears with count numbers that exceed the control limit. We now want to determine the true proportion exceeding the upper limit within an error $E = 0.02$ with 80% confidence. What sample size is needed to do this if
   a) we assume that $p = 0.05$?
   b) we do not know the value of p?

7.19 An alpha source was counted 10 times over 5-min intervals, yielding the following count numbers: 19, 23, 29, 15, 22, 29, 28, 23, 22, 26. Assume that the sample came from a Poisson distribution.
   a) Estimate the rate parameter λ in counts per minute.
   b) Obtain an approximate 95% confidence interval for λ.

7.20 Two types of filters, A and B, were compared for their ability to reduce the activity of a radioactive aerosol. A fixed activity of 1000 Bq was drawn through a fresh filter, and the amount that passed through was measured. Ten type A and eight type B filters were tested. The average activity that passed through type A was 250 Bq, while the average for type B was 242 Bq. Assume that the variances

in the transmitted activities for the two filter types were $\sigma_A^2 = 25$ Bq^2 and $\sigma_B^2 = 49$ Bq^2.
   a) Obtain a 95% confidence interval for the difference between the true means of the two groups.
   b) Which filter type is more effective, based on these tests? Why?

7.21 Assume that the variances in the last problem are not known, but that measurements from the experiment give the estimates, $s_A^2 = 25$ Bq^2 and $s_B^2 = 49$ Bq^2. Assume, further, that the variances are equal.
   a) Determine the best estimate of the common variance.
   b) Obtain a 95% confidence interval for the difference in the means, assuming that the variances are equal, but unknown.

7.22 In the last problem, assume that the variances are unknown and unequal.
   a) Find a 95% confidence interval for the difference in means.
   b) Write down the 95% confidence intervals for the last two problems. Why should the length of the interval get larger as one proceeds from there to the present problem?

7.23 Two ointments are being tested for possible use in skin decontamination, in order to see whether they produce a rash. Ointment A was applied to 200 persons, and 15 developed a rash. Ointment B was used with 250 persons, and 23 developed a rash.
   a) What percentages of subjects developed a rash with each ointment?
   b) Use the central limit theorem to obtain a 95% confidence interval on the difference between the two percentages.
   c) Are the proportions of rashes different for the two ointments?

7.24 For each of the following, construct a 95% confidence interval for the population variance:
   a) $n = 15$, $s^2 = 40.1$;
   b) $n = 25$, $s^2 = 25.6$;
   c) $n = 11$, $s^2 = 10.2$.

7.25 A random sample of size 11 is taken from a normal population. The sample mean is 12.6, and the sample standard deviation is 4.2. Find 95% confidence intervals for the population mean and variance.

7.26 The amount of potassium in mg per kg of body weight for a random sample of five adult males is reported to be 1740, 1820, 1795, 1910, and 1850.
   a) What is the sample standard deviation?
   b) Obtain a 95% confidence interval on the variance, assuming a normal population.

7.27 Is the quantity Q given by Eq. (7.75) a statistic? Why or why not?

7.28 Starting with Eqs. (7.81) and (7.82), verify the result (7.88).

7.29 A new procedure is to be tested for performing a certain assay at a laboratory. Ten technicians are randomly chosen and divided into two groups, A and B, with five members each. Group A is taught the new procedure, while group B continues to use the standard method. A test preparation is made with a known amount of material, and each technician independently performs the assay. The objective is to see whether the new method reduces the variability of results among the technicians. The results of the testing for the two groups are as follows:

Group A: 6.12, 6.08, 6.15, 6.08, 6.10.
Group B: 5.95, 6.25, 5.80, 6.05, 6.01.

The true amount material used in the test preparation is 6.00.
a) Obtain the sample mean and sample variance for each group.
b) Obtain a 90% confidence interval for the ratio of variances (A to B).
c) Is the new method better than the old? Why or why not?
d) Knowing that the true value is 6.00, what is the significance of the results for the two procedures? Which method would you use and why?

7.30 For each of the following, construct a 90% confidence interval for the ratio of the variances (1 to 2):
a) $n_1 = 20$, $s_1^2 = 25.6$; $n_2 = 15$, $s_2^2 = 16.8$;
b) $n_1 = 10$, $s_1^2 = 5.40$; $n_2 = 25$, $s_2^2 = 8.90$.

7.31 Uniform random soil samples were collected from two sites, A and B. The number of alpha counts measured in 5 min from the collected samples were as follows:

Site	Sample size	Sample mean	Sample standard deviation
A	6	3.2	2.5
B	8	4.5	4.9

a) Calculate the coefficient of variation for each site.
b) Calculate a 90% confidence interval on the ratio of the variance at site A to the variance at site B.
c) Does the variability appear to be different at the two sites?

7.32 Let $x_1, x_2, \ldots, x_n$ denote n values obtained by randomly sampling from an exponential distribution with density

$$f(x; \theta) = \frac{1}{\theta} e^{-x/\theta}, \quad x > 0, \quad \theta > 0.$$

a) Show that the likelihood function for θ is given by

$$L(\theta) = \frac{1}{\theta^n} e^{-\sum x_i/\theta},$$

where the summation goes from $i = 1$ to n.
b) Show that the maximum likelihood estimator for θ is given by $\hat{\theta} = \bar{x}$.
c) Verify that $\hat{\theta}$ is unbiased for θ; that is, $E(\hat{\theta}) = \theta$.
d) Show that the variance of $\hat{\theta}$ is given by θ^2/n.

7.33 Show that the value of $\hat{p}$ given by Eq. (7.99) maximizes, rather than minimizes, $L(p)$.

7.34 The maximum likelihood estimator for the variance when sampling from a normal population is given by Eq. (7.115). Using this result and the fact that

$(n-1)S^2/\sigma^2 \sim \chi^2_{n-1}$, find the expected value of the MLE for σ^2. Use this information to adjust $\hat{\sigma}^2$ to be unbiased.

7.35 For a random variable X, show that the second moment about the mean μ is equal to the difference between the second moment about an arbitrary point b and the square of the first moment about b. That is, show that

$$E[(X-\mu)^2] = E[(X-b)^2] - [E(X-b)]^2.$$

7.36 Let $X_1, X_2, \ldots, X_n$ be the disintegration times in seconds of n individual, identical atoms. Assume that these random variables are independent and identically distributed as exponential distributions with unknown parameter λ.
   a) Find a method of moments estimator for λ.
   b) Determine the variance of the exponential and use this information to obtain another moment estimator for λ. Is the moment estimator for λ unique?
   c) Find the probability $\Pr(X_1 \geq 1)$ that the first atom will "live" at least 1 s before decaying.
   d) Obtain a method of moments estimator for the probability in part (c).

7.37 Show that the moment generating function for the normal probability density function is that given in Table 7.3.

7.38 Use the generating function from the last problem to calculate the first two moments of the normal distribution about the origin.

7.39 Use moment generating functions to show that, if $X_1, X_2, \ldots, X_n$ are independent, normal random variables with mean μ and variance σ^2, then $Y = X_1 + X_2 + \cdots + X_n$ is normal with mean $n\mu$ and variance $n\sigma^2$.

7.40 If X has the moment generating function $M_x(t)$, show that $Y = aX + b$ has the moment generating function $M_y = e^{bt} M_x(at)$.

7.41 Use the results of the last two problems to show that $Z = \bar{X} = (X_1 + X_2 + \cdots + X_n)/n$ is normal with mean μ and variance σ^2/n.
   (Hint: From the problem before last, we know the moment generating function of $Y = X_1 + X_2 + \cdots + X_n$. Note that $Z = \bar{X} = Y/n$, and apply the results of the last problem.)

8
Propagation of Error

8.1
Introduction

In addition to statistical errors associated with measurements of a random variable, one often needs to assess the error in a quantity that is derived from a combination of independent random variables. For example, the activity of a radioactive source can be inferred by subtracting background counts observed in the absence of the source from counts with the source present. Both counts are independent random variables. Given their individual errors, how does one obtain the error for the count difference and hence the activity of the source?

Uncertainty in a derived quantity is a combination of all errors in the component parts. Systematic errors, if known, can be used to correct the values of the components before calculating the derived quantity. Independent random errors, on the other hand, act together to contribute to the overall random variation in a composite quantity. The way in which random errors in individual variables are combined to estimate the resulting random error in a derived quantity is commonly called *error propagation* – the subject of this chapter.

8.2
Error Propagation

We consider a quantity $Q(X_1, X_2, \ldots, X_n)$, which is a function of n independent random variables X_i, each, respectively, having a mean μ_i and variance σ_i^2, with $i = 1, 2, \ldots, n$. We wish to determine how random errors in the X_i propagate into the error in the derived quantity Q. This goal is accomplished by calculating the variance of Q in terms of the variances σ_i^2. The analysis is greatly simplified if Q is a linear function of the X_i. If Q is nonlinear, then we approximate it in linear form by using a Taylor series expansion, keeping only the zero- and first-order terms. Nonlinear aspects will be considered in an example presented in Section 8.4. We shall see that, when the $\sigma_i \ll \mu_i$ for all i, the linear approximation works well.

Statistical Methods in Radiation Physics, First Edition. James E. Turner, Darryl J. Downing, and James S. Bogard.
© 2012 Wiley-VCH Verlag GmbH & Co. KGaA. Published 2012 by Wiley-VCH Verlag GmbH & Co. KGaA.

We expand Q through first order in the X_i about the point $\mu = (\mu_1, \mu_2, \ldots, \mu_n)$ in n-dimensional space by writing

$$Q(X_1, X_2, \ldots, X_n) \cong Q(\mu) + \sum_{i=1}^{n} \left(\frac{\partial Q}{\partial X_i}\right)_\mu (X_i - \mu_i)$$
$$\equiv Q^*(X_1, X_2, \ldots, X_n). \tag{8.1}$$

We thus approximate the exact function Q by the function Q^*, which is linear in the X_i. The partial derivatives are understood to be evaluated at the point μ. It is straightforward to calculate the mean and variance of the function Q^*. (If Q is linear to begin with, then, of course, $Q = Q^*$.)

The expected value of the sums in Eq. (8.1) is equal to the sum of the individual expected values (Section 4.2), and so it follows that (Problem 8.1)

$$E(Q^*) = E[Q(\mu)] + \sum_{i=1}^{n} \left(\frac{\partial Q}{\partial X_i}\right)_\mu [E(X_i) - E(\mu_i)] = Q(\mu). \tag{8.2}$$

Since $E(X_i) = \mu_i$, each term in brackets in Eq. (8.2) is identically zero. Also, the expected value of the constant $Q(\mu)$, which remains, is just the constant itself. Therefore, Eq. (8.2) shows that the expected value of Q^* is just $Q(\mu) = Q(\mu_1, \mu_2, \ldots, \mu_n)$, or Q evaluated at the means of the X_i.

The variance of Q is given approximately by $\text{Var}(Q^*)$. Since the mean is $Q(\mu)$, we have from Eqs. (4.44) and (8.1)

$$\text{Var}(Q^*) = E\left\{[Q^*(X_1, X_2, \ldots, X_n) - Q(\mu)]^2\right\}$$
$$= E\left\{\left[\sum_{i=1}^{n} \left(\frac{\partial Q}{\partial X_i}\right)_\mu (X_i - \mu_i)\right]^2\right\}. \tag{8.3}$$

Separating the squared and the cross terms in the square of the sum, one can write

$$\text{Var}(Q^*) = E\left[\sum_{i=1}^{n} \left(\frac{\partial Q}{\partial X_i}\right)_\mu^2 (X_i - \mu_i)^2\right] + E\left[\sum_{i \neq j=1}^{n} \left(\frac{\partial Q}{\partial X_i}\right)_\mu \left(\frac{\partial Q}{\partial X_j}\right)_\mu (X_i - \mu_i)(X_j - \mu_j)\right] \tag{8.4}$$

$$= \sum_{i=1}^{n} \left(\frac{\partial Q}{\partial X_i}\right)_\mu^2 E\left[(X_i - \mu_i)^2\right] + \sum_{i \neq j=1}^{n} \left(\frac{\partial Q}{\partial X_i}\right)_\mu \left(\frac{\partial Q}{\partial X_j}\right)_\mu E\left[(X_i - \mu_i)(X_j - \mu_j)\right]. \tag{8.5}$$

In the first summation, $E[(X_i - \mu_i)^2] = \sigma_i^2$, the variance of X_i. In the second summation, each term $E[(X_i - \mu_i)(X_j - \mu_j)]$ is the covariance σ_{ij} (Eq. (4.82)), which vanishes because the X_i are independent. We are left with the result (Problem 8.2),

$$\text{Var}(Q^*) = \sum_{i=1}^{n} \left(\frac{\partial Q}{\partial X_i}\right)_\mu^2 \sigma_i^2. \tag{8.6}$$

Equations (8.2) and (8.6) give the mean and variance of the (linear) approximation Q^* to the function Q. The latter equation shows explicitly how the variance of the derived quantity Q^* depends on the standard errors σ_i in the individual, independent random variables.

In practice, the value of the derived quantity is often inferred from repeated determinations of the random variables. If one measures a total of m_i values x_{ik}, $k = 1, 2, \ldots, m_i$, for the independent random variables X_i, then useful estimates of μ_i and σ_i^2 are

$$\hat{\mu}_i = \frac{\sum_{k=1}^{m_i} x_{ik}}{m_i} = \bar{x}_i \tag{8.7}$$

and

$$\hat{\sigma}_i^2 = \frac{\sum_{k=1}^{m_i} (x_{ik} - \bar{x}_i)^2}{m_i - 1} = s_i^2, \tag{8.8}$$

in which x_{ik} is the value obtained in the kth measurement of X_i. These estimates from the sample are then substituted for μ_i and σ_i^2 in Eqs. (8.2) and (8.6) in order to estimate the mean and standard error of Q^*, which approximates Q when the latter is nonlinear.

We can use these results to calculate confidence intervals. If we assume that the X_i are normally distributed with means μ_i and variances σ_i^2, then the approximate $100(1 - \alpha)\%$ confidence interval for $E(Q)$ in Eq. (8.2) is

$$Q(\boldsymbol{\mu}) \pm z_{\alpha/2} \sqrt{\sum_{i=1}^{n} \left(\frac{\partial Q}{\partial X_i}\right)_{\boldsymbol{\mu}}^2 \sigma_i^2}, \tag{8.9}$$

where $z_{\alpha/2}$ is the quantity that cuts off an area of size $\alpha/2$ to the right under the standard normal distribution. If we obtain sample estimates of μ_i and σ_i^2, given by Eqs. (8.7) and (8.8), then an approximate $100(1 - \alpha)\%$ confidence interval for $E(Q)$ is given by

$$Q(\bar{x}) \pm t_{\nu,\alpha/2} \sqrt{\sum_{i=1}^{n} \left(\frac{\partial Q}{\partial X_i}\right)_{\bar{x}}^2 s_i^2}, \tag{8.10}$$

in which $t_{\nu,\alpha/2}$ is defined in Table A.5 and $\bar{x} = (\bar{x}_1, \bar{x}_2, \ldots, \bar{x}_n)$, where the $\bar{x}_i$ are the sample averages. The degrees of freedom are calculated by using the Satterthwaite approximation to the degrees of freedom (Satterthwaite, 1946; Anderson and McLean, 1974),

$$\nu = \frac{\left[\sum_{i=1}^{n} (\partial Q/\partial X_i)_{\bar{x}}^2 s_i^2\right]^2}{\sum_{i=1}^{n} (1/(m_1 - 1))(\partial Q/\partial X_i)_{\bar{x}}^4 s_i^4}. \tag{8.11}$$

The degrees of freedom, which are nonintegral, can be truncated to the next lowest integer (in order to be conservative), so that Student's t-tables can be used.

8.3
Error Propagation Formulas

One can use Eq. (8.6) to compute the variance of any derived quantity Q in the linear approximation. A number of functional forms occur frequently in practice, and we next develop explicit formulas for estimating their variances.

8.3.1
Sums and Differences

When Q consists of sums and differences of random variables, one has the general form

$$Q(X_1, X_2, \ldots, X_n) = \sum_{i=1}^{n} a_i X_i, \qquad (8.12)$$

where the a_i are constants. In this case, Q itself is linear in the X_i and, therefore, identical with Q^*. The error computation is then exact. The partial derivatives, obtained from Eq. (8.12), are $\partial Q/\partial X_i = a_i$. Equation (8.6) then gives

$$\text{Var}(Q) = \text{Var}(Q^*) = \sum_{i=1}^{n} a_i^2 \sigma_i^2. \qquad (8.13)$$

In the example cited in the first paragraph of this chapter, the net number of counts from a source is the difference in the two direct count measurements: X_1 = source plus background (gross counts) and X_2 = background alone. Comparing with Eq. (8.9), we can write for the net number of counts $Q = X_1 - X_2$, with $n = 2$, $a_1 = 1$, and $a_2 = -1$. Applying Eq. (8.6) shows that the variances in the two individually measured count numbers add in quadrature:

$$\text{Var}(Q) = a_1^2 \sigma_1^2 + a_2^2 \sigma_2^2 = \sigma_1^2 + \sigma_2^2. \qquad (8.14)$$

The standard deviation in the net number of counts is thus equal to the square root of the sum of the variances in the gross and background count numbers. As mentioned, this result is exact, because Q is a linear function of the X_i in Eq. (8.12).

8.3.2
Products and Powers

For all other functions, we substitute the linear approximation Q^* for Q. For a combination of products of n random variables X_i raised to any power p_i, the most general form for Q can be written as

$$Q = X_1^{p_1} X_2^{p_2} \cdots X_n^{p_n}. \qquad (8.15)$$

Taking the partial derivative of Q with respect to X_i introduces a multiplicative factor of p_i and reduces the power of X_i by one unit. Therefore, we may write (Problem 8.6)

$$\frac{\partial Q}{\partial X_i} = \frac{p_i Q}{X_i}. \tag{8.16}$$

According to Eq. (8.6), with the derivatives evaluated at the point $\boldsymbol{\mu} = (\mu_1, \mu_2, \ldots, \mu_n)$ we obtain

$$\mathrm{Var}(Q^*) = \sum_{i=1}^{n} \left(\frac{p_i Q}{X_i}\right)_{\boldsymbol{\mu}}^{2} \sigma_i^2 = Q^2(\boldsymbol{\mu}) \sum_{i=1}^{n} \left(\frac{p_i \sigma_i}{\mu_i}\right)^2. \tag{8.17}$$

This equation can be written conveniently in dimensionless form:

$$\frac{\mathrm{Var}(Q^*)}{Q^2(\boldsymbol{\mu})} = \sum_{i=1}^{n} \left(\frac{p_i \sigma_i}{\mu_i}\right)^2. \tag{8.18}$$

8.3.3
Exponentials

We consider a function of the form $Q = e^W$, where the exponent

$$W = X_1^{p_1} X_2^{p_2} \cdots X_n^{p_n} \tag{8.19}$$

is the same as Q in Eq. (8.15). The partial derivatives are

$$\frac{\partial Q}{\partial X_i} = \frac{\partial e^W}{\partial X_i} = \frac{\partial W}{\partial X_i} e^W = \frac{p_i W}{X_i} Q. \tag{8.20}$$

With the quantities evaluated at $\boldsymbol{\mu}$, we obtain from Eq. (8.6) (Problem 8.7)

$$\mathrm{Var}(Q^*) = \sum_{i=1}^{n} \left(\frac{p_i W Q}{\mu_i}\right)^2 \sigma_i^2. \tag{8.21}$$

In dimensionless form,

$$\frac{\mathrm{Var}(Q^*)}{Q^2(\boldsymbol{\mu})} = W^2(\boldsymbol{\mu}) \sum_{i=1}^{n} \left(\frac{p_i \sigma_i}{\mu_i}\right)^2. \tag{8.22}$$

8.3.4
Variance of the Mean

The formalism here can be applied to calculate the variance of the mean. For a set of n independent measurements X_i of a random variable X, the mean is, by definition, the linear function

$$Q(X_1, X_2, \ldots, X_n) = \frac{\sum_{i=1}^{n} X_i}{n}. \tag{8.23}$$

Here the X_i can be regarded as n independent random variables, all having the same variance $\sigma_i^2 = \sigma^2$ as the random variable X. The function (8.23) is then a special case of Eq. (8.12), with $Q^* = Q$ and $a_i = 1/n$ for all i. Applying Eq. (8.6), we find that

$$\text{Var}(Q) = \sum_{i=1}^{n} \frac{\sigma^2}{n^2} = n\left(\frac{\sigma^2}{n^2}\right) = \frac{\sigma^2}{n}. \tag{8.24}$$

The result was given earlier (Eq. (6.35)) without proof.

■ *Example*

The specific gross count rate of a source is defined as the count rate of the sample plus background divided by the mass of the source, and it can be expressed as $N/(TM)$ in $s^{-1} g^{-1}$. An unknown source is placed in a counter, and a total of $n = \hat{\mu}_N = 1999$ counts is registered in a time $t = \hat{\mu}_T = 60.00 \pm 0.05$ s. The mass of the source is $m = \hat{\mu}_M = 3.04 \pm 0.02$ g. (The usual notation, showing one standard deviation, will be used in this example.) Find

a) the specific gross count rate and
b) its standard error.

Solution

a) The specific gross count rate is given by

$$Q(n, t, m) = \frac{\hat{\mu}_N}{\hat{\mu}_T \hat{\mu}_M} = \frac{1999}{(60.00 \text{ s})(3.04 \text{ g})} = 11.0 \text{ s}^{-1} \text{g}^{-1}. \tag{8.25}$$

The given values of $\hat{\mu}_N$, $\hat{\mu}_T$, and $\hat{\mu}_M$ serve as estimates of the individual means, which are to be used in accordance with Eq. (8.1). The result (8.25) has been rounded to three significant figures, equal to the number in the least precise measurement – that of the mass.

b) The function Q in Eq. (8.25) has the form (8.15) with each p_i equal to either $+1$ or -1. The variance is given by Eq. (8.18):

$$\text{Var}(Q^*) = Q^2(\hat{\mu}_N, \hat{\mu}_T, \hat{\mu}_M)\left[\left(\frac{\sigma_N}{\mu_N}\right)^2 + \left(\frac{-\sigma_T}{\mu_T}\right)^2 + \left(\frac{-\sigma_M}{\mu_M}\right)^2\right], \tag{8.26}$$

where the quantities σ are the standard deviations in the respective quantities. The estimated errors are given for T and M. We shall assume that the number of counts is Poisson distributed and use as an estimate $\hat{\sigma}_N \cong \sqrt{1999} = 44.7$. Substitution of the numerical values into (8.26) gives for the estimated variance of Q^* and hence Q,

$$\hat{\sigma}_Q^2 \cong \hat{\sigma}_{Q^*}^2 = (11.0 \text{ s}^{-1} \text{g}^{-1})^2\left[\left(\frac{44.7}{1999}\right)^2 + \left(\frac{-0.05}{60.00}\right)^2 + \left(\frac{-0.02}{3.04}\right)^2\right] \tag{8.27}$$

$$= 121(5.00 \times 10^{-4} + 6.94 \times 10^{-7} + 4.33 \times 10^{-5})$$
$$= 6.58 \times 10^{-2} \text{ s}^{-2} \text{ g}^{-2}. \tag{8.28}$$

The estimated standard error is, therefore, approximately, $\hat{\sigma}_Q = 0.257 \text{ s}^{-1} \text{ g}^{-1}$. We report the specific gross count rate with its one standard error as $\hat{Q} = 11.0 \pm 0.3 \text{ s}^{-1} \text{ g}^{-1}$, to three significant figures. We see from Eq. (8.28) that most of the uncertainty is due to the random error in the number of counts in this example. (Time measurements almost always have negligible error compared with other factors in a counting experiment.)

■ **Example**

The activity Q of an airborne radionuclide deposited on the filter of a constant air monitor is related to the deposition rate D (activity per unit time), the decay constant Λ of the nuclide, and the collection time T. If there is initially no activity on the filter and the deposition rate D is constant, then the activity of the material on the filter at time T is given by the following expression (Cember, 1996, p. 571; Turner et al., 1988, p. 32):

$$Q = \frac{D}{\Lambda}(1 - e^{-\Lambda T}). \tag{8.29}$$

If $d = 115 \pm 7$ Bq h^{-1}, $\lambda = 0.301 \pm 0.004$ h^{-1}, and $t = 8.00 \pm 0.03$ h, where the standard errors are indicated, find

a) the activity on the filter at time t and
b) its standard error.

Solution

a) The activity expressed by Eq. (8.29) is a function $Q(D, \Lambda, T)$ of the three variables shown. Using their given values as estimates of the means, we find for the estimated activity

$$\hat{Q} = \frac{115 \text{ Bq h}^{-1}}{0.301 \text{ h}^{-1}}(1 - e^{-(0.301 \text{ h}^{-1})(8.00 \text{ h})}) = 348 \text{ Bq}. \tag{8.30}$$

b) The function (8.29) is not one of the "standard" forms for variance worked out in this section, and so we employ the general expression (8.6) to calculate the variance. The partial derivatives are to be evaluated at the point μ determined by the means of the three variables, estimated with the given values. The exponential term, $e^{-\lambda t} = e^{-0.301 \times 8.00} = 0.0900$, calculated in Eq. (8.30), will occur in all of the partial derivatives. We estimate the σ_i in Eq. (8.6) by using the given standard deviations. Thus, we find that

$$\left(\frac{\partial Q}{\partial D}\right)_{\hat{\mu}} \hat{\sigma}_D = \left(\frac{1-e^{-\lambda t}}{\lambda}\right)\hat{\sigma}_D = \frac{1-e^{-0.301\times 8}}{0.301\text{ h}^{-1}} \times 7\text{ Bq h}^{-1}$$

$$= 21.2\text{ Bq}, \qquad (8.31)$$

$$\left(\frac{\partial Q}{\partial \lambda}\right)_{\hat{\mu}} \hat{\sigma}_\lambda = d\left(-\frac{1-e^{-\lambda t}}{\lambda^2} + \frac{te^{-\lambda t}}{\lambda}\right)\hat{\sigma}_\lambda \qquad (8.32)$$

$$= (115\text{ Bq h}^{-1})\left[-\frac{1-e^{-0.301\times 8}}{(0.301\text{ h}^{1-})^2} + \frac{(8\text{ h})\times e^{-0.301\times 8}}{0.301\text{ h}^{-1}}\right](0.004\text{ h}^{-1})$$

$$= -3.52\text{ Bq}, \qquad (8.33)$$

and

$$\left(\frac{\partial Q}{\partial t}\right)_{\hat{\mu}} \hat{\sigma}_t = \left[\frac{d}{\lambda}(\lambda e^{-\lambda t})\right]\hat{\sigma}_t$$

$$= (115\text{ Bq h}^{-1})(e^{-0.301\times 8})(0.03\text{ h}) = 0.310\text{ Bq}. \qquad (8.34)$$

Adding the squares from Eqs. (8.31), (8.33), and (8.34), we obtain from Eq. (8.6) the estimated variance,

$$\hat{\sigma}_Q^2 \cong (21.2)^2 + (-3.52)^2 + (0.310)^2 = 462\text{ Bq}^2. \qquad (8.35)$$

The estimated standard error for the activity on the filter is, therefore,

$$\hat{\sigma}_Q \cong \sqrt{462\text{ Bq}^2} = 21.49\text{ Bq} = 20\text{ Bq}. \qquad (8.36)$$

Since the standard errors for the random variables all carry only one significant figure, in writing Eq. (8.36) we have rounded off the calculated error in the derived activity accordingly. The filter activity together with its one standard error can be reported as

$$\hat{Q} = 348 \pm 20\text{ Bq}. \qquad (8.37)$$

■ **Example**

A certain nuclide has a radiological half-life T_R with estimated mean $\hat{\mu}_{T_R} = 8.0 \pm 0.1$ d and a metabolic half-life T_M with estimated mean $\hat{\mu}_{T_M} = 100 \pm 5$ d, where both uncertainties are one standard deviation.

a) Calculate the effective half-life, given by

$$T_E = \frac{T_R T_M}{T_R + T_M}. \qquad (8.38)$$

b) What is the standard error in the effective half-life?

Solution

a) Inserting the given data into Eq. (8.38), we obtain

$$\hat{T}_E = \frac{(8.0\,\text{d}) \times (100\,\text{d})}{(8.0\,\text{d}) + (100\,\text{d})} = 7.4\,\text{d}. \tag{8.39}$$

b) The expression (8.38) is not one of the standard forms from earlier in this section, and so we employ Eq. (8.6), writing

$$\text{Var}(T_E) \cong \left(\frac{\partial T_E}{\partial T_R}\right)_{\boldsymbol{\mu}}^2 \sigma_R^2 + \left(\frac{\partial T_E}{\partial T_M}\right)_{\boldsymbol{\mu}}^2 \sigma_M^2. \tag{8.40}$$

The partial derivatives are to be evaluated at the point $\boldsymbol{\mu}$ determined by using the given mean estimates $\hat{\mu}_{T_R}$ and $\hat{\mu}_{T_M}$. Similarly, the given values can be employed to estimate σ_R and σ_M. Using Eq. (8.38) in Eq. (8.40) gives (Problem 8.13)

$$\text{Var}(T_E) = \sigma_E^2 \cong \left(\frac{T_M}{T_R + T_M}\right)_{\boldsymbol{\mu}}^4 \sigma_R^2 + \left(\frac{T_R}{T_R + T_M}\right)_{\boldsymbol{\mu}}^4 \sigma_M^2. \tag{8.41}$$

Putting in the numerical values, we find

$$\hat{\sigma}_E^2 \cong \left(\frac{100\,\text{d}}{108\,\text{d}}\right)^4 (0.1\,\text{d})^2 + \left(\frac{8.0\,\text{d}}{108\,\text{d}}\right)^4 (5\,\text{d})^2$$
$$= 8.10 \times 10^{-3}\,\text{d}^2. \tag{8.42}$$

Thus, the estimated standard error in the effective half-life is $\hat{\sigma}_E \cong 0.09$ d, to the appropriate number of significant figures.

8.4
A Comparison of Linear and Exact Treatments

It is not our intent to go beyond an introductory presentation of the conditions under which error computation by means of the linear approximation is satisfactory. Suffice it to say that the formulas in the last section are sufficiently accurate for many purposes when the standard deviations of the random variables are much smaller than their means, that is, when $\sigma_i \ll \mu_i$ for all i. Looking back at the three examples presented in the last section, one can see that this condition was met. Use of the linear function Q^* thus gave a good approximation to the propagated error.

The condition $\sigma_i \ll \mu_i$ for the validity of the linear approximation can be understood from Eq. (8.1), which represents the lowest order expansion of the random variables about their means. Higher order terms are not important when the spread of the variables about their mean values is relatively small. To illustrate, we next present an example in which the propagated error is computed in the linear approximation and then compared with the exact result.

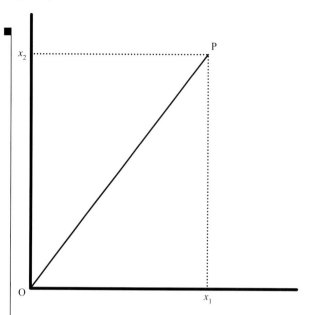

Figure 8.1 See Eq. (8.43) and example in Section 8.4.

Example

The distance of a fixed point P from the origin O of the orthogonal coordinate axes in Figure 8.1 is determined by measuring the displacements X_1 and X_2 of the projections of P along the two perpendicular axes. The square of the distance of P from O is given by

$$Q(X_1, X_2) = X_1^2 + X_2^2. \tag{8.43}$$

The random variables X_1 and X_2 have means μ_1 and μ_2 and variances σ_1^2 and σ_2^2. Find the mean and variance of Q

a) approximately, by using the linear approximation Q^* for Q, and
b) exactly, given that X_1 and X_2 have normal distributions.

Solution

a) From Eqs. (8.2) and (8.43) it follows that the mean value for the square of the distance from O to P is, in the linear approximation,

$$E(Q) \cong E(Q^*) = E(\mathbf{\mu}) = \mu_1^2 + \mu_2^2. \tag{8.44}$$

($Q(\mathbf{\mu}) = Q(\mu_1, \mu_2)$ is the value of Q at the point $\mathbf{\mu} = (\mu_1, \mu_2)$.) For the variance, the partial derivatives in Eq. (8.6) are, from Eq. (8.43),

$$\left(\frac{\partial Q}{\partial X_i}\right)_{\boldsymbol{\mu}} = 2X_i|_{\boldsymbol{\mu}} = 2\mu_i, \tag{8.45}$$

where $i = 1, 2$. Thus, we find from Eq. (8.6) that

$$\text{Var}(Q) \cong \text{Var}(Q^*) = \sum_{i=1}^{2} 4\mu_i^2 \sigma_i^2 = 4\mu_1^2 \sigma_1^2 + 4\mu_2^2 \sigma_2^2, \tag{8.46}$$

which is the estimate of Var(Q) in the linear approximation.

b) We next deal with Q exactly. The mean is

$$E(Q) = E(X_1^2 + X_2^2) = E(X_1^2) + E(X_2^2). \tag{8.47}$$

From Eq. (4.48),

$$E(Q) = \mu_1^2 + \sigma_1^2 + \mu_2^2 + \sigma_2^2. \tag{8.48}$$

This exact result can be compared with Eq. (8.44). The error in our estimating the mean of Q by the linear approximation is seen to amount to neglecting $\sigma_1^2 + \sigma_2^2$ compared with $\mu_1^2 + \mu_2^2$. As pointed out at the beginning of this section, use of the linear approximation for error propagation is accurate to the extent that the standard deviations are negligible compared with the means of the random variables. For the exact computation of the variance, we have

$$\text{Var}(Q) = \text{Var}(X_1^2 + X_2^2). \tag{8.49}$$

Using Eq. (4.48), we write in place of Eq. (8.49)

$$\text{Var}(Q) = E[(X_1^2 + X_2^2)^2] - [E(X_1^2 + X_2^2)]^2. \tag{8.50}$$

The expected value in the second term is given by Eqs. (8.47) and (8.48). The expected value in the first term can be expanded:

$$E[(X_1^2 + X_2^2)^2] = E(X_1^4) + 2E(X_1^2)E(X_2^2) + E(X_2^4). \tag{8.51}$$

After a lengthy, term-by-term evaluation one finds from Eq. (8.50) that (Problem 8.21)

$$\text{Var}(Q) = 4\mu_1^2 \sigma_1^2 + 4\mu_2^2 \sigma_2^2 + 2\sigma_1^4 + 2\sigma_2^4. \tag{8.52}$$

Comparison with Eq. (8.46) shows that the linear approximation for the variance neglects the last two terms in the exact expression (8.52) in comparison with the first two. As with the mean, the linear approximation for variance is good when $\sigma_i^2 \ll \mu_i^2$ for all i.

8.5
Delta Theorem

The results in Sections 8.1–8.3 have been presented without much distributional underpinning. There is a theorem referred to as the *delta method* that proves that, under certain conditions, the distribution of $Q(X_1, X_2, \ldots, X_n)$ as expressed by Eq. (8.1) is normal with mean $Q(\mu)$ and variance given by Eq. (8.6). For this to be true, one must assume that each $X_i \sim N(\mu_i, \sigma_i)$. The Taylor series gives a linear approximation, and the sum of normal random variables is normal. There are many situations where the X_i will not be normally distributed, and the question of how to handle them arises. Generally, good estimates of the μ_i are needed for the Taylor approximation to work well. In most instances, if it is possible to sample m_i times from the populations X_i, then the sample means $\bar{X}_{im_i}$ will be good estimators of the μ_i. We also know that under rather general conditions the central limit theorem holds and, therefore, $\bar{X}_{im_i} \sim N(\mu_i, \sigma_i/\sqrt{m_i})$. If this condition holds for each $\bar{X}_{im_i}$, $i = 1, 2, \ldots, n$, then

$$Q(\bar{X}_{1m_1}, \bar{X}_{2m_2}, \ldots, \bar{X}_{nm_n}) \overset{.}{\sim} N\left(Q(\mu), \sqrt{\operatorname{Var} Q^*}\right) \tag{8.53}$$

where, from Eq. (8.6),

$$\operatorname{Var}(Q^*) = \sum_{i=1}^{n} \left(\frac{\partial Q}{\partial X_i}\right)_{\bar{X}}^2 \frac{\sigma_i^2}{m_i}. \tag{8.54}$$

The delta theorem is discussed by Bishop, Fienberg, and Holland (1975).

Problems

8.1 Verify Eq. (8.2).

8.2 Show that Eq. (8.6) follows from Eq. (8.3).

8.3 A technician takes a series of five 30-s counts with a long-lived source. The results are 72, 49, 55, 63, and 51.
   a) Find the mean and standard deviation of the measurements.
   b) What is the standard deviation of the mean?

8.4 A student takes five 10-min measurements with a long-lived radioactive source. The numbers of counts observed are 120, 108, 131, 117, and 137.
   a) What is the mean count number?
   b) What is the standard deviation of the count number?
   c) What is the standard deviation of the mean count number?
   d) State the value of the mean count rate ± its standard error in cpm.

8.5 A ^{90}Mo generator is to be milked for the daughter ^{99m}Tc with an efficiency of $91.2 \pm 0.8\%$. Before milking, the ^{99m}Tc content of the generator is 844 ± 8 mCi. After milking, how much ^{99m}Tc is expected? State the result with its standard error.

8.6 Verify Eq. (8.16).

8.7 Show that Eq. (8.21) follows from Eq. (8.6) for the exponential function $Q = e^W$ with W defined by Eq. (8.19).

8.8 a) How does the standard deviation of the measurements in Problem 8.3 compare with the square root of the mean number of counts?
b) Why are the two numbers not the same?

8.9 For two independent random variables X_1 and X_2 show that the standard deviations of $a_1 X_1 + a_2 X_2$ and $a_1 X_1 - a_2 X_2$ are the same.

8.10 In what sense are the quantities D, λ, and t in Eq. (8.29) independent random variables?

8.11 A fresh air monitor filter begins collecting activity of a radionuclide at a constant rate of 97.2 ± 10.3 Bq min^{-1}. The decay constant of the nuclide is given as 2.22 ± 0.46 h^{-1}.
a) Show that the activity on the filter and its standard error when the filter is removed after 1.25 h is 2460 ± 490 Bq. The time measurement has negligible error.
b) State the value of the half-life and its standard error.

8.12 In the last problem, what is the activity on the filter and its standard error 90.00 ± 0.10 min after it is removed?

8.13 Verify Eq. (8.41).

8.14 Show that the variance of the effective half-life (8.41) can be written in the compact form
$$\text{Var}(Q_E) = T_E^4 \left(\frac{\sigma_R^2}{T_R^4} + \frac{\sigma_M^2}{T_M^4} \right).$$

8.15 A pressurized ion chamber is used to determine the typical penetrating radiation background in a green field location adjacent to an operating radiological facility. The results of eight observations are provided below. A series of such measurement sets is planned, each consisting of eight readings. What numerical value would you expect the standard deviation of the averages of these sets to have?

Background measurement results (μrem)							
175	163	171	182	185	170	180	173

8.16 An ion chamber is used to determine the exposure rate $\dot{E}$ (R h^{-1}) from an X-ray machine according to the relation
$$\dot{E} = (I_{\text{Beam}} - I_{\text{BKg}}) \frac{fT}{P}.$$

With units given below, I_{Beam} and I_{Bkg} are, respectively, the ion chamber currents with the X-ray beam on and off, f is a conversion factor, T is the ambient temperature, and P is the atmospheric pressure.

a) What is the exposure rate and its standard error for the following observed values with their standard errors?

$$f = (1.49 \pm 0.02) \times 10^{10} \text{ R Torr min C}^{-1} \text{ K}^{-1} \text{ h}^{-1};$$

$$I_{Beam} = (3.513 \pm 0.165) \times 10^{-11} \text{ C min}^{-1};$$

$$I_{Bkg} = (6.000 \pm 5.000) \times 10^{-13} \text{ C min}^{-1};$$

$$T = 24.3 \pm 0.2 \,°C;$$

$$P = 752 \pm 1 \text{ Torr}.$$

b) Which of the variables makes the largest contribution to the error in the exposure rate?

8.17 Current measurements with an ion chamber, given in the table below, are used to determine the net air kerma rate from a radiation source. A set of readings, I_S and I_B, were taken with the source present and background alone, respectively. The data were taken at a temperature $T = 22.0\,°C$ and pressure $P = 743$ Torr. The ion chamber was calibrated at standard temperature and pressure ($T_0 = 0\,°C$ and $P_0 = 760$ Torr) to obtain the conversion factor $f = (9.807 \pm 0.098) \times 10^7$ Gy C^{-1}. The net air kerma rate $\dot{K}$ at ambient temperature and pressure is given by

$$\dot{K} = f \bar{I}_{Net} \left(\frac{P_0}{P}\right)\left(\frac{T}{T_0}\right),$$

where $\bar{I}_{Net} = \bar{I}_S - \bar{I}_B$.

Ion chamber current (C min^{-1})	
$I_S \times 10^8$	$I_B \times 10^{13}$
3.630	6.0
3.612	8.8
3.624	8.0
3.618	5.2
3.620	7.6
3.622	9.4
3.618	5.5
3.616	4.8
3.624	2.7
3.626	6.3

a) What is the best estimate of average net air kerma rate $\dot{K}$ from these data?

b) If temperature and pressure in the laboratory can be measured within ± 0.5 °C and ± 1 Torr, determine the total uncertainty in the estimate of $\dot{K}$ in part (a).

c) What variable contributes the greatest uncertainty in the estimate of $\dot{K}$?

8.18 A radiation worker submits a monthly urine sample, which is analyzed for the particular radioisotope to which she is occupationally exposed. Average background B in the counter used to determine activity in the processed sample is obtained from several individual measurements in the standard time t. Gross activity G in the sample is based on a single measurement for the same time t. Daily urinary excretion Q (Bq d^{-1}) of the radioisotope is estimated from the analytical result according to the expression

$$Q = \frac{R}{f},$$

where $R = G - B$ is the net activity in the sample. The conversion factor f includes a chemical recovery factor, counter efficiency, sample volume, and daily urinary excretion rate. Values and associated standard errors for the quantities described above are

$f = 0.3041 \pm 0.0742$ d;
$G = 244$ s^{-1}.

Individual background measurements (s^{-1})	
175	169
178	182
170	185
160	182

a) What is the average background and its standard error?
b) Determine the best estimate of Q and its standard error.

8.19 Californium-252 decays both by alpha emission (96.91% of transitions) and by spontaneous fission (3.09%). A ^{252}Cf source has a total neutron emission rate of 7.63×10^9 s^{-1}, with a standard error of 1.5%. Answer the questions below, giving the value requested and its standard error.

a) What is the neutron fluence rate (cm^{-2} s^{-1}) from this source at a distance of 50.0 ± 0.3 cm?

b) The half-life of ^{252}Cf is 2.645 ± 0.212 y. What is the neutron fluence (cm^{-2}) in 30 d from the source under the conditions in part (a)? (Take into account the decrease in the activity of the source during this time.)

8.20 The specific activity A of radium in soil was determined by measuring the activity of a sample with a well counter and applying the relationship

$$A = \frac{G - B}{m \varepsilon t}.$$

Here,

G = number of gross counts with sample = 15 000;
B = number of background counts = 350 ± 15;
m = mass of soil = 150 ± 1 g;
ε = counter efficiency = 0.250 ± 0.005;
t = count time = 4 h (exactly).

Assume that fluctuations in G are due solely to the random nature of radioactive decay. Find
a) the specific activity of the sample in Bq g^{-1} and
b) its standard error.

8.21 Verify Eq. (8.52).

8.22 The fraction of monoenergetic, normally incident photons that traverse a uniform shield of thickness x without collision is given by $e^{-\mu x}$, where μ is the linear attenuation coefficient for the shield material (cf. Cember, 1996, pp. 134–138; Turner, 2007, pp. 187–190). In a certain experiment, $\mu = 0.90$ cm^{-1} and $x = 2.0$ cm.
  a) If the shield thickness is increased by 5%, differentiate and use the linear approximation with $dx/x = 0.05$ to determine how much the fraction of uncollided photons will change.
  b) Use the exact exponential function to answer (a).

9
Measuring Radioactivity

9.1
Introduction

Analyzing samples for radioactive content plays an important role in health physics. Procedures are needed to determine whether an unknown sample should be treated as "radioactive" and, if so, how much activity is present. In the simplest procedure, two measurements can be made and compared: (1) a background count with a blank in place of a sample and (2) a gross count of the sample plus background. The difference between the two observed count rates, called the net count rate, can be used to infer the possible presence of radioactivity in the sample itself. Because of statistical fluctuations in both the number of counts due to background and the possible number of disintegrations in a source, such procedures often do not yield a simple "yes" or "no" answer to the question of whether radioactive material is present. In principle, the level of activity, if any, can be described only in statistical terms. That is to say, measurements indicate only that the activity of a sample probably lies in a specified range with a certain degree of confidence.

In this chapter, we develop some concepts and descriptions that apply to the counting and characterization of samples for radioactivity. Formal protocols and criteria that provide decision tools for judging activity are given in the next chapter.

We first treat long-lived radionuclide sources, having negligible change in activity over the time of observation. Short-lived sources will be dealt with in Section 9.6. It is assumed that background radiation does not change and that it is characterized by a Poisson distribution of counts N_b in a fixed measurement time t_b. The number of gross counts N_g obtained in a given time t_g will also be treated as Poisson distributed. Both random variables are also assumed to be approximated well by normal distributions, having means μ_b and μ_g, respectively, and standard deviations $\sqrt{\mu_b}$ and $\sqrt{\mu_g}$. Time intervals are determined with negligible error compared with random fluctuations in the count numbers.

Statistical Methods in Radiation Physics, First Edition. James E. Turner, Darryl J. Downing, and James S. Bogard
© 2012 Wiley-VCH Verlag GmbH & Co. KGaA. Published 2012 by Wiley-VCH Verlag GmbH & Co. KGaA.

9.2
Normal Approximation to the Poisson Distribution

Section 6.5 described how the normal distribution can be used to approximate the exact binomial distribution for radioactive decay when n is large (Eq. (6.45)). That the normal distribution can also be used to approximate the Poisson for long counting times (provided the change in activity is negligible) can be seen in the following way. Let X represent the number of counts observed in time t. Since this random variable is Poisson distributed, we write $X \sim P(\lambda t)$, where the rate parameter λ is counts per unit time. With λ expressed in s^{-1}, for example, we can think of X as the sum of the counts Y_i in each of the $i = 1, 2, \ldots, t$ seconds of the observation. The observed count rate over time t can then be expressed as

$$\frac{X}{t} = \frac{1}{t}\sum_{i=1}^{t} Y_i, \tag{9.1}$$

where the right-hand side represents the *average* number of counts per second. We recall from the central limit theorem (Eq. (6.36)) that averages tend to have normal distributions. Thus,

$$X \sim P(\lambda t) \,\dot{\sim}\, N(\lambda t, \sqrt{\lambda t}) \tag{9.2}$$

tends asymptotically with large λt toward the normal distribution with mean λt and standard deviation $\sqrt{\lambda t}$. The normal approximation to the Poisson distribution can be generally used when $\lambda t \geq 10$.

9.3
Assessment of Sample Activity by Counting

It is important to distinguish between the *number* of counts and an associated *count rate*. To this end, we shall begin the subscript of a symbol for a random variable that represents a *rate* with the letter, lowercase "r". Thus, with N_g counts obtained in time t_g, the gross count rate is given by $R_{rg} = N_g/t_g$. Similarly, for the background rate measurement, $R_{rb} = N_b/t_b$.

Operationally, a "measurement" of activity in a sample begins with the observation of the numbers of gross and background counts, with and without the sample present, made under a specified set of standard conditions. The difference, or net value $R_{rn} = R_{rg} - R_{rb}$, of the resulting count rates is assumed to be proportional to the sample activity A. Thus,

$$A = \gamma R_{rn} = \gamma(R_{rg} - R_{rb}) = \gamma\left(\frac{N_g}{t_g} - \frac{N_b}{t_b}\right). \tag{9.3}$$

The constant of proportionality γ is determined numerically by calibration of the instrument. It can depend on sample preparation, self-absorption in the sample, and other factors. In the present discussion, γ in Eq. (9.3) can be considered as the reciprocal of the counter efficiency, $\varepsilon = R_{rn}/A$, which relates the net count rate (counts per unit time, cps) and the sample activity (disintegrations per unit time, dps). More generally, the calibration constant γ can be evaluated to yield other quantities, such as activity concentration, based on count rates or count numbers. Under the assumptions given in Section 9.1, the means of the gross and background count rates in Eq. (9.3) are, respectively, $\mu_{rg} = \mu_g/t_g$ and $\mu_{rb} = \mu_b/t_b$. The standard deviations of the two rates are

$$\sigma_{rg} = \frac{\sqrt{\mu_g}}{t_g} \quad \text{and} \quad \sigma_{rb} = \frac{\sqrt{\mu_b}}{t_b}. \tag{9.4}$$

The difference of the two functions R_{rg} and R_{rb} in Eq. (9.3) forms a distribution that describes the net count rate R_{rn} and the corresponding activity A when repeated measurements are taken. The means μ_A of the activity and μ_{rn} of the net count rate are given by

$$\mu_A = \gamma \mu_{rn} = \gamma(\mu_{rg} - \mu_{rb}). \tag{9.5}$$

By the central limit theorem (Section 6.4), A will be normally distributed about the true activity μ_A, which is the object of the measurements. If no activity is present, then the distribution is centered about the value $\mu_A = 0$.

In the terminology of Section 7.3, the two sample values n_g and n_b for N_g and N_b in times t_g and t_b can be used to estimate the sample activity. We write for the two count rates

$$\hat{\mu}_{rg} = \frac{n_g}{t_g} \quad \text{and} \quad \hat{\mu}_{rb} = \frac{n_b}{t_b}. \tag{9.6}$$

From Eq. (9.5) the estimate for the mean activity is

$$\hat{\mu}_A = \gamma \hat{\mu}_{rn} = \gamma(\hat{\mu}_{rg} - \hat{\mu}_{rb}) = \gamma \left(\frac{n_g}{t_g} - \frac{n_b}{t_b} \right). \tag{9.7}$$

9.4
Assessment of Uncertainty in Activity

The uncertainty associated with the determination of the activity can be expressed in terms of the standard deviation of the distribution for A in Eq. (9.3). The variance σ_{rn}^2 of the net rate R_{rn} in Eq. (9.3) is obtained (exactly) from Eqs. (8.12) and (8.13) for the linear combination (difference) of the two independent random variables. It follows that

$$\sigma_{rn}^2 = \sigma_{rg}^2 + \sigma_{rb}^2 = \frac{\mu_g}{t_g^2} + \frac{\mu_b}{t_b^2}. \tag{9.8}$$

Since $\mu_{rg} = \mu_g/t_g$ and $\mu_{rb} = \mu_b/t_b$, we arrive at the alternative forms for the standard deviation of the net count rate,

$$\sigma_{rn} = \sqrt{\frac{\mu_g}{t_g^2} + \frac{\mu_b}{t_b^2}} = \sqrt{\frac{\mu_{rg}}{t_g} + \frac{\mu_{rb}}{t_b}}. \tag{9.9}$$

As before, one can use the sample count numbers and rates as estimators for the indicated quantities. The estimates of the standard deviations are (Problem 9.7)

$$\hat{\sigma}_{rn} = \sqrt{\frac{n_g}{t_g^2} + \frac{n_b}{t_b^2}} = \sqrt{\frac{\hat{\mu}_{rg}}{t_g} + \frac{\hat{\mu}_{rb}}{t_b}} \qquad (9.10)$$

and

$$\hat{\sigma}_A = \gamma \hat{\sigma}_{rn} \qquad (9.11)$$

(see Problem 9.8). Either of the alternative forms in Eq. (9.10) can be used for the net rate standard deviation. One sees, incidentally, from the last equality in Eq. (9.10) that the standard deviation of an individual count rate decreases as the square root of the counting time.

■ *Example*

Measurement with a long-lived radioactive source gives 123 gross counts in 1 min.

a) Estimate the mean gross count rate and its standard deviation.
b) Based on this measurement, how long would the sample have to be counted in order to obtain the gross count rate to a precision of ±5% with 95% confidence?

Solution

a) With $n_g = 123$ in the time $t_g = 1$ min, the estimated mean gross count rate is, from Eq. (9.6), $\hat{\mu}_{rg} = 123/(1 \text{ min}) = 123 \text{ min}^{-1}$. From Eq. (9.4), the estimated standard deviation of the gross count rate is

$$\hat{\sigma}_{rg} = \frac{\sqrt{n_g}}{t_g} = \frac{\sqrt{123}}{1 \text{ min}} = 11.1 \text{ min}^{-1}. \qquad (9.12)$$

b) We use a confidence interval as described by (7.14) with $n = 1$ for the single measurement. For 95% confidence, $\alpha = 0.050$. The symmetric limits that leave this amount of area outside their boundaries in the tails of the standard normal distribution are $z_{0.025} = 1.96$ (Table 6.1 or A.3). With the use of the sample values, the required condition is $0.05\hat{\mu}_{rg} = 1.96\hat{\sigma}_{rg}$. The new counting time is found from the relation

$$0.05 \times 123 \text{ min}^{-1} = 1.96\sqrt{\frac{123 \text{ min}^{-1}}{t_g}}, \qquad (9.13)$$

giving $t_g = 12.5$ min for the estimated time needed. One can also solve this part of the example by first estimating the *number of counts* n_g that would be needed. For the count *number* one writes in place of Eq. (9.13)

$$0.05 n_g = 1.96\sqrt{n_g}, \qquad (9.14)$$

where $\sqrt{n_g}$ is the estimated standard deviation of the count number. Solution gives $n_g = 1.54 \times 10^3$, and so the counting time is

$$t_g = \frac{n_g}{\hat{\mu}_{rg}} = \frac{1.54 \times 10^3}{123 \text{ min}^{-1}} = 12.5 \text{ min.} \quad (9.15)$$

It is important to emphasize that the statistical precision of the count rate is governed solely by the *number of counts* that are observed, and not by the magnitude of the rate itself. A larger number of counts increases confidence in the result obtained for the count rate.

This example can also be treated in a probabilistic context, as we show next.

■ *Example*

In the last example, solve part (b) by specifying that there is a 0.95 probability that the estimated gross count rate will be within ±5% of the mean.

Solution

Using the estimated mean gross rate $\hat{\mu}_{rg} = 123 \text{ min}^{-1}$, we want to find the time t_g such that

$$\Pr\left(\left|\frac{N_g}{t_g} - \mu_{rg}\right| \leq 0.05\mu_{rg}\right) = 0.95. \quad (9.16)$$

Multiplying both sides of the inequality by $t_g/\sqrt{\mu_{rg}t_g}$ inside the parentheses, we may write

$$\Pr\left(\frac{|N_g - \mu_{rg}t_g|}{\sqrt{\mu_{rg}t_g}} \leq \frac{0.05\mu_{rg}t_g}{\sqrt{\mu_{rg}t_g}}\right) = 0.95. \quad (9.17)$$

The term on the left-hand side of the inequality in the parentheses has the form of the standard normal variable Z, since N_g has mean $\mu_{rg}t_g$ and standard deviation $\sqrt{\mu_{rg}t_g}$. Therefore, the term on the right-hand side of the inequality has the value $z_{0.025} = 1.96$, corresponding to a probability of 0.95, as reflected on the right-hand side of Eq. (9.17). With the estimate $\hat{\mu}_{rg} = 123 \text{ min}^{-1}$ from the last example, we obtain

$$\frac{0.05\hat{\mu}_{rg}t_g}{\sqrt{\hat{\mu}_{rg}t_g}} = 0.05\sqrt{123 \text{ min}^{-1}t_g} = 1.96, \quad (9.18)$$

giving $t_g = 12.5$ min, as before.

■ *Example*

At a certain facility, a 60-min background reading gives 4638 counts. Measurement with a sample yields 3217 counts in 15 min. The counter efficiency is 18%.

a) Estimate the mean net count rate and its standard deviation.
b) Estimate the activity of the sample and its standard deviation.
c) How might one express numerically this measurement of activity and its associated uncertainty?

Solution

a) Using the given data, we write for the estimated gross and background mean count rates

$$\hat{\mu}_{rg} = \frac{n_g}{t_g} = \frac{3217}{15 \text{ min}} = 214 \text{ cpm} \tag{9.19}$$

and

$$\hat{\mu}_{rb} = \frac{n_b}{t_b} = \frac{4638}{60 \text{ min}} = 77.3 \text{ cpm}. \tag{9.20}$$

The estimated mean net count rate is, therefore,

$$\hat{\mu}_{rn} = \hat{\mu}_{rg} - \hat{\mu}_{rb} = 214 - 77.3 = 137 \text{ cpm}. \tag{9.21}$$

Using the first equality in Eq. (9.10),[1] we find for the estimated standard deviation of the net count rate

$$\hat{\sigma}_{rn} = \sqrt{\frac{3217}{(15 \text{ min})^2} + \frac{4638}{(60 \text{ min})^2}} = 3.95 \text{ cpm}. \tag{9.22}$$

b) With $\gamma = 1/0.18$, we find for the estimated activity

$$\hat{\mu}_A = \gamma \hat{\mu}_{rn} = \frac{137 \text{ min}^{-1}}{0.18} = 761 \text{ dpm} = 12.7 \text{ dps} = 12.7 \text{ Bq}. \tag{9.23}$$

The estimated standard deviation is

$$\hat{\sigma}_A = \gamma \hat{\sigma}_{rn} = \frac{3.95 \text{ min}^{-1}}{0.18} = 21.9 \text{ dpm} = 0.366 \text{ Bq}. \tag{9.24}$$

c) There are various ways to express the result. Using the standard error (one standard deviation of the estimate) as a measure of the precision of the result, one could report the estimated activity as $\hat{\mu}_A = 12.7 \pm 0.4$ Bq, with an implied confidence level of 68% (Section 7.5). Alternatively, using the probable error, $0.675\hat{\sigma}_A = 0.675 \times 0.366 = 0.2$ Bq, one could report the activity as 12.7 ± 0.2 Bq, at a confidence level of 50%. At the 95% confidence level ($\pm 1.96\hat{\sigma}_A$), the activity could be reported as 12.7 ± 0.7 Bq. Specification of a higher confidence level is accompanied by a larger interval of uncertainty.

---

1) Use of the first, rather than second, equality in Eq. (9.10) is slightly preferable here to the extent that it depends directly on the information as given. The second equality involves calculated rates.

Example

a) With only the single background measurement in the last example, how long would the sample have to be counted in order to obtain the activity to within ±5% of its expected value with 95% confidence?
b) Repeat part (a) for a precision of ±3% with 95% confidence.
c) Without remeasuring background, what is the greatest precision that one could obtain for the activity itself at the 95% confidence level?

Solution

a) Our estimate of the mean net count rate from Eq. (9.21) is $\hat{\mu}_{rn} = 137$ min^{-1}. For the normal distribution, 95% confidence corresponds to the interval ±1.96 standard deviations centered about the mean. For this interval to span ±5% about the estimated mean requires that

$$0.05 \hat{\mu}_{rn} = 1.96 \hat{\sigma}_{rn}. \tag{9.25}$$

Employing the second equality from Eq. (9.10), letting t_g represent the new gross counting time in minutes, and inserting the other numerical values, we write in place of Eq. (9.25),

$$0.05 \times 137 = 1.96 \sqrt{\frac{214}{t_g} + \frac{77.3}{60}}. \tag{9.26}$$

The solution is $t_g = 19.6$ min.

b) The 95% confidence interval corresponds to ±1.96 standard deviations centered about the mean. In place of Eq. (9.26) we have, for a precision of ±3%,

$$0.03 \times 137 = 1.96 \sqrt{\frac{214}{t_g} + \frac{77.3}{60}}, \tag{9.27}$$

giving the time $t_g = 68.8$ min.

c) Given the single background measurement that was made, greater precision in the net count rate and activity can be obtained by increasing the gross counting time. The limiting precision is found by letting $t_g \to \infty$. If P represents the limiting precision in place of 0.03 on the left-hand side of Eq. (9.27), that is, $P \times 137 = 1.96\sqrt{(214/t_g) + (77.3/60)}$, then when $t_g \to \infty$, we have for the 95% confidence level (1.96 standard deviations),

$$\lim_{t_g \to \infty} P = \lim_{t_g \to \infty} \frac{1.96}{137} \sqrt{\frac{214}{t_g} + \frac{77.3}{60}} = \frac{1.96}{137} \sqrt{\frac{77.3}{60}} = 0.016. \tag{9.28}$$

Since our estimate of the activity from part (b) of the last example is $\hat{\mu}_A = 12.7$ Bq, the greatest precision obtainable for A at the 95% confidence level without additional background counting is $0.016 \times 12.7 = 0.203$ Bq. In essence, letting t_g become very large in Eq. (9.27) reduces the variance of the gross count rate until it is negligible compared with the variance for background. Apart from any uncertainty in the value of the counter efficiency, the precision in the measured activity is then limited only by the relative magnitude of the random fluctuations in the number of background counts during its measurement. We also see the law of diminishing returns, in that as we reduce P we increase t_g, and we see that going from a $P=0.05$ to a $P=0.03$ costs an additional 49.2 min of counting time. Decreasing P even further requires an even longer counting time. So the question is how much precision is necessary? One can look at the interplay between P and t_g and pick the pair that gives us the best precision for the cost (in time) that we can afford.

9.5
Optimum Partitioning of Counting Times

One can see from Eq. (9.9) how the standard deviation of the net count rate changes when the gross and background counting times are changed. It is sometimes important to know just how to partition a fixed total counting time, $t_t = t_g + t_b$, between t_g and t_b in such a way that the standard deviation of the net rate has its smallest possible value. To this end, we can express the second equality in Eq. (9.9) as a function of either time variable alone, t_g or t_b, and then minimize σ_{nr} by differentiation. It is simpler, moreover, to apply this procedure to the variance, rather than the standard deviation, thus avoiding differentiation of the square root function. Substituting $t_b = t_t - t_g$ in Eq. (9.9), we write for minimization of the variance

$$\frac{d\sigma_{rn}^2}{dt_g} = \frac{d}{dt_g}\left(\frac{\mu_{rg}}{t_g} + \frac{\mu_{rb}}{t_t - t_g}\right) = 0, \tag{9.29}$$

or

$$-\frac{\mu_{rg}}{t_g^2} + \frac{\mu_{rb}}{(t_t - t_g)^2} = 0. \tag{9.30}$$

Replacing $t_t - t_g$ by t_b then gives

$$\frac{\mu_{rg}}{t_g^2} = \frac{\mu_{rb}}{t_b^2}, \tag{9.31}$$

so that

$$\frac{t_g}{t_b} = \sqrt{\frac{\mu_{rg}}{\mu_{rb}}}. \tag{9.32}$$

When the gross and background counting times are in the same ratio as the square roots of the two respective rates, then the standard deviation of the net count rate has its smallest possible value for the total counting time.

■ **Example**
In the next-to-last example, a total counting time of 75 min was partitioned as $t_b = 60$ min and $t_g = 15$ min. The standard deviation of the net count rate was estimated to be 3.95 cpm.

a) What values of t_b and t_g, totaling 75 min, would minimize the standard deviation of the net count rate?
b) What is the estimated minimum value?

Solution
a) From Eq. (9.32) with the estimates $\hat{\mu}_{rg}$ and $\hat{\mu}_{rb}$ given by Eqs. (9.19) and (9.20), one finds

$$t_g = t_b \sqrt{\frac{214}{77.3}} = 1.66 t_b. \tag{9.33}$$

Since $t_g + t_b = 75$ min, we have

$$t_g = 1.66(75 - t_g), \tag{9.34}$$

giving $t_g = 46.8$ min. It follows that $t_b = 28.2$ min.
b) Substitution of these two times into the second equality in Eq. (9.10) gives

$$\min \hat{\sigma}_{rn} = \sqrt{\frac{214 \text{ min}^{-1}}{46.8 \text{ min}} + \frac{77.3 \text{ min}^{-1}}{28.2 \text{ min}}} = 2.70 \text{ cpm}, \tag{9.35}$$

as compared with 3.95 cpm found before (Eq. (9.22)) for the same total counting time.

One can use Eqs. (9.32) and (9.9) to show that the minimum standard deviation of the net count rate is, in general, given by (Problem 9.22)

$$\min \sigma_{rn} = \sqrt{\frac{\mu_{rg} + \mu_{rb} + 2\sqrt{\mu_{rg}\mu_{rb}}}{t_g + t_b}}. \tag{9.36}$$

9.6
Short-Lived Radionuclides

The discussions thus far in this chapter have been based on the use of Poisson and normal statistics to describe radioactive decay. The specific underlying mathematical assumptions are stated in Section 9.1. When the time t for counting a pure radionuclide source with decay constant λ is very short compared with the mean life ($=1/\lambda$), then $\lambda t \ll 1$. The probability p for a given atom to decay in time t is then very small, and the distribution of the number of disintegrations during t is nearly Poisson. When the counting time is not short compared with the mean life, the condition $\lambda t \ll 1$ does not hold. One must then revert to the binomial distribution, which describes radioactive decay exactly (Chapter 2). We show next how the binomial distribution is compatible with the formalism employed thus far in this chapter when $\lambda t \ll 1$. We then examine the consequences when λt is not small.

The expected number of atoms that decay in time t from a pure radionuclide source, containing n atoms at time $t = 0$, is seen from Eq. (2.22) to be, exactly,

$$\mu = np = n(1 - e^{-\lambda t}). \tag{9.37}$$

When $\lambda t \ll 1$, the exponential term is approximated well by a Taylor's series with only the lowest-order term retained: $e^{-\lambda t} \cong 1 - \lambda t$. Equation (9.37) then gives

$$\mu \cong n\lambda t. \tag{9.38}$$

The mean, or expected value, of the disintegration rate, or activity A, is

$$\mu_A = \frac{\mu}{t} \cong n\lambda, \tag{9.39}$$

as we have seen before (e.g., Eq. (2.3)). The standard deviation for the binomial distribution is (Eq. (5.15))

$$\sigma = \sqrt{np(1-p)} = \sqrt{\mu e^{-\lambda t}}. \tag{9.40}$$

For very small λt, the exponential term is close to unity, and so

$$\sigma \cong \sqrt{\mu}. \tag{9.41}$$

This important approximation from the binomial distribution when $\lambda t \ll 1$ is exactly true for the Poisson distribution. It enables a single measurement to provide estimates of both the expected value of the count number and its standard deviation, as utilized in the earlier sections of this chapter.

When λt is large, Eqs. (9.39) and (9.41), which do not depend on the time, are no longer valid approximations for radioactive decay. On the other hand, Eqs. (9.37) and (9.40) for the binomial distribution describe the decay exactly. They apply over any time period.

■ **Example**

a) Show that the expected value of the number of disintegrations of a pure radionuclide in a time period equal to the half-life, $t = T_{1/2} = (\ln 2)/\lambda$, is $n/2$, where n is the number of atoms initially present.
b) What is the standard deviation of this number?
c) How does this standard deviation compare with the Poisson value? Which should be larger? Why?

Solution
a) The expected value of the number of disintegrations is given exactly by Eq. (9.37) for all times t. For the specified time $t = (\ln 2)/\lambda$, we have

$e^{-\lambda t} = e^{-(\ln 2)} = 1/e^{\ln 2} = 1/2$. Equation (9.37) gives

$$\mu = n\left(1 - \frac{1}{2}\right) = \frac{1}{2}n. \tag{9.42}$$

The expected number of disintegrations in one half-life is, of course, one-half the original number n of atoms.

b) For the standard deviation, Eq. (9.40) gives

$$\sigma = \sqrt{\frac{1}{2}n \times \frac{1}{2}} = \frac{1}{2}\sqrt{n}. \tag{9.43}$$

c) The Poisson distribution with parameter μ would give for the standard deviation $\sigma_P = \sqrt{\mu} = \sqrt{n/2} = \sigma\sqrt{2}$. Thus, σ_P is larger than the binomial value σ by the factor $\sqrt{2} = 1.414$. The larger standard deviation for the Poisson approximation can be understood in the following way. The number of disintegrations represented by the binomial function spans the finite closed interval $[0, n]$. The Poisson distribution, on the other hand, has a broader spread, spanning the infinite interval $[0, \infty]$.

An interesting situation occurs when the observation time is so long that the original nuclide decays completely away. We assume that the background is zero. When $\lambda t \to \infty$, the exact Eqs. (9.37) and (9.40) give, respectively, $\mu = n$ and $\sigma = 0$. In this case, the expected number of disintegrations is thus equal to the original number of atoms, and the standard deviation of this number is zero. If the experiment is repeated, the result will always be the same: exactly n atoms decay.

If the counter efficiency ε is less than unity, then the expected number of counts registered from a source that decays completely away in time t will generally be less than the number of disintegrations. Repetition of the experiment will result in a distribution of values for the count number. Not registering every atom that disintegrates introduces uncertainty in the knowledge of the number initially present. A given atom might not decay, or it might decay and escape detection when $\varepsilon < 1$. Since the decay probability for a given atom in time t is $p = 1 - e^{-\lambda t}$, the probability that a given atom decays and is detected, thus registering a count, is

$$p^* = \varepsilon(1 - e^{-\lambda t}). \tag{9.44}$$

The probability for not registering a count when a given atom decays is, then,

$$q^* = 1 - p^* = 1 - \varepsilon + \varepsilon e^{-\lambda t}. \tag{9.45}$$

The expected number of counts during any time t is thus

$$\mu^* = np^* = \varepsilon n(1 - e^{-\lambda t}). \tag{9.46}$$

For the standard deviation of the count number, one has

$$\sigma^* = \sqrt{np^* q^*} = \sqrt{\mu^* q^*} = \sqrt{n\varepsilon(1 - e^{-\lambda t})(1 - \varepsilon + \varepsilon e^{-\lambda t})}. \tag{9.47}$$

Both Eqs. (9.46) and (9.47) are exact. If a sample decays completely away ($\lambda t \to \infty$), then the expected number of counts (9.46) is

$$\mu^* = \varepsilon n \qquad (9.48)$$

and the standard deviation (9.47) is

$$\sigma^* = \sqrt{n\varepsilon(1-\varepsilon)} = \sqrt{\mu^*(1-\varepsilon)}. \qquad (9.49)$$

The standard deviation of the number of atoms initially present, from Eqs. (9.48) and (4.101), is σ^*/ε.

■ *Example*
A short-lived radionuclide source gives 3 212 675 counts before dying away completely. Background is zero. Determine the number of atoms initially present in the sample and the standard deviation of this number if the counter efficiency ε is (a) 1.00 or (b) 0.24.

Solution
a) If $\varepsilon = 1.00$, then there were exactly $\mu = 3\,212\,675$ atoms present initially, and the standard deviation is $\sigma = 0$.
b) From the single measurement, the best estimate $\hat{\mu}^*$ for μ^* in Eq. (9.48) is the observed number of counts. The estimated number of atoms originally present is

$$\frac{\hat{\mu}^*}{\varepsilon} = \frac{3\,212\,675}{0.24} = 13\,386\,146. \qquad (9.50)$$

From Eq. (9.49), the estimated standard deviation of the number of counts is

$$\hat{\sigma}^* = \sqrt{\hat{\mu}^*(1-\varepsilon)} = \sqrt{3\,212\,675(1-0.24)} = 1563. \qquad (9.51)$$

The estimated standard deviation of the number of atoms initially present is $\hat{\sigma}^*/\varepsilon = 1563/0.24 = 6513$.

Problems

9.1 A source gives 385 gross counts in 5 min.
   a) Estimate the standard deviation of the gross count rate.
   b) Estimate how much longer the sample would have to be counted in order to reduce the standard deviation of the gross count rate by a factor of 10.
9.2 How long would the source in the last problem have to be counted to obtain the gross count rate to a precision of $\pm 2\%$ with 90% confidence?
9.3 How many counts are needed to obtain a coefficient of variation of 1%?

9.4 For a certain radioisotope, having a half-life of 4.12 d, the expected number of disintegrations in 1 wk is 58.8.
   a) How many atoms of the radioisotope are initially present?
   b) What is the standard deviation of the number of disintegrations in 1 wk?
   c) What is the standard deviation for 2 d?

9.5 a) In the last problem, at what time does the maximum value of the standard deviation of the number of disintegrations occur?
   b) What is the maximum value of the standard deviation?
   c) Make a sketch of the standard deviation of the number of disintegrations as a function of the time.

9.6 The number of decays in any time period from a radioactive source is an integral number. Justify the fact that the expected number of 58.8 disintegrations in 1 wk in the last two problems is not an integer.

9.7 Verify Eqs. 9.8–9.11.

9.8 While perhaps "obvious," show that $\sigma_A = \gamma \sigma_{rn}$ and hence Eq. (9.11) follow from the relation $A = \gamma R_{rn}$.
   (*Hint*: Use propagation of error Eq. (8.17) for a product.)

9.9 Measurements of 10 min each give gross and background count rates, respectively, of 72 and 54 cpm, for a net count rate of 18 cpm. What are the standard deviations of the following count rates:
   a) gross?
   b) background?
   c) net?

9.10 The true count rate of a long-lived source is $12.0\,\mathrm{s}^{-1}$.
   a) What is the standard deviation of the number of counts for a 1-min measurement?
   b) The sample is counted for 3 min. What is the probability that the measured count rate will be within $\pm 5\%$ of the true count rate?

9.11 A long-lived source is reported to have a count rate of 127 cpm.
   a) Technician A obtains 140 counts in 1 min, implying a count rate of 140 cpm. What is the probability that such a measurement would differ from the expected value by no more than ± 13 cpm, if the true rate is 127 cpm? Does A's measurement tend to confirm the reported count rate?
   b) Technician B makes a 30-min measurement and obtains 4180 counts, giving him a measured rate of 139 cpm, close to that of A. Does B's measurement tend to confirm the reported count rate? Give a quantitative argument to support your conclusion.
   c) Give your estimate of the true count rate and its uncertainty with some specified confidence level.

9.12 A source registers 1050 counts in 10 min. A 1-h measurement with a blank yields 4800 background counts.
   a) What are the net count rate and its standard deviation?
   b) If the counter efficiency is 40%, estimate the activity of the sample and its standard deviation?

9.13 If the efficiency (40%) of the counter in the last problem is known with a precision of only ±1%, what is the precision of the standard deviation in the determination of the sample activity?

9.14 In Problem 9.12, estimate the probability that a second measurement would give 1100 or more gross counts in 10 min.

9.15 Without remeasuring the background in Problem 9.12, how long would one have to count the source in order to determine the activity to within ±10% with a confidence of 90%?

9.16 Of two available counters, the one with the better precision for determining net count rate is to be chosen for making a long-term measurement with a certain source. In 15-min runs with the source present, a NaI detector gives 3277 counts and an HPGe detector gives 1213 counts. In 30-min measurements of background, the NaI and HPGe instruments give, respectively, 952 and 89 counts. Calculate the coefficient of variation for the net count rate for the
a) NaI detector.
b) HPGe detector.
c) Based on these data, which counter would you choose?

9.17 With the HPGe detector in the last problem, 1213 gross counts were obtained in 15 min and 89 background counts in 30 min. New measurements are to be made with this detector, keeping the relative background and gross counting times in the ratio 2 : 1, as before.
a) How much total counting time is needed to determine the net count rate to within ±5% with 95% confidence?
b) What is the standard deviation for the new measurement of the net count rate?

9.18 a) What would be the optimum division of the total counting time in the last problem in order to minimize the standard deviation for the new measurement of the net count rate?
b) What is the value of the minimum standard deviation?

9.19 The gross count rate with a sample is observed to be 73 cpm in a counter that has a background rate of 58 cpm. What is the optimum division of a total counting time of 1 h for gross and background counts in order to obtain the minimum standard deviation of the net rate?

9.20 A background measurement yields 4442 counts in 1 h with a certain counter. A long-lived sample is then placed in the counter, and 1888 counts are registered in 10 min.
a) What is the net count rate and its standard deviation?
b) What is the minimum value of the standard deviation of the net count rate obtainable for the total counting time of 1 h + 10 min?
c) Without redoing the background measurement, how long would the sample have to be counted to obtain the net count rate to within ±3% of its true value with 95% confidence?

9.21 A long-lived radionuclide gives 714 counts in 5 min. A 10-min background reading yields 1270 counts. The counter has an efficiency of 45%.

a) What is the standard deviation of the source activity in Bq?

b) How should a total counting time of 2 h be divided between gross and background counting in order to minimize the standard deviation of the measured activity?

9.22 Derive Eq. (9.36).

9.23 a) What is the expected value of the fraction of the atoms that decay in a pure radionuclide source in 2.7 half-lives?

b) What is the standard deviation of this fraction?

9.24 A short-lived radioactive source produces 121 497 counts before decaying away completely. The counter efficiency is 18%.

a) Estimate the initial number of atoms of the radionuclide in the source.

b) Calculate the standard deviation of the initial number.

10
Statistical Performance Measures

10.1
Statistical Decisions

Chapter 7 treated random sampling, the estimation of population parameters, estimator properties, and associated confidence intervals. We focus now on a different, but related, aspect of *statistical inference*. In practice, one is frequently called upon to use sample data to reach a specific conclusion about a *population characteristic or parameter*. Is the uncertainty in results from two different gas proportional counters the same? Will no more than 1% of dosimeters fabricated by a particular process fail acceptance testing? Is the uncertainty in results from a liquid scintillation counter adequately described by the statistics of radioactive decay alone? Such questions can be approached by first forming an answer as a hypothesis, or conjecture, and then using data from random sampling as evidence to either support or not support the hypothesis.

10.2
Screening Samples for Radioactivity

A procedure to routinely screen samples for radioactivity affords an example of applying statistical inference. The number of net counts from a sample, determined under standard conditions as the difference between observed gross and background counts, is compared with a preset *critical value,* L_C. If the observed net value is equal to or greater than L_C, then the decision is made that radioactivity is present in the specimen. Otherwise, the sample is considered as having "no significant activity." Calibration provides the relationship between L_C and the corresponding activity A_I, which is called the *minimum significant measured activity*. Operationally, A_I is the smallest measured value that is interpreted as meaning that activity is present in a sample.

We shall assume that all measured counts are Poisson distributed. Unless otherwise stated, we also assume that count numbers are sufficiently large to be

Statistical Methods in Radiation Physics, First Edition. James E. Turner, Darryl J. Downing, and James S. Bogard.
© 2012 Wiley-VCH Verlag GmbH & Co. KGaA. Published 2012 by Wiley-VCH Verlag GmbH & Co. KGaA.

adequately represented by normal distributions with mean and variance equal to the expected number of counts.

▪ **Example**

A protocol is being assessed to screen for sample activity. It specifies taking a gross count, N_g, for 2 min and a background count, N_b, for 10 min. The detector efficiency has been determined to be $\gamma = 1.74$ disintegrations per count. A selected sample registers $n_g = 64$ gross counts in time $t_g = 2$ min, compared with $n_b = 288$ background counts in time $t_b = 10$ min.

a) Estimate the expected value of the net count rate and its standard deviation.
b) What is the implied sample activity?
c) What is the probability that a sample with zero true activity ($A_T = 0$) would give an activity measurement exceeding that found in (b)? Is the measurement in (b) consistent with $A_T = 0$ for this sample?
d) Assume that $A_T = 0$. What is the smallest net rate value, r_{rnc}, over a 2-min period that would be exceeded with a probability no greater than $\alpha = 0.05$? (As in Chapter 9 and elsewhere, the leading subscript "r" is used when a quantity describes a rate.)
e) Using this value of r_{rnc} as the critical value, what would be the minimum significant measured activity for the protocol?

Solution

a) From the relations (9.7), the estimate of the mean net count rate based on the sample values is, with time in minutes,

$$r_{rn} = \frac{n_g}{t_g} - \frac{n_b}{t_b} = \frac{64}{2} - \frac{288}{10} = 3.20 \text{ cpm}. \tag{10.1}$$

The estimated standard deviation of the net count rate is, from Eq. (9.10),

$$S_{rn} \sqrt{\frac{n_g}{t_g^2} + \frac{n_b}{t_b^2}} = \sqrt{\frac{64}{4} + \frac{288}{100}} = 5.91 \text{ cpm}. \tag{10.2}$$

b) Multiplying the net count rate by the counter efficiency implies a measured activity of $A = r_{rn}\gamma = (1.74)(3.20 \text{ min}^{-1})/(60 \text{ s min}^{-1}) = 0.0928 \text{ s}^{-1}$ = 0.0928 Bq.
c) With assumed zero activity, the variable $Z = R_{rn}/\sigma_{rn}$ is approximately distributed as the standard normal distribution. By what amount is the observed value of $r_{rn} = 3.20$ cpm greater than zero, our assumed value? To check this, we can calculate the probability that we would observe $R_{rn} \geq 3.20$ given that $\mu_{rn} = 0$. Using the normal approximation and standardizing, we have $\Pr(R_{rn}/\sigma_{rn} \geq 3.20/\sigma_{rn}) = \Pr(Z \geq 3.20/5.91) = \Pr(Z \geq 0.542) = 0.294$ (Table A.3). Therefore, the probability is 0.294 that a sample with zero activity would show a count rate greater than or equal to the observed 3.20 cpm in (a). This result is consistent with the true sample

activity being zero, in which case the observed net rate $r_{rn} = 3.20$ is due solely to random fluctuations of the background.

d) With $A_T = 0$, the value $z_\alpha = z_{0.05} = 1.645$ leaves the area 0.05 to its right under the standard normal curve approximating the standardized net rate distribution. That is, $\Pr(R > r_{rnc}) = \Pr(Z > r_{rnc}/\sigma_{rn}) = 0.05$ implies that $r_{rnc}/\sigma_{rn} = 1.645$, or $r_{rnc} = 1.645\sigma_{rn}$. Now r_{rnc} and σ_{rn} are not independent since $r_{rnc} = r_{rg} - r_{rb}$ and σ_{rn} is a function of r_{rg} and r_{rb}. Since $r_{rnc} = r_{rg} - r_{rb}$, we can substitute $r_{rg} = r_{rnc} + r_{rb}$ in Eq. (9.10) for s_{rn}. Therefore,

$$r_{rnc} = z_\alpha \sqrt{\frac{r_{rnc} + r_{rb}}{t_g} + \frac{r_{rb}}{t_b}} = 1.645\sqrt{\frac{r_{rnc} + 28.8}{2} + \frac{28.8}{10}}. \quad (10.3)$$

The resulting quadratic equation, $r_{rnc}^2 - 1.36 r_{rnc} - 47.0 = 0$, has the solution $r_{rnc} = 7.57$ cpm (Problem 10.1). The required number of counts in 2 min would be $7.57 \times 2 = 15.1$. However, the actual count number has to be an integer – in this case, either 15 or 16. With 15, the probability would exceed 0.05, and so the answer is 16 net counts.

e) The minimum significant measured activity corresponding to the critical value $L_C = 16$ net counts in 2 min is $A_I = L_C\gamma = (16)(1.74\,\text{Bq})/(2\,\text{min})$ $(60\,\text{s min}^{-1}) = 0.232\,\text{s}^{-1} = 0.232\,\text{Bq}$.

10.3
Minimum Significant Measured Activity

This section discusses the significance and implications of using the minimum significant measured activity as a decision tool. As seen from Eq. (10.3), the net rate r_{rnc} depends on the background rate, the gross and background counting times, and the value of z_α. The latter, which is selected to fix L_C at the desired level, determines the probability that the measured activity of a sample will be larger than A_I when the true mean activity $A_T = 0$ (see Figure 10.1). When this happens, the wrong conclusion that the true mean activity $A_T > 0$ is unknowingly reached. The resulting false positive is referred to as a *type I error*. The value of α represents an "acceptable" type I error rate that is chosen as one element of a screening protocol. This error means that we are declaring a sample as having activity when it does not. This might trigger compensatory responses unnecessarily, costing money and time. Thus, type I errors can be expensive to a laboratory in terms of false alarms and consequent actions that have to be taken. Errors will be discussed more fully as we proceed.

One can work out the explicit dependence of A_I on the various factors mentioned in order to gain further insight into the decision level. Starting with the first equality in Eq. (10.3) and solving for r_{rnc}, one finds (Problem 10.2)

$$r_{rnc} = \frac{z_\alpha^2}{2t_g} + \frac{z_\alpha}{2}\sqrt{\frac{z_\alpha^2}{t_g^2} + 4r_{rb}\left(\frac{t_g + t_b}{t_g t_b}\right)}. \quad (10.4)$$

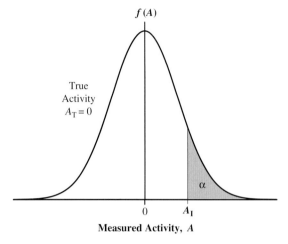

Figure 10.1 Probability density function $f(A)$ for measured activity when true activity is zero. The probability of a type I error (false positive) does not exceed the area α.

When the gross and background counting times are the same, this general expression reduces to a simpler form. With $t_g = t_b = t$, the critical net count becomes (Problem 10.4)

$$L_C = r_{rnc} t = z_\alpha \sqrt{2n_b} \left(\frac{z_\alpha}{\sqrt{8n_b}} + \sqrt{1 + \frac{z_\alpha^2}{8n_b}} \right) \quad \text{(equal counting times)}. \quad (10.5)$$

The detector efficiency γ relates L_C to the minimum significant measured activity,

$$A_I = \frac{\gamma L_C}{t}. \quad (10.6)$$

If, in addition, the number of background counts is large, so that $z_\alpha/\sqrt{n_b} \ll 1$, Eq. (10.5) reduces to

$$L_C \cong z_\alpha \sqrt{2n_b} \quad \text{(equal times and } z_\alpha/\sqrt{n_b} \ll 1\text{)}. \quad (10.7)$$

In many instances, the background is stable. The number B of background counts over an extended time can then be determined with greater precision than that obtained from a single measurement of N_g over the prescribed time t_g. With zero activity ($A = 0$), the standard deviation of the number of net counts is simply $\sqrt{B}$. It follows that the minimum significant net count difference is

$$L_C \cong z_\alpha \sqrt{B} \quad \text{(background stable and well known)}. \quad (10.8)$$

Comparison with Eq. (10.7) shows that the minimum significant measured activity is lower by approximately a factor of $\sqrt{2}$ when the background is stable and well known.

In addition to false positive (type I) errors that can occur when using L_C as a screening tool, false negative (type II) errors are also possible. This happens when the measured net count is less than L_C, but the true value is $A_T > 0$. A type II error poses potential risks of a different nature from those associated with type I. Falsely declaring zero activity may imply an unrecognized hazard, possibly leading to radiation exposure and the resulting health issues. We next develop a decision tool that specifically applies when $A_T > 0$ and thus addresses type II errors.

10.4
Minimum Detectable True Activity

Figure 10.2 shows the probability density function for the measured sample activity when the true activity has a value $A_T = A_{II} > 0$. Also shown for comparison to its left is the density function from Figure 10.1 for $A_T = 0$ with the associated minimum significant measured activity A_I (Eq. (10.6)). With $A_T = A_{II} > 0$, it is more likely than before that the measured sample activity will be greater than A_I. Applying the decision tool $A > A_I$ or equivalently $N_n > L_C$, for the net count, then correctly gives the decision that the sample has activity (recall that $A = N\gamma/t$). However, when $A_T > 0$ there is also a possibility that the measured result could be $N_n < L_C$. When this happens, the false negative decision is reached, indicating that the sample has no significant activity when, in fact, it does. The probability that this type II error will be committed when the true activity is A_{II} is equal to the area β under the distribution centered at A_{II} left of A_I in Figure 10.2. We write this probability symbolically as $\Pr(A < A_I | \mu_A = A_{II}) = \beta$.

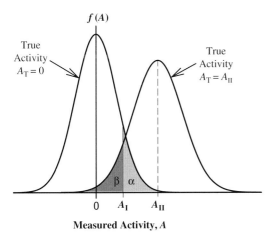

Figure 10.2 Probability density function for true activity $A_T = A_{II} > 0$ and density function from Figure 10.1 for $A_T = 0$. With α chosen, setting a protocol value for β fixes A_{II}, which is then called the minimum detectable true activity.

Like α, the value of β can be selected to set an "acceptable" risk, related now to type II errors, as part of the screening protocol. Since the value of A_I is fixed once α is set, the value of β will be changed only by moving A_II to the left or right. Therefore, a particular numerical choice of β determines a particular value for the activity A_II. With β specified, this value of A_II is called the *minimum detectable true activity*. Any smaller mean activity A_T, where $0 < A_\text{T} < A_\text{II}$, gives a type II error rate (false negative) with probability greater than β. A_II is the smallest mean activity that will be declared undetected ($N_\text{n} < L_\text{C}$ or $A < A_\text{I}$) with a probability no larger than β.

We next find the mean value R_rnd of the net count rate when the true value of the sample activity is A_II. With μ_A denoting the mean activity level, the mean rate μ_rnd is related to the mean activity by the equation $\mu_\text{A} = \mu_\text{rnd}\gamma$, where γ is the detector efficiency. Because γ is a fixed constant, we can use the count rate in place of the activity to determine if any activity is present. Since the count rate is easier to calculate, we employ it rather than activity. Using Figure 10.3, we see that r_rnd is the count rate corresponding to A_II, and what we want to determine is the value of r_rnd for which $\Pr(R_\text{rn} < r_\text{rnc} | \mu_\text{rnd} = r_\text{rnd}) = \beta$. Once we determine r_rnd, we can easily determine the corresponding activity. Also, note that r_rnc has been previously determined as the critical value for which we declare falsely with probability α that there is activity when the true mean activity is zero. It is important to note that r_rnc has been previously computed via Eq. (10.4) and it is this value that we are using in our calculations. As long as the number of counts is sufficiently large, then R_rn will be approximately normally distributed with mean $\mu_\text{rnd} = r_\text{rnd}$ and standard deviation σ_m given by Eq. (9.9). We want to determine the value of r_rnd when $\Pr(R_\text{rn} < r_\text{rnc} | \mu_\text{rnd} = r_\text{rnd}) = \beta$. Standardizing the distribution of R_rn, we find that $\Pr[R_\text{rn} < r_\text{rnc} | \mu_\text{rnd} = r_\text{rnd}] = \Pr[(R_\text{rn} - \mu_\text{rnd})/\sigma_\text{m} < (r_\text{rnc} - r_\text{rnd})/\sigma_\text{m}]$. Continuing, this probability is

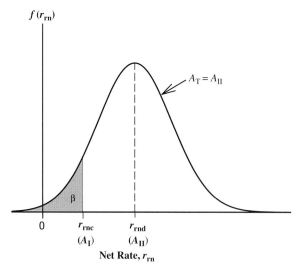

Figure 10.3 Probability density function $f(r_\text{rn})$ for net count rate with true activity $A_\text{T} = A_\text{II}$. See Eq. (10.10).

$\Pr[Z < (r_{rnc} - r_{rnd})/\sigma_{rnd}] = \beta$, where Z is the standard normal random variable. The value that cuts off the fraction β of the area to the left of the standard normal distribution is denoted by $-z_\beta$. Hence, we have $(r_{rnc} - r_{rnd})/\sigma_{rn} = -z_\beta$, which we can solve for r_{rnd}. The result is

$$r_{rnd} = r_{rnc} + z_\beta \sigma_{rn}. \tag{10.9}$$

Recall that $r_{rnc} = r_{rg} - r_{rb}$ has been determined already by Eq. (10.4). We can use this result to express the value of r_{rnd} in terms of r_{rnc}, r_{rb}, z_β, t_g, and t_b. Equation (10.9) can be rewritten as

$$r_{rnd} = r_{rnc} + z_\beta \sqrt{\frac{r_{rnc}}{t_g} + r_{rb}\left(\frac{t_g + t_b}{t_g t_b}\right)}. \tag{10.10}$$

When the counting times are equal, $t = t_g = t_b$, one finds for the minimum detectable true net count number with activity A_{II},

$$L_D = r_{rnd} t = L_C + z_\beta \sqrt{2 n_b} \left(1 + \frac{z_\alpha^2}{4 n_b} + \frac{z_\alpha}{\sqrt{2 n_b}} \sqrt{1 + \frac{z_\alpha^2}{8 n_b}}\right)^{1/2}, \tag{10.11}$$

where L_C is given by Eq. (10.5) (Problem 10.8). Finally, use of the detector efficiency γ gives the minimum detectable true activity,

$$A_{II} = \frac{\gamma L_D}{t}. \tag{10.12}$$

This and other related quantities are discussed in the literature under various headings, such as the *lower limit of detection* (LLD) or the *minimum detectable amount* (MDA) (Currie, 1968; HPS, 1996).

As before with L_C, often $z_\alpha/\sqrt{n_b} \ll 1$. Then Eq. (10.11) reduces to the simple result

$$L_D \cong L_C + z_\beta \sqrt{2 n_b} = (z_\alpha + z_\beta)\sqrt{2 n_b}, \tag{10.13}$$

in which the approximation (10.7) has been substituted for L_C. When $\alpha = \beta$, it follows that $L_D \cong 2 L_C$.

When the background is stable and accurately known, Eq. (10.11) can be written (Problem 10.9) as

$$L_D = \sqrt{n_b}\left(z_\alpha + \frac{z_\beta^2}{2\sqrt{n_b}} + z_\beta \sqrt{1 + \frac{z_\alpha}{\sqrt{n_b}} + \frac{z_\beta^2}{4 n_b}}\right). \tag{10.14}$$

In addition, Eq. (10.13) becomes

$$L_D \cong (z_\alpha + z_\beta)\sqrt{n_b}. \tag{10.15}$$

As with the minimum significant measured activity, a well-characterized and stable background lowers the minimum detectable true activity by approximately $\sqrt{2}$. The relationship $L_D \cong 2L_C$ also holds when $\alpha = \beta$ with a well-known background.

■ **Example**

Measurement of a sample and background for the same length of time yields $n_g = 333$ gross and $n_b = 296$ background counts. The calibration constant is $\gamma = 4.08$ Bq per count. The maximum risks for both type I and type II errors are set at $\alpha = \beta = 0.05$. Determine

a) the critical count number;
b) whether the sample has activity;
c) the measured sample activity;
d) the minimum significant measured activity;
e) the minimum detectable true activity.

Solution

a) With $k_\alpha = 1.645$ in Eq. (10.5), the critical count number is

$$L_C = 1.645\sqrt{2 \times 296}\left(\frac{1.645}{\sqrt{8 \times 296}} + \sqrt{1 + \frac{1.645^2}{8 \times 296}}\right)$$
$$= 41.4. \qquad (10.16)$$

We round to the next larger integer, using $L_C = 42$ as the smallest net count number for declaring activity to be present, with a probability not exceeding 0.05 for a false positive.

b) The net count number is

$$n_n = n_g - n_b = 333 - 296 = 37 < L_C. \qquad (10.17)$$

Therefore, the sample is judged as having no significant activity.

c) The measured activity is $A = \gamma n_n = 4.08$ Bq $\times 37 = 151$ Bq.

d) From Eqs. (10.6) and (10.16), the minimum significant measured activity is

$$A_I = \gamma L_C = 4.08 \text{ Bq} \times 41.4 = 169 \text{ Bq}. \qquad (10.18)$$

e) The expected net count when the true activity is A_{II} is given here by combining Eqs. (10.11) and (10.16):

$$L_D = 41.4 + 1.645\sqrt{2 \times 296}\left(1 + \frac{1.645^2}{4 \times 296} + \frac{1.645}{\sqrt{2 \times 296}}\sqrt{1 + \frac{1.645^2}{8 \times 296}}\right)^{1/2}$$
$$= 82.8.$$
$$(10.19)$$

Therefore, the minimum detectable true activity is $A_{II} = \gamma L_D = 4.08$ Bq $\times 82.8 = 338$ Bq.

10.4 Minimum Detectable True Activity

■ **Example**

Pertaining to the last example, a technician subsequently finds that the background radiation level at the facility has been extensively documented. A large number of separate measurements, based on more than 50 000 total counts, show the background to be stable with an estimated mean number of 290 counts for the length of time during which a sample is counted. Repeat the previous analysis, using the improved background data.

Solution

a) With $n_b = 290$, Eq. (10.8) yields $L_C = 1.645\sqrt{290} = 28.01$. The critical net count number, accordingly, is $L_C = 29$. (Note that the single measurement $n_b = 296$ in the last example is consistent with a stable background.)

b) The measured net count is now $n_n = n_g - n_b = 333 - 290 = 43$. Since it is greater than L_C, we declare the sample to have activity, that is, $A_T > 0$. This conclusion contradicts the previous one, based on the less certain background data.

c) The measured sample activity is $A = \gamma n_n = 4.08 \text{ Bq} \times 43 = 175 \text{ Bq}$.

d) The minimum significant measured activity is

$$A_\text{I} = \gamma L_C = 4.08 \text{ Bq} \times 28.01 = 115 \text{ Bq}. \quad (10.20)$$

e) From Eqs. (10.12) and (10.14), the minimum detectable true activity is

$$A_\text{II} = \gamma L_D = 4.08 \text{ Bq}$$

$$\times \sqrt{290}\left(1.645 + \frac{1.645}{2\sqrt{290}} + 1.645\sqrt{1 + \frac{1.645}{\sqrt{290}} + \frac{1.645^2}{4 \times 290}}\right)$$

$$= 240 \text{ Bq}. \quad (10.21)$$

Table 10.1 summarizes the last two examples. In essence, with the smaller coefficient of variation for the background counts compared with the single gross count, the

Table 10.1 Summary of data from examples in the text.

Quantity	With single background count	With extensive background count
Gross counts, n_g	333	333
Background counts, n_b	296	290
Net counts, n_n	37	43
(a) Critical level, L_C	42	29
(b) Zero true activity?	Yes, $A_T = 0$	No, $A_T > 0$
(c) Measured activity, A	151 Bq	175 Bq
(d) A_I	169 Bq	115 Bq
(e) A_II	338 Bq	240 Bq

sensitivity for A_I and A_{II} is improved by about $\sqrt{2}$, as borne out by the table. Moreover, the decision itself about the presence of activity is reversed in the two examples. It is more likely that the conclusion in the second example with the much larger number of total background counts is correct, and that a type II error was made in the first example.

10.5
Hypothesis Testing

The two types of questions, whether activity is really present and the likelihood that we will detect it at a given level, can be addressed statistically through *hypothesis testing*. The usual setting involves deciding between two conjectures, or hypotheses, concerning characteristics of one or more populations. Frequently, a hypothesis is a claim about the value of a population parameter, such as a mean, variance, or proportion. It might also be a general assertion, for example, that a new drug is more effective that an existing one for curing an illness. Although our discussion will deal primarily with statements about a single population parameter, the methodology has general applicability.

Hypothesis testing has the following formal structure. Any stated statistical assertion or conjecture that we wish to test about a population is called the *null hypothesis* and is denoted by H_0. Rejection of H_0 implies acceptance of an *alternative hypothesis*, denoted by H_1. We decide whether to reject the null hypothesis on the basis of data from random sampling of the population. For example, in order to test the assertion that a population parameter θ has a particular set of values ω, the hypotheses can be stated symbolically:

$$H_0 : \theta \in \omega \quad \text{versus} \quad H_1 : \theta \in \Omega - \omega. \tag{10.22}$$

Here Ω represents the *whole space*, or the set of all possible values that θ can take on. The *null space* ω refers to the values that θ can assume under the null hypothesis. The alternative H_1 states that θ is in the remainder of the whole space, exclusive of ω. In the usual hypothesis testing model, a null hypothesis H_0 and an alternative H_1 are stated. If the null and alternative hypotheses specify single values, for example, $H_0: \theta = \theta_0$ versus $H_1: \theta = \theta_1$, then the hypotheses are described as *simple versus simple*. If one hypothesis specifies a single value while the other gives an interval, for example, $H_0: \theta = \theta_0$ versus $H_1: \theta > \theta_0$, then they are referred to as *simple versus composite*. Composite versus composite hypotheses are possible as well, for example, $H_0: \theta \leq \theta_0$ versus $H_1: \theta > \theta_0$. To judge between the two hypotheses, we calculate some suitable test statistic from a random sample from the population and base our decision on its value. Usually, we look at extremes of the test statistic. If its value falls in some range C, called the *critical region* or the *rejection region*, then we reject H_0. The complement of C is the *acceptance region*.

The screening procedure from the foregoing sections provides an example of hypothesis testing. The measured activity A of a sample (gross minus background) is compared with an a priori established minimum significant measured activity A_I, which depends on the background and the value chosen for the parameter α. The true activity A_T of a given sample is judged to be zero or greater than zero according to how A compares with A_I. There are two possibilities:

$$\left.\begin{array}{l} H_0 : A_T = 0 \text{ (null hypothesis)}, \\ H_1 : A_T > 0 \text{ (alternative hypothesis)}. \end{array}\right\} \quad (10.23)$$

The test statistic is the single measurement of A. Compared with Eq. (10.22), the whole space Ω in Eq. (10.23) is the set of all nonnegative numbers ($A_T \geq 0$), the null space is the single value $\omega = A_T = 0$, and $\Omega - \omega$ is the set of values $A_T > 0$. Hypothesis (10.23) is an example of simple versus composite.

The general relationship between errors and the wrong decision is shown in Table 10.2. One either accepts or rejects H_0, while H_0 is either actually true or false. The rejection of H_0 when it is true results in a type I error. On the other hand, acceptance of H_0 when it is false (i.e., H_1 is true) is a type II error. The implications of the two types of error can be quite different (Problem 10.11).

The probability α associated with type I errors in a test of hypothesis is referred to as the *size*, or *significance level*, of the test. The probability associated with a type II error is commonly referred to as β. The probability $(1 - \beta)$ of the complement for type II errors is called the *power* of the test. Note that the power is the probability of rejecting H_0 when H_1 is true, a correct decision. Ideally, one would like to design a test with a rejection region so as to minimize both α and β. However, this is not generally feasible, because of their complicated relationship. The significance level α is set beforehand to some specified value, which acts to determine the decision level for the test statistic. Similarly, β can be arbitrarily chosen, but it enters the test procedure in a different way. The probability of making a type II error depends additionally on the level of the true activity A_T.

The situation is represented schematically in Figure 10.4 by a so-called *power curve* for the test. The probability $\Pr(A > A_I | \mu_A = A_T)$ that the measured activity A will be greater than A_I is shown plotted against the true activity A_T. Starting at $A_T = 0$, the probability α has been assigned in the test procedure for a type I error. As A_T starts out from zero, the probability $\Pr(A > A_I | \mu_A = A_T)$ increases gradually at first and then climbs more steeply. When $A_T = A_I$, $\Pr(A > A_I | \mu_A = A_I) \cong 0.5$. (Provided the count numbers are sufficiently large to allow the normal approximation to be valid, it is about equally probable that A will lie on either side of A_I.) When A_T reaches the particular value A_{II}, we have $\Pr(A \leq A_I | \mu_A = A_{II}) = \beta$, which stipulates the probability for a type II error at that mean activity. Thus, the complementary probability $\Pr(A > A_I | \mu_A = A_{II}) = 1 - \beta$, called the power of the test, is shown in Figure 10.4.

Table 10.2 "Truth" table showing possible outcomes of hypothesis testing with relations (10.23).

	H_0 is true	H_0 is false
Accept H_0	Correct decision	Type II error
Reject H_0	Type I error	Correct decision

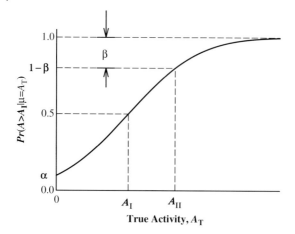

Figure 10.4 Power curve showing the relationship of the relevant quantities. The power, $1-\beta$, is the probability of correctly stating that the true activity is greater than zero when $A_T = A_{II}$.

Thereafter, $\Pr(A > A_I | \mu_A = A_T)$ approaches unity as A_T increases. The power of a test can be particularly useful in assessing how sensitive some tests are in distinguishing small differences in numerical values of population parameters.

In optimizing the test procedure, if we reduce α, then we move A_I to the right and cause β to increase (e.g., see Figure 10.2). The classical approach to designing a test procedure is to first fix α, usually at some standard level (e.g., $\alpha = 0.05$ or 0.01), and then look for a test that minimizes β. If the smallest possible value of β for that α is unacceptably large, then we should increase α, increase the sample size, or try another test. Additional discussion on optimization of the test procedure is given in Section 10.8.

It should be apparent that acceptance of a statistical hypothesis usually means that the sample data are not sufficiently strong to refute the hypothesis. Rejection, on the other hand, implies that the sample data would result only with a small probability if the hypothesis were true. There is thus a strong qualitative difference between acceptance and rejection of the null hypothesis. Generally, the null hypothesis H_0 expresses the status quo, while the alternative H_1 requires a burden of proof. The test statistic is the random variable whose numerical value determines the decision.

■ *Example*

When first received from the manufacturer, the response of a thermoluminescence dosimeter (TLD) to a 5-mGy dose from ^{137}Cs photons is measured several times. In terms of the charge Q measured by the reader for this exposure, an acceptable dosimeter should show a normal distribution with mean $\mu = 117.2$ nC and standard deviation $\sigma = 1.5$ nC. If a new dosimeter checks out, it is assigned to a radiation worker. After issuance, the dosimeter is periodically retrieved and checked by criteria specified below to see whether the response has changed. If so, the unit is removed from service.

a) To check the response of a retrieved dosimeter, it is decided to test the hypothesis that $\mu = 117.2$ nC, the standard setting, versus the alternative

10.5 Hypothesis Testing

that $\mu > 117.2$ nC. Formally state H_0 and H_1 and whether the hypothesis is simple or composite.

b) What are the values of ω and Ω as defined by Eq. (10.22)?
c) The value $\alpha = 0.05$ is chosen as the significance level of the test. Assume that the reading is normally distributed with the mean and standard deviation given above. Obtain the critical value that defines the test of size $\alpha = 0.05$.
d) What is the acceptance region for this test?
e) Suppose that the alternative hypothesis is H_1: $\mu = 122.0$ nC. What is the probability β of a type II error?
f) What is the power for this alternative hypothesis?

Solution

a) The null hypothesis states that the mean equals the standard setting. Hence, H_0: $\mu = 117.2$ nC. The alternative is simply that the mean has shifted to some larger value. Thus, we write H_1: $\mu > 117.2$ nC. The null hypothesis, referring to a single value, is a simple hypothesis. The alternative, which involves an interval of values, is composite.
b) $\omega = 117.2$ nC and $\Omega = [117.2, \infty)$.
c) Recall that $\alpha = \Pr(\text{reject } H_0 | H_0 \text{ is true})$. That is, α equals the probability that we reject H_0 given that H_0 is true. Letting Y denote the reading from the retrieved dosimeter and Y_C the critical value, we write

$$\Pr(Y > Y_C | H_0 \text{ is true}) = \alpha = 0.05. \quad (10.24)$$

Under H_0, $\mu = 117.2$ nC, and we are given $\sigma = 1.5$ nC. Converting to the standard normal distribution, we write

$$\Pr\left(\frac{Y-\mu}{\sigma} > \frac{Y_C - 117.2}{1.5}\right) = 0.05. \quad (10.25)$$

Thus, with $Z = (Y - \mu)/\sigma \sim N(0, 1)$,

$$\Pr\left(Z > \frac{Y_C - 117.2}{15}\right) = 0.05. \quad (10.26)$$

From Table 6.1 or from Table A.3, we find $\Pr(Z > 1.645) = 0.05$. Therefore,

$$\frac{Y_C - 117.2}{1.5} = 1.645, \quad (10.27)$$

giving $Y_C = 119.7$ nC. With this critical value, a type I error (rejection of H_0 when it is true) can be expected 5% of the time.

d) The acceptance region is the complement of the critical, or rejection, region C. The latter is the set of all values equal to or greater than the critical value: $C = [Y_C = 119.7 \text{ nC}, \infty)$. Therefore, the acceptance region is $C' = [0, 119.7 \text{ nC})$. Note that, since H_1: $\mu > 117.2$ nC, we will reject H_0 only when we observe larger values of the response. Also, because negative readings do not occur, the lower bound of the acceptance region is zero.

Table 10.3 Power curve, $(1-\beta)$ versus μ, for example in the text.

μ (nC)	118	119	120	121	122	123
β	0.8715	0.6796	0.4207	0.1931	0.0626	0.0139
$1-\beta$	0.1285	0.3204	0.5793	0.8069	0.9374	0.9861

e) A type II error occurs when H_0 is accepted when H_1 is true. With the critical value Y_C we write, analogous to Eq. (10.24),

$$\Pr(Y < Y_C | H_1 : \mu = 122.0 \text{ is true}) = \beta. \tag{10.28}$$

Applying the standard normal variable $Z = (Y-\mu)/\sigma$ with $\mu = 122.0$, $Y_C = 119.7$, and $\sigma = 1.5$ gives

$$\Pr\left(\frac{Y-\mu}{\sigma} < \frac{119.7-122.0}{1.5}\right) = \Pr(Z < -1.53) = \beta. \tag{10.29}$$

From Table A.3, $\beta = 0.063$.

f) The power is the probability $1 - \beta = 0.937$ (the complement for a type II error) that we reject H_0 given that H_1 is true. So if the dosimeter has shifted to a higher mean value of 122.0 nC, we would have a 93.7% chance of detecting this change and rejecting the null hypothesis. We can calculate a power curve by choosing different values of μ for H_1. Table 10.3 gives some values and their corresponding powers in the test above in which the critical value is $Y_C = 119.7$ nC. The resulting power curve is shown in Figure 10.5.

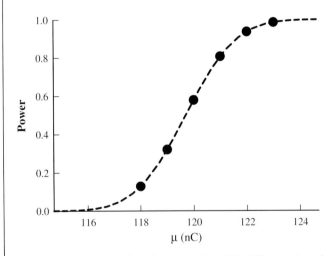

Figure 10.5 Power curve from the quantity $(1-\beta)$ for different values of μ in Table 10.3 (example in the text).

Table 10.4 Critical values z_α and $z_{\alpha/2}$ for one- and two-tailed tests, respectively, for different levels of significance α with the standard normal distribution.

α	One-tailed, z_α	Two-tailed, $z_{\alpha/2}$
0.100	1.282	1.645
0.050	1.645	1.960
0.010	2.326	2.576
0.005	2.576	2.810
0.002	2.880	3.080
0.001	3.080	3.295

The discussion until now has dealt only with a one-sided, or one-tailed, alternative hypothesis. The critical region for rejection then lies in the right or left tail of the distribution for the test statistic, depending on whether the alternative is greater than or less than the null hypothesis, respectively. Hypothesis tests can be two-tailed. A simple null hypothesis will always be stated as an equality to a single value. The alternative can then be either one- or two-tailed. For example,

One-tailed test
$H_0 : \theta = \theta_0$
$H_1 : \theta > \theta_0$

Two-tailed test
$H_0 : \theta = \theta_0$
$H_1 : \theta \neq \theta_0$

In a two-tailed test, we might reject H_0 if our test statistic is either too small or too large. The significance level α then must be split to account for either of these errors if H_0 is true. Generally, there is no reason to suspect that one error might occur more often than the other. One then simply uses $\alpha/2$ for each. Table 10.4 lists the critical values z_α and $z_{\alpha/2}$, respectively, for one- and two-tailed tests for different levels of significance α for the standard normal distribution. One-tailed tests are used to establish performance measures to evaluate radioactivity and radiation dose, for which results giving less than background have no physical significance.

■ *Example*

Consider again the definition of an acceptable thermoluminescence dosimeter in the last example. We shall now use different criteria, involving a two-tailed test, to judge whether the response of a retrieved dosimeter has changed. Specifically, a dosimeter will be discarded if it reads too far above or below the mean, $\mu = 117.2$ nC, at a significance level $\alpha = 0.05$.

a) Determine the acceptance region for the response of a retrieved dosimeter.
b) State the null and alternative hypotheses for the dosimeter test.
c) If the acceptance region is within $\pm 4\%$ of the mean, what is the probability of a false positive?

Solution

a) With a significance level $\alpha = 0.05$, the acceptance region in this two-tailed test is the interval within $z_{\alpha/2} = z_{0.025} = 1.960$ standard deviations on either side of the mean (Table 10.4). Inserting the given values of μ and σ, we find that the acceptance region, $\mu \pm 1.960\sigma$, spans the interval (114.3 nC, 120.1 nC).

b) The null hypothesis states that the response of a retrieved dosimeter has not changed. That is, the measured charge Q has mean equal to 117.2 nC, so that $H_0: \mu = 117.2$ nC. The alternative hypothesis is that Q has changed, or $H_1: \mu \neq 117.2$ nC. The hypothesis test, which is two-tailed, is conveniently stated in terms of the standard normal test statistic, $Z = (Q - \mu)/\sigma$. The acceptance and rejection regions are, respectively, given by

$$|Z| < 1.960 \quad \text{and} \quad |Z| \geq 1.960. \tag{10.30}$$

This is an example of a simple versus a composite test.

c) With $Q = \mu \pm 0.04\mu = \mu(1 \pm 0.04)$, the standard normal variable has the values

$$Z = \frac{\mu - \mu(1 \pm 0.04)}{\sigma} = \pm \frac{0.04\mu}{\sigma} = \pm \frac{0.04 \times 117.2 \text{ nC}}{1.5 \text{ nC}}$$
$$= \pm 3.1. \tag{10.31}$$

The probability of a false positive is the probability that we reject H_0 given that H_0 is true. That is, $\Pr(Z < -3.1) + \Pr(Z > 3.1) = 0.001 + 0.001 = 0.002$.

■ Example

The activity concentration of uranium in soil around a proposed munitions plant has a mean $\mu = 1.017$ pCi g^{-1} and standard deviation $\sigma = 0.108$ pCi g^{-1}. The plant will manufacture depleted uranium armor penetrators. When in operation, soil samples will be taken periodically around the site to monitor the efficacy of practices for the containment of the uranium. The goal is to maintain a state of "no detectable uranium above background" in the environment. What measured level of uranium concentration in soil would indicate a failure of the containment controls, given an acceptable probability $\alpha = 0.01$ of a false positive?

Solution

This is a one-tailed test of hypothesis, since we are interested only in results that exceed the mean background. From Table A.3, for $\alpha = 0.01$ one has $z_{0.01} = 2.326$. The critical level for the soil concentration X is thus

$$L_C = \mu + z_\alpha \sigma = 1.017 + 2.326(0.108) = 1.268 \text{ pCi g}^{-1}. \tag{10.32}$$

The null hypothesis states that there is no activity concentration of uranium above that of background. The alternative hypothesis is that there is increased activity. Symbolically,

$$H_0 : \mu = 1.017 \text{ pCi g}^{-1} \quad \text{versus} \quad H_1 : \mu > 1.017 \text{ pCi g}^{-1}. \quad (10.33)$$

A measurement in excess of the amount L_C given by (10.32) is the basis for rejection of the null hypothesis. It is to be considered as indicative that uranium levels have exceeded background, presumably because of inadequate containment at the plant.

In many applications we do not know the population variance, and so we must estimate it from sampled observations. Unless the number of observations is large (usually $n > 30$), our test statistic will be the Student's t-distribution (Section 6.9) rather than the normal. When $n > 30$, the normal distribution provides an adequate approximation to the t-distribution, and so it can be used.

■ *Example*
In the last example, calculate the critical level if the given values, $\bar{x} = 1.107 \text{ pCi g}^{-1}$ and $s = 0.108 \text{ pCi g}^{-1}$, are estimates of μ and σ obtained from

a) $n = 4$ measurements and
b) $n = 10$ measurements.
c) Plot the probability density functions for the uranium activity concentrations in the last example and in parts (a) and (b) here.

Solution
a) In this case we do not know the true population mean and standard deviation, for which we have only the sample estimates. Rather than the normal distribution, we employ Student's t-distribution (Section 6.9). Instead of z_α we use $t_{\nu,\alpha} = t_{3,0.01} = 4.541$ from Table A.5 for $\nu = n-1 = 3$ degrees of freedom. In place of Eq. (10.32) we write

$$L_C = \bar{x} + t_{3,0.01}s = 1.017 + 4.541(0.108) = 1.507 \text{ pCi g}^{-1}. \quad (10.34)$$

b) With the same estimates from the larger sample, $t_{\nu,\alpha} = t_{9,0.01} = 2.821$, and the critical level is

$$L_C = \bar{x} + t_{9,0.01}s = 1.017 + 2.821(0.108) = 1.322 \text{ pCi g}^{-1}. \quad (10.35)$$

c) Figure 10.6 shows the three probability density functions, $f(x)$ – the Student's t-distributions for $\nu = 3$ and $\nu = 9$ degrees of freedom – and the standard normal distribution $N(0, 1)$, which is the limit approached by the t-distribution as $n \to \infty$. Note that the critical level (10.34) is largest for the broadest distribution ($\nu = 3$) and becomes progressively smaller as the number of measurements for the estimates of μ and σ increases.

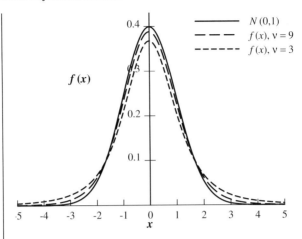

Figure 10.6 Three probability density functions $f(x)$ for example in the text: Student's t-distributions for $\nu = 3$ and $\nu = 9$ degrees of freedom and the standard normal distribution $N(0, 1)$.

10.6
Criteria for Radiobioassay, HPS N13.30-1996

The analysis described in Sections 10.3 and 10.4 for radiobioassay largely follows that given by Altshuler and Pasternack (1963) and discussed further by Turner (2007). This section reviews guidance furnished by the American National Standard, HPS N13.30 (HPS, 1996), building on original work by Currie and others. Unless otherwise stated, it will be assumed that count numbers are Poisson distributed and sufficiently large to be represented by a normal distribution with equal mean and variance.

The American National Standard, HPS N13.30-1996, *Performance Criteria for Radiobioassay*, (HPS, 1996) is in widespread use today. It presents a protocol that defines a *decision level* (L_C) and *minimum detectable amount* for measurements of a radioactive analyte in a sample or a person. These quantities play the same role as their related namesakes in our previous discussions, but differ from the former in the way "background" is assessed.[1] The critical level formalized in Section 10.3 is applied to the net count above background measured with a subject under analysis. The background count is typically made with the subject replaced by an *appropriate blank*, radiometrically identical with the subject, but containing no added radioactivity.[2] We denote the standard deviation of the count number n_{B0} from the appropriate blank

---

1) The same symbol, L_C, which we have employed in this chapter for the *critical level*, is used in N13.30 for the so-called *decision level*.

2) Examples of appropriate blanks include synthetic urine for *in vitro* radiobioassay and anthropomorphic phantoms for *in vivo* analysis (body counting).

over a time t_b by $s_{B0} = \sqrt{n_{B0}}$. An additional quantity is required by N13.30 – namely, "… the count n_{B1} of a subject, by the routine measurement procedure, where the subject contains no actual analyte activity above that of an appropriate blank." The standard deviation $s_{B1} = \sqrt{n_{B1}}$ of this count number is applied to the gross time, t_g, used to count the subject. Rather than comparing the subject count with a single background count over the same time t_g, as in Eq. (10.7), N13.30 compares the subject count with the equivalent *net count* over time t_g derived from n_{B1} and n_{B0} as follows. According to Eq. (9.10), the standard deviation of this *net rate* is

$$s_{r0} = \sqrt{\frac{n_{B1}}{t_g^2} + \frac{n_{B0}}{t_b^2}}. \tag{10.36}$$

The standard deviation in the number of net counts during the time t_g that a subject is counted is

$$s_0 = s_{r0} t_g = \sqrt{n_{B1} + \left(\frac{t_g}{t_b}\right)^2 n_{B0}} = \sqrt{s_{B1}^2 + \left(\frac{t_g}{t_b}\right)^2 s_{B0}^2}. \tag{10.37}$$

Whereas the subject count time t_g for the procedure is fixed at some standard value, the time t_b for the appropriate blank can have any value. Letting $\varrho = t_b/t_g$ be the ratio of the two times, we have

$$s_0 = \sqrt{s_{B1}^2 + \frac{1}{\varrho^2} s_{B0}^2}. \tag{10.38}$$

When the counting times are the same ($\varrho = 1$), the two standard deviations in Eq. (10.38) are approximately equal. Then $s_0 \cong \sqrt{2 s_{B0}^2} = \sqrt{2 n_b}$, which is essentially the same as Eq. (10.7).

For the type I error probability α, the decision level L_C in N13.30 is defined with the help of Eq. (10.38) as

$$L_C = \Delta_B B + z_\alpha s_0. \tag{10.39}$$

Here, B ($= n_{B0}$) is the total count of the appropriate blank (standard deviation $= s_{B0}$). The factor Δ_B equals the "maximum expected fractional systematic uncertainties bound in the appropriate blank B. Δ_B is the maximum fractional difference between the background of the subject being counted and the background for the subject estimated from the appropriate blank." As further described in N13.30, the use of Eq. (10.39) assumes that any systematic bias, such as background radiation attenuation by the sample matrix, is relatively constant for the two counts and is accounted for by the term $\Delta_B B$. Because Δ_B is a measure of the systematic uncertainty in the appropriate blank, it cannot be reduced by replicate measurements. Using an appropriate blank that is not radiometrically identical to the uncontaminated sample will also bias the mean of the net background and, with it, the decision level. Systematic error can be minimized only by repeated calibrations, by background and quality assurance measurements, and by adding to blanks known amounts of radionuclides that do not interfere with the measurement, but that provide information about chemical recoveries and detector efficiencies. In good laboratories, Δ_B in Eq. (10.39) will be close to zero. N13.30 recommends neglecting the first term in Eq. (10.32) when it does not exceed one-tenth of the value of the second term. We shall not consider systematic errors further, assuming for the decision level, $L_C = z_\alpha s_0$.

The minimum detectable true activity described in Section 10.4 is similar to Currie's *detection limit* and the *minimum detectable amount* in N13.30. That is, the mean of the distribution of net results from replicate analyses of a sample with the MDA level of added radioactivity is such that the fraction β of the net distribution lies below the critical level, in the same way as depicted in Figure 10.2. However, the net count in N13.30 refers to that between a subject under analysis and the net count that defines s_0 (Eq. (10.38)). For simplicity, we assume that systematic uncertainties are negligible. We also choose equal counting times and select $\alpha = \beta = 0.05$ ($z_\alpha = z_\beta = 1.645$). In place of Eq. (10.13), N13.30 then employs for the paired blank

$$L_D = (z_\alpha + z_\beta)s_0 = 3.29\, s_0 = 3.29\sqrt{s_{B1}^2 + s_{B0}^2} = 4.65\, s_B. \tag{10.40}$$

Here $s_B \equiv s_{B0} \cong s_{B1}$ is introduced to reflect the near equality of the two terms under the radical that represent the background counts. We can reduce the component of the variance given by a well-known blank if we take a sample of background measurements and average them, rather than using a single sample. As we know from Eq. (8.24), the variance of the mean of m observations that are independently distributed with constant variance s_{B0}^2 is s_{B0}^2/m. Hence, for the mean of m independent samples, with similar assumptions as before, we have

$$L_D = 3.29 s_B \sqrt{1 + \frac{1}{m}}. \tag{10.41}$$

The expressions (10.39) and (10.41) provide the appropriate operational count numbers that determine the critical level and the MDA, according to the explicit criteria set out by the protocol. They establish the required measurement time through the estimate of the number of background counts required. Of more relevance for exposure monitoring than the numbers *per se* is their expression in some appropriate units. For example, the MDA is usually expressed in derived units, such as nCi of uranium in the lung, dpm per day urinary excretion of ^{241}Am, and so on, rather than in units of the actual measurement (counts). In general, count numbers are converted to the appropriate derived analytical units by dividing by KT, where K is a calibration factor (e.g., expressing count rate per unit analytical amount) and T is the counting time for analysis. The quantity K may be a combination of factors, such as counting efficiency, chemical recovery, and urinary excretion rate. Thus, for the paired blank, the minimum detectable amount in appropriate analytical units is, from Eq. (10.40),

$$\text{MDA} = \frac{4.65 s_B}{KT}. \tag{10.42}$$

In applying derived units, it should be remembered that L_C and MDA are restricted to integral values by definition.

These expressions from N13.30 for L_C and MDA were derived under the assumption that total counts are large enough to be adequately approximated by normal distributions. While this condition is met in many types of radiobioassay (e.g., whole-body counts for ^{137}Cs against a ^{40}K background in body tissues), some modern applications entail very small count numbers, which are not well approximated by the normal distribution. The background in a well-maintained alpha spectroscopy counting chamber might be only one or two counts over several days. N13.30 does not deal with this situation at length, but adds a semi-empirical term to L_D in Eqs. (10.40) and (10.41) in order to assure that $\beta \leq 0.05$ in very low backgrounds. With a Poisson distribution, an MDA of three counts satisfies this condition. The N13.30 MDA (10.42) for the paired blank then becomes

$$\text{MDA} = \frac{4.65 s_B + 3}{KT}. \tag{10.43}$$

For the mean of m samples of a well-known blank (see Eq. (10.41)),

$$\text{MDA} = \frac{3.29 s_B \sqrt{1 + (1/m)} + 3}{KT}. \tag{10.44}$$

A well-maintained alpha-particle spectrometer might register on average a single event in the energy range of interest for a 1000-min count, even though the detection chamber contains a blank sample with no activity. The Poisson distribution for the number of counts with a mean background $\mu_B = 1$ over the counting interval is shown in Figure 10.7. The probability of observing zero counts is $\Pr(X=0) = 0.368$, the same as that for a single count. Although there is no radioactivity in the sample, the probability for registering one or more counts is $\Pr(X \geq 1) = 0.632$.

Setting a critical level for the measurement of samples with the spectrometer from Figure 10.7 can be approached in the following way. We relate the probability

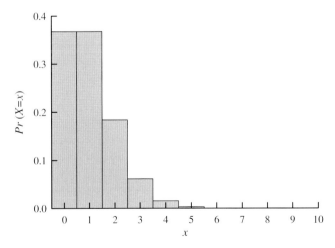

Figure 10.7 Poisson distribution $\Pr(X=x)$ for count number X with mean background $\mu_B = 1$ over the counting interval.

α associated with a type I error to the Poisson distribution with parameter μ by writing

$$\alpha = \Pr(X > L_C) = 1 - \Pr(X \le L_C) = 1 - \sum_{x=0}^{L_C} p(x;\mu) = 1 - P(L_C;\mu). \quad (10.45)$$

Here $P(X;\mu)$ is the cumulative distribution of the Poisson probability $p(x;\mu)$ tabulated in Table A.2 in the Appendix. In contrast to using $L_C = z_\alpha s_0$ as before, under the assumption of normal statistics, we deal now with α having noncontinuous values. If we wanted to set $\alpha = 0.05$, for example, we see from Table A.2 with $\mu = 1$ for our spectrometer that $\Pr(X \le 2) = 0.920$ and $\Pr(X \le 3) = 0.981$. Thus, choosing $L_C = 2$ or 3, respectively, gives $\alpha = 0.080$ or 0.019. In order to assure that the type I error probability is no greater than 0.05, we need $L_C = 3$, for which $\alpha = 0.019$. A measurement that yields four or more counts in the 1000-min interval with the spectrometer from Figure 10.7 is to be interpreted as meaning that there is activity in the sample. In this case, we reject H_0 with actual significance level $\alpha = 0.019$.

The MDA and L_D for the spectrometer depend on L_C and the value selected for the probability β for a type II error. We choose the MDA as the smallest mean activity level above background that results in three or fewer counts in 1000 min with a probability $\beta = 0.05$. The L_D is the mean Poisson count for which $\Pr(X \le 3 | \mu = L_D) = \beta = 0.05$. In Table A.2 we look for the value of $\mu = L_D$ for which $P(3,\mu) = 0.05$, that is, for which the cumulative Poisson probability is 0.05 for $X = 3$ counts. We see that this occurs between $\mu = 7.5$ and $\mu = 8.0$. Iterative calculation on μ gives $\beta = 0.05$ when $\mu = L_D = 7.75$. (Simple linear interpolation gives $\mu = 7.76$.) Figure 10.8 shows the distributions for a

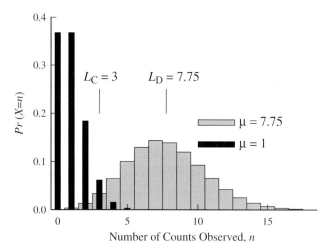

Figure 10.8 A detection level $L_D = 7.75$ counts gives a probability of false negative $\beta = 0.05$ for an alpha-particle spectrometer with background mean of 1 count in the counting interval and the probability of false positive, $\alpha \le 0.05$. This means that replicate measurements of a sample containing activity at the detection level would yield a distribution of results (gray bars) that, 5% of the time, lie below the critical level ($L_C = 3$) defined by the background distribution (black bars) and the value of α. The Poisson distribution is assumed for count results obtained for both background and sample with activity.

background with mean $\mu = 1$ count and for a sample with a mean at the detection level $L_D = 7.75$ counts. By convention, the critical level L_C is expressed in the actual units of the measurement (in this case, counts). The detection level and MDA are typically expressed in derived units (in this case, activity). For this reason, we have not restricted L_D to be an integer.

■ Example

A radiobioassay laboratory evaluates ^{241}Am in urine with a procedure that involves chemical separation and electroplating, followed by alpha spectrometry. A total of $m = 21$ reagent blanks (having everything that would be in an appropriate blank except real or artificial urine) were obtained and counted to characterize the background distribution. The count numbers x varied from zero to a maximum of five. The numbers of blanks n_x with x counts were distributed as follows:

Count number, x	0	1	2	3	4	5
Occurrence, n_x	3	7	5	3	1	2

Conversion of the measured count number into derived units of disintegrations per second (dps, or Bq) is made with the calibration factor $K = 0.27$ counts per disintegration and a counting time $T = 60\,000$ s.

a) What are the mean and standard deviation for the number of counts?
b) What is the HPS N13.30 critical level?
c) Estimate the MDA from HPS N13.30 and show graphically the resulting β probability for a false negative.

Solution
a) The mean count number is

$$\bar{x} = \sum_{x=0}^{5} \frac{x n_x}{m} = 1.905, \qquad (10.46)$$

and the standard deviation is

$$s_{B0} = \left[\sum_{x=0}^{5} \frac{n_x (x - \bar{x})^2}{m - 1} \right]^{1/2} = 1.480. \qquad (10.47)$$

Note that, for the Poisson distribution, the mean and the variance should be equal. However, we found the two to have different values in Eqs. (10.46) and (10.47). The difference is due to the fact that we have used two different estimations in Eqs. (10.46) and (10.47) for the same quantity. This condition is described as *overdispersion* (the variance estimate is larger than the sample mean) or *underdispersion* (the variance estimate is smaller than the sample mean). The theory surrounding this circumstance is beyond the scope of this text. We suggest using the sample mean if the

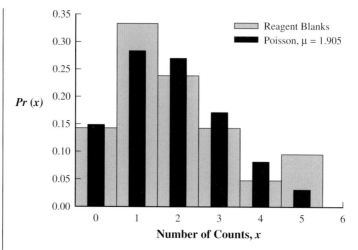

Figure 10.9 Distribution of alpha-spectrometer results (counts) in the ^{241}Am region of interest from reagent blanks (gray bars) in the example of Section 10.6, together with a superimposed Poisson distribution with the same mean, $\mu = 1.905$ (black bars), for comparison.

difference between the estimators is not too great. If the difference is quite large, then the population from which the sample is taken might not be Poisson distributed. Figure 10.9 shows the distribution of the reagent results, together with a superimposed Poisson distribution with the same mean, $\mu = 1.905$, for comparison.

b) We assume the default value $\alpha = 0.05$ with $s_{B0} = 1.38$. With $\Delta_B = 0$, Eq. (10.39) and Eq. (10.41) with $m = 21$ then give the HPS N13.30 critical count number for the well-known blank,

$$L_C = 1.645 s_{B0} \sqrt{1 + \frac{1}{m}} = 1.645(1.38)\sqrt{1 + \frac{1}{21}} = 2.3. \quad (10.48)$$

As discussed in connection with Eq. (10.45), one needs to adjust L_C to be an integer, thus changing the attainable value of α. In this case, $L_C = 3$.

c) The MDA for low background and well-known blank, Eq. (10.44), should be used. Substituting the given information yields

$$\text{MDA} = \frac{3.29(1.38)\sqrt{1 + (1/21)} + 3}{(0.27)(60\,000\,\text{s})} = 4.7 \times 10^{-4}. \quad (10.49)$$

Figure 10.10 shows the relationship between the MDA determined in this way (converted to counts) and the critical level, with the distributions being represented as normal. Note the result, $\beta \cong 0$. This is an artifact of adding $3/KT$ to the MDA in going from Eq. (10.41) to Eq. (10.44) to ensure that $\beta \lesssim 0.05$ even for very low backgrounds.

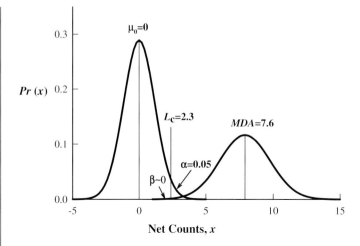

Figure 10.10 Normal approximation to the distribution of net background in the example of Section 10.6, showing the critical level L_C, MDA, and probabilities of false positive (α) and false negative (β). Note the artificially small value of β, a coincidental (in this case) result of the 3/KT term in the HPS N13.30 MDA formula.

The formulas provided in N13.30 are intended to be used as guidance in the absence of other information. Radiobioassay laboratories may be required to use the formulas in some cases – in intercomparison studies with other laboratories or as part of the quality control criteria of regulatory agencies. A better picture of an installation's performance is usually obtained by carefully controlled studies, using appropriate blanks, designed to provide empirically derived measures. Spiked test samples can be used to estimate standard deviations and biases.

Low-level counting has received additional attention since N13.30 was published. The reader is referred to an evaluation of eight decision rules for low-level radioactivity counting carried out by Strom and MacLellan (2001).

10.7
Thermoluminescence Dosimetry

Performance measures for external dosimetry, which utilizes integrating devices such as thermoluminescence dosimeters (TLDs) and radiosensitive film to measure radiation doses from sources outside the body, have been developed in the United States by the Department of Energy for their Laboratory Accreditation Program (DOELAP) for personnel dosimetry (DOE, 1986). A statistical model developed by Roberson and Carlson (1992) derives the formulas used for estimating a critical level and a lower limit of detectability. We describe this model and its parameters.

TLDs and film provide data that are continuously distributed, namely, light output from a TLD and light transmission through a film, respectively. Although we shall concentrate on TLD measurements in the discussions that follow, much of the work applies also to film dosimeters.

Some crystalline materials are thermoluminescent, absorbing and storing energy from incident radiation by the promotion of valence electrons to higher quantum states with relatively long lifetimes. When an exposed thermoluminescent crystal is later heated under controlled conditions in a reader, the promoted electrons make transitions back to the lower energy states, releasing their stored energy in the form of light. The light thus emitted is detected and registers an amount of electric charge, which ideally is proportional to the radiation energy that was absorbed in the crystal. Calibration of the equipment enables measurement of the charge to be converted into absorbed dose. The reading process also "zeroes out" the stored energy and restores the dosimeter to its approximate original state for use again (unlike a film dosimeter).

A personnel TLD usually comprises several elements, each consisting of a separate thermoluminescent crystal, or "chip." Chips can be fabricated from different materials. Some chips might be provided with different filters to help identify the type of radiation in mixed fields and to provide crude spectral information. For instance, chips made of pure ^{6}LiF or pure ^{7}LiF have the same response to gamma radiation, but respond differently to neutrons, which are absorbed by ^{6}Li but not by ^{7}Li. Comparing the response of these two isotopically different LiF chips in the same dosimeter gives an indication of the separate gamma and neutron doses to the wearer. TLDs can be calibrated to provide information on deep dose, shallow beta and low-energy photon doses, and other information. Individual dose components from a multi-element TLD are determined by means of unfolding algorithms that analyze and interpret combinations of the responses from the individual chips.

We shall assume that normal statistics can by employed to describe the response of a single TLD chip exposed repeatedly to a given radiation dose and then read under fixed conditions and, similarly, to describe the response of a collection of identical chips. However, the results unfolded from the combination of outputs from the same multichip TLD might not be normally distributed. The discussions below pertain to the response of an individual chip. We generally follow the approach of Roberson and Carlson (1992) and Currie (1968).

For personnel monitoring of external exposure, one is interested in a net dosimeter reading, $X_H = X_T - \bar{X}_B$, which is the difference between a total dosimeter reading X_T and average background $\bar{X}_B$. The latter is the mean value determined from $i = 1, 2, \ldots, n$ dosimeter measurements X_{Bi}, each having variance σ_B^2:

$$\bar{X}_B = \frac{1}{n} \sum_{i=1}^{n} X_{Bi}. \tag{10.50}$$

The respective net, total, and background variances satisfy the relation (see Eq. (8.24))

$$\sigma_H^2 = \sigma_T^2 + \frac{1}{n} \sigma_B^2. \tag{10.51}$$

10.7 Thermoluminescence Dosimetry

If the true net signal is zero, then $\mu_H = 0$, and we can assume that $\sigma_T = \sigma_B$. We denote the standard deviation of this net signal as $\sigma_H = \sigma_0$, with variance

$$\sigma_0^2 = \sigma_B^2 \left(1 + \frac{1}{n}\right). \tag{10.52}$$

Like Eq. (10.39) with $\Delta_B = 0$, the critical level is given by

$$L_C = z_\alpha \sigma_0. \tag{10.53}$$

As before, measurements due only to background fluctuations that exceed L_C cause type I errors (false positive) when the null hypothesis is (incorrectly) rejected.

■ **Example**

The penetrating radiation background for a calendar quarter was measured with 100 single-chip TLDs. Randomly distributed in the geographical area of interest, their mean value with standard error was 15.0 ± 5.0 mrem.

a) What is the critical level for net results obtained with this background, if one accepts $\alpha = 0.05$ for the probability of a false positive in deciding whether there is added dose?
b) What is the probability of a false positive in part (a) if $L_C = 5.0$ mrem?
c) What is the critical level in part (a) when a single background dosimeter is used with the same background as stated above (paired blank)?

Solution

a) Combining Eqs. (10.52) and (10.53) with $n = 100$, $z_\alpha = 1.645$, and the sample estimate $\hat{\sigma} = 5.0$ mrem gives

$$L_C = z_\alpha \hat{\sigma}_B \sqrt{1 + \frac{1}{n}} = 1.645(5.0 \text{ mrem}) \sqrt{1 + \frac{1}{100}}$$
$$= 8.3 \text{ mrem}. \tag{10.54}$$

b) Setting $L_C = 5.0$ mrem in the last equation, we can solve for z_α:

$$z_\alpha = \frac{L_C}{\hat{\sigma}_B \sqrt{1 + (1/n)}} = \frac{5.0 \text{ mrem}}{(5.0 \text{ mrem}) \sqrt{1 + (1/100)}} = 0.995$$
$$= 1.0, \tag{10.55}$$

or one standard deviation. From Table A.3, $1 - \alpha = 0.84$, and so $\alpha = 0.16$. Alternatively, this part of the example can be solved by starting with $\Pr(X > L_C | \mu_n = 0)$ (see Problem 10.25).
c) In this case we repeat the calculation of part (a) with $n = 1$. The result is 12 mrem.

When we have to estimate the variance and the sample size is <30, then we should use Student's t-distribution rather than the normal. This is accomplished by replacing in Eq. (10.54) z_α with $t_{\alpha,\nu}$, where $\nu = n - 1$, the number of degrees of freedom. For part (a) of the last example, $t_{\alpha,\nu} = t_{0.05,99}$, giving again $L_C = 8.3$. Comparison with

Eq. (10.54) shows that the normal distribution is a good approximation for this sample size ($n = 100$). For $n = 10$, on the other hand, the result from Student's t-distribution is $L_C = 9.2$, while for $n = 3$, $L_C = 15$ (Problem 10.26). Student's t-distribution is not applicable in the case of a paired blank ($n = 1$), because the variance cannot be estimated.

The detection level is given by

$$L_D = L_C + t_{\beta, n-1} \sigma_D. \tag{10.56}$$

Here σ_D is the standard deviation of the signal at the exposure level L_D and β is the probability that the reading will be less than L_C. Thus, L_D is the minimum mean dose for which the probability of a type II error (incorrectly accepting the null hypothesis) is no greater than β (cf. Figure 10.2), where L_D corresponds to A_{II}. Our task next is to find how L_D can be determined from the measured parameters.

The average background $\bar{X}_B$ in Eq. (10.50) is contributed by all signals not due to the radiation exposure X_H being evaluated. These include background radiation, reader noise, and any other factors that might arise from treatment or handling of the dosimeters (e.g., fogging of film, fading of TLDs). We can single out the part X_N (having standard deviation σ_N) of $\bar{X}_B$ that is due to reader noise, which is not attributable to the dosimeter. The total dosimeter reading, X_T, is equal to the sum of the net dosimeter reading, X_H, and the average background reading, $\bar{X}_B$. Hence,

$$X_T = X_H + \bar{X}_B = (X_H + \bar{X}_B - X_N) + X_N = (X_T - X_N) + X_N, \tag{10.57}$$

in which the expression in parentheses describes the dosimeter signal alone. The standard deviation of the dosimeter signal is assumed to be a constant fraction k of the signal $X_T - X_N$ itself. Equation (10.57) then implies for the variance that

$$\sigma_T^2 = k^2 (\mu_T - \mu_N)^2 + \sigma_N^2, \tag{10.58}$$

in which k is the relative standard deviation of the dosimeter reading. Similarly, for the background,

$$\sigma_B^2 = k^2 (\mu_B - \mu_N)^2 + \sigma_N^2. \tag{10.59}$$

Substitution of Eq. (10.58) into Eq. (10.51) gives

$$\sigma_H^2 = k^2 (\mu_T - \mu_N)^2 + \sigma_N^2 + \frac{1}{n} \sigma_B^2. \tag{10.60}$$

Taking expectations in Eq. (10.57), we find that $\mu_T = \mu_H - \mu_B$. Using this and σ_N^2 from Eq. (10.59) and collecting terms leads to the following (Problem 10.22):

$$\sigma_H^2 = k^2 \left[\mu_H^2 + 2\mu_H (\mu_B - \mu_N) \right] + \sigma_B^2 \left(1 + \frac{1}{n} \right). \tag{10.61}$$

At the detection level, $\mu_H = L_D$, and so the variance is $\sigma_H^2 = \sigma_D^2$, where the latter is defined by Eq. (10.56). Introducing the notation $\mu_{B'} = \mu_B - \mu_N$ for the background mean, excluding system noise, and remembering Eq. (10.52), we obtain (Problem 10.23)

$$\sigma_D^2 = k^2(L_D^2 + 2L_D \mu_{B'}) + \sigma_0^2. \tag{10.62}$$

On the left, σ_D^2 depends on L_D. Using Eq. (10.56) for L_D and substituting Eq. (10.53) for L_C (assuming n is large so that $t_{\beta,n-1} \approx z_\beta$) gives

$$\sigma_D = \frac{L_D - z_\alpha \sigma_0}{z_\beta}. \tag{10.63}$$

Employing this in Eq. (10.62) and rearranging terms gives the following quadratic equation for L_D:

$$(1 - z_\beta^2 k^2) L_D^2 - 2(z_\alpha \sigma_0 + z_\beta^2 k^2 \mu_{B'}) L_D + (z_\alpha^2 - z_\beta^2) \sigma_0^2 = 0. \tag{10.64}$$

With $z_\alpha = z_\beta = z$, the detection level is given by (Problem 10.24)

$$L_D = \frac{2(z\sigma_0 + z^2 k^2 \mu_{B'})}{1 - z^2 k^2}. \tag{10.65}$$

This final expression shows the explicit dependence of the detection level on the various components of the measurement. The dominant factor is almost always the term $z\sigma_0$ in the numerator. The second term in the numerator gives the contribution of background. When the background and relative standard deviation are small, $L_D \cong 2z\sigma_0 \cong 2L_C$.

System noise must be determined in some way, since it has to be subtracted from background in order to obtain $\mu_{B'}$. In one method, a plot is made of background accumulation in a set of TLDs as a function of time. This plot should, if there is no system error, yield a straight line with its intercept at the origin. The value of a nonzero intercept is taken as the measure of the system noise. In a study at one site, TLDs were distributed to 50 employees, each of whom stored the dosimeters in a suitable location at home. A subset of the dosimeters was retrieved from each employee at monthly intervals and read, and the accumulated background signal thus determined over the course of a year. As shown in Figure 10.11, the resulting regression curve (Chapter 14) has an intercept at 12 mrem, which can be used as an estimator for μ_N, introduced in Eq. (10.57).

In applications of Eq. (10.65), Student's t-values replace the values of z, and sample standard deviations can replace the σ values. In accordance with Eqs. (10.52) and (10.53), the first value of z in the numerator of Eq. (10.65) should be replaced by $t_{n-1,\alpha}$, since there are n background dosimeter readings. We replace the z^2 terms in the numerator and denominator. The estimated relative standard deviation, $k = \hat{\sigma}_0/\bar{x}$, is determined by m dosimeter readings at large dose compared with background, and since $\alpha = \beta$, the replacement is $t_{m-1,\alpha}$. Finally, $\mu_{B'}$ is replaced by

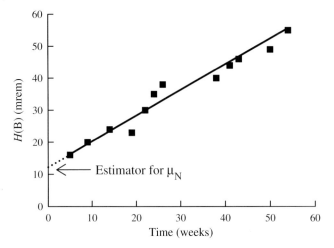

Figure 10.11 Regression method for estimating system noise (Sonder and Ahmed, 1991). H_B is the time-dependent accumulated background signal plus system noise (corresponding to $\bar{H}_B$ in the discussion leading to Eq. (10.66)).

$\bar{H}_{B'} = \bar{H}_B - N$ (average background signal with system noise removed). With these substitutions, the detection level becomes

$$L_D \cong \frac{2(t_{n-1,\alpha}\hat{\sigma}_0 + t^2_{m-1,\alpha}k^2\bar{H}_{B'})}{1-t^2_{m-1,\alpha}k^2} \tag{10.66}$$

As pointed out already, the formulas developed here apply to the response of a single TLD element. They can be used for the dose response of multiple elements combined in such a way that the response algorithm is a function of the element readings without discontinuities, as long as both background dosimeters and field dosimeters (measuring possible added dose) are analyzed in identical ways. The formulas are not suitable for mixed fields when the dosimetric ratios of the field components (e.g., beta/gamma) are not known.

In practice, the assumption of normally distributed responses should be checked. Figure 10.12 shows the distributions of responses for each of the four elements that comprise the TLDs used to determine system noise in Figure 10.11. The bars represent the empirical data (number of dosimeters giving a particular dose response in steps of 1 mrem), and the continuous curves are the normal distributions with means and standard deviations equivalent to those of the actual distributions. At best, the data appear to be only very roughly normally distributed. Although formulas developed above probably give reasonable approximations of L_C and L_D for many purposes, a more rigorous treatment should be considered for critical applications.

Care should also be taken in the way L_C and L_D are used. The critical level, by definition, determines whether a result is significantly different from background. The detection level should not be considered as the limit for reporting net dose as either "zero" or "positive." As employed in personnel dosimetry programs, the

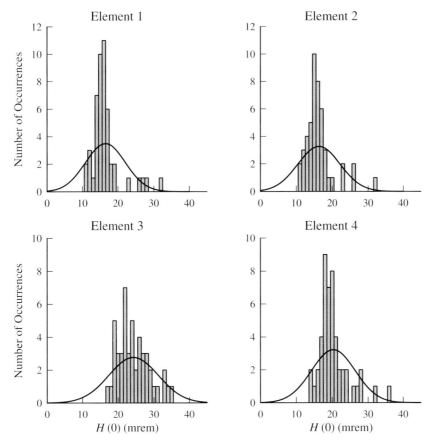

Figure 10.12 Typical distributions of environmental background signals (bars) recorded in the vicinity of a nuclear laboratory. Each graph is for a single element of a four-element thermoluminescence dosimeter. A normal distribution with the same mean and standard deviation as the bar graph is superimposed in each case. The figure shows that assuming a normal distribution for these results is problematic and illustrates the importance of verifying assumptions about the type of distribution in establishing a well-characterized personnel dosimetry program.

detection level is most useful as a standard measure by which the performance of different dosimetry laboratories may be compared. The value of the critical level is best determined empirically, rather than by the ideal represented by the standard formulas. In radiation protection programs, the presence or absence of added doses at low levels can have significant impacts on operations in addition to potential implications for worker health. Values selected for α and β may be based on any reasonable criteria, such as program costs and regulatory requirements. The generally accepted assignment $\alpha = \beta = 0.05$ is useful when there are no other imposing considerations. Using $\beta = 0.05$ can impose a large sample size, and in many practical situations using $\beta = 0.10$ is acceptable.

10.8
Neyman–Pearson Lemma

In Section 10.5, we touched on optimization of hypothesis testing. As ordinarily practiced, α is set at some agreed upon level and β is independently chosen. The test that minimizes β for fixed α is called the *most powerful test*, or *best test*, of size α. In this section, we describe the Neyman–Pearson lemma, which allows one to find such a best test.

Neyman–Pearson Lemma
Let $X_1, X_2, \ldots, X_n$ form a random sample from a distribution with probability density function $f(x, \theta)$. We wish to carry out the following test of hypothesis regarding the numerical value of the parameter θ:

$$H_0 : \theta = \theta_0 \quad \text{versus} \quad H_1 : \theta = \theta_1. \tag{10.67}$$

If $L(\theta)$ is the likelihood function, then the best test of size α of H_0 versus H_1 has a critical (or rejection) region of the form

$$\frac{L(\theta_1)}{L(\theta_0)} \geq A \tag{10.68}$$

for some positive constant A.

We shall not present a proof of the lemma, which is given, for example, in Hogg and Tanis (1993). We show here how it can be used to derive the best test for a simple versus composite example.

■ *Example*
Let $X_1, X_2, \ldots, X_n$ be a random sample from a normal population with known variance σ^2 and unknown mean μ. We wish to test

$$H_0 : \mu = \mu_0 \quad \text{versus} \quad H_1 : \mu = \mu_1, \quad \text{with } \mu_1 > \mu_0. \tag{10.69}$$

Use the Neyman–Pearson lemma to find the best test of size α.

Solution
We know that $\bar{X}$ is a useful estimator of μ. The likelihood function for the normal distribution has been given earlier by Eq. (7.109). Since here we treat μ as the only unknown parameter, we write

$$L(\mu) = (2\pi\sigma^2)^{-(n/2)} e^{-(1/2\sigma^2)\sum (X_i - \mu)^2}, \tag{10.70}$$

in which the sum is carried out over $i = 1, 2, \ldots, n$. By taking the logarithms on both sides of Eq. (10.68), the Neyman–Pearson lemma implies that best test of H_0 versus H_1 has a critical region of the form $(A > 0)$

$$\ln\left[\frac{L(\mu_1)}{L(\mu_0)}\right] \geq \ln A. \tag{10.71}$$

Combining the last two relations and taking logarithms, one finds

$$\sum_{i=1}^{n}[2X_i(\mu_1-\mu_0)+\mu_0^2-\mu_1^2] \geq 2\sigma^2 \ln A. \tag{10.72}$$

With $\mu_1 > \mu_0$, this is equivalent to

$$\bar{X} \geq \frac{\sigma^2 \ln A}{n(\mu_1-\mu_0)} + \frac{\mu_0+\mu_1}{2} \equiv B, \tag{10.73}$$

where B is a positive constant (Problem 10.28). Hence, by virtue of the lemma, we find that the best test is a function of the sample mean $\bar{X}$. We can readily determine the critical region by knowing that $\bar{X} \sim N(\mu, \sigma/\sqrt{n})$ and that we want the probability of a type I error to be α. That is, we stipulate

$$\alpha = \Pr(\text{reject } H_0 | H_0 \text{ is true}) = \Pr(\bar{X} > B | H_0 \text{ is true}). \tag{10.74}$$

Now, under H_0, $\bar{X} \sim N(\mu_0, \sigma/\sqrt{n})$ and so we can write in place of the last equality

$$\alpha = \Pr\left(\frac{\bar{X}-\mu_0}{\sigma/\sqrt{n}} > \frac{B-\mu_0}{\sigma/\sqrt{n}}\right). \tag{10.75}$$

Since the function $(\bar{X}-\mu_0)/(\sigma/\sqrt{n}) \sim N(0,1)$, the quantity $(B-\mu_0)/(\sigma/\sqrt{n})$ must be equal to the value of z_α that cuts off the fraction α to the right of the standard normal distribution. It follows, therefore, that the critical value is

$$B = \mu_0 + \frac{z_\alpha \sigma}{\sqrt{n}}, \tag{10.76}$$

and the critical region (Section 10.5) is

$$C = \left\{(X_1, X_2, \ldots, X_n) : \bar{X} > B = \mu_0 + \frac{z_\alpha \sigma}{\sqrt{n}}\right\}. \tag{10.77}$$

This methodology can be applied to any hypothesis testing situation. The interested reader is referred to Hogg and Tanis (1993) or Garthwaite, Jolliffe, and Jones (2002).

10.9
Treating Outliers – Chauvenet's Criterion

Data from a sampled population might contain one or more values that do not appear to be consistent with the others in the sample as a whole. The question then arises whether to exclude such *outliers* from the analysis of the data, as not belonging to the population. *Chauvenet's criterion*, described in the next paragraph, can be used as an aid in considering such a decision, especially when there appear to be no objective reasons for rejecting the outlier.

For a random variable X, having a sample mean $\bar{x}$ and standard deviation s, we consider a value $X = x$, typical of an outlier. Multiplying the probability that $X \geq x$ by the number of measurements n, including the suspect points, used in determining the sample mean and standard deviation, one forms the product

$$\eta = n \Pr(X \geq x). \tag{10.78}$$

Chauvenet's criterion states that a result may be considered for rejection if the expected number η of such results in the sample is less than 0.5. Consider, for example, a set of normally distributed data with $n = 1000$ measurements, a mean $\bar{x} = 10$, and a standard deviation $s = 1$. We can use the criterion to decide whether to consider for exclusion an individual measurement result of $x = 13$, for example. This value is three standard deviations beyond the mean. We find from Table A.3 that, in a random sample of this size, one would expect to observe about $\eta = (1000)(0.0013) = 1.3$ instances in which a value $x \geq 13$ occurred. Finding a result of $x = 13$ in this size population is not an unusual event, and Chauvenet's criterion indicates that there is no justification to reject it as an outlier. With the same mean and standard deviation, but with a sample size of $n = 100$, one finds $\eta = 0.13$. This value is considerably less than 0.5 as specified by the criterion, and so the number of such examples expected in this smaller sample is significantly less than unity. The sample member $x = 13$ can thus be considered for exclusion as an outlier, with a new mean and standard deviation estimated by using the remaining data.

The expression for η with normally distributed data, having mean μ and standard deviation σ, is

$$\eta = n \left(\frac{1}{\sqrt{2\pi}} \int_z^\infty e^{-(1/2)t^2} \, dt \right) = n \left(1 - \frac{1}{\sqrt{2\pi}} \int_{-\infty}^z e^{-(1/2)t^2} \, dt \right), \tag{10.79}$$

where $z = |x - \mu|/\sigma$ is the standard normal variable tabulated in Table A.3. The evaluation of normally distributed results for sample sizes $n \lesssim 30$ should be carried out by using Student's t-distribution with appropriate degrees of freedom (Table A.5).

■ *Example*

The dose rate from a certain source of beta radiation has been measured with the following results (mrad h^{-1}): $x_i = 179, 181, 180, 176, 181, 182,$ and 180.

a) Should the result, 176 mrad h^{-1}, be considered for exclusion as an outlier?
b) If this result is excluded, what is the effect on the estimated mean and standard deviation?

Solution
a) With $n = 7$, the mean of the original data set is

$$\bar{x} = \frac{1}{n} \sum_{i=1}^n x_i = 179.9. \tag{10.80}$$

The standard deviation is

$$s = \sqrt{\frac{n\sum_{i=1}^{n} x_i^2 - \left(\sum_{i=1}^{n} x_i\right)^2}{n(n-1)}} = 1.95. \tag{10.81}$$

We use Student's t-distribution to calculate η, since the sample is small with $n = 7 < 30$. The percentile t, which expresses the difference between the datum under consideration and the estimated mean divided by the estimated standard deviation, is

$$t = \frac{|x - \bar{x}|}{s} = \frac{|176 - 179.9|}{1.95} = 2.00. \tag{10.82}$$

This value corresponds to an area of approximately 0.05 to either the left of $\bar{x} - ts$ or the right of $\bar{x} + ts$. (From Table A.5, $t = 1.943$ with $n - 1 = 6$ degrees of freedom.) The number of such observations expected for this sample size is

$$\eta = n \Pr(t \geq 2.0) \approx 7(0.05) = 0.35. \tag{10.83}$$

Thus, we may consider rejecting the measurement 176 mrad h^{-1} from this sample on the basis of Chauvenet's criterion, since $\eta < 0.5$.

b) If this measurement is excluded from the sample, then the new estimate of the mean is $\bar{x} = 180.5$ mrad h^{-1} (a change of only 0.3%), and the new (unbiased) estimate of the standard deviation is $s = 1.05$ mrad h^{-1} (a change of almost 50%). Removing an outlying datum usually does not alter the sample mean very much, but tends instead to significantly reduce the estimate of variability in the remaining data.

More than a single suspect datum might exist in a data set, in which case there are at least two applicable strategies for using Chauvenet's criterion. One approach is to evaluate η for the suspect datum closest to the mean and consider rejecting it and other further outlying data if $\eta < 0.5$. Another is to use multiples of 0.5 as the rejection criterion if all the suspect data lie close together. Rejection may be considered for two suspect data lying about the same distance from the mean, for example, if $\eta < 2(0.5) = 1$. Chauvenet's criterion is not an absolute indicator for rejecting even a single suspect result, however, and justification for rejecting multiple measurement results becomes rapidly more problematic and less desirable as a strategy for statistical decision making. There is virtually no support for using Chauvenet's criterion a second time to evaluate data that remain after one or more have been rejected and a new estimate of the mean and standard deviation determined.

10 Statistical Performance Measures

The decision to exclude data from a sample should be made with care. The presence of outliers might be a signal that something unsuspected and important is going on. Also, reducing sample size increases uncertainty in estimates of population parameters. Additional measurements should be made, if possible, to support decisions about retaining or rejecting outliers. Besides that of Chauvenet, other outlier rejection criteria can be found in the literature (e.g., Barnett and Lewis, 1994).

Problems

10.1 Show that Eq. (10.3) yields the quadratic equation, $r_{rnc}^2 - 1.36 r_{rnc} - 47.0 = 0$.

10.2 Show that the first equality in Eq. (10.3) leads to the solution given by Eq. (10.4) for r_{rnc}.

10.3 Find the solution yielded by Eq. (10.4) when $t_b \to \infty$.

10.4 For equal gross and background counting times, show that the critical level is given by Eq. (10.5).

10.5 Samples will be counted for 5 min in a screening facility and compared with a background count of 10 min. The background rate is 33 cpm, and the calibration constant is 3.44 disintegrations per count.
   a) If the maximum risk for a type I error is to be 0.05, what is the minimum significant measured activity?
   b) If the gross counting time is increased to 10 min, what is the value of the minimum significant measured activity?

10.6 The quarterly penetrating background is determined to be 12.0 ± 5.0 mrem for a particular geographical area by employing seven dosimeters.
   a) What is the critical level, given that a $\leq 2.5\%$ probability of a false positive is acceptable? (Use Student's t-distribution for small sample size.)
   b) A personnel dosimeter from this area indicates a response of 24.5 rem. What dose should be reported for the employee?
   c) What would have been the fractional change in the critical level if the same background had been determined by using 60 dosimeters?

10.7 Determine in each part below which of the two changes proposed would have the greater impact on the critical level in the last problem.
   a) Doubling the background standard deviation or using only 2 dosimeters to measure background.
   b) Halving the background standard deviation or using 120 dosimeters to measure background.

10.8 Use Eq. (10.10) to show that Eq. (10.11) follows when $t_g = t_b = t$.

10.9 Show that Eq. (10.14) applies when the background is stable and accurately known.

10.10 Derive an expression from Eq. (10.64) for the detection level, L_D, when $z_\alpha \neq z_\beta$.

10.11 Describe and contrast the implications making type I and type II errors in radiation protection measurements.

10.12 Assume the conditions of Altshuler and Pasternack (1963). Two counting systems are being considered for routine use. The calibration constant for counter 1 is 0.0124 nCi per count, and the background $B_1 = 7928$ counts is accurately known. The corresponding data for counter 2 are 0.00795 nCi per count and $B_2 = 15160$ counts, also accurately known. Counting times are the same for evaluating both systems.
   a) At a given level of risk for a type I error, what is the ratio of the minimum significant measured activities for the two counters?
   b) Additional shielding can by placed around counter 1 to reduce its background. It is decided that the acceptable risks for type 1 and type 2 errors are both to be 10%. If only the shielding of counter 1 is changed, what number of background counts B_1 would then be required to achieve a minimum detectable true activity of 1.0 nCi?
   c) What factors determine the value of the calibration constant?

10.13 A 4-min background count and a 16-min gross count are taken with specimens being assessed for activity. The calibration constant is 2.36 Bq per net count. A selected sample registers 120 background and 584 gross counts.
   a) Estimate the expected value of the net count rate and its standard deviation.
   b) What is the implied sample activity?
   c) What is the probability that a sample with zero true activity ($A_T = 0$) would give an activity measurement exceeding that found in (b)? Is the measurement in (b) consistent with zero true activity for this sample?
   d) Assume that $A_T = 0$. What are the smallest net count numbers, L_C, over a 4-min period that would be exceeded with a probability no greater than 0.05 or 0.10?
   e) With these values of L_C, what would be the minimum significant measured activities?

10.14 Measurements of a sample and background, taken over the same length of time, yield, respectively, 395 and 285 counts. The calibration constant is 3.15 Bq per net count. If the maximum risks for both type I and type II errors are 0.05, determine
   a) the critical count number;
   b) whether the sample has activity;
   c) the measured sample activity;
   d) the minimum significant measured activity;
   e) the minimum detectable true activity.

10.15 Ten randomly selected dosimeters give the following readings (in mrem): 4.70, 4.89, 5.18, 4.57, 5.41, 5.11, 4.28, 4.90, 5.19, and 5.42.
   a) Calculate the sample mean and standard deviation.
   b) Determine a 95% confidence interval for the true mean value.
   c) An acceptable reading for dosimeters from this population is 5.1 mrem. Use the t-test to determine whether the mean response from this group differs significantly from the acceptable value with $\alpha = 0.05$. (Recall from Section 6.9 that $t = (\bar{x}-\mu)/(s/\sqrt{n}) \cong t_{n-1}$.)

d) By considering the confidence interval, could one infer that this sample differed significantly from the acceptable value?

10.16 State the (a) null and (b) alternate hypotheses, both symbolically and in words, for deciding whether a net bioassay result indicates the presence of radioactivity in a subject. Are these hypotheses simple or composite? One-tailed or two-tailed, and why?

10.17 The activity concentration of uranium in soil around a processing plant has a mean $\mu = 3.2$ pCi g^{-1} and standard deviation $\sigma = 0.2$ pCi g^{-1}. Soil samples are collected monthly to monitor for possible contamination. The goal is to maintain a state of "no detectable uranium above background" in the soil.
   a) What measured level of uranium in soil would indicate a failure of the containment controls, given an acceptable probability $\alpha = 0.01$ of a false positive?
   b) Suppose that we use $L_C = 3.7$ pCi g^{-1} as the critical value for our test. That is, we will reject H_0: $\mu = 3.2$ pCi g^{-1} if a sample shows a concentration greater than L_C. What is the probability that we will accept H_0 when the true value is 3.2 pCi g^{-1} (a type II error, β)?
   c) Repeat (b) for $\mu = 3.4$, 3.6, 3.8, 4.0, and 4.1 pCi g^{-1}.
   d) Plot the values obtained in (b) and (c), showing μ as the abscissa and $(1 - \beta)$ as the ordinate.
   e) From your plot, determine the power when $\mu = 3.2$ pCi g^{-1}.

10.18 Suppose we estimate μ and σ in the last problem from a sample of size n. Answer the following questions using the corresponding average and sample standard deviation.
   a) Assume H_0: $\mu = \mu_0$ and show that the expression for the critical value L_C, using $\bar{X}$ with a significance level of α is $L_C = \mu_0 + (s/\sqrt{n})t_{n-1,\alpha}$.
   b) Calculate L_C if $s = 0.2$, $\alpha = 0.05$, and $n = 10$ using the expression in (a).
   c) Calculate β using the critical value from (b) when the true mean is $\mu = 3.4$ pCi g^{-1}.
   d) For fixed α, the text mentioned that β can be reduced by increasing the sample size. With everything else remaining the same, consider a sample size of 15. Given $\mu = 3.4$ pCi g^{-1}, calculate β. (Note that you need to recalculate L_C in order to do this.)

10.19 Analysis of a uranium worker's lung count results shows that there are 118×10^3 counts in the region of interest for ^{235}U. The worker's background in the region of interest was established by three lung counts prior to his beginning work with uranium. The results were 45×10^3, 65×10^3, and 80×10^3 counts. All lung counts were performed for a standard 20-min count time. Activity of ^{235}U in the lung for this body counter is determined by using the conversion factor $K = 10$ s^{-1} nCi^{-1}. Answer the following questions, stating all assumptions.
   a) Does the analytical result indicate that there is ^{235}U activity in the worker's lung?
   b) What is the best estimate of ^{235}U activity in the worker's lung?

c) What is the N13.30 MDA for ^{235}U in the worker's lung, treating the pre-uranium-work background counts as measurements of a well-known blank? How does this MDA compare with the ^{235}U activity estimate in part (b)?

d) What would you do next to assess this worker's exposure?

10.20 The ANSI N13.30 formula for calculating MDA contains a semi-empirical term, $3/KT$, for assurance that the $\beta \leq 0.05$ probability of a type II error is assured, even under low-background conditions when the assumption of a normal distribution of results may not be valid. To illustrate this, consider an alpha spectrometer, used for measuring actinides in excreta for a radio-bioassay program, that registers no counts (over equal time intervals) from either background or a paired blank. Ignore the conversion factor, $1/KT$, in the ANSI MDA formula and answer the questions below, considering only the counts recorded in an appropriate region of interest.

a) What is the critical level for an $\alpha \leq 0.05$ probability of a type I error?
b) What is the ANSI N13.30 detection level, L_D (in counts), using Eq. (10.43)?
c) Use the critical level from (a) and assume that counting results in any particular region of interest are Poisson distributed to calculate the probability of a type II error for $L_D = 2$, for $L_D = 3$, and for $L_D = 4$.
d) What would be the value of L_D using the ANSI N13.30 formula without the semi-empirical term (i.e., MDA $= 4.65 s_B$)?
e) Explain why the $3/KT$ term is included in the ANSI N13.30 formula, considering the answers to (c) and (d).

10.21 Twenty dosimeters receive a calibrated dose of 10 000 mGy. The response mean and standard deviation are 10 100 ± 500 mGy. Assume that any difference between the mean response and the known delivered dose is due to system noise. Estimate the relative standard deviation k of the dosimetric response with noise removed.

10.22 Verify Eq. (10.61).

10.23 Verify Eq. (10.62).

10.24 Show that L_D is given by Eq. (10.65) when $z_\alpha = z_\beta = z$.

10.25 Starting with $\Pr(X > L_C | \mu_n = 0)$ after Eq. (10.55), solve part (b) of the example.

10.26 When the Student's t-distribution is used in the example considered in the last problem, show that, for $n = 10$, $L_C = 9.2$ and, for $n = 3$, $L_C = 15$.

10.27 A personnel dosimetry program retrieves dosimeters from its radiation workers every 90 days. The average background accumulation (including system noise) over the 90-day period, determined using five dosimeters from this population, is $\bar{H}_B = 22.0 \pm 5.5$ mrem, and a regression of the background accumulation with time shows an intercept at $H_{B,t=0} = 10.0$ mrem. Calibrated irradiations of three dosimeters drawn from this population show that the uncertainty (one standard deviation) in response to a 500-mrem delivered dose is ±35 mrem.

a) Estimate the critical level at the end of the 90-day assignment period.

b) Estimate the detection level, L_D, at the end of the assignment period, using Eq. (10.66).
c) What are the critical level and decision level near the beginning of the assignment period, if $\bar{H}_B = 11.2 \pm 5.5$ mrem and all other parameters are unchanged?
d) What are the critical level and decision level if 35 dosimeters drawn from this population show the uncertainty in response to a 500-mrem delivered dose to be ± 35 mrem?
e) What are the critical level and decision level at the end of the assignment period if the uncertainty in $\bar{H}_B$ is ± 2.5 mrem, but all other parameters are unchanged?

10.28 Verify Eq. (10.73).

10.29 a) A sample of n observations is taken from a Poisson population with parameter μ. Use the Neyman–Pearson lemma to show that the critical region for testing H_0: $\mu = \mu_0$ versus H_1: $\mu = \mu_1$ ($<\mu_0$) is given by

$$\sum_{i=1}^{n} x_i \leq \frac{\log A + n(\mu_1 - \mu_0)}{\log \mu_1 - \log \mu_0}.$$

In the Neyman–Pearson lemma, the best critical region is such that the ratio of the two likelihoods, one under the alternative hypothesis and the other under the null hypothesis, is less than some constant, which we refer to here as A.

(Hint: $f(x, \mu) = e^{-\mu} \mu^x / x!$ and $L(\mu_i) = \prod_{j=1}^{n} e^{\mu_i} \mu_i^{x_j} / x_j!$, $i = 0, 1$.)

b) If $\mu_0 = 10$, $\mu_1 = 8$, and the sample size is $n = 5$, find the value of A that yields a significance level α closest to 0.05. (Recall that $Y = \sum X_i$ is a Poisson random variable with mean $n\mu$. Use the normal approximation to find k such that $\Pr(Y \leq k) = 0.05$, then use this result to solve for A.)

10.30 A laboratory calibrates ion chambers by recording their responses to a well-characterized source of radiation under the same measurement conditions each day for 5 d. The ionization responses (nC) from one such instrument are 5.224, 5.535, 5.339, 4.980, and 4.516.
a) Which result lies furthest from the mean, and by what amount?
b) What is the *expected* number of results η lying at least as far from the mean as the one in part (a)?
c) Can the value lying furthest from the mean be excluded, and a new mean and variance calculated by using Chauvenet's criterion?

11
Instrument Response

11.1
Introduction

Detection and quantitative measurements are basic to assessment, control, and protection practices in dealing with ionizing radiation. It is essential, therefore, that instrument readings be understood fully and interpreted correctly. A counter can give a misleading result due to the malfunction of a component or due to an incorrect setting. In addition, the responses of many devices also reflect random errors inherent in the atomic processes being monitored. As distinct from systematic errors, such as a wrong setting, effects of purely statistical fluctuations on instrument readings will be the focus of this chapter. We thus deal with certain irreducible limits imposed on precision by the fluctuations in quantum physics, apart from any other sources of uncertainty.

Only a few topics from this rather broad aspect of instrument response will be addressed here. We consider the energy resolution attainable from pulse height measurements with scintillation counters and ionization devices, chi-square testing to check proper functioning of a count rate meter, and dead time corrections for count rate measurements.

11.2
Energy Resolution

A variety of instruments are available for measuring the energy spectra of alpha particles, gamma photons, and other types of radiation. Many devices depend on the collection of a number of charge carriers[1] that result from the complete absorption of a single incident particle or photon in the detector. In a scintillation counter, the energy of the absorbed particle is partially converted into a burst of low-energy

---

1) For example, electrons or, in the case of semiconductors, electron–hole pairs.

(scintillation) photons, some of which liberate photoelectrons that are collected from the cathode of a photomultiplier tube. Ideally, the total charge of the electrons collected in a pulse from the photomultiplier tube – the pulse height – is proportional to the initial energy of the particle absorbed in the scintillator. Depending on the amount of charge collected, the pulse height is registered in the proper energy channel, as determined by independent calibration of the instrument. In a gas proportional counter, electrons liberated directly by absorption of the incident particle are accelerated to produce additional ionizations. Gas multiplication thus occurs, providing a pulse of size dependent on the particle energy.

In all of these "energy proportional" detectors, repeated pulse height measurements with the absorption of single, monoenergetic particles or photons will result in a distribution of recorded energies. The distribution is called the response function of the detector. A hypothetical example is shown in Figure 11.1. The response function is characterized by a peak centered about a mean value E_o, which is the energy of the monoenergetic radiation emitted by the source. The width of the peak reflects the extent of statistical variations in the energy measurement. The narrower the peak, the better the resolution of the counter – that is, its ability to distinguish radiation of one energy in the presence of another. The peak can be characterized quantitatively by its full width at half its maximum height (FWHM). The energy resolution of the detector at energy E_o is then defined as the dimensionless ratio

$$R = \frac{\text{FWHM}}{E_o}, \qquad (11.1)$$

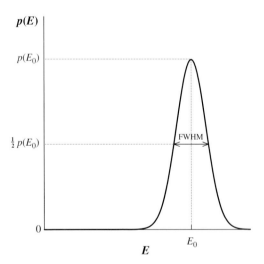

Figure 11.1 Response function for detector absorbing monoenergetic particles or photons with energy E_o. The quantity $p(E)dE$ represents the probability that a given pulse falls within the interval from E to $E + dE$. For a normal distribution, the full width of the peak at one-half the maximum value is equal to 2.35 standard deviations: FWHM $= 2.35\sigma$.

often expressed as a percentage. As we shall assume throughout this chapter, the response function in Figure 11.1 can often be approximated by a normal curve, for which FWHM $= 2.35\sigma$ in terms of the standard deviation σ (Problem 11.1). The resolution can then be written as

$$R = \frac{2.35\sigma}{E_o}. \qquad (11.2)$$

In many detectors, the energy of an absorbed particle or photon is registered according to the amount of charge, or number of charge carriers (e.g., electrons), collected in a pulse. This number is a discrete random variable. Other factors being the same, resolution can be associated with the average energy needed to produce an electron that is collected in the pulse. When this energy is small, numerically large samples of charge carriers per pulse result from the absorption radiation, characteristic of good resolution. We shall see below how the average energy needed to produce a charge carrier compares for different detector types.

Some insight into energy resolution can be gained by assuming initially that fluctuations in the number of collected charge carriers are described by Poisson statistics when monoenergetic radiation is absorbed. For detectors with a linear energy response (i.e., a linear conversion of pulse height into channel number for a pulse of any size), the mean pulse amplitude E_o is proportional to the mean number μ of charge carriers. We write $E_o = k\mu$, where k, the constant of proportionality, is the mean energy needed to produce a collected charge carrier. For a Poisson distribution, the standard deviation in the pulse height is then given by $\sigma = k\sqrt{\mu}$. Applying Eq. (11.2), we write for the resolution with Poisson statistics

$$R_P = \frac{2.35k\sqrt{\mu}}{k\mu} = \frac{2.35}{\sqrt{\mu}}. \qquad (11.3)$$

The estimator for μ is the average number of charge carriers $\bar{n}$. Substituting $\bar{n}$ for μ in Eq. (11.3) gives the estimate for R_P that can be determined by measurement. In this approximation, one sees that the resolution of a detector improves as the reciprocal of the square root of the average number of charge carriers collected from the absorption of a particle or photon. Therefore, the resolution improves as the reciprocal of the square root of the energy of the particle or photon.

Measurements show that a number of radiation detector types have considerably better resolution than that implied by the Poisson value (11.3). Therefore, the processes that produce the individual charge carriers that are collected are not independent in such detectors. The departure of the response of an instrument from Poisson statistics is quantitatively expressed by means of the Fano factor, defined as the ratio of the actual, or observed, variance σ_o^2 in the number of charge carriers and the Poisson variance σ_P^2:

$$F = \frac{\sigma_o^2}{\sigma_P^2}. \qquad (11.4)$$

Substitution of the sample variance s_o^2 as the estimator for σ_o^2 gives the estimator for F, which can be determined by measurement. The observed resolution R_o is related to the Fano factor as follows. Using Eqs. (11.2) and (11.4), one can write

$$R_o = \frac{2.35 \sigma_o}{E_o} = \frac{2.35 \sigma_P \sqrt{F}}{E_o}. \tag{11.5}$$

Since $E_o = k\mu$ and $\sigma_P = k\sqrt{\mu}$, one obtains

$$R_o = 2.35 \sqrt{\frac{F}{\mu}} = R_P \sqrt{F}, \tag{11.6}$$

where Eq. (11.3) has been used for the last equality. Reported values of the Fano factor are in the range from about 0.05 to 0.14 for semiconductors and about 0.05 to 0.20 for gases. The Fano factor is close to unity for many scintillation counters.

■ *Example*

a) Interpret the physical meaning for the two limiting values of the Fano factor, $F = 0$ and $F = 1$, applied to a gas proportional counter.
b) Give a physical reason, based on energy conservation, to explain why gas ionization cannot strictly be a Poisson process.
c) How is energy expended in a gas by a charged particle without producing ionization?

Solution

a) According to the definition (11.4) of the Fano factor, $F = 0$ would mean that there are no fluctuations observed in the number of electrons collected for a given amount of absorbed energy. The resolution would be precise, and the response function (Figure 11.1) would be a delta function at the energy E_o, the FWHM being zero. The value $F = 1$, on the other hand, would mean that the distribution of the number of electrons is consistent with Poisson statistics.
b) A minimum amount of energy, called the *ionization potential*, is always required to free an electron from an atom or a molecule in the gas. This minimum is equal to the binding energy E_{min} of the most loosely bound electron. Theoretically, because of energy conservation, the maximum number of ion pairs that could be produced by absorption of a particle of energy E_o is E_o/E_{min}. Thus, energy conservation and electron binding prevent gas ionization from rigorously obeying Poisson statistics. The latter implies that there is a nonzero probability for the formation of any number of ions.
c) As just described, energy is spent in overcoming the binding energy of electrons. In addition, gas atoms and molecules undergo transitions to discrete, bound, excited states by absorbing energy from incident radiation without ionization. The excited states can then relax (i.e., lose their excess energy) by photon emission, molecular dissociation, or collisions with other molecules.

The discussions in this section have addressed principally effects that the statistical nature of radiation interaction has on resolution when energy measurements are made. A number of additional factors can also affect the overall resolution of a detector. In a gas proportional counter, for example, electronic stability, geometrical nonuniformity in structural parts and in the sensitive volume, and gas purity play a role. Each of these independent sources of error, which add in quadrature, can affect resolution. For most gas and scintillation counters, the principal limitation on resolution arises from the statistical fluctuations. For semiconductors, the same is true at high energies; at low energies, other phenomena can become more important.

11.3
Resolution and Average Energy Expended per Charge Carrier

A charged particle, passing through a gas, loses energy by ionizing and exciting the atoms or molecules of the gas. Some of the secondary electrons it liberates through ionization have enough kinetic energy themselves to cause additional ionizations and excitations in the gas. The total number of electrons thus produced can be collected and measured for different particles of known initial energy that stop in the gas. The average energy spent to produce an ion pair (i.e., a free electron and a positive gas ion) when a particle stops in a gas is called the W value and is usually expressed in eV per ion pair (eV ip^{-1}). Examples of measured W values for alpha and beta particles in several gases are shown in Table 11.1. The values of W are numerically the same, whether expressed in eV ip^{-1} or J C^{-1} (Problem 11.3). W values for alpha particles and other heavy charged particles in polyatomic gases are generally in the range of 30–35 eV ip^{-1}. They are essentially independent of the initial particle energy above several hundred keV, but can be considerably larger for particles of lower initial energy. For a given gas, the W value for beta particles is somewhat smaller than that for alpha particles, and it remains constant down to very low energies. The fact that W values are practically independent of the initial energy of energetic particles has important implications for the use of ionization to measure radiation dose.

Table 11.1 Average energies, W_α and W_β, needed to produce and ion pair (eV ip^{-1}) for alpha and beta particles in several gases.[a]

Gas	W_α (eV ip^{-1})	W_β (eV ip^{-1})
He	43	42
H$_2$	36	36
CO$_2$	36	32
CH$_4$	29	27
Air	36	34

Note: 1 eV ip^{-1} = 1 J.

■ Example
What is the average number of ion pairs produced when a 5-MeV alpha particle stops in air? Does the alpha particle itself produce all of the ion pairs?

Solution
According to Table 11.1, an alpha particle expends an average of 36 eV to produce an ion pair in air. The average number of ionizations is, therefore,

$$\frac{5.00 \times 10^6 \text{ eV}}{36 \text{ eV}} = 1.39 \times 10^5. \tag{11.7}$$

The alpha particle does not produce all of these ion pairs by itself. Some of its ionizing collisions provide secondary electrons with enough energy to ionize additional air molecules. A typical energy loss by an energetic alpha particle in a single ionizing collision is at most a few tens of eV.

For ionization in semiconductors, the W values for producing an electron–hole pair (at 77 K) are 3.76 eV for Si and 2.96 eV for Ge. Compared with a gas, the absorption of a given amount of energy in a semiconductor produces about 10 times as many charge carriers initially, thus providing for inherently better resolution. As we shall see in the next section, the average energy required to produce an electron at the photocathode in a scintillation detector is several hundred eV. For comparison, the energy resolution for 662-keV gamma rays from ^{137}Cs is about 0.3% for high-purity germanium (HPGe), in the neighborhood of 2% for a gas proportional counter, and in the range of 6–10% for the best resolution with a NaI scintillation counter.

11.4
Scintillation Spectrometers

Radiation detectors utilize both organic and inorganic scintillating materials in a number of varied applications. Figure 11.2 shows an example of a pulse height spectrum measured with a sodium iodide crystal scintillator exposed to the 662-keV gamma photons from ^{137}Cs. Various features of the measured spectrum are interpreted as follows. An incident photon that does not escape from the crystal gives rise to an event with an energy registered under the total energy peak (light shading), which is also called the photopeak. Such a photon undergoes complete absorption in the crystal, either producing a photoelectron directly or after one or more Compton scatterings in the crystal. Scintillation photons associated with these processes as well as with any subsequent Auger electrons, characteristic X-rays, or bremsstrahlung rapidly combine and give rise to a single pulse in the region of the total energy peak, centered at 662 keV. The resolution of this particular counter is seen to be about 8%.

Other incident gamma photons, not photoelectrically absorbed, undergo single or multiple Compton scatterings before escaping from the crystal. Such photons produce an event that is registered under the continuous Compton distribution (dark shading) in Figure 11.2. The Compton edge at 478 keV marks the maximum

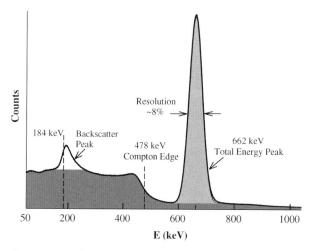

Figure 11.2 Pulse height spectrum from a 2 in. × 2 in. NaI(Tl) scintillation counter exposed to 662-keV gamma photons from ^{137}Cs. The resolution under the total energy peak is ~8%.

energy that a 662-keV gamma ray can transfer to an electron by a single Compton scattering. Some gamma rays from the source enter the crystal only after being scattered into it from surrounding objects. Many of these are reflected in the backward direction from objects beyond the crystal, giving rise to the backscatter peak (unshaded). Since most are not reflected at exactly 180°, their average energy is somewhat larger than the minimum possible after Compton scattering, namely, larger than $662 - 478 = 184$ keV.

As mentioned in the last section, the Fano factor is close to unity for many scintillator systems. We shall assume for analysis here that the resolution of the detector in Figure 11.2 is determined solely by the distribution of charge carriers produced by the Poisson process. A charge carrier in this case is a photoelectron liberated from the cathode at the first stage of the photomultiplier tube. Electron multiplication in the tube is assumed to add negligible variance to that associated with the distribution of the number of photoelectrons that initiate a pulse. The following example illustrates how the inherent resolution of a scintillator arises from the underlying statistical processes.

■ *Example*

A scintillation crystal, like that used for Figure 11.2, is exposed to monoenergetic, 420-keV, gamma rays. The crystal has an efficiency of 8.1% for the conversion of absorbed radiation energy into scintillation photons, which have an average energy of 2.83 eV. An average of 52% of the scintillation photons produced by the absorption of a gamma ray reach the cathode of the photomultiplier, where the efficiency for producing a photoelectron in the initiating pulse is 13%.

a) Calculate the average number of photoelectrons produced per pulse.
b) Calculate the resolution (Poisson) for the 420-keV gamma rays.

c) What is the average energy needed to produce a charge carrier (electron) collected from the cathode of the photomultiplier tube?

d) If the relationship between light yield and absorbed energy is independent of the gamma-ray energy, the detector is said to have a linear response. With assumed linearity, what would be the resolution of the detector for 750-keV gamma rays?

Solution

a) Given the absorption of a 420-keV gamma ray, we trace the various processes sequentially to find the average number $\bar{n}$ of photoelectrons produced. The total energy of the scintillation photons, created with an efficiency of 8.1% when a gamma photon is absorbed, is $4.20 \times 10^5 \text{ eV} \times 0.081 = 3.40 \times 10^4 \text{ eV}$. With an average energy of 2.83 eV, the average number of scintillation photons is $(3.40 \times 10^4 \text{ eV})/(2.83 \text{ eV}) = 1.20 \times 10^4$. Of these, the average number that reach the photocathode is $(1.20 \times 10^4) \times 0.52 = 6.25 \times 10^3$. The average number of photoelectrons produced per pulse is thus $\bar{n} = (6.25 \times 10^3) \times 0.13 = 813$.

b) The resolution, assumed to be determined by Poisson statistics, is found from Eq. (11.3):

$$R_P = \frac{2.35}{\sqrt{\bar{n}}} = \frac{2.35}{\sqrt{813}} = 0.082, \qquad (11.8)$$

or 8.2%.

c) The average energy needed to produce a single photoelectron from the cathode of the photomultiplier tube in the detector is

$$\frac{420\,000 \text{ eV}}{813} = 517 \text{ eV}. \qquad (11.9)$$

d) Under the assumption of a linear energy response for the detector, the above conversion efficiencies are the same for absorption of a 750-keV gamma photon as for a 420-keV photon. It follows from Eq. (11.9) that a 750-keV gamma ray will produce an average of

$$\bar{n}' = \frac{750\,000 \text{ eV}}{517 \text{ eV}} = 1.45 \times 10^3 \qquad (11.10)$$

photoelectrons. In place of Eq. (11.8), we have for the resolution at the higher energy

$$R'_P = \frac{2.35}{\sqrt{\bar{n}'}} = \frac{2.35}{\sqrt{1.45 \times 10^3}} = 0.062, \qquad (11.11)$$

or 6.2%. The collection of a larger number of electrons in a pulse at the higher gamma-ray energy results in improved resolution. One can see that the resolution improves as the inverse square root of the incident photon energy. As an alternative way of solution, one finds directly from Eq. (11.8) that

$$R'_P = R_P \sqrt{\frac{420 \text{ keV}}{750 \text{ keV}}} = 0.082 \sqrt{\frac{420}{750}} = 0.061, \qquad (11.12)$$

which agrees with Eq. (11.11) to within roundoff.

11.5
Gas Proportional Counters

We consider next the effects of statistical fluctuations in energy loss and ionization on the resolution of a gas proportional counter. The pulse of charge collected following the energy lost by a charged particle or photon results from two independent processes. First, the radiation interacts directly with the gas to produce an initial number of secondary electrons, this number being a discrete random variable. These electrons are produced before charge collection begins. For a given amount of energy deposited, this initial number will be distributed about some mean value, $\bar{n}$. When a charged particle of energy E stops in the gas, $\bar{n} = E/W$, as illustrated by the last example. Second, the initial electrons are accelerated by a strong collecting field and can acquire enough energy to produce additional ionizations, which in turn can produce still more, leading to an avalanche. Gas multiplication of the initial charge thus occurs. Under proper, ideal operating conditions, the number of electrons collected in the pulse will be proportional to the original number of secondary electrons produced by the radiation and hence proportional to the energy deposited in the gas.

In an oversimplified picture of what actually takes place, each initial secondary electron produces an avalanche with its own number of additional electrons, which is another discrete random variable. If $\bar{m}$ is the average multiplication factor for an initial electron, then the average charge collected in pulses from the deposition of a given amount of energy in the gas is

$$\bar{Q} = e\bar{n}\bar{m}, \qquad (11.13)$$

where e is the magnitude of the charge of the electron. The distribution of the charge from otherwise identical events is shown directly by the response function of the detector, as in Figure 11.1.

Statistical fluctuations embodied in the response function of a gas proportional counter are due to variations from pulse to pulse in both the initial number of electrons and the individual electron multiplication factors. One can apply error propagation analysis to study the relative contribution of each to the variance of the charge collected (pulse amplitude). Investigations show that fluctuations in gas multiplication typically contribute much more to the spread of the pulse size distribution than fluctuations in the initial number of electrons. To a good approximation, with linearity the resolution for different radiation energies is still inversely proportional to the square root of the energy.

11.6
Semiconductors

Semiconductors, particularly high-purity germanium, give the best energy resolution of any detector. As pointed out in Section 11.3, their excellent resolution is associated with a high yield in the number of charge carriers per unit of energy expended in them. At high energies, the resolution is governed primarily by the statistical fluctuations in the number of charge carriers (electron–hole pairs) produced. Incomplete charge collection, which occurs at all energies, becomes more important at low energies, where the number of charge carriers is relatively small.

Figure 11.3 shows a comparison of measurements made on the same source of 93% enriched uranium using scintillation detectors of thallium-doped sodium iodide (NaI(Tl)) and cerium-activated lanthanum bromide (LaBr$_3$(Ce)), and also using solid-state detectors of cadmium zinc telluride (CdZnTe) and high-purity germanium. The absolute scintillation efficiency of NaI(Tl) is around 13%, whereas that for cerium-activated lanthanum bromide is around 21% (160% as efficient as NaI(Tl)). The greater scintillation efficiency of LaBr$_3$(Ce) results in better resolution than that

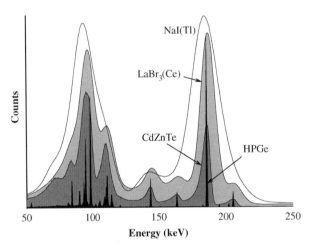

Figure 11.3 Comparison of gamma spectra from 93% enriched uranium measured with a NaI(Tl) scintillation counter, LaBr$_3$(Ce) scintillator, a CdZnTe wide-bandgap semiconductor detector, and a high-purity germanium semiconductor detector. The improved resolution of LaBr$_3$(Ce), compared with NaI(Tl), is due to its superior light output, and that of high-purity germanium, compared with CdZnTe, because of the creation of larger numbers of electron–hole pairs per unit energy absorbed. (Spectral data courtesy of Steven E. Smith, Oak Ridge National Laboratory, U.S. Department of Energy.)

obtained using NaI(Tl), as seen in Figure 11.3. High-purity germanium has superior energy resolution because of the ease of ion-pair formation ($\sim 3\,\text{eV ip}^{-1}$) in the semiconductor crystal. CdZnTe is a semiconductor detector material that can be used at room temperature because of its wide bandgap (HPGe must be cooled to around 77 K), but about 4.6 eV is required to produce an ion pair, so the energy resolution is not as good as with HPGe.

Most of the closely spaced photopeaks, evident in the HPGe spectrum, cannot be resolved at all by the NaI(Tl) and can only be inferred by the presence of asymmetry in the primary photopeaks centered around 98 and 186 keV. CdZnTe and $LaBr_3$(Ce) exhibit better resolution than NaI(Tl), but not as good as that of HPGe.

11.7 Chi-Square Test of Counter Operation

As discussed in Sections 6.6–6.8, chi-square tests are designed to see how well a set of values fits an assumed statistical distribution or model. An important example of their application in radiation protection and in nuclear physics is provided by a quality control procedure used to check whether a counting system is operating properly. A series of repeated counts are made over time intervals of fixed duration. If the counter is functioning as it should, the observed fluctuations in the number of counts are expected to be random and consistent with Poisson statistics. The observance of abnormally large or small fluctuations would indicate the possible malfunction of some component of the counting system. The chi-square test provides a numerical measure for comparison of the observed and expected fluctuations. An example will illustrate this test.

■ *Example*

A GM counter is to be checked for proper operation. Twenty independent, 1-min readings are taken with the counter under identical conditions. The observed count numbers, n_i ($i = 1, 2, \ldots, 20$), are shown in the first two columns of Table 11.2.

a) Compute the value of χ^2 for these data.
b) What conclusion can be drawn about how the counter appears to be functioning?

Solution

a) To determine χ^2 (Eq. (6.93)), we need to compute the mean count number, which will serve as our estimate of the mean μ, and the sum of the squares of the deviations from the mean. The number of degrees of freedom is 19. From the data in column 2 of Table 11.2, we find that

$$\bar{n} = \frac{1}{20} \sum_{n=1}^{20} n_i = \frac{192}{20} = 9.60. \tag{11.14}$$

11 Instrument Response

Table 11.2 Count numbers n_i observed in 1-min intervals from example in the text.

i	n_i	$n_i - \bar{n}$	$(n_i - \bar{n})^2$
1	11	1.40	1.96
2	12	2.40	5.76
3	5	−4.60	21.16
4	13	3.40	11.56
5	10	0.40	0.16
6	11	1.40	1.96
7	7	−2.60	6.76
8	13	3.40	11.56
9	3	−6.60	43.56
10	12	2.40	5.76
11	6	−3.60	12.96
12	11	1.40	1.96
13	9	−0.60	0.36
14	13	3.40	11.56
15	9	−0.60	0.36
16	5	−4.60	21.16
17	6	−3.60	12.96
18	13	3.40	11.56
19	14	4.40	19.36
20	9	−0.60	0.36
Total	192	0.00	202.80

From columns 3 and 4,

$$\sum_{i=1}^{20} (n_i - \bar{n})^2 = 202.80. \qquad (11.15)$$

(The estimated variance of the sample is thus

$$s^2 = \frac{1}{19} \sum_{i=1}^{20} (n_i - \bar{n})^2 = \frac{202.80}{19} = 10.67. \qquad (11.16)$$

This value is close to the sample mean, $\bar{n} = 9.60$, consistent with the Poisson distribution.) It follows that

$$\chi^2 = \sum_{i=1}^{20} (n_i - \bar{n})^2 / \bar{n} = \frac{202.80}{9.60} = 21.13. \qquad (11.17)$$

b) Using Table A.3, we find that the probability of observing a χ^2_{19} value as large as 21.13 or larger is at least 0.3. Thus, this event is not rare or

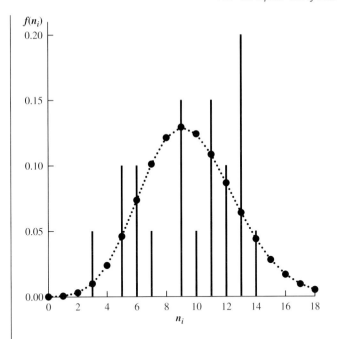

Figure 11.4 Plot of probability distribution $f(n_i)$, shown by bars, and Poisson distribution (filled circles) for 20 values of n_i given in Table 11.3. See example in the text. Both distributions have the mean $\bar{n} = 9.60$. The filled circles are connected by a dotted line for ease of visualization.

significant, and we may conclude that the counter is operating properly. Figure 11.4 shows a plot of the sample frequency distribution of the n_i compiled from Table 11.3 and the Poisson distribution with the same mean, $\bar{n} = 9.60$.

Table 11.3 Data analysis for example in the text.

n	$f(n)$	$nf(n)$	$n - \bar{n}$	$(n - \bar{n})^2 f(n)$
3	0.05	0.15	−6.60	2.18
4	0	0	−5.60	0
5	0.10	0.50	−4.60	2.12
6	0.10	0.60	−3.60	1.30
7	0.05	0.35	−2.60	0.34
8	0	0	−1.60	0
9	0.15	1.35	−0.60	0.50
10	0.05	0.50	0.40	0.01
11	0.15	1.65	1.40	0.29
12	0.10	1.20	2.40	0.58
13	0.20	2.60	3.40	2.31
14	0.05	0.70	4.40	0.97
Total	1.00	9.60	0.00	10.14

11.8
Dead Time Corrections for Count Rate Measurements

A count rate meter registers an individual particle or photon that interacts with it. However, immediately following an event, the counter needs a certain minimum length of time, called the dead time, in order to recover and thus be able to record the next event. Another particle or photon, interacting during this dead interval, will not be registered. When counting a radioactive sample, it is important, therefore, to be aware of any dead time corrections that should be made to the observed count rate, especially with an intense source. The count rate indicated by the detector might be substantially smaller than the rate of events taking place in the detector, which is the relevant quantity.

Two idealized models can be used to approximate the behavior of counters. A nonparalyzable detector is inert for a fixed time τ following an event, irrespective of any other events that occur during τ. A *paralyzable detector*, on the other hand, is unable to respond again until a time τ has passed following any event, even when the event occurs during a dead interval. Whereas the nonparalyzable counter simply ignores events that happen during the downtime τ, the start of the recovery period τ is reset in the paralyzable counter each time an event happens, irrespective of whether that event is registered.

The behavior of the two models is illustrated in Figure 11.5. The top line shows the occurrence of nine events, distributed in time according to the position shown along the horizontal line. The middle and bottom lines indicate how the two types of detectors would respond, given the same dead time τ. Both counters register events 1,

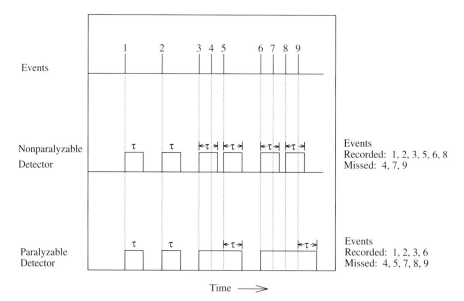

Figure 11.5 Example of events registered by nonparalyzable and paralyzable counter models. See the text.

2, and 3, but miss event 4, which occurs in a time less than τ after event 3. The nonparalyzable counter just ignores event 4. The paralyzable instrument needs at least a time τ following event 4 in order to be able to respond again. Event 5, which follows 3 by a time greater than τ and 4 by less than τ, is registered by the nonparalyzable, but not the paralyzable, counter. As seen from Figure 11.5, the four events 6, 7, 8, and 9 give two counts and one count, respectively, with the two detectors. The events recorded and missed are shown on the right in Figure 11.5. In this example, the nonparalyzable instrument would register 2/3 and the paralyzable counter only 4/9 of the actual events. Most real counting systems exhibit behavior intermediate to these two models. It is interesting to note, as Figure 11.5 illustrates, that radiation counters actually count the number of *intervals* that occur between the events they respond to, rather than the number of events themselves. That is, they register the number of time periods during which the instrument is not responding.

To analyze the response of the two types of counters to radiation fields of different fluence rates, we let r_t be the mean event rate and r_c the mean count rate as registered by the instrument. Both rates are assumed to be constant in time. When the true event rate r_t is small, both detector types in Figure 11.5 will register almost the same count rate r_c. Even though a few events might be missed by both counters, $r_c \cong r_t$. As the event rate increases somewhat, the count rate from both instruments goes up. For the nonparalyzable detector, the count rate will be a little higher than that for the paralyzable detector. For both instruments, though, $r_c < r_t$ because of dead time. If the event rate becomes very large, the nonparalyzable counter will be triggered almost immediately after each recovery time τ. Its count rate will approach the limiting value $1/\tau$, which is the maximum reading that such an instrument is capable of giving. With the paralyzable counter, on the other hand, one can see from Figure 11.5 that increasing r_t to ever larger values will eventually cause the count rate to decrease. In the limiting case of very large r_t, the paralyzable counter never has a chance to recover, and so the count rate approaches zero.

It is straightforward to work out relationships between r_t, r_c, and τ for the models, which enable one to make dead time corrections in order to convert an observed count rate into an estimated true event rate. If a measurement is made over a long time t with a nonparalyzable counter, having a dead time τ, then the number of counts registered, $r_c t$, implies that the counter was unresponsive for a total length of time $r_c t \tau$. The amount of time during which it was responsive, therefore, was $t - r_c t \tau = (1 - r_c \tau) t$. Thus, the fraction of the time t that it was "alive" is $1 - r_c \tau$, which is the fraction of the true events that are registered:

$$1 - r_c \tau = \frac{r_c}{r_t}. \tag{11.18}$$

Solving for the true event rate, we obtain

$$r_t = \frac{r_c}{1 - r_c \tau} \quad \text{(nonparalyzable)}. \tag{11.19}$$

When r_t gets very large, then $r_c\tau \to 1$ in these equations, and so $r_c \to 1/\tau$, as pointed out in the last paragraph. When the event rate is small, $r_c\tau \ll 1$, Eq. (11.19) gives, for low event rates,

$$r_t \cong r_c(1 + r_c\tau), \tag{11.20}$$

where we have used the approximation $(1-x)^{-1} \cong 1 + x$ for small x.

The analysis for the paralyzable counter is a little more involved. We can see from the third line in Figure 11.5 that this counter registers only the number of time intervals that are of length τ or greater between successive events. To see the effect of the dead time, we need to consider the distribution of such time intervals that occur at a mean event rate r_t. Since these events are random, they obey Poisson statistics. The mean number of events that take place in a time t is $r_t t$. Therefore, the probability that no event occurs in a time interval between 0 and t is given by Eq. (5.27) with $\mu = r_t t$ and $x = 0$:

$$p_o = e^{-r_t t}. \tag{11.21}$$

The probability that an event does occur in the time interval between t and $t + dt$ is equal to $r_t\, dt$. Thus, the probability that the duration of a particular time interval, void of any event, will end between t and $t + dt$ is

$$p(t)dt = p_o r_t\, dt = e^{-r_t t} r_t\, dt. \tag{11.22}$$

That is, the probability is the product of (1) the probability $e^{-r_t t}$ that no event has occurred in t and (2) the independent probability $r_t\, dt$ that an event will occur in dt (Eq. (3.50)). The probability that a time interval T longer than τ will occur without an event happening is

$$\Pr(T > \tau) = \int_\tau^\infty p(t)dt = r_t \int_\tau^\infty e^{-r_t t}\, dt = -e^{-r_t t}\Big|_\tau^\infty = e^{-r_t \tau}. \tag{11.23}$$

(The exponential distribution was discussed in Section 6.7.) The observed count rate r_c is the product of the true event rate and the probability (11.23):

$$r_c = r_t e^{-r_t \tau} \quad \text{(paralyzable)}. \tag{11.24}$$

Unlike Eq. (11.19), this transcendental equation cannot be solved in closed form for r_t as a function of r_c. The dead time corrections must be dealt with numerically. For small event rates and $r_t\tau \ll 1$, we can use the exponential series approximation, $e^x = 1 + x + (x^2/2!) + \cdots$ with Eq. (11.24) to show

$$r_c \approx r_t(1 - r_t\tau). \tag{11.25}$$

Additionally, when $r_c\tau \ll 1$, this equation also leads, after some manipulation, to the same Eq. (11.20) for both models (Problem 11.19).

The relationships between count rates r_c and event rates r_t for the paralyzable and nonparalyzable models are shown graphically in Figure 11.6. If there were no dead

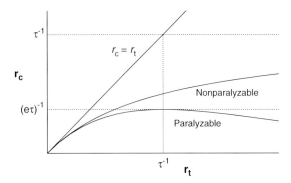

Figure 11.6 Behavior of observed count rate r_c as a function of the true event rate r_t for paralyzable and nonparalyzable counters with dead time τ.

time ($\tau = 0$), then $r_c = r_t$, and the response of both counters would be identical and linear, as illustrated. When $\tau \neq 0$, the count rate for the nonparalyzable detector increases with increasing r_t and tends toward its maximum value of $1/\tau$. The count rate for the paralyzable detector also rises, but then passes through a maximum at the value $1/\tau$, after which it decreases toward zero at large r_t, as explained earlier. The value of r_t that makes r_c a maximum and the resulting maximum value of r_c can be found by differentiation from Eq. (11.24):

$$\frac{dr_c}{dr_t} = (1 - r_t\tau)e^{-r_t\tau} = 0. \tag{11.26}$$

It follows that the maximum count rate occurs when the event rate is $r_t = 1/\tau$. (Note that this event rate is numerically the same as the maximum count rate for the nonparalyzable counter.) From Eq. (11.24), the maximum count rate for the paralyzable counter is then

$$\max r_c = \frac{1}{\tau}e^{-1} = \frac{1}{e\tau}. \tag{11.27}$$

We note in Figure 11.6 for the paralyzable counter that, except at the maximum, there are always *two* values of r_t that correspond to a given instrument reading r_c. A low count rate found with a paralyzable system could be the response to a very intense radiation field.

■ **Example**
A nonparalyzable counter, having a dead time of 1.40 μs, shows a count rate of $1.10 \times 10^5 \, \text{s}^{-1}$.

a) What fraction of the true events is being counted?
b) What is the maximum count rate that the instrument can register?

Solution

a) The true event rate can be found from Eq. (11.19) with $r_c = 1.10 \times 10^5$ s^{-1} and $\tau = 1.40 \times 10^{-6}$ s:

$$r_t = \frac{1.10 \times 10^5 \text{ s}^{-1}}{1 - (1.10 \times 10^5 \text{ s}^{-1}) \times (1.40 \times 10^{-6} \text{ s})}$$
$$= 1.30 \times 10^5 \text{ s}^{-1}. \tag{11.28}$$

The fraction of events being counted is thus

$$\frac{r_c}{r_t} = \frac{1.10 \times 10^5 \text{ s}^{-1}}{1.30 \times 10^5 \text{ s}^{-1}} = 0.846, \tag{11.29}$$

or about 85%. (A more direct solution is obtained from Eq. (11.18).)

b) The maximum count rate that the instrument can record is

$$\frac{1}{\tau} = \frac{1}{1.40 \times 10^{-6} \text{ s}} = 7.14 \times 10^5 \text{ s}^{-1}. \tag{11.30}$$

■ *Example*

Apply the data given in the last example ($\tau = 1.40$ μs and $r_c = 1.10 \times 10^5$ s^{-1}) to a paralyzable counter.

a) Find the true event rate.
b) At what event rate will the instrument show its maximum count rate?
c) What is the maximum reading that the counter will give?

Solution

a) Substitution of the given information into Eq. (11.24) gives

$$1.10 \times 10^5 = r_t \, e^{-1.40 \times 10^{-6} r_t}, \tag{11.31}$$

in which the time unit, s, is implied. As pointed out after Eq. (11.27), we expect two solutions for r_t. These have to be found numerically, as can be accomplished by trial and error. To this end, rather than using Eq. (11.31) with its exponential term, we take the natural logarithm of both sides and write

$$1.40 \times 10^{-6} r_t = \ln \frac{r_t}{1.10 \times 10^5}. \tag{11.32}$$

This form is handier to work with than Eq. (11.31). As a first attempt at solution, we try the result $r_t = 1.30 \times 10^5$ s^{-1} found previously for the nonparalyzable counter in the last example. Substitution of this value into Eq. (11.32) gives 0.182 on the left-hand side, compared with the smaller value, 0.167, on the right. This is apparently close to the actual solution, for which r_t should be somewhat larger (see Figure 11.6). Accordingly, we try 1.35×10^5 s^{-1}, which gives 0.189 on the left and 0.205 on the right in Eq. (11.32). Since the value on the right is now the larger of the two, the solution must lie between these two trial values of r_t. The actual solution is

$r_t = 1.32 \times 10^5 \text{ s}^{-1}$, as can be verified by substitution. To locate the second solution, we arbitrarily try order-of-magnitude steps in the direction of larger r_t. Starting with $r_t = 10^6 \text{ s}^{-1}$, we find that the left-hand side of Eq. (11.32) has the smaller value, 1.40, compared with 2.21 on the right-hand side. With $r_t = 10^7 \text{ s}^{-1}$, the two sides have the respective values 14.0 and 4.51, the left now being the larger of the two. Therefore, the equation has a root in the interval, $10^6 < r_t < 10^7$. Closer inspection shows that $r_t = 3.45 \times 10^6 \text{ s}^{-1}$ satisfies Eq. (11.32) and is the desired solution. In this example, the paralyzable counter gives the same reading in one radiation field as in another where the true event rate is more than 25 times greater.

b) From Eq. (11.26), the maximum count rate occurs when the event rate is

$$r_t = \frac{1}{\tau} = \frac{1}{1.40 \times 10^{-6} \text{ s}} = 7.14 \times 10^5 \text{ s}^{-1}. \tag{11.33}$$

c) The maximum count rate is, from Eq. (11.27),

$$\frac{1}{e\tau} = \frac{1}{2.72 \times 1.40 \times 10^{-6} \text{ s}} = 2.63 \times 10^5 \text{ s}^{-1}. \tag{11.34}$$

We have dealt with the random occurrence of true events as a Poisson process with constant mean rate r_t per unit time. Equation (11.22) gives the probability that the duration of a particular time interval between two successive events will lie between t and $t + dt$. The quantity

$$p(t) = r_t e^{-r_t t} \tag{11.35}$$

gives the probability density for the length of time t between successive randomly spaced events, commonly called the exponential distribution (Section 6.7). This function plays a role in a number of Poisson counting processes. For a large number N of intervals observed over a long time t, the number n that have a length between t_1 and t_2 is

$$n = N \int_{t_1}^{t_2} p(t') dt' = N r_t \int_{t_1}^{t_2} e^{-r_t t'} dt' = N(e^{-r_t t_1} - e^{-r_t t_2}). \tag{11.36}$$

For the response of a paralyzable counter with dead time τ, n is simply the number of intervals for which $t_1 = \tau$ and $t_2 \to \infty$:

$$n = N e^{-r_t \tau}. \tag{11.37}$$

Dividing both sides by the observation time t, we can make the replacements $n/t = r_c$ and $N/t = r_t$, from which Eq. (11.24) follows.

Are the intervals between the successive n counts generated by a Poisson process? One sees from Eq. (11.35) that short time intervals have a higher probability than long intervals between randomly distributed true events. It follows that the counts registered by a paralyzable counter are not a random subset of the Poisson distributed true events, since the shorter intervals are selectively suppressed by the locking mechanism of the instrument. The subset of events registered as counts by a nonparalyzable counter, on the other hand, is Poisson distributed. The distribution of interval lengths between successive counts is exponentially distributed, like Eq. (11.35), with probability density

$$p_c(t) = r_c e^{-r_c t}. \tag{11.38}$$

For a large number of counts, a given event is registered with a fixed probability r_c/r_t. In random time intervals of fixed length t, the mean number of events is $r_t t$, and the mean number of counts is $r_c t$.

Most gas proportional counters, self-quenching GM instruments, and scintillation counters are of the nonparalyzable type. Dead times for GM tubes and other gas-filled detectors are in the range of 50–250 µs. Scintillation counters are much faster, having dead times on the order of 1–5 µs. Charge collection in semiconductor detectors is extremely fast, and system dead time typically depends on factors such as the preamplifier rise time.

Commercial spectral software packages commonly report dead time in terms of the fractional difference (as a percent) between real elapsed time and the system live time during a measurement. The corresponding system response time τ, as presented in this section, is related to this fractional difference by noting that the apparent count rate r_c is the ratio of the total number of events recorded by the detector system and the elapsed time, whereas the true count rate r_t is approximately the ratio of events to the system live time. The relationship can be expressed as

$$\tau \approx \frac{f}{R_c}, \tag{11.39}$$

where $f =$ (real time – live time)/real time. Care should be used, however, in inferring the detector dead time τ from system dead time f, which may depend on several factors.

Problems

11.1 Show that FWHM $= 2.35\sigma$ for the normal distribution (Figure 11.1).

11.2 The resolution of a certain proportional counter, having a linear energy response, is 9.10% for 600-keV photons. What is the resolution for 1.50-MeV photons?

11.3 For the average energy needed to produce an ion pair, show that $1\,\text{eV ip}^{-1} = 1\,\text{J C}^{-1}$.

11.4 Calculate the Poisson energy resolution (FWHM in keV) for the absorption of a 5.61-MeV alpha particle in an HPGe detector ($W = 2.96\,\text{J C}^{-1}$).

11.5 What is the energy resolution of the detector in the last problem if the Fano factor is 0.07?

11.6 If the resolution of a gamma scintillation spectrometer is 10.5% at 750 keV, what is the FWHM in keV at this energy?

11.7 What percentage energy resolution is needed for a gamma-ray spectrometer to resolve two peaks of comparable intensity at 621 and 678 keV?

11.8 With a certain NaI gamma spectrometer, an average of 726 eV is needed to produce an electron collected from the cathode of the photomultiplier tube. If other sources of error are negligible, determine the energy resolution in keV for the total energy peak of the 1.461-MeV gamma ray from ^{40}K.

11.9 Can the instrument in the last problem resolve the two ^{60}Co photons, emitted with 100% frequency at energies of 1.173 and 1.332 MeV?

11.10 A thick sodium iodide crystal with an efficiency of 12% is used in a spectrometer exposed to 580-keV gamma rays. The scintillation photons have an average wavelength of 4130 Å, and 8.2% of them produce a photoelectron that is collected from the cathode of the photomultiplier tube. Calculate the energy resolution in percent for the total energy peak of the spectrometer at 580 keV. Assume that the resolution is determined solely by random fluctuations in the number of photoelectrons produced following the complete absorption of a photon.

11.11 Absorption of a 500-keV beta particle in an organic scintillator produces, on average, 12 400 photons of wavelength 4500 Å. What is the efficiency of the scintillator?

11.12 A series of ten 1-min background readings with a GM counter under identical conditions give the following numbers of counts: 21, 19, 26, 21, 26, 20, 21, 19, 24, 23.
   a) Calculate the mean and its standard deviation.
   b) Compute χ^2.
   c) Interpret the result found in (b).

11.13 Repeat the last problem for test results that give the following numbers of counts: 21, 17, 30, 24, 29, 16, 18, 17, 31, 26.

11.14 Apply the χ^2 test to determine whether the following set of independent count numbers, obtained under identical conditions, shows fluctuations consistent with Poisson statistics: 114, 129, 122, 122, 130, 134, 127, 141.

11.15 A nonparalyzable counter has a dead time of 12 µs.
   a) What is the true event rate when it registers 22 300 cps?
   b) What is the maximum count rate that this instrument can give?

11.16 Repeat the last problem for a paralyzable counter.

11.17 A nonparalyzable counter has a dead time of 27 µs. What count rate will it register when the true event rate is
   a) $20\,000\,\text{s}^{-1}$?
   b) $60\,000\,\text{s}^{-1}$?
   c) $600\,000\,\text{s}^{-1}$?

11.18 Repeat the last problem for a paralyzable counter.

11.19 Show that Eq. (11.24) for the paralyzable counter leads to the same approximate expression, Eq. (11.20), that holds for the nonparalyzable counter, when the event rate is small.

11.20 A point source gives a count rate of $21\,200\,\text{s}^{-1}$ with a nonparalyzable counter. With a second, identical point source added to the first, the count rate is

$38\,700\,\text{s}^{-1}$. What is the dead time of the counter? Background and self-absorption are negligible.

11.21 A nonparalyzable counter, having a dead time of $12.7\,\mu\text{s}$, shows a reading of 24 500 cps.
   a) Calculate the true event rate.
   b) What fraction of the events is not being registered?

11.22 In the last problem, what is the count rate if 10% of the true events are not being registered?

11.23 A paralyzable counter has a dead time of $2.4\,\mu\text{s}$. If the instrument shows a reading of 1.27×10^5 cps, what are the two possible values of the true event rate?

11.24 A thin source of a long-lived radionuclide is plated onto a disk. This source is placed in a counter, and 224 622 counts are registered in 10 s. A second such disk source of the radionuclide is made, having an activity exactly 1.42 times that of the first disk. When the first source is replaced by the second, 461 610 counts are registered in 15 s. If the counter is of the nonparalyzable type, what is its dead time? (Background and self-absorption of radiation in the sources are negligible.)

11.25 As a paralyzable counter is brought nearer to a source, its reading increases to a maximum of 62 000 cps and decreases thereafter as the source is approached even closer. What is the dead time of the instrument?

11.26 A series of gamma photons randomly traverse the sensitive volume of a GM counter at an average rate of $620\,\text{s}^{-1}$.
   a) What is the probability that no photons will traverse the sensitive volume in 1 ms?
   b) What is the probability that a time equal to or greater than 0.010 ms will pass without a photon traversal?

11.27 A series of events occur randomly in time at a constant mean rate of $6.10\,\text{min}^{-1}$ as the result of a Poisson process.
   a) What is the probability that no events will occur during a randomly chosen interval of 1 min?
   b) What is the probability that no events will occur during a randomly chosen interval of 12 s?
   c) What is the relative number of intervals between successive events that have a duration between 20 and 30 s?
   d) What is the median length of time between successive events?

12
Monte Carlo Methods and Applications in Dosimetry

12.1
Introduction

A Monte Carlo procedure is a method of numerical analysis that uses random numbers to construct the solution to a physical or mathematical problem. Its very name and its dependence on random numbers indicate a close association with statistics. The use of sampling from random processes to solve deterministic or other problems was begun by Metropolis and Ulam (1949). Von Neumann and Ulam coined the phrase "Monte Carlo" to refer to techniques employing this idea. It was so named because of the element of chance in choosing random numbers in order to play suitable games for analysis. Monte Carlo procedures provide an extremely powerful and useful approach to all manner of problems – particularly ones that are not amenable to accurate analytical solution. Such statistical techniques are employed in a number of areas in radiation physics and dosimetry, as will be brought out in this chapter.

In the next section we discuss the generation of random numbers, which are at the foundation of any Monte Carlo calculation. In Section 12.3, we illustrate the use of statistical algorithms and random numbers to determine the known numerical answers to two specific problems. The remainder of the chapter deals with applications in radiation physics and dosimetry, including photon and neutron transport. Monte Carlo transport calculations are widely used to determine dose, dose equivalent, and shielding properties. Random sampling is used to simulate a series of physical events as they might occur statistically in nature at the atomic level. Monte Carlo models and computer codes are thus used to calculate radiation penetration through matter. They provide computer-generated histories for a number of charged particles, photons, or neutrons incident on a target. Individual particle histories are generated by a fixed algorithm, enabling the random selection of flight distances, energy losses, scattering angles, and so on to be made as each particle and its secondaries are transported. With the help of a sequence of random numbers, selections for all of these events are made one after another from statistical distributions provided as input to the computer programs. Ideally, the input distributions themselves are obtained directly from experimental measurements or

Statistical Methods in Radiation Physics, First Edition. James E. Turner, Darryl J. Downing, and James S. Bogard.
© 2012 Wiley-VCH Verlag GmbH & Co. KGaA. Published 2012 by Wiley-VCH Verlag GmbH & Co. KGaA.

reliability theory. A large set of particle histories can then be analyzed to compile any information desired, such as the absorbed dose and LET distributions in various volume elements in the target. In a shielding calculation, the relative number, type, and energy spectrum of particles that escape from the target, as well as possible induced radioactivity in it, are of interest. The number of histories that need to be generated for such computations will depend on the specific information wanted and the statistical precision desired, as well as the computer time required.

12.2
Random Numbers and Random Number Generators

Early generators relied on physical methods, such as tossing a die or drawing numbered beads from a bag, to produce a series of "random" numbers. Before the era of electronic computers, tables of random numbers were published in book form. Today, random number generators are available on pocket calculators, personal computers, and sophisticated mainframe machines. Typically, an arbitrary random number "seed" is used to produce a second number, which, in turn, acts as the seed for the next number, and so on. The sequence of random numbers r_i thus produced is designed to span uniformly the semiclosed interval $0 \leq r_i < 1$, or $[0, 1)$.

An example of a random number generator is provided by the equation[1]

$$r_i = \text{FRAC}[(\pi + r_{i-1})^5]. \tag{12.1}$$

Here FRAC[x] is the remainder of x, after the integer portion of the number is truncated (e.g., FRAC[26.127] = 0.127). To illustrate how Eq. (12.1) works, we use an arbitrary seed number, $r_o = 0.534$, and generate the first five numbers, rounding off each to three significant figures before calculating the next r_{i+1}. To begin, we have

$$r_1 = \text{FRAC}[(\pi + 0.534)^5] = \text{FRAC}[670.868] = 0.868. \tag{12.2}$$

The next numbers are

$$\begin{aligned} r_2 &= 0.338, \\ r_3 &= 0.084, \\ r_4 &= 0.179, \\ r_5 &= 0.718. \end{aligned} \tag{12.3}$$

One can see that such a computer-generated random number sequence will only go a finite number of steps without repeating itself, commencing when one of the numbers appears for the second time. Since there are only $10^3 = 1000$ different numerical entries possible in the three-place sequence exemplified by Eq. (12.2), no such sequence can have a period greater than 1000 entries before repeating itself. Also, since there are only 1000 possible seed numbers, there can be no more than

---

1) This generator was utilized with the Hewlett-Packard model HP-25 hand calculator.

1000 independent random number sequences. While hand calculators and electronic computers, especially with double precision, carry many more than three significant figures, their precision is nevertheless finite, and such generators can pose a practical limitation in some Monte Carlo studies. Other types of generators are based on different algorithms, such as use of some aspect of the computer's internal clock reading at various times during program execution. Whereas the sequences thus generated are not periodic in the sense of Eq. (12.1), there are still only a finite number of different entries that the sequences in a computer can contain. Because of these and other considerations, when speaking about random numbers and random number generators in Monte Carlo work, one should perhaps apply the more accurate terminology, "pseudorandom."

Various tests are possible to determine how well a given sequence of pseudorandom numbers approaches randomness. The distribution of truly random numbers is uniform over the interval $0 \leq r_i < 1$. Their mean, or expected value, is, therefore, $E(r_i) = 1/2$, exactly. Since different entries in a random number sequence are completely uncorrelated, the expected value of the product of two different entries is $E(r_i r_j) = 1/4$, $i \neq j$. On the other hand, $E(r_i^2) = 1/3$.

Given a series of numbers in the semiclosed interval [0, 1), how can one test for randomness? There are a number of both simple and sophisticated criteria. Any set of pseudorandom numbers can be expected to pass some tests and fail others. The r_i should be uniformly distributed over the interval. Thus, the fraction of the numbers that lie in any subinterval within [0, 1] should be proportional to the width of that subinterval. The average value of a large set of random numbers, $0 \leq r_i < 1$, should be exactly 1/2. The numbers should also be completely independent of one another. Since the mean value of each r_i is 1/2, the mean of the product $r_i r_j$, with $i \neq j$, of the series should be

$$E(r_i r_j) = E(r_i) E(r_j) = \frac{1}{2} \times \frac{1}{2} = \frac{1}{4}. \tag{12.4}$$

On the other hand,

$$E(r_i r_i) = E(r_i^2) = \frac{1}{3}. \tag{12.5}$$

Many other tests for randomness have been developed. To mention only one, groups of five numbers are used to select poker hands. The resulting frequencies of the various hands are then compared with the known probabilities for randomly drawing them.

For additional information on Monte Carlo techniques, random numbers, and random number generators, the reader is referred to the following publications in the Bibliography at the end of the book: Atkinson (1980), Carter and Cashwell (1975), Kalos and Whitlock (1986), Kennedy and Gentle (1980), Knuth (1980), Metropolis and Ulam (1949), Newman and Barkema (1999), and Ulam (1983). The paper by Turner, Wright, and Hamm (1985) is a Monte Carlo primer written especially for health physicists.

12.3
Examples of Numerical Solutions by Monte Carlo Techniques

In this section we obtain the numerical solutions to two problems by the use of Monte Carlo techniques. For both problems, the Monte Carlo results can be compared with the exact solutions, which are known.

12.3.1
Evaluation of $\pi = 3.14159265\ldots$

In Figure 12.1, a quadrant of a circle of unit radius is inscribed in a unit square. The area of the circular quadrant is $\pi/4$, which is also just the fraction of the area of the square inside the quadrant. If one randomly selects a large number of points uniformly throughout the square, then the fraction of points that are found to lie within the circular quadrant would have the expected value $\pi/4$. An estimate of π can thus be obtained from a random sample of a large number of points in the square. Statistically, one would expect that the precise value of π would be approached ever more closely by using more and more points in the sample.

A simple computer program can perform the needed computations. Pairs of random numbers, $0 \leq x_i < 1$ and $0 \leq y_i < 1$, determine the coordinates (x, y) of points in the square, as shown in Figure 12.1. A numerical check is made for each

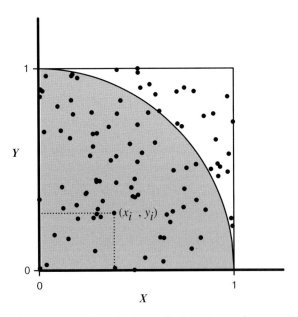

Figure 12.1 Quadrant of circle inscribed in unit square has area $\pi/4$, which is the expected value of the fraction of randomly chosen points (x_i, y_i) in the square that fall within the quadrant. Inside the circle, $x_i^2 + y_i^2 \leq 1$. See Table 12.1.

Table 12.1 Value of π (=3.14159265...) obtained by Monte Carlo procedure.

Total number of points	Number of points within circle	Estimated value of π	Percent error
10	5	2.000000	36.3
10^2	75	3.000000	4.51
10^3	783	3.132000	0.305
10^4	7 927	3.170800	0.930
10^5	78 477	3.139080	0.0800
10^6	785 429	3.141716	0.00393
10^7	7 855 702	3.142281	0.0219
10^8	78 544 278	3.141771	0.00568
10^9	785 385 950	3.141544	0.00155

point to see whether $x_i^2 + y_i^2 \leq 1$. When this condition holds, the point is tallied; otherwise, it is not tallied. After selecting a large number of points, the ratio of the number of tallied points and the total number of points used gives the estimated value of $\pi/4$.

Numerical results calculated in this way are shown in Table 12.1 for samples with sizes ranging from 10 to 10^9 points. In a sample of just 1000 random points, for example, 783 were found to lie within the circular quadrant, implying that $\hat{\pi} = 4 \times (783/1000) = 3.13200$. This estimate differs from the true value by 0.305%.

One sees from the table that using more points does not always give a more accurate computed value of $\hat{\pi}$, although there is general improvement as one goes down in Table 12.1. One must remember that the numbers of tallied points in column 2 of Table 12.1 are themselves random variables. With a different sample of 1000 points, chosen with a different set of random numbers, one would probably not obtain exactly 783 tallies again. The expected number of tallies for a sample of 1000 is, in fact, $(\pi/4)(1000) = 785.3982$. The number of tallies is a binomial random variable with $p = \pi/4 = 0.7854$. For $N = 1000$, we know that its coefficient of variation (Section 7.3) is $\sigma/\mu = \sqrt{(1-p)/Np} = \sqrt{0.2146/785.4} = 0.0165$, or about 1.65%. The particular sample used for Table 12.1 had very nearly the expected value of tallies, and the actual error in the calculated value of π, shown in the last column of Table 12.1, is only 0.3%. One sees from the last column of the table that the percent error decreases with sample size, but not monotonically, because of random fluctuations.

12.3.2
Particle in a Box

Figure 4.3 shows the quantum-mechanical probability density (Eqs. (4.15) and (4.16)) for a particle confined to a one-dimensional box in its state of lowest energy (ground state). As determined analytically by Eq. (4.24), the probability of finding the particle between $X = 0$ and $X = a/4$ is 0.409. We now evaluate this probability in two different ways by means of Monte Carlo algorithms.

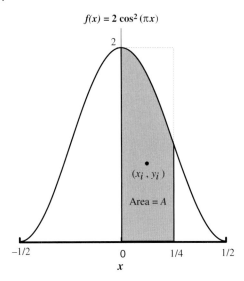

Figure 12.2 Probability of finding particle in the box in a location between $x=0$ and $x=1/4$ is equal to the shaded area A under the curve $f(x)$.

Method 1
As indicated in Figure 12.2, the desired probability is equal to the area A within the heavy border under the curve $f(x)$, defined by Eq. (4.15). (Since the width a of the box is arbitrary, we have set $a=1$ in Figure 12.2. Alternatively, if one carries a explicitly through the computations, it drops out identically later.) As we did to find the value of π in the last problem, we can use pairs of random numbers to select a large number of points (x_i, y_i) uniformly within the rectangle completed with the dotted lines in Figure 12.2. The expected value of the fraction F of points in a large sample that fall under the curve $f(x)$ is equal to the ratio of (1) the area A, being sought, and (2) the area of the rectangle, which is $2 \times 1/4 = 1/2$. Thus, $F=2A$, and so $A=F/2$. Finding the fraction by choosing random points thus determines the magnitude of A. (Both this and the last problem are examples of doing numerical integration by Monte Carlo methods.) One random number $0 \leq r_i < 1$ provides a value $x_i = r_i/4$ between $x=0$ and $x=1/4$, as desired. A second random number r'_i picks $y_i = 2r'_i$ between 0 and 2. For each point chosen in the rectangle, we test the inequality

$$y_i < 2\cos^2 \pi x_i. \tag{12.6}$$

When a point with coordinates (x_i, y_i) satisfies Eq. (12.6), it is under the curve $f(x)$ in Figure 12.2, and is, therefore, tallied. Otherwise, the point is not tallied. At the end of a calculation, the fraction F of the points that are tallied provides the estimate of the area: $A = F/2$. This procedure was carried out for samples ranging in size from 10 to 10^9 points, and the results for this method are shown in column 2 of Table 12.2. They are in good agreement with the known value of the area, which is, from Eq. (4.24), $A = 0.409155$.

Table 12.2 Value of area (A = 0.409155) under curve in Figure 12.2 calculated by two Monte Carlo methods.

Total number of points	Area by method 1	Area by method 2
10	0.350000	0.361441
10^2	0.400000	0.400124
10^3	0.410500	0.406790
10^4	0.413400	0.410627
10^5	0.408445	0.409148
10^6	0.409145	0.409473
10^7	0.409259	0.413869
10^8	0.409170	0.083886
10^9	0.409141	0.008389

Method 2

In this method of solution, random values of x_i, $0 \leq x_i < 1/4$, are selected, just as in the first method. Now, however, for each x_i one computes the value of the ordinate on the curve, $y_i = 2 \cos^2 \pi x_i$. The average value $\bar{y}$ of the ordinate is then evaluated for a large number of points N:

$$\bar{y} = \frac{1}{N} \sum_{i=1}^{N} y_i. \tag{12.7}$$

The area A in Figure 12.2 is given by the product of the average value $\bar{y}$ of y and the width $1/4$ of the interval: $A = \bar{y}/4$. Results for calculated area are presented in the last column of Table 12.2. The estimated area at first improves as more points are used, but then becomes progressively worse when more and more points are used. This failure of method 2 occurs because of the limited numerical precision of the computer. The problem is one of roundoff. A real (floating point) number like $\bar{y}$ in Eq. (12.7) is stored in a register with a fixed number of bytes, giving its sign, decimal point location (exponent), and a certain number of significant figures. The computer cannot add a small number to a very large number beyond the precision that this system imposes. For instance, with a precision of six significant figures, the following sum would be thus evaluated: $4.29846 \times 10^9 + 12.7 = 4.29846 \times 10^9$. Roundoff occurs. In the present problem, as seen from Figure 12.2, the largest value of y_i that can contribute to the sum in Eq. (12.7) is 2.0, and so roundoff occurs once the accumulated sum reaches a certain size, which is beyond this precision. We see from the last two entries in the last column of Table 12.2 that the sum in Eq. (12.7) is the same for $N = 10^9$ points as for 10^8. (The terminal value of the sum, with the computer used for these computations, turns out to be 3.355443×10^7.) Roundoff error can be made to occur later in such a calculation (i.e., for larger N) when the computations are performed in double precision, as is often done in Monte Carlo work. In contrast to method 2 for solving the same numerical problem, method 1 calculates the area as the ratio of two *integers*, which are stored with greater precision as binary sequences in the computer. Thus, knowledge of how the computer stores numbers and performs numerical analysis is very important in the proper use of Monte Carlo methods.

12.4
Calculation of Uniform, Isotropic Chord Length Distribution in a Sphere

Spherical, tissue-equivalent, gas proportional counters find widespread applications in dosimetry and microdosimetry. Often called Rossi counters, after the pioneering work of Harald H. Rossi and coworkers, they were used early on for the task of determining LET spectra from energy proportional pulse height measurements. In an idealized concept, a charged particle (produced in the wall of a Bragg–Gray chamber) traverses the sensitive volume of a proportional counter with constant LET, L, along a chord of length η. With energy-loss straggling ignored (in reality, it is large), the energy registered by the counter would be $\varepsilon = L\eta$. A pulse of a given amplitude ε could be due to either a particle of relatively high LET traversing a short chord or a particle of low LET traversing a long chord. If the radiation is isotropic, then, since the distribution of isotropic chord lengths in a spherical cavity is known (see the next section), one can assign a statistical probability for chord length to each pulse. This assignment, in turn, implies a statistical probability for the LET of the particle that produced it. Mathematically, the unfolding of the LET spectrum from an observed pulse height spectrum and a known chord length distribution is equivalent to solving a convolution integral equation. Technically, the experimental procedure presents formidable difficulties, but it has led to important and useful information on LET spectra. Today, spherical, tissue-equivalent proportional counters are employed in a number of different ways in research and routine monitoring. In addition, with proper selection of sphere size and gas pressure, one can employ such chambers to simulate unit density spheres that have relevant biological sizes, for example, that of a cell, cell nucleus, or other microscopic structure. The subject of microdosimetry deals with the distribution of physical events on a scale of microns and smaller and the use of such distributions to interpret and understand the biological effects of radiation (Rossi and Zaider, 1996).

Monte Carlo calculations can be performed to find the distribution of chord lengths, having any spatial arrangement in a cavity of any shape. To show some details of this kind of a Monte Carlo calculation, we now compute the isotropic chord length distribution for a sphere.

Figure 12.3a shows schematically a sphere of radius R traversed by a parallel, uniform beam of tracks. The distribution of chord lengths in the sphere with this geometry is the same as that for a sphere immersed in a uniform, isotropic field of tracks. Because of spherical symmetry, the relative orientation of the sphere and the tracks does not matter, as long as the tracks are uniform in space. To obtain the desired chord length distribution by Monte Carlo means, we can select a large number of incident tracks at random from the parallel beam, compute the resulting chord length for each, and compile the results. The random selection of the tracks can be made with reference to Figure 12.3b, which is a view at right angles to that in Figure 12.3a, along the direction of the tracks. The projection of the sphere presents a circle of radius R to the beam. The intersection of the tracks is uniform over the plane of the circle. We can select a random point of intersection (x_i, y_i) by choosing two random numbers, as we did in the computation of π (Figure 12.1). If the point is

12.4 Calculation of Uniform, Isotropic Chord Length Distribution in a Sphere

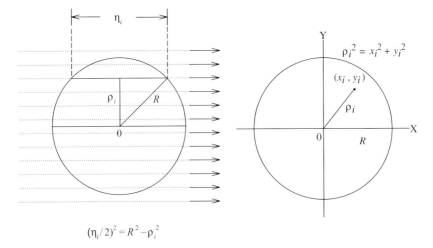

Figure 12.3 (a) Broad, uniform, parallel beam of tracks traversing sphere of radius R centered at O. Chord length distribution in sphere is same as for uniform, isotropic tracks. (b) Random point (x_i, y_i) in circle of radius R can be used to select random chord, having length η_i in sphere.

outside the circle of radius R, then we simply ignore it here and choose the next one. If the point is inside the circle, then its displacement from the center is

$$\varrho_i = \sqrt{x_i^2 + y_i^2}, \tag{12.8}$$

and so the resulting chord length is

$$\eta_i = 2\sqrt{R^2 - \varrho_i^2}. \tag{12.9}$$

Because of the symmetry in this problem, no generality is lost by always choosing the random point (x_i, y_i) in the first quadrant of the circle in Figure 12.3b.

It is convenient to express the chord lengths as dimensionless fractions of the longest chord, whose length is equal to the sphere diameter, $2R$. Dividing both sides of Eq. (12.9) by $2R$, we write

$$\frac{\eta_i}{2R} = \sqrt{1 - \frac{\varrho_i^2}{R^2}} = \sqrt{1 - 4\left(\frac{\varrho_i}{2R}\right)^2}. \tag{12.10}$$

Letting $\eta'_i = \eta_i/2R$ and $\varrho'_i = \varrho/2R$ denote the dimensionless quantities, we have

$$\eta'_i = \sqrt{1 - 4\varrho'^2_i}. \tag{12.11}$$

Since, in units of $2R$, x_i and y_i vary in magnitude between 0 and 1/2, their random values can be selected by writing

$$x_i = 0.5 r_i \quad \text{and} \quad y_i = 0.5 r'_i, \tag{12.12}$$

where $0 \leq r_i < 1$ and $0 \leq r'_i < 1$ are random numbers from the uniform distribution.

Table 12.3 Data for 25 random chords in sphere (see the text).

i	x_i	y_i	ϱ'_i	η'_i
1	0.2172	0.2272	0.3143	0.7778
2	0.0495	0.1633	0.1707	0.9399
3	0.3518	0.1144	0.3699	0.6728
4	0.3088	0.0398	0.3114	0.7824
5	0.3359	0.1289	0.3598	0.6944
6	0.1692	0.2132	0.2722	0.8389
7	0.0585	0.1762	0.1856	0.9285
8	0.2657	0.2890	0.3926	0.6193
9	0.4330	0.0067	0.4331	0.4998
10	0.4394	0.0561	0.4430	0.4636
11	0.2194	0.4386	0.4904	0.1945
12	0.3115	0.1170	0.3328	0.7463
13	0.2503	0.1046	0.2713	0.8400
14	0.1618	0.0334	0.1652	0.9438
15	0.1374	0.0667	0.1528	0.9522
16	0.3379	0.3452	0.4831	0.2580
17	0.0366	0.3832	0.3850	0.6381
18	0.1966	0.3598	0.4100	0.5724
19	0.1612	0.4135	0.4438	0.4607
20	0.4017	0.2500	0.4731	0.3235
21	0.0570	0.0663	0.0874	0.9846
22	0.1483	0.2940	0.3293	0.7525
23	0.2448	0.3278	0.4091	0.5749
24	0.0220	0.2917	0.2925	0.8111
25	0.2408	0.1282	0.2728	0.8380

To show some details involved in obtaining such a distribution by Monte Carlo means, we produce and analyze a small sample of 25 chords. Table 12.3 shows the lengths of 25 chords obtained with the help of a random number generator and Eq. (12.12). The values of x_i, y_i, ϱ'_i, and the 25 chord lengths η'_i are given explicitly. It is convenient to display such Monte Carlo results in the form of a histogram, showing how many times the random values η'_i of the chord lengths fall within different "bins," or specified ranges of η'_i. We arbitrarily choose 10 bins of equal width, $\Delta\eta'_i = 0.10$, for an initial display. Tabulation by bin of the sample of 25 chords from Table 12.3 is presented in the first two columns of Table 12.4. The relative frequency with which chords occur in each bin (the number in column 2 divided by 25) is given in column 3. The last column shows the relative frequency per unit bin width, which is obtained by dividing each of the numbers in column 3 by the width of its bin, which in this example is 0.10 for each. (It is not necessary for the bins to have the same size. In any case, plotting the *relative frequency per unit bin* width forces the area of the histogram to be unity.) The histogram thus obtained from the first and last columns of Table 12.4 is displayed in Figure 12.4. Since its area has been made to equal unity, the histogram gives directly our sample's approximation to the actual *probability density function*, $p(\eta'_i)$, which we seek to find by Monte Carlo means. The actual probability density is a function of the continuous random variable η' such that

12.4 Calculation of Uniform, Isotropic Chord Length Distribution in a Sphere

Table 12.4 Tabulation of chords from Table 12.3 into 10 bins of width 0.10 (see the text).

Chord length, η'_i	Number of chords	Relative frequency	Probability density, $p(\eta'_i)$
0.0000–0.1000	0	0.0000	0.0000
0.1000–0.2000	1	0.0400	0.4000
0.2000–0.3000	1	0.0400	0.4000
0.3000–0.4000	1	0.0400	0.4000
0.4000–0.5000	3	0.1200	1.2000
0.5000–0.6000	2	0.0800	0.8000
0.6000–0.7000	4	0.1600	1.6000
0.7000–0.8000	4	0.1600	1.6000
0.8000–0.9000	4	0.1600	1.6000
0.9000–1.0000	5	0.2000	2.0000

$p(\eta')d\eta'$ gives the *probability* that the length of a random isotropic chord η' falls between η' and $\eta' + d\eta'$ (Eqs. (4.6, 4.7, 4.8)). Since chord lengths in Figure 12.4 are represented as fractions of the sphere diameter, the distribution is numerically the same as that for chord lengths in a sphere of unit diameter.

The histogram in Figure 12.4, although the result of a sample of limited size, suggests that the chord length distribution is a monotonically increasing function of η', as can be surmised from inspection of Figure 12.3. There is some scatter in the results, reflecting the fact that each bin has only a few (≤ 5) representatives. Statistical scatter can generally be reduced in Monte Carlo histograms of a fixed sample size by using larger bins. In so doing, however, one suppresses some information by

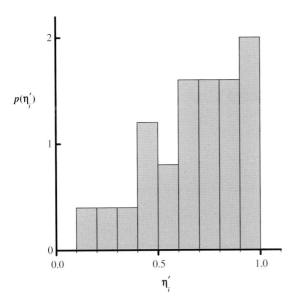

Figure 12.4 Sample approximation to probability density function $p(\eta'_i)$ for isotropic chord length distribution in sphere of unit diameter. The histogram is formed by sorting the 25 chord lengths in the sample into 10 equal bins of width $\Delta\eta'_i = 0.10$. See Table 12.4.

Table 12.5 Tabulation of chords from Table 12.3 into five bins of width 0.20 (see the text).

Chord length, η'_i	Number of events	Relative frequency	Probability density, $p(\eta'_i)$
0.0000–0.2000	1	0.0400	0.2000
0.2000–0.4000	2	0.0800	0.4000
0.4000–0.6000	5	0.2000	1.0000
0.6000–0.8000	8	0.3200	1.6000
0.8000–1.0000	9	0.3600	1.8000

lowering the resolution with which the independent variable, in this case η', is portrayed. Using five bins of equal width, $\Delta\eta'_i = 0.20$, for our sample of 25 chords presents the data shown in Table 12.5 and plotted in Figure 12.5. The new presentation of the same sample of chord lengths conveys a different impression of how the probability density distribution might look. The monotonic behavior of $p(\eta'_i)$ is more evident here than in the previous figure. In fact, drawing a smooth curve from the origin through the midpoints at the top of each bin gives a good approximation to the actual continuous distribution function (see Figure 12.10). Since there is no requirement that bins be of equal size, it is sometimes advantageous to combine small bins into larger ones along the abscissa only in regions where the number of events is small. In any case, the height of the histogram for a given bin must show the relative number of events there divided by the width of that bin in order that normalization be preserved to represent the probability density function for the continuous random variable.

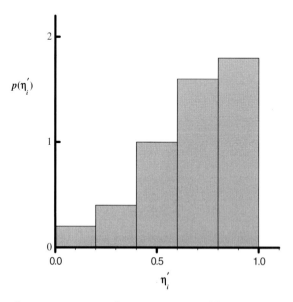

Figure 12.5 Same sample as in Figure 12.4 with histogram constructed by using five bins of equal width $\Delta\eta'_i = 0.20$. See Table 12.5.

12.4 Calculation of Uniform, Isotropic Chord Length Distribution in a Sphere

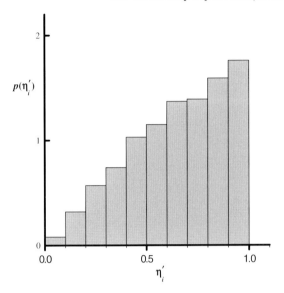

Figure 12.6 Probability density function $p(\eta'_i)$ for isotropic chord length distribution obtained from 10^3 random chords, sorted into 10 equal bins of width $\Delta\eta'_i = 0.10$, in sphere of unit diameter.

A more definitive representation of the probability density function for the isotropic chord length distribution can be generated with a large sample. As examples, Figures 12.6 and 12.7 show, respectively, results from tallying 10^3 random chords in 10 uniform bins and 10^5 chords in 100 uniform bins.

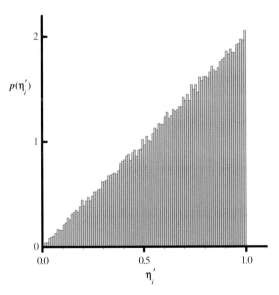

Figure 12.7 Probability density function $p(\eta'_i)$ for isotropic chord length distribution obtained from 10^5 random chords sorted into 100 equal bins of width $\Delta\eta'_i = 0.01$, in sphere of unit diameter.

One sees that the chord length distribution is linear, as we shall show analytically in Section 12.5.

12.5
Some Special Monte Carlo Features

Some special devices can be employed to improve the efficiency of Monte Carlo computations and to assess their statistical uncertainties. We briefly mention several here.

12.5.1
Smoothing Techniques

Figures 12.4 and 12.5 exemplify how the same set of raw data can have a different appearance and possibly even convey more than one impression when presented in different ways. We mentioned that, in general, grouping data into fewer, but larger, bins gives "better statistics" in the sense of a smoother looking histogram, but at the expense of poorer resolution of any structure in the probability density function for the independent variable. Smoothing routines are sometimes employed to reduce fluctuations. Instead of plotting the individual values of the probability density as the ordinate of each bin, one plots a weighted average of that density and the densities from several (e.g., two, four, or more) adjacent bins. The resulting histogram will generally show less fluctuation than the original raw numbers. As an example, the data for the 25 chords were smoothed as follows. We arbitrarily modified the numbers in the last column of Table 12.4 by weighted averages. Except for the two end bins, each value of $p(\eta'_i)$ in the last column was replaced by the average comprised of 1/2 times that value plus 1/4 times the sum of the values in the adjacent bins on either side. For instance, between $\eta'_i = 0.4000$ and $\eta'_i = 0.5000$, the probability density 1.200 was replaced by the smoothed value $(1/2)(1.200) + (1/4)(0.4000 + 0.8000) = 0.9000$. For the two end bins, the original value in Table 12.4 was weighted by 3/4 and added to 1/4 the value from the single adjacent bin. The smoothed data are shown by the histogram in Figure 12.8. In this instance, we obtain a much closer resemblance to Figure 12.7 than that afforded by Figure 12.4, even though both are based on the same sample of 25 events.

12.5.2
Monitoring Statistical Error

Monte Carlo calculations that produce data like we have been considering here can be repeated exactly, except with a different set of random numbers. For each independent computation, the number of events that occur in any given bin and its distribution can be compiled. Generation of this number satisfies the criteria for Bernoulli trials. The number in each bin of the Monte Carlo histogram is, then, a binomial random variable (Sections 5.3 and 5.4). For large samples, such as those

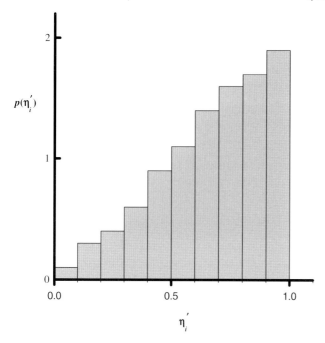

Figure 12.8 Re-plot of Figure 12.4 after simple smoothing of raw data for the 25 chords, as described in the text.

used for Figures 12.6 and 12.7, the number of events in each bin can also be described by Poisson statistics. The standard deviation of the number in each bin can, therefore, be estimated as the square root of the number obtained in the calculation. To illustrate, Table 12.6 gives the number of events, the estimated standard deviation, and the coefficient of variation, or relative error (Section 7.3), for the first and last three bins plotted in Figure 12.7 from the sample of 10^5 isotropic chords. The coefficients of variation are, of course, largest (15–30%) in the first bins, where there are the fewest events. Combining the first three bins, which contain a total of

Table 12.6 Estimated coefficients of variation for first and last three bins of histogram in Figure 12.7.

Chord length	Number of events	Estimated standard deviation	Coefficient of variation
0.0000–0.0100	11	3.32	0.30
0.0100–0.0200	39	6.24	0.16
0.0200–0.0300	42	6.48	0.15
...			
0.9700–0.9800	1973	44.4	0.023
0.9800–0.9900	1964	44.3	0.026
0.9900–1.0000	2054	45.3	0.022

92 chords, into a single bin reduces its coefficient of variation to $1/\sqrt{92} = 0.10$. As a rule in Monte Carlo work, depending on the information desired, one makes a trade-off between the resolution needed, the amount of random fluctuation that can be tolerated, and the computer time available. In any case, statistical uncertainties in Monte Carlo results can be estimated from the data generated.

12.5.3
Stratified Sampling

Variances in the estimated values of π given in Table 12.1 are due to both limited sample sizes and the random nature of the selection of points (x_i, y_i). As for the latter, at the end of a calculation, the values of the x_i and y_i in the sample would, ideally, be found to be uniformly distributed over the semiclosed interval [0, 1). The variance in the values estimated for π can, in principle, be reduced without bias by simply choosing one of the independent variables systematically before selecting the other randomly. For example, the following algorithm selects N values, spaced systematically and with a uniform separation of $1/N$, along the X-axis between $x = 0$ and $x = 1$:

$$x_i = \frac{2i - 1}{2N}, \quad i = 1, 2, \ldots, N. \tag{12.13}$$

For each of these x_i, taken sequentially, y_i can be chosen at random, as before, to determine the random point (x_i, y_i). By thus reducing the variance of the x_i, the overall variance in the results is reduced. Such a technique of selecting a random variable with minimum variance is called *stratified sampling*.

■ **Example**

a) The algorithm (12.12) was used with a random number generator to select the 25 random chords described in Table 12.3. Would you expect that more than 25 random number pairs are needed to produce the 25 chord lengths by means of this algorithm?

b) In place of $x_i = 0.5 r_i$ in Eq. (12.12), write an algorithm that selects 25 values of x_i uniformly by stratified sampling over the proper interval.

c) With the stratified sample of values of x_i in (b), the random selection of y_i for a given x_i could give $\varrho'_i > 0.5$. Should one then select another random value of y_i for that x_i, or should one skip that value of x_i altogether and proceed to x_{i+1}?

Solution

a) The point (x_i, y_i) in Figure 12.3b must be within the circle to be used. Therefore, we would expect that more than 25 random points selected by the algorithm (12.12) would be needed in order for 25 to be inside. In connection with Figure 12.1 and Table 12.1, we found that the probability is 0.785 for a randomly chosen point to fall inside. The probability that 25 successive points are all within the circle is $(0.785)^{25} = 0.00235$.

b) A stratified sample of values between zero and unity is given by Eq. (12.13) with $N = 25$. The chord lengths in Eq. (12.12) are expressed in units of their maximum, which is $2R$. Since we require that the x_i be distributed uniformly between 0.0 and 0.5, we write in place of Eq. (12.13), with $N = 25$,

$$x_i = \frac{2i-1}{4N} = \frac{2i-1}{100}, \quad i = 1, 2, \ldots, 25. \tag{12.14}$$

The values of the x_i are $1/100$, $3/100$, ..., $49/100$.

c) As noted in (a), it is unlikely that 25 randomly chosen points will fall inside the circle when the y_i are randomly selected. Previously, when $\varrho'_i > 0.5$, we simply ignored such a point and kept randomly sampling for both x_i and y_i in pairs until we obtained 25 points inside the circle. Now, however, we can use each x_i only once in the unbiased stratified sample. If a point is not in the circle, we drop it and go to the next value, x_{i+1}. In contrast to the algorithm (12.12), the algorithm (12.14) will probably furnish fewer than 25 actual chords.

12.5.4
Importance Sampling

As Figure 12.7 strongly suggests, the chord length distribution $p(\eta'_i)$ passes through the origin (0, 0). One might wish to focus on the detailed structure of the distribution in this region. The calculation as we have presented it has the fewest events and thus the largest coefficient of variation in the first few bins in this region. To improve the statistical uncertainty there, one could simply make calculations with a larger sample of chords. However, this procedure would be very inefficient – we see from Table 12.6 that only 11 chords fell within the first bin out of 10^5 selected randomly. Much more efficient is a technique of *importance sampling*, in which the chord selection process is forced to sample repeatedly from the range of η'_i values of particular interest, in order to reduce the variance there. Special bin sizes can also be utilized to such an end. Results obtained by importance sampling must then be given proper statistical weighting when combined with the rest of the probability density function.

Some other special techniques, used to increase the efficiency of Monte Carlo calculations of radiation transport, will be pointed out in Section 12.11.

12.6
Analytical Calculation of Isotropic Chord Length Distribution in a Sphere

In the last section, we reasoned that the distribution of uniform, isotropic chords in a sphere is the same as that for the parallel, uniform beam shown in Figure 12.3. We now present an analytical solution for that distribution, which was found numerically to a good approximation by Monte Carlo means, as shown by Figure 12.7.

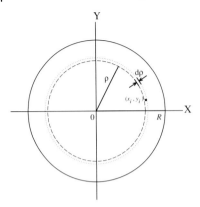

Figure 12.9 Same circle as in Figure 12.3b, looking in the direction of the beam. The probability that a chord will have a length in the interval $d\varrho$ is equal to the ratio of the areas of the annular ring, $2\pi\varrho\,d\varrho$, and the circle, πR^2.

Figure 12.9 shows the same view of the sphere as in Figure 12.3b, looking in the direction of the beam. The probability $w(\varrho)d\varrho$ that a given chord will pass at a distance between ϱ and $\varrho + d\varrho$ from the center of the circle that the sphere presents is equal to the ratio of the differential annular area $2\pi\varrho d\varrho$ and the total area πR^2 of the circle:

$$w(\varrho)d\varrho = \frac{2\pi\varrho\,d\varrho}{\pi R^2} = \frac{2\varrho}{R^2}d\varrho. \tag{12.15}$$

Transforming to the variable η and using Eq. (4.124), we write for the probability density function $p(\eta)$ for the chord lengths,

$$p(\eta) = w[\varrho(\eta)]\left|\frac{d\varrho(\eta)}{d\eta}\right|. \tag{12.16}$$

From Eq. (12.9),

$$\varrho = \sqrt{R^2 + \frac{\eta^2}{4}}. \tag{12.17}$$

Using Eqs. (12.15) and (12.17), we obtain in place of Eq. (12.16),

$$p(\eta)d\eta = \frac{\eta}{2R^2}d\eta. \tag{12.18}$$

The probability density function for isotropic chord lengths in a sphere is, therefore,

$$p(\eta) = \frac{\eta}{2R^2} \tag{12.19}$$

when $0 \leq \eta < 2R$, and $p(\eta) = 0$, otherwise. The function is shown in Figure 12.10. Its normalization to unit area is easily checked (Problem 12.13).

Comparison of Figures 12.7 and 12.10 illustrates the power of the Monte Carlo technique. The former figure gives an accurate numerical solution for the isotropic chord lengths, in this case also known analytically for the sphere, as shown in the

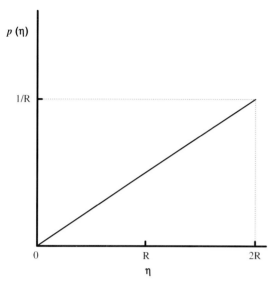

Figure 12.10 Analytical probability density function, $p(\eta) = \eta/(2R^2)$, for isotropic chord lengths η in a sphere of radius R.

latter figure. Monte Carlo techniques have been used to compute accurate chord length distributions in other geometries, such as cylinders and spheroids, for which no analytical solutions are known. Part of the beauty that Monte Carlo possesses is its complete generality.

■ *Example*
The *Cauchy theorem* states that, for any convex body, the mean value μ_η of the lengths of uniform, isotropic chords is given by

$$\mu_\eta = \frac{4V}{S}, \tag{12.20}$$

where V is the volume and S is the surface area of the body. To test the theorem for a sphere, use the probability density function (12.19) to compute μ_η and then compare the result with that obtained from Eq. (12.20).

Solution
With the help of the probability density function (12.19), we find that

$$\mu_\eta = \int_0^{2R} \eta p(\eta) d\eta = \frac{1}{2R^2} \int_0^{2R} \eta^2 d\eta = \frac{1}{2R^2} \times \frac{1}{3}\eta^3 \Big|_0^{2R} = \frac{4}{3}R. \tag{12.21}$$

The Cauchy theorem (12.20) gives

$$\mu_\eta = \frac{4 \times (4/3)\pi R^3}{4\pi R^2} = \frac{4}{3}R. \tag{12.22}$$

The mean isotropic chord length in a "square" cylinder, having both diameter and height equal to $2R$, is also $4R/3$ (Problem 12.17). Thus, the mean isotropic chord length in a "square" cylinder is identical with that in its inscribed sphere, although the two distributions are different. The Cauchy theorem, Eq. (12.20), is a remarkable finding.

12.7
Generation of a Statistical Sample from a Known Frequency Distribution

Many Monte Carlo applications begin with a random sample taken from a known probability distribution. For example, one might wish to generate a number of transport histories for individual fission neutrons released in a target. The initial neutron energies for the computations need to be selected at random from the known fission neutron energy spectrum. Given such a frequency distribution, which can be expressed either numerically or analytically, we now show how to generate a random sample from it by Monte Carlo techniques.

The upper curve in Figure 12.11 shows a hypothetical probability density function $f(x)$ for a continuous random variable x. Given $f(x)$, it is desired to generate a random sample of values of x having this frequency distribution. The curve right below shows the corresponding cumulative function $F(x)$, which, when differentiated, yields $f(x)$ (Eq. (4.14)). The probability that an event occurs with a value of x in a small interval dx, as indicated on the upper curve, is given by $f(x)dx$. In terms of the cumulative

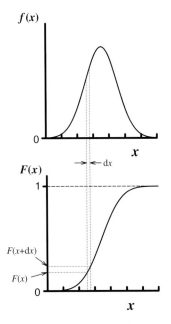

Figure 12.11 Example of a hypothetical probability density function $f(x)$ and cumulative distribution $F(x)$. A series of random numbers $0 \le r < 1$, chosen along the ordinate of $F(x)$, determine values of x distributed as $f(x)$.

distribution, this probability is also given by the difference $F(x+dx) - F(x)$ shown on the bottom curve. This difference, in turn, is just the fraction of the line interval between 0 and 1 that is determined by dx. The difference represents, therefore, the probability that a random number, $0 \leq r < 1$, would fall within the interval corresponding to dx. The random number, set equal to the value of the cumulative function $F(x)$, thus selects a value of x, which in turn determines $f(x)$ with its appropriate statistical probability density.

In this way, a random sample of values of x can be generated from the known distribution $f(x)$ by choosing random numbers. We note from the equality

$$f(x)dx = F(x+dx) - F(x) \tag{12.23}$$

that, in the limit $dx \to 0$, $f(x) = dF(x)/dx$. This relationship is just that given by Eq. (4.14).

As we have seen, data are often generated and treated in the form of histograms in Monte Carlo work. As stated in Section 4.1 (footnote 2), the probability that a continuous random variable has exactly the value x is zero. One deals, instead, with the probability that x has values within specified finite intervals. In generating a sample as just described, one can compile a histogram, showing the relative frequency with which values of x fall within various ranges.

■ **Example**

The probability density function for isotropic chord lengths in a sphere of radius R is given by Eq. (12.19). Using this function, derive an algorithm that gives the length η_i of a random chord as a function of a random number, $0 \leq r_i < 1$. Write the algorithm for the chord length also expressed in units of the sphere diameter.

Solution

According to the procedure just developed, one equates a random number to the cumulative probability distribution function. Using Eq. (12.19) with the dummy variable of integration λ, we obtain for the cumulative function

$$P(\eta) = \int_0^\eta p(\lambda)d\lambda = \frac{1}{2R^2}\int_0^\eta \lambda\, d\lambda = \frac{\eta^2}{4R^2}. \tag{12.24}$$

The cumulative function for $\eta \geq 0$ is shown in Figure 12.12; $P(\eta) = 0$ for $\eta < 0$. As indicated in Figure 12.12, a random number, $0 \leq r_i < 1$, determines a value η_i of the chord length. With $P(\eta_i) = r_i$, Eq. (12.24) can be conveniently rewritten in the form

$$\eta_i = 2R\sqrt{P(\eta_i)} = 2R\sqrt{r_i}. \tag{12.25}$$

If we express the chord length in units of the sphere diameter, $\eta' = \eta/(2R)$, which is the length of the longest chord, then Eq. (12.25) becomes

$$\eta'_i = \sqrt{r_i}. \tag{12.26}$$

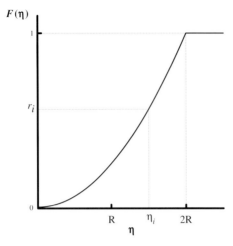

Figure 12.12 Cumulative analytical distribution for isotropic chord lengths η in a sphere of radius R.

Figure 12.13 shows a sample of 1000 values obtained by Eq. (12.26) and sorted into 20 bins of equal size, $\Delta\eta' = 0.05$. The average chord length found for this sample is 0.6767; the true mean, Eq. (12.21), is 2/3.

In many problems, stratified sampling can improve the statistical precision of the results. In the example just presented, choosing 1000 points uniformly along the ordinate

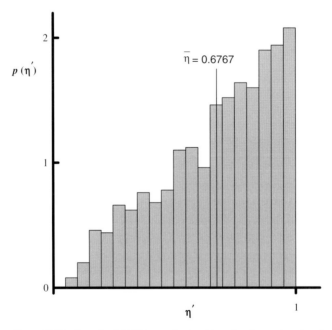

Figure 12.13 Sample of 1000 isotropic chord lengths chosen by using the algorithm $\eta'_i = \sqrt{r_i}$ (Eq. (12.26)) for a sphere of unit diameter. Uniform bin width is 0.05. Average value of chord length in sample is 0.6767, compared with expected value of 2/3.

in the interval [0, 1) should reduce the scatter in Figure 12.13. On the other hand, if we wanted to simulate examples of individual particle tracks in a proportional counter, then the "unstratified" algorithm in the last example would be more appropriate.

12.8
Decay Time Sampling from Exponential Distribution

The exponential function (Section 6.7) plays an important role in radiation protection. It is fundamental to radioactive decay and, as we shall see later in this chapter, to radiation transport in matter. As another application, we use Monte Carlo methods to show how one can generate a sample of decay times for a radionuclide and calculate the mean life and standard deviation. To this end, we begin by finding the cumulative distribution function for exponential decay.

Starting at time $t = 0$, we let T denote the time of decay of an atom. As discussed in Section 2.4, the probability of survival of a given atom past time t is $\Pr(T > t) = e^{-\lambda t}$, where λ is the decay constant. The cumulative probability of decay at time t is

$$F(t) = \Pr(T \leq t) = 1 - \Pr(T > t) = 1 - e^{-\lambda t}, \quad (12.27)$$

and the corresponding probability density function is (Eq. (4.14))

$$f(t) = \frac{dF(t)}{dt} = \lambda e^{-\lambda t}. \quad (12.28)$$

The function (12.27) is shown in Figure 12.14. A random number, $0 \leq r_i < 1$, chosen along the ordinate for $F(t)$ determines a value t_i for the decay time of an atom.

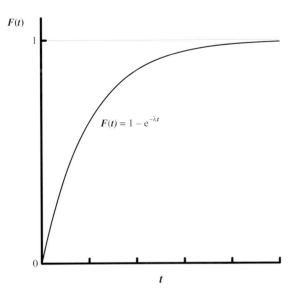

Figure 12.14 Cumulative probability $F(t)$ for decay time t of a given atom in a radionuclide source (Eq. (12.27)).

With $F(t_i) = r_i$ in Eq. (12.27), it follows that

$$t_i = -\frac{1}{\lambda}\ln(1 - r_i). \qquad (12.29)$$

Equation (12.29) was used with a random number generator to produce samples t_1, t_2, ..., t_N of decay times for ^{42}K atoms. The decay constant of this nuclide is $\lambda = 0.05608\,\text{h}^{-1}$, and the mean life is $\tau = 1/\lambda = 17.83\,\text{h}$. Data for samples of different sizes N were analyzed. For each sample, the estimator, $\hat{\tau}$, of the mean life was taken to be the sample average, $\bar{t}$:

$$\hat{\tau} = \frac{1}{N}\sum_{i=1}^{N} t_i = \bar{t}. \qquad (12.30)$$

The estimator $\hat{\sigma}$ for the standard deviation was the sample standard deviation s, given by

$$\hat{\sigma} = s = \sqrt{\frac{1}{N-1}\sum_{i=1}^{N}(t_i - \bar{t})^2}. \qquad (12.31)$$

To carry out the actual computations, the variance for each sample was evaluated according to Eq. (4.48) in place of Eq. (12.31):

$$\hat{\sigma}^2 = \frac{1}{N-1}\sum_{i=1}^{N} t_i^2 - \frac{N\bar{t}^2}{N-1}. \qquad (12.32)$$

Table 12.7 shows some results of calculations of the mean life (column 2) for sample sizes N ranging from 10 to 10^6 random decay times. Columns 3 and 4 give the calculated standard deviations and coefficients of variation, and column 5 shows the standard error of the estimator $\hat{\tau}$. The variance of the estimator is obtained from Eq. (7.6),

$$\text{Var}(\hat{\tau}) = \text{Var}\left(\frac{1}{N}\sum_{i=1}^{N} T_i\right) = \frac{1}{N^2}\sum_{i=1}^{N}\text{Var}(T_i). \qquad (12.33)$$

Table 12.7 Results of Monte Carlo calculations for decay of ^{42}K atoms in samples of different sizes.

Number of decays, N	Mean life, $\hat{\tau}$ (h)	Standard deviation, $\hat{\sigma}$ (h)	Coefficient of variation, $\hat{\sigma}/\hat{\tau}$	Standard error, $\hat{\sigma}/\sqrt{N}$, in $\hat{\tau}$
10	18.33	13.40	0.7312	4.237
10^2	16.01	14.21	0.8871	1.421
10^3	19.01	18.83	0.9907	0.5955
10^4	17.55	17.72	1.010	0.1772
10^5	17.85	17.95	1.005	0.0568
10^6	17.85	17.86	1.001	0.0179

Because the decay times are observations of independent and identically distributed exponential random variables (Eq. (12.28)), they each have the same variance, $1/\lambda^2$. Thus, Eq. (12.33) yields

$$\text{Var}(\hat{\tau}) = \frac{1}{N^2}\left(N\frac{1}{\lambda^2}\right) = \frac{1}{N\lambda^2}. \tag{12.34}$$

Since $\hat{\sigma}^2$ is an unbiased estimator for $1/\lambda^2$ (Problem 12.28), we can substitute it in Eq. (12.34) to obtain an estimate for the $\text{Var}(\hat{\tau})$. Thus,

$$\sqrt{\widehat{\text{Var}(\hat{\tau})}} = \frac{\hat{\sigma}}{\sqrt{N}} \tag{12.35}$$

is the standard error associated with $\hat{\tau}$. Notice that the estimate of the standard deviation $\hat{\sigma}$ in Table 12.7 converges to $1/\lambda = 17.83$ h, as it should, while the standard error of our estimate gets smaller as the sample size increases.

12.9 Photon Transport

Monte Carlo procedures are used extensively to calculate the transport of photons and other kinds of radiation through matter. Such calculations permit assessments of doses and shielding under a variety of conditions. In principle, all manner of complicated target materials, geometrical configurations, and radiation fields can be handled by Monte Carlo methods. However, in very complex situations, involved statistical alternatives and algorithms become necessary, and computer capacity and computing time can become excessive. Detailed results can require an inordinately large number of particle histories. We shall describe the basics of photon and neutron transport, interaction, and dose calculation in these last sections of this chapter.

The linear attenuation coefficient, or macroscopic cross section, is the numerical parameter that statistically determines the distribution of flight distances that photons of a given energy travel in a specified material before having an interaction. This quantity can be measured experimentally under conditions of "good geometry," as shown in Figure 12.15 (Turner, 1995, pp. 186–187). A pencil beam of monoenergetic photons is directed normally onto a slab of the material, having variable thickness x. A count rate detector is located some distance behind the slab. It is far enough removed so that any photon, scattered in the slab, has a negligible chance of reaching it and being counted. Under these "good geometry" conditions, only uncollided photons – that is, only those that pass through the slab without interacting – are detected. The ratio of count rates, $\dot{D}$ with the slab present and $\dot{D}_o$ with the slab absent, is observed to decrease exponentially with the slab thickness x. As indicated in Figure 12.15, the best fit to measurements of $\dot{D}/\dot{D}_o$ versus x furnishes the measured value of the linear attenuation coefficient μ, having the dimensions of inverse length:

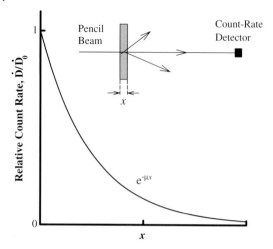

Figure 12.15 Attenuation coefficient μ is measured under conditions of "good geometry" with a pencil beam of normally incident, monoenergetic photons. Relative count rate is given by $\dot{D}/\dot{D}_o = e^{-\mu x}$ for different slab thicknesses x.

$$\frac{\dot{D}}{\dot{D}_o} = e^{-\mu x}. \qquad (12.36)$$

The mass attenuation coefficient, μ/ϱ, where ϱ is the density of the material, is the quantity more often given in tables and graphs. Whereas μ depends upon temperature and pressure, especially for a gas, μ/ϱ does not.

Equation (12.36) gives the relative number of photons that traverse a slab of thickness x without interacting. The quantity $e^{-\mu x}$ on the right-hand side is, therefore, the probability that a given photon, normally incident on the slab, will pass through it without interacting. More generally, if we let X be the depth at which an incident photon has its first interaction in a uniform, infinitely thick target, then $e^{-\lambda x}$ is the probability that the photon will travel at least a distance x without interacting:

$$\Pr(X > x) = e^{-\mu x}. \qquad (12.37)$$

Conversely, the probability than an interaction occurs at a depth less than or equal to x is

$$\Pr(X \leq x) = 1 - \Pr(X > x) = 1 - e^{-\mu x}. \qquad (12.38)$$

Like Eq. (12.27), this last expression gives the usual cumulative distribution function, in this case for the depth of penetration for a photon's first interaction. It follows from Eq. (12.37) that the probability for the interaction to occur at a depth between x_1 and x_2 (with $x_2 > x_1$) is

$$\Pr(x_1 \leq X < x_2) = \Pr(X > x_1) - \Pr(X > x_2) = e^{-\mu x_1} - e^{-\mu x_2}. \tag{12.39}$$

In analogy with the decay constant λ discussed in Section 2.2 as the probability per unit time that an atom will decay, μ represents the probability per unit distance that a photon will interact. Also, just as $1/\lambda$ is the mean life in Eqs. (2.7) and (4.36), $1/\mu$, called the mean free path, is the average distance that a photon travels before it interacts.

Until a photon interacts, it penetrates a target "without memory." That is, the probability that the photon will interact in its next segment of travel is always independent of whether the photon has already traveled a long distance in the target or is only just starting out. This property can be formulated in terms of a conditional probability (Section 3.5). Given that a photon has reached a depth x, the probability that it will travel at least an additional distance s without interaction is (Problem 12.32)

$$\Pr(X > x + s | X > x) = \Pr(X > s). \tag{12.40}$$

The following example shows how the penetration of photons in matter is determined statistically by the linear attenuation coefficient.

■ **Example**

A parallel beam of 200-keV gamma rays is normally incident on a thick slab of soft tissue. The linear attenuation coefficient is $0.137\,\text{cm}^{-1}$.

a) What is the probability that an incident photon will reach a depth of at least 1 cm without interacting? What is the probability for 5 cm?
b) What is the probability that an incident photon will have its initial interaction at a depth between 4 and 5 cm?
c) What is the probability that an incident photon, having reached a depth of 10 cm without interacting, will travel at least an additional 1 cm without interacting?

Solution

a) With $\mu = 0.137\,\text{cm}^{-1}$, the probability that a 200-keV gamma ray will penetrate to a depth of at least $x = 1$ cm without interacting is, from Eq. (12.37),

$$\Pr(X > 1) = e^{-0.137 \times 1} = 0.872. \tag{12.41}$$

Distances in this example will be expressed in cm, without writing the unit explicitly. Note that the same length unit has to be employed for both μ and x, because the exponent must be dimensionless. The probability of penetrating at least 5 cm without interacting is

$$\Pr(X > 5) = e^{-0.137 \times 5} = 0.504. \tag{12.42}$$

Alternatively, one can regard Eq. (12.42) as giving the probability of traversing each of the five 1-cm distances consecutively. Thus,

$$\Pr(X > 5) = [\Pr(X > 1)]^5 = (0.872)^5 = 0.504, \qquad (12.43)$$

which is the same as Eq. (12.42). The equivalence of Eqs. (12.42) and (12.43) is due to the special property of the exponential distribution, called *independent increments*. Exponents are added in multiplication. As expressed by Eq. (12.40), a photon penetrates a target without memory, the probability of interaction being independent from one interval to the next.

b) As with part (a), there are different ways of answering. The difference between $\Pr(X > 4)$ and $\Pr(X > 5)$, according to Eq. (12.39), is the probability that a given incident photon will have its first interaction at a depth between 4 and 5 cm. Thus,

$$\Pr(4 < X < 5) = e^{-0.137 \times 4} - e^{-0.137 \times 5} = 0.578 - 0.504$$
$$= 0.0740 \qquad (12.44)$$

is the desired probability. We can also consider the probability that an incident photon will reach at least a depth of 4 cm, but will travel at most an additional 1 cm without interaction. Since, from Eq. (12.38), the latter probability is $\Pr(X \leq 1) = 1 - 0.872 = 0.128$, we have for the probability of interacting between 4 and 5 cm,

$$\Pr(4 < X < 5) = \Pr(X > 4)\Pr(X \leq 1) = 0.578 \times 0.128$$
$$= 0.0740, \qquad (12.45)$$

as before. The last solution applies the property of independent increments, mentioned after Eq. (12.43).

c) As an uncollided gamma photon penetrates a uniform target, its interaction probability per unit distance of travel remains equal to the linear attenuation coefficient μ, regardless of how deep the photon may have already penetrated. Applying Eq. (12.40) with $x = 10$ and $s = 1$, we write

$$\Pr(X > 10 + 1 | X > 10) = \Pr(X > 1) = 0.872. \qquad (12.46)$$

The attenuation of X-rays and gamma rays in matter occurs by means of four major processes: the photoelectric effect, Compton scattering, pair production, and photonuclear reactions. Each contributes additively to the numerical value of the attenuation coefficient. In the modeling of photon transport, the specific types of interaction can be determined statistically, as the next example illustrates.

■ *Example*

A parallel beam of 400-keV gamma rays is normally incident on a crystal of sodium iodide (density, $\varrho = 3.67 \text{ g cm}^{-3}$). The mass attenuation coefficients for the photoelectric effect and for Compton scattering (the only significant physical interaction mechanisms) are, respectively, $\tau/\varrho = 0.028 \text{ cm}^2 \text{ g}^{-1}$ and $\sigma/\varrho = 0.080 \text{ cm}^2 \text{ g}^{-1}$.

a) Calculate the probability that a normally incident, 400-keV gamma ray will produce a photoelectron at a depth between 2.4 and 2.6 cm in the crystal.
b) How thick a crystal is needed in order to be 90% efficient in detecting these photons?

Solution

a) The answer is given by the product of (1) the probability that there is an interaction in the specified depth interval and (2) the (independent) probability that the interaction is photoelectric absorption, rather than Compton scattering. In terms of conditional and independent probabilities, discussed in Section 3.5, we let A be the event that an interaction occurs in the interval and B the event that the interaction (given A) is photoelectric. Then, by Eq. (3.31), the probability that both independent events happen is

$$P(A \cap B) = P(A)P(B|A), \quad (12.47)$$

as just stated. To express $P(A)$, we must use the linear attenuation coefficient. With the given values of μ/ϱ and ϱ, we have

$$\mu = \left(\frac{\mu}{\varrho}\right) \times \varrho = (0.108 \text{ cm}^2 \text{ g}^{-1}) \times (3.67 \text{ g cm}^{-3})$$
$$= 0.396 \text{ cm}^{-1}. \quad (12.48)$$

The probability that an incident photon has its first encounter between 2.4 and 2.6 cm is, therefore, by Eq. (12.39),

$$P(A) = \Pr(2.4 < X < 2.6) = e^{-0.396 \times 2.4} - e^{-0.396 \times 2.6}$$
$$= 0.3866 - 0.3572 = 0.0294. \quad (12.49)$$

Given that the interaction takes place, the probability for a photoelectric, rather than Compton, event is equal to the photoelectric fraction of the total mass (or linear) attenuation coefficient. Thus,

$$P(B|A) = \frac{\tau/\varrho}{\mu/\varrho} = \frac{0.028 \text{ cm}^2 \text{ g}^{-1}}{0.108 \text{ cm}^2 \text{ g}^{-1}} = 0.259. \quad (12.50)$$

It follows from Eqs. (12.47), (12.49), and (12.50) that the probability for an incident photon to undergo photoelectric absorption at a depth between 2.4 and 2.6 cm is

$$P(A)P(B|A) = 0.0294 \times 0.259 = 0.00761. \quad (12.51)$$

b) When a 400-keV gamma ray interacts in the crystal by either process, it produces a secondary electron, which in turn produces scintillation photons that are detected electronically. (Multiple Compton events from a single incident gamma ray are detected in a single pulse.) To detect 90% of the normally incident, 400-keV gamma rays, the thickness x must be such that only 10% of the incident photons are still uncollided as they reach this depth. Thus,

$$\Pr(X > x) = e^{-\mu x} = e^{-0.396x} = 0.100, \quad (12.52)$$

giving $x = 5.81$ cm for the required thickness.

Monte Carlo computations can be made to simulate photon transport and to generate individual histories for a number of photons incident on a target and any secondary particles that the photons produce. The accumulated statistical data from a number of histories can be compiled to assess the information desired, such as the dose at various locations in a target or the radiation leakage through a shield. The random selection of flight distances to the sites of interaction for the photons is central to these computations. The cumulative distribution for random flight distances, which is governed by Eq. (12.38), is mathematically the same as Eq. (12.27) for the survival probability for radioactive decay, treated in the last section. The use of a random number r_i to select a flight distance x_i to the site of interaction for a photon can be made in complete analogy with Eq. (12.29):

$$x_i = -\frac{1}{\mu}\ln(1 - r_i). \tag{12.53}$$

In a Monte Carlo simulation of photon transport and interaction in matter, the fates of each photon and any secondary particles they produce are determined by random numbers. The simulation can be carried out in any desired detail. The following example shows some of the elements of a Monte Carlo transport calculation for photons.

■ **Example**

A broad, uniform beam of 40-keV photons is normally incident on a soft tissue slab, 12 cm thick, as shown in Figure 12.16. The linear attenuation coefficient, $\mu = 0.241 \text{ cm}^{-1}$, is the sum of the coefficients $\tau = 0.048 \text{ cm}^{-1}$ for the photoelectric effect and $\sigma = 0.193 \text{ cm}^{-1}$ for the Compton effect. (Other processes occur to a negligible extent.) For an incident photon, use one random number to select a flight distance to the site of its first interaction. (The photon might, of course, traverse the slab completely without interacting.) Let a second random number decide whether the interaction is

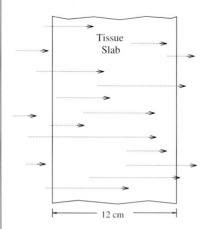

Figure 12.16 Uniform, broad beam of 40-keV photons incident normally on a 12-cm soft tissue slab. See the text for description of Monte Carlo calculation of dose per unit fluence as a function of depth in slab.

photoelectric or Compton. To this end, we partition the relative interaction probabilities along the unit interval according to their magnitudes. We have $\tau/\mu = 0.048/0.241 = 0.199$ and $\sigma/\mu = 0.193/0.241 = 0.801$. If the second random number r falls in the interval $0 \leq r < 0.199$, the interaction is photoelectric absorption; if $0.199 \leq r < 1.000$, the photon is Compton scattered. Use sequentially the random numbers from Eqs. (12.2) and (12.3) in Section 12.2 to determine the depth of the first collision and the type of interaction that takes place for two incident photons.

Solution

The flight distance of an incident photon is selected by means of Eq. (12.53). With the given value of μ and $r_1 = 0.868$ from Eq. (12.2), we obtain

$$x_1 = -\frac{1}{0.241} \ln(1 - 0.868) = 8.40 \text{ cm} \qquad (12.54)$$

for the depth at which the first photon interacts. The second random number $r_2 = 0.338$ from Eq. (12.3) determines, according to the specified algorithm, that the photon is Compton scattered at this depth. The flight distance for the second photon is, with the help of r_3 from Eq. (12.3),

$$x_2 = -\frac{1}{0.241} \ln(1 - 0.084) = 0.364 \text{ cm}. \qquad (12.55)$$

With $r_4 = 0.179$ from Eq. (12.3), this photon undergoes photoelectric absorption.

12.10
Dose Calculations

Using the two photon histories started in the last example as illustrations, we show how absorbed dose in the tissue slab can be calculated. Specifically, we outline a procedure for determining the absorbed dose per unit fluence from a uniform, broad beam as a function of depth in the slab. For analysis of the Monte Carlo generated data, we divide the slab into 12 subslabs of thickness 1 cm, as shown in Figure 12.17. (The number and thicknesses of the subslabs, which are used for accumulating energy deposition, are arbitrary.) We let the first photon enter the slab at some arbitrary point P. It travels to the point Q_1 at a depth of 8.40 cm in the ninth subslab, as shown in Figure 12.17, where it is Compton scattered. For the dose calculation, one needs to select a value of the energy deposited in the subslab from this collision, based on the known probability density function for Compton scattering. There are various ways to compute this quantity with the help of random numbers. One way is to randomly select a scattering angle θ, which then uniquely determines the energy of the Compton electron. The cumulative probability function for Compton scattering at an angle θ with respect to the incident direction is known for photons of all energies from the Klein–Nishina formula. This function can be stored numerically in the computer as a

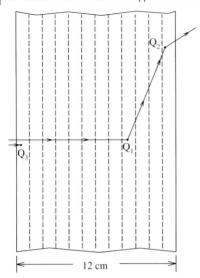

Figure 12.17 Schematic histories of two photons. First photon is Compton scattered at Q_1 and then at Q_2 before escaping from slab. Second photon undergoes photoelectric absorption at Q_3. See the text.

function of photon energy as part of the input data of the code for the dose calculation. A new random number thus selects θ from the cumulative distribution (just as r_i determined η_i in Figure 12.11) and, with it, the energy of the Compton electron. The physical ranges of the secondary electrons that can be produced by the photons are much shorter than the 1-cm dimension of the subslabs. Therefore, the energy lost by the photon in this collision is tallied as energy deposited in the ninth subslab.

The Compton scattered photon must also be tracked through the slab from the collision point Q_1. Its direction of travel is fixed by the polar angle θ and a second, azimuthal, angle ϕ distributed uniformly over the interval $0° \leq \phi < 360°$ about the original line of travel. An additional random number r is used to pick the azimuthal angle: $\phi = 360r$. The scattered photon has an energy equal to the original photon energy minus the energy lost to the Compton electron. The linear attenuation coefficient for soft tissue at the new photon energy is found from data stored as input to the dose code. One now starts the scattered photon from Q_1 along the new direction of travel at the reduced energy and chooses a flight distance by means of Eq. (12.53), thus continuing the transport and interaction simulation as before. Eventually, each incident photon will disappear from the slab either by photoelectric absorption or by escaping through its surface. In this example, we have shown another Compton scattering at a point Q_2 in the 12th subslab in Figure 12.17, after which the scattered photon escapes the slab. The collision at Q_2 is handled as before, and the energy absorbed is tallied in that subslab.

We let the second photon (from the last example) be incident at the same point P on the slab as the first photon. Its track is shown schematically in Figure 12.17 somewhat below that of the first photon for clarity. This photon disappears by photoelectric absorption at a point Q_3 at a depth of 0.364 cm. The entire photon energy, 40 keV, is tallied as absorbed in the first subslab. The calculation of the complete histories for two photons and the energy absorbed at different depths in the slab have now been completed by Monte Carlo techniques.[2] The histories of a large number of additional photons, incident at P, can be similarly generated, until the absorbed energy in the various subslabs, plotted as a histogram, shows acceptably small statistical scatter. The number of histories needed to obtain a smooth representation of the absorbed energy throughout the slab will depend on various factors, principally the number of subslabs into which the slab is divided for analysis and the statistical precision with which the absorbed dose is needed.

The first stage of the dose computation thus consists of statistically generating the histories of a large number N of photons, normally incident at a point P on the slab, and then compiling the total energies E_i absorbed in each subslab. It remains to convert this information into the absorbed dose in each subslab from a broad beam of unit fluence. It is evident from Figure 12.17 that the total absorbed dose in a given subslab is distributed very nonuniformly in lateral directions away from the perpendicular line of incidence through P. The absorbed dose will generally get smaller as one moves away from this line. On the other hand, the lateral distribution of the absorbed dose in each subslab will be constant for the uniform, broad beam in which we are interested. Instead of N photons incident at the single point P, consider a fluence, N/A, of photons uniformly incident over the entire (infinite) surface of the slab. (The area A over which the N photons are spread can have any size.) Then the absorbed energy in each volume element with lateral area A perpendicular to the beam direction in the ith subslab must be equal to the *total* energy E_i absorbed in that subslab as a result of N photons incident at the *single* point P. If the slab density is ϱ, then the mass of the lateral element is $M_i = \varrho A \Delta z_i$, where Δz_i is the subslab thickness ($=1$ cm in Figure 12.17). Dividing the total absorbed energy E_i by the mass, we obtain for the absorbed dose in the ith subslab

$$D_i = \frac{E_i}{\varrho A \Delta z_i}. \tag{12.56}$$

Since this is the dose for a fluence N/A, we find for the absorbed dose per unit fluence in the ith subslab

$$\hat{D} = \frac{D_i}{N/A} = \frac{E_i}{N \varrho \Delta z_i}. \tag{12.57}$$

---

[2] In a more careful analysis, a distinction is to be made between the energy lost by a photon and the energy absorbed locally in a subslab. In both photoelectric absorption and Compton scattering, the initial kinetic energy of the struck electron is somewhat less than the energy lost by the photon because of the energy spent in overcoming the binding of the electron. The binding energy is regained when the parent ion is neutralized, but partly in the form of fluorescence radiation, which can escape from the immediate vicinity of the collision site. In addition, the struck electron can produce some bremsstrahlung, also taking energy away from the site. These differences are embodied in the contrasting definitions of the attenuation, energy transfer, and energy absorption coefficients. In the present computation of a depth–dose curve for 40-keV photons over 1-cm subslabs, the distinction is of no practical importance.

Note that the units in Eq. (12.57) are those of dose times area. In SI units, the dose per unit fluence is thus expressed in Gy per photon per m^2, or Gy (m^{-2})$^{-1}$ = Gy m^2, and in CGS units, rad cm^2.

■ **Example**
With the geometry shown in Figure 12.17, histories are generated by Monte Carlo means for 2500 photons normally incident on a soft tissue slab at P with an energy of 1 MeV. The accumulated energy absorbed in the third subslab, having a thickness of 1 cm, is 15.9 MeV. What is the absorbed dose at this depth per unit fluence for a uniform, broad beam of incident 1-MeV photons? Express the answer in both SI and CGS units.

Solution
Using the second equality in (12.57) with $E_3 = 15.9$ MeV, $N = 2500$, $\varrho = 1$ g cm^{-3}, and $\Delta z_3 = 1$ cm, we obtain for the dose per unit fluence in the third subslab

$$\hat{D}_3 = \frac{15.9 \text{ MeV}}{2500 \times 1 \text{ g cm}^{-3} \times 1 \text{ cm}} = 9.36 \times 10^{-3} \text{ MeV g}^{-1} \text{ cm}^2. \quad (12.58)$$

Since 1 MeV = 1.60×10^{-6} erg and 1 rad = 100 erg g^{-1}, we write

$$\hat{D}_3 = 9.36 \times 10^{-3} \frac{\text{MeV cm}^2}{\text{g}} \times 1.6 \times 10^{-6} \frac{\text{erg}}{\text{MeV}} \times \frac{1}{100 \text{ erg g}^{-1}} \frac{\text{rad}}{}, \quad (12.59)$$

giving

$$\hat{D}_3 = 1.02 \times 10^{-10} \text{ rad cm}^2 \quad (12.60)$$

in CGS units. Since 1 rad = 0.01 Gy and 1 cm^2 = 10^{-4} m^2, we have, in SI units,

$$\hat{D}_3 = 1.02 \times 10^{-16} \text{ Gy m}^2. \quad (12.61)$$

Although we have dealt with relatively simple problems, having a high degree of symmetry, it should be apparent that Monte Carlo methods can be undertaken with complete generality. In the dose calculation just described, the incident photons in a Monte Carlo simulation need not be monoenergetic or incident from a single direction. They can be accompanied by neutrons and other kinds and energies of radiation. The target might contain regions of different tissues, such as bone and lung, and can be of finite extent. Monte Carlo calculations of dose and shielding provide an extremely valuable tool for radiation protection, in which statistics plays an important role. In another application, the formalism developed by the Medical Internal Radiation Dose (MIRD) Committee of the Society of Nuclear Medicine for internal emitters utilizes the concept of "absorbed fraction." This quantity is defined as the fraction of the energy emitted as a specific radiation type from a given body tissue or organ, called the source organ, that is absorbed in a target tissue or organ (which may be the same as the source). For example, the thyroid (target organ) will be irradiated by gamma rays from an emitter lodged in the lung (source organ). Absorbed fractions for numerous organ pairs, photon energies, and radionuclide spectra have been calculated for anthropomorphic phantoms by Monte Carlo methods. They are an essential part of the basis for risk assessment through internal dosimetry. Dose and committed dose models have been developed for determining annual limits of intake and derived air concentrations. Used by regulatory agencies, many of these related quantities have been enacted into law.

12.11
Neutron Transport and Dose Computation

Monte Carlo calculations of neutron transport and dose have many similarities with computations for photons. They can proceed in analogous fashion. A flight distance to the site of first interaction for an incident neutron can be selected with a random number from an equation identical with Eq. 12.29. The quantity μ for neutrons is usually referred to as the macroscopic cross section (a term used for photons, too) and is also the inverse of the mean free path. Its numerical value is determined by the atomic composition of the target and the values of the energy-dependent neutron cross sections for the elements therein. Neutrons interact with matter through the short-range strong force, which is exerted by nucleons. The nuclear interaction can be either elastic or inelastic. When an interaction occurs, additional random numbers can be used to determine the type of nucleus struck, the type of interaction, and the energy deposited. A scattered neutron is transported, just as one transports a Compton scattered photon. If one is computing dose equivalent, then the linear energy transfer (LET) of the struck recoil nucleus can be inferred from its identity and energy for the assignment of a quality factor with which to multiply the absorbed energy from the event. Except for neutrons at high energies, the recoil nuclei have ranges that are short on a scale of 1 cm in condensed matter, and so one can usually treat their energy as locally absorbed in a dose calculation.

Fast neutrons deposit most of their energy in soft tissue through elastic scattering, principally with hydrogen. For neutron energies up to about 10 MeV, the elastic scattering with hydrogen is isotropic in the center-of-mass coordinate system. This fact has the interesting consequence that the energy-loss spectrum for neutron collisions with hydrogen is uniform in the laboratory system. Any energy loss between zero and the full energy of the neutron is equally probable. A neutron thus loses one-half of its energy, on average, in a hydrogen collision. Hydrogen is unique for neutrons in having the same mass. It is only with hydrogen that a neutron can have a head-on, billiard-ball-like collision and lose its entire energy. If the collision is not head-on, then, because of their equal masses, the neutron and recoil proton are scattered at right angles to one another. Slow and thermal neutrons, on the other hand, have a high probability of capture by hydrogen or nitrogen in tissue. Thermal neutron capture by hydrogen results in the release of a 2.22-MeV gamma ray, which should also be transported in a dose calculation. These basic aspects of neutron physics, as they affect dosimetry, are discussed in Chapter 9 of Turner (2007).

■ **Example**

a) What is the probability that a 6-MeV neutron will lose more than 1.2 MeV in an elastic collision with hydrogen?
b) What is the probability that it will lose between 100 keV and 1.0 MeV in such a collision?
c) What is the average energy lost by a 6-MeV neutron in a collision with hydrogen?

Solution

a) We let X denote the energy loss of the neutron. It has the uniform distribution $p(x)$ shown in Figure 12.18. The probability that the energy loss will be greater than 1.2 MeV is equal to the fraction of the rectangular area above this energy:

$$\Pr(X > 1.2) = \int_{1.2}^{6.0} \frac{1}{6} dx = \frac{6.0 - 1.2}{6} = 0.80. \qquad (12.62)$$

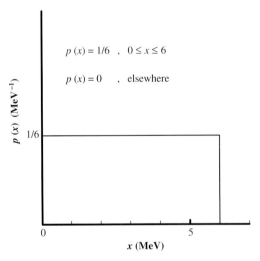

Figure 12.18 Probability density function $p(x)$ for energy loss x by a 6-MeV neutron in a collision with hydrogen. See example in the text.

b) The probability is equal to the fraction of the spectral area between the two given values:

$$\Pr(0.10 < X < 1.0) = \int_{0.1}^{1.0} \frac{1}{6} dx = \frac{1.0 - 0.1}{6} = 0.15. \tag{12.63}$$

c) The average energy lost is at the midpoint of the flat spectrum in Figure 12.18,

$$E(X) = \int_0^6 xp(x)dx = \frac{x^2}{12}\Big|_0^6 = 3.0 \text{ MeV}. \tag{12.64}$$

Some different aspects of Monte Carlo calculations, not yet discussed, are important in many shielding calculations, particularly for thick shields. A fast neutron in a thick target can have a very large number of collisions before becoming thermalized or escaping. By its nature, a shielding calculation places primary importance on the relatively few neutrons that do get through – their relative number and their energy and angular distributions. The size of the sample in which one is interested gets increasingly small with depth. Doing neutron transport in the straightforward manner we have been describing would be inefficient for determining the properties of a thick shield. One would find that a large fraction of the incident neutrons are absorbed before they come out of the other side, after much computer time is spent on their transport. In addition, other neutrons would tend to be scattered away from the back surface after penetrating to a considerable depth. (Collisions with heavy nuclei can scatter neutrons backward.) We mention two techniques that improve computational efficiency for deep penetration problems.

First, there is *splitting and Russian roulette*. As in Figure 12.17, one specifies a series of imaginary boundaries z_i at different depths inside a target. A statistical weight is assigned to each incident neutron. Every time a neutron crosses a boundary z_i, it is split into v_i identical neutrons, each being given a reduced statistical weight w_i/v_i, where w_i is the weight assigned to a neutron that crosses z_i. Weight is preserved thereby, and a larger sample of neutrons with reduced weight is obtained. Splitting itself is made more efficient by playing Russian roulette with neutrons that wander into less desirable parts of a target, like those that tend to return toward the surface of incidence. When Russian roulette is played, a neutron is given a random chance with probability p of continuing to survive or $(1 - p)$ of being killed at that point in the calculation. If it survives, its weight is increased by $1/p$. This technique decreases the number of less desirable histories that would otherwise be continued in the computations.

Second, *exponential transformations* can be performed in the selection of flight distances. One artificially decreases the macroscopic cross section μ in regions of the target that are of less interest and increases it in regions where greater sampling is desired. The appropriate weighting factors can be calculated and used to compensate for the biases thus introduced.

Problems

12.1 a) Devise a (pseudo-)random number generator.
b) How well do its numbers fit the criteria (12.4) and (12.5)?
c) Specify at least one additional criterion and use it to test the generator.

12.2 Consider a random number generator, such as that given by Eq. (12.1). Starting with a given seed, it generates a sequence, $0 \leq r_i < 1$, in which each successive number r_i acts as the seed for the next number.
a) With roundoff to three significant figures at each step, what is the longest possible sequence of numbers that can be generated before the sequence begins to repeat itself?
b) How many independent, three-digit sequences can be constructed without repetition of one of the numbers in a sequence?

12.3 Prove Eqs. (12.4) and (12.5).

12.4 Evaluate the integral

$$\int_0^2 e^{-0.5x} \, dx$$

by using a Monte Carlo technique. Compare your result with that obtained by analytical evaluation.

12.5 a) Use a Monte Carlo technique to calculate the area of a regular octagon, having a side of length b.
b) What is the analytical formula for the area as a function of b?

12.6 When the computations for Table 12.1 are repeated with a different sequence of random numbers, the values in the second column are generally different from those given there. The numbers in column 2 are themselves random variables, which show fluctuations with repeated, independent samples.
a) What kind of statistics do the numbers in column 2 of Table 12.1 obey?
b) When the total number of points is 10^5, what is the expected value of the number in column 2?
c) What is its standard deviation?

12.7 In applying method 2 for finding the area, is it possible for the sum in Eq. (12.7) not to be exact even after adding only the first two terms? Explain.

12.8 The 25 random chord lengths η_i' in Table 12.3 were tabulated into 10 bins of uniform size in Table 12.4.
a) Redo the tabulation, using bins with the following boundaries: 0.00, 0.15, 0.30, 0.45, 0.60, 0.70, 0.85, 0.95, 1.00.
b) Show that the new probability density, $p(\eta_i')$, is normalized.
c) Plot the new histogram and compare with Figures 12.4 and 12.5.

12.9 What is the expected value of the total number of points (x_i, y_i) that one would have to sample in order to obtain 25 chords, as in Table 12.3, that entered the sphere?

Table 12.8 Data for Problem 12.10.

Time (s)	Number of observations
0–5	61
5–10	43
10–15	72
15–20	28
20–25	36
25–30	0
Total	240

12.10 In a certain experiment, the times between successive counts registered by an instrument are recorded. Table 12.8 shows the data for 240 observations, with the times tabulated into six bins.
   a) Plot the data from the table directly as a histogram.
   b) Use the smoothing technique described in Section 12.5 and plot the resulting histogram.

12.11 Estimate the coefficient of variation for each of the six entries in Table 12.8.

12.12 A Monte Carlo analysis requires the selection of random angles, θ_i, distributed uniformly over the interval $0 \leq \theta_i < 90°$.
   a) Write an algorithm that determines an angle θ_i from the value of a random number, $0 \leq r_i < 1$.
   b) A calculation using 1000 random angles θ_i is to be made. In place of the algorithm in (a), write a formula, like Eq. (12.13), that selects the 1000 angles by stratified sampling.

12.13 Show that the probability density function for isotropic chord lengths, Eq. (12.19), is normalized.

12.14 A spherical proportional counter, having a radius of 2.20 cm, is traversed uniformly by isotropic chords.
   a) What is the probability that a randomly chosen chord has a length between 2.90 and 3.10 cm?
   b) What is the expected value of the chord length?
   c) Write the cumulative probability function for the chord length.

12.15 In the last problem, what is the probability that a random chord will pass the center of the sphere at a distance between 1.5 and 2.0 cm? (See Figure 12.9.)

12.16 Isotropic chords uniformly traverse a sphere of diameter 50 cm.
   a) What is the mean chord length?
   b) What is the mean square chord length?

12.17 Use the Cauchy theorem, Eq. (12.20), to show that the mean isotropic chord length for a "square" cylinder (diameter = height) is the same as that for the sphere inscribed in the cylinder.

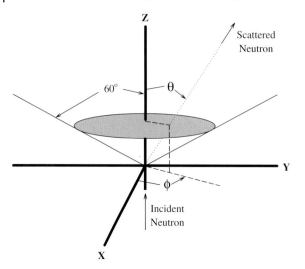

Figure 12.19 See Problem 12.20.

12.18 Use the Cauchy theorem to find the mean isotropic chord length in
  a) a parallelepiped, having edges of length a, b, and c;
  b) a cube with edges of length a.

12.19 What is the median chord length in Problem 12.14?

12.20 In a Monte Carlo calculation, a neutron, incident along the Z-axis in Figure 12.19, is scattered at the origin of the XYZ axes. The neutron has an equal likelihood of being scattered in any forward direction (i.e., isotropically) within a 60° cone, as indicated in Figure 12.19. The probability of being scattered outside the cone is zero.
  a) Write equations for computing the polar and azimuthal scattering angles (θ, ϕ) within the cone from the values of two random numbers, $0 \leq r < 1$.
  b) Given the "random" numbers 0.88422 and 0.01731, calculate a pair of angles, θ and ϕ, in either order, according to your algorithm in (a).

12.21 A fast neutron (n) is scattered from a proton (p) at an angle θ with respect to its original direction of travel (Figure 12.20). The probability density for scattering at the angle θ (in three dimensions) is $p(\theta) = 2 \sin \theta \cos \theta$, corresponding to isotropic scattering in the center-of-mass reference system.
  a) What is the cumulative probability function $P(\theta)$ for scattering at an angle θ?
  b) Plot $P(\theta)$ versus θ.
  c) Write a formula that determines a random scattering angle θ from the value of a random number $0 \leq r < 1$.
  d) What is the scattering angle θ when $r = 0.21298$?

12.22 Give an algorithm that uses a random number, $0 \leq r < 1$, to select a value of z from the standard normal distribution.

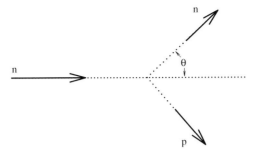

Figure 12.20 See Problem 12.21.

12.23 One can use Monte Carlo techniques to simulate the decay of a radioactive source. Apply the algorithm given in Section 12.8 to a sample of ^{32}P (beta emitter, half-life = 14.3 d).
   a) What two numerical random number choices determine whether a given atom will decay during the time period between $t=0$ d and $t=2$ d?
   b) From these choices, determine $\Pr(0 \leq T < 2\,\mathrm{d})$.

12.24 Repeat the last problem for the time period between $t=20$ d and $t=22$ d.

12.25 Why is the probability for the decay of an atom over a 2-day period different in the last two problems?

12.26 Show that the random numbers found in Problem 12.24 satisfy the conditional probability
$$\Pr(0\,\mathrm{d} \leq T < 2\,\mathrm{d}) = \Pr(T < 22\,\mathrm{d}|T > 20\,\mathrm{d}).$$

12.27 Verify the entries in the last two columns of Table 12.7.

12.28 Show that $\hat{\sigma}^2$ is an unbiased estimator of λ^2 in Eq. (12.34).

12.29 The mass attenuation coefficient for 2-MeV photons in iron (density = 7.86 g cm^{-3}) is 0.0425 cm^2 g^{-1}.
   a) What is the probability that a 2-MeV photon will travel at least 2.14 cm in iron without interacting?
   b) What is the probability that it will have its first interaction after traveling between 2.14 and 3.09 cm?
   c) What is the mean free path of a 2-MeV photon in iron?

12.30 Repeat the last problem for a 2-MeV photon in air at 20 °C and 752 Torr pressure (density = 1.293 kg m^{-3} at 0 °C and 760 Torr). The mass attenuation coefficient for air is 0.0445 cm^2 g^{-1}.

12.31 The linear attenuation coefficient for an 80-keV X-ray in soft tissue is 0.179 cm^{-1}.
   a) What is the probability that a normally incident, 80-keV photon will reach a depth of at least 10 cm in soft tissue without having an interaction?

b) What is the probability that a normally incident photon will first interact in the tissue at a depth between 10 and 11 cm?

c) If a photon reaches a depth of 10 cm without interacting, what is the probability that it will continue to a depth of at least 11 cm without interacting?

12.32 Prove Eq. (12.40).

12.33 A pencil beam of 500-keV photons is normally incident on a 3.5-cm thick aluminum slab, which is followed immediately by a 1.6-cm lead slab. The linear attenuation coefficient for Al is $0.227\,\text{cm}^{-1}$ and that for Pb is $1.76\,\text{cm}^{-1}$.

  a) Give a Monte Carlo procedure to determine the first collision depth of successive photons, based on a random number sequence.

  b) What is the probability that an incident photon will traverse both slabs without interacting?

  c) Use the random number sequence (12.2)–(12.3) in Section 12.2 to determine the first collision depths for two photons.

12.34 a) In the last problem, what is the median depth of travel of a photon before the first interaction?

  b) Of the photons that do not penetrate both slabs, what is the average distance of travel to the depth of the first collision?

  c) What is the mean free path of the incident photons?

12.35 Histories are generated in a Monte Carlo calculation for 50 000 photons, normally incident on a broad slab, having a density of $0.94\,\text{g\,cm}^{-3}$. For analysis, the target is divided uniformly into subslabs of 0.50-cm thickness, similar to the slab in Figure 12.17. From the histories it is found that a total of 78.4 MeV was deposited in one of the subslabs. From these data, determine the average absorbed dose in this subslab per unit fluence from a uniform, broad beam of these photons.

12.36 Neutrons of 5 MeV are scattered by hydrogen.

  a) What is the average energy lost by a neutron in a single collision?

  b) What is the probability that a neutron will lose more than 4 MeV in a single collision?

  c) What is the probability that a neutron will lose between 1.0 and 1.5 MeV in a single collision?

12.37 a) For the last problem, write the probability density function for neutron energy loss.

  b) Write the cumulative energy-loss probability function.

  c) Sketch both functions from (a) and (b).

12.38 For the scattering of a neutron of energy E by a proton (hydrogen nucleus), as illustrated in Figure 12.20, the conservation of energy and momentum requires that $x = E \sin^2 \theta$, where x is the energy lost by the neutron.

a) From the uniform energy-loss distribution, as illustrated in Figure 12.18, show that the probability density $w(\theta)$ for scattering at the angle θ is $w(\theta) = \sin 2\theta$.
b) The angular scattering probability density function is thus independent of the neutron energy. How is this result related to fact that the scattering is isotropic in the center-of-mass coordinate system?

13
Dose–Response Relationships and Biological Modeling

13.1
Deterministic and Stochastic Effects of Radiation

Radiation damages the cells of living tissue. If the damage is not sufficiently repaired, a cell might die or be unable to reproduce. On the other hand, it might survive as a viable cell, altered by the radiation. The two alternatives can have very different implications for an individual who is exposed to radiation. The distinction is manifested in the description of biological effects due to radiation as either *stochastic* or *deterministic* (also previously called *nonstochastic*). We consider deterministic effects first.

At relatively small doses of radiation, the body can typically tolerate the loss of a number of cells that are killed or inactivated, without showing any effect. A certain minimum, or threshold, dose is necessary before there is a noticeable response, which is then characterized as a deterministic effect. Reddening of the skin and the induction of cataracts are examples of deterministic effects of radiation. In addition to there being a threshold, there is also a direct cause and effect relationship between the severity of a deterministic effect and the dose that produced it. The larger the dose to the skin, for instance, the greater the damage, other conditions being the same.

Stochastic effects, on the other hand, arise from cells that survive and reproduce, but also carry changes induced by the radiation exposure. Leukemia, bone cancer, and teratogenic effects in irradiated fetuses are but a few examples of such effects. They are associated with alterations of *somatic* cells, and are manifested in the irradiated individual. Genetic changes due to radiation represent another important example of stochastic effects. Alterations of the *germ* cells of an irradiated individual manifest themselves in his or her descendents. All somatic and genetic stochastic effects known to be caused by radiation also occur with natural incidence. In principle, since a single energy-loss event by radiation in a cell can produce a molecular change, the argument is often made that there is no threshold dose needed to produce stochastic effects. Presumably, even at very small doses, the probability for a stochastic effect is not zero, in contrast to a deterministic effect. The *probability* for producing a stochastic disease, such as cancer, increases with dose. However, the severity of the

---

Statistical Methods in Radiation Physics, First Edition. James E. Turner, Darryl J. Downing, and James S. Bogard.
© 2012 Wiley-VCH Verlag GmbH & Co. KGaA. Published 2012 by Wiley-VCH Verlag GmbH & Co. KGaA.

disease in an individual does not appear to be dose related, also in contrast to the response for deterministic effects.

The problem of recognizing and quantitatively assessing the probabilities for radiation-induced stochastic effects, especially at low doses, is complicated by the fact that, as already pointed out, these effects also occur statistically at natural, or spontaneous, rates of incidence. In a population that has received a high radiation dose, such as those among the survivors of the atomic bombs at Hiroshima and Nagasaki, the excess incidence of a number of maladies is evident. While the probability for a person in the population to develop leukemia, for example, might be increased by the radiation exposure, one cannot determine whether a given case would have occurred spontaneously in the absence of the radiation. At low doses, any real or presumed added incidence of a stochastic effect due to radiation cannot be distinguished from fluctuations in the normal incidence. The estimation of radiation risks for stochastic effects at low doses – at levels in the range of those used as limits for the protection of workers and the public – remains an important and controversial unresolved issue in radiation protection today.

Modeling of biological effects on a wide variety of systems is carried out to try to understand the mechanisms of radiation action and to make quantitative predictions of results from radiation exposures. In this chapter, we examine the role of statistics in some simple aspects of biological modeling for stochastic effects. The discussions are limited to just a few specific examples to introduce and illustrate this important subject.

13.2
Dose–Response Relationships for Stochastic Effects

Biological responses can be conveniently represented in the form of dose–response curves.

Figure 13.1 is adapted from a report in the literature on a study of the incidence of breast cancer in women (Boise et al., 1979).

The data serve to illustrate the general nature of many dose–response relationships – what can be learned and what problems and uncertainties accompany their interpretation, especially at low doses.

A higher-than-normal incidence of breast cancer was observed in 1047 women with tuberculosis who were examined fluoroscopically over a number of years between 1930 and 1954 in Massachusetts. The average number of examinations per patient was 102, with an accompanying X-ray dose estimated to be, on the average, 0.015 Gy to each breast per exam. Figure 13.1a shows a plot of the observed incidence of breast cancer versus the estimated total dose to the breasts. The incidence is expressed as the number of cases per 100 000 woman years (WY) at risk, averaged over all ages of the women and all exposure regimes. The sense of this unit is such that it applies, for example, to a group of 1000 women over 10 years or to 10 000 women over 1 year. The error bars represent the 90% confidence interval. Results from a different study in New York are shown in Figure 13.1b. In this case, irradiation

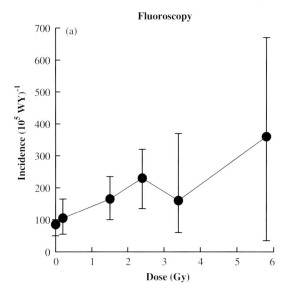

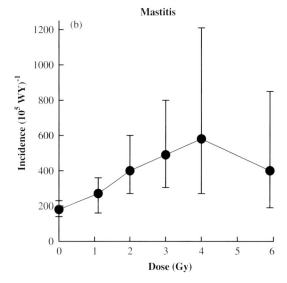

Figure 13.1 Examples of dose–response relationships for breast cancer: (a) fluoroscopic examinations in a Massachusetts study; (b) treatment of mastitis in a New York study. The observed incidence of breast cancer per 10^5 WY (women years, see the text) is shown as a function of the dose to the breasts.

of the breasts by X-rays occurred in 571 females who were treated for mastitis. As in Figure 13.1a, the error bars denote the 90% confidence level. The data points in Figure 13.1a and b are arbitrarily connected by straight lines to form continuous curves. Ideally, the resulting dose–response functions should at least approximately represent the risk for breast cancer as a function of dose for X-rays.

It is instructive to examine the data sets in some detail. The two curves in Figure 13.1 leave little doubt that substantial doses of X-rays increase the risk of breast cancer in women. As with any stochastic effect, the numbers of cases found in various dose groups are subject to statistical fluctuations. The ranges of the expected variations are shown by the error bars.

As is typical, the shape of the dose–response function for breast cancer based on Figure 13.1 is not clearly established by the available data. In Figure 13.1a, the data appear to be compatible with a straight line drawn from the point at the highest dose to the intercept at the ordinate. One might assume this linear response as a working hypothesis in considering, for example, the establishment of risk and acceptable radiation limits needed for workers and for the general public. Thus, a most important aspect of any dose–response function is its shape at low doses. The annual whole-body occupational limit for X-rays is 0.050 Gy, for instance. One sees from Figure 13.1 that, under the linear hypothesis, any increased incidence of breast cancer at such a level of dose to the breasts is small compared with random fluctuations in both the irradiation data and the normal incidence at zero dose. The data, in fact, do not rule out the existence of a threshold dose for causing breast cancer. The situation encountered here illustrates the uncertainty inherent in estimating the risk for any stochastic effect at low dose levels.

A critical factor in these and similar studies is the assignment of specific values to the doses that individuals might have received. The exposure of the breasts in the fluoroscopic examinations was not the primary focus of attention, but was incidental to the main procedure being carried out. Reconstructed doses for Figure 13.1a were based on medical records, interviews with physicians and patients, later laboratory measurements with similar equipment, and even Monte Carlo calculations. In Figure 13.1b, calibrated therapy units were used, and the doses are well documented. Generally, there can be considerable uncertainty in doses assigned retrospectively to individuals in such studies. Moreover, even given the total dose that a person received, the response is known to depend markedly on the way it was distributed in fractions during the time of the repeated examinations.

Several other points about dose–response relationships can be noted. The appearance of the plotted data can be affected to some extent by the particular groupings of the raw numbers into specific cohort groups. We met this circumstance in dealing with Monte Carlo histograms in the previous chapter. The raw data are represented by grouping numbers of cases into selected dose intervals. Also, the control group, representing unexposed individuals for comparison, should be the same in all other respects except for the radiation. This ideal is not often attainable. Still another concern in the example shown in Figure 13.1b is whether the condition of mastitis predisposes one to breast cancer in the first place.

13.3 Modeling Cell Survival to Radiation

In the paper from which Figure 13.1 is adapted, the authors discuss these matters as well as others that we shall not attempt to address here. The paper is highly recommended to the reader interested in statistics and dose–response relationships.

13.3 Modeling Cell Survival to Radiation

One type of dose–response relationship that lends itself to statistical modeling and interpretation is that for killing cells of a specified type irradiated *in vitro*. The data are usually represented by means of a cell survival curve, showing the fraction of a population of cells that survives a given dose of radiation. When plotted in semilog fashion, the survival curve for a single, acute dose often closely approximates one of two general shapes, shown in Figure 13.2. It can either be linear, as is typical for the response to high-LET radiation, or have a shoulder at low doses before becoming linear at high doses, characteristic of the response to low-LET radiation. (On a linear, rather than semilog, plot, the two types of survival curves are exponential and sigmoidal.) In Figure 13.2, S represents the number of cells from an original irradiated population of S_o that survive a dose D. The ratio S/S_o, plotted as the ordinate, can be interpreted as the survival probability for a given cell in the irradiated population as a function of the dose D, given by the abscissa.

Before discussing statistical models as a basis for interpreting observed cell survival curves, we describe briefly what is meant by "cell killing" in the present context. For some nonproliferating cells, like nerve and muscle, irradiation can lead

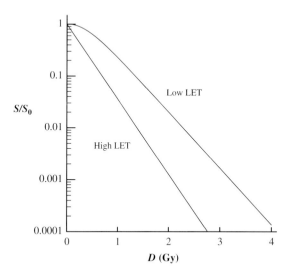

Figure 13.2 Two general classes of cell survival curves are common. Linear survival on a semilog plot is characteristic of the response to high-LET radiation, such as neutrons and alpha particles. A response with a shoulder at low doses is characteristic of low-LET radiation, such as gamma rays and beta particles.

to loss of function, which can be considered as "death" for them. For proliferating cells, the relevant end point is "reproductive death" – that is, loss of the ability of a cell to divide indefinitely and thus produce a large colony, or clone. An irradiated cell might continue to "live," in the sense of having metabolic activity and even undergoing mitosis a few times. But if it cannot produce a large colony, the cell is considered to be "dead," by definition. A surviving cell must be clonogenic, as measured by specific techniques that involve further handling, incubation, and comparison with similarly treated unirradiated controls.

13.4
Single-Target, Single-Hit Model

We consider first a single-target, single-hit model for cell killing by radiation. The irradiated sample is considered to consist of S_0 identical, independent, cells suspended in solution. According to the model, there is only one mechanism for killing a cell. Each cell contains a single critical target that, when acted upon, or "hit," by radiation, leads to cell death. The interaction of the radiation with the target is described in terms of a cross section, σ, which has the dimensions of area and is defined like that for other processes of radiation interaction. That is, when S_0 targets are traversed by a monoenergetic, uniform, parallel beam of radiation, having fluence φ, then the expected number of hits in targets is $\varphi \sigma S_0$. In general, σ will depend on the type of radiation employed and its energy. The absorbed dose is proportional to φ. It follows that the average number of hits per target in the population is

$$\mu = \frac{\varphi \sigma S_0}{S_0} = \varphi \sigma. \tag{13.1}$$

The hits are assumed to be randomly distributed in the targets of the cell population according to Poisson statistics. Therefore, if X represents the number of hits in the target of a given cell, then the probability that exactly n hits occur there is

$$\Pr(X = n) = \frac{\mu^n e^{-\mu}}{n!}. \tag{13.2}$$

The survival probability S/S_0 for a given cell, under the assumptions of the model, is the probability that there are no hits in its target:

$$\frac{S}{S_0} = \Pr(X = 0) = e^{-\mu} = e^{-\varphi \sigma}, \tag{13.3}$$

where Eq. (13.1) has been used to write the last equality. Since the fluence φ and dose D are proportional, we may write

$$D = k\varphi, \tag{13.4}$$

13.4 Single-Target, Single-Hit Model

where k is the constant of proportionality. Thus, Eq. (13.3) can also be written as

$$\frac{S}{S_o} = e^{-D\sigma/k} = e^{-D/D_o}, \qquad (13.5)$$

where the constant

$$D_o = \frac{k}{\sigma}. \qquad (13.6)$$

has been introduced in place of k. Comparison with Eq. (13.3) shows that

$$\mu = \frac{D}{D_o} \qquad (13.7)$$

gives the mean number of hits per target in terms of D_o.

The survival curve obtained with the single-target, single-hit model, Eq. (13.5), is exponential. As Figure 13.2 indicates, this type of response is often found for densely ionizing radiations. The slope of the response curve on the semilog plot is $-1/D_o$. When the dose D is equal to D_o, Eq. (13.5) gives for the survival

$$\frac{S}{S_o} = e^{-1} = 0.37. \qquad (13.8)$$

For this reason, D_o is often referred to as the "D_{37}" dose. Also, in analogy with the mean life of a radionuclide (Eq. (4.38)) and the mean free path for radiation interaction (Section 12.9), D_o in Eq. (13.5) represents the average dose for killing a cell (Problem 13.3). D_o is thus also called the "mean lethal dose."

Equation (13.6) shows the formal relationship in the model between the target "size" σ and the mean lethal dose. When the target is small, D_o is large, and vice versa, as one would expect. The quantity $k = D/\varphi$, introduced in Eq. (13.4), represents the absorbed dose per unit fluence for the radiation. This quantity can be measured or calculated for the radiation used in an experiment. Knowledge of k, coupled with Eq. (13.6) and the observed slope of the cell survival curve, $1/D_o$, allows numerical evaluation of the target cross section φ. To the extent that such a model might be realistic, one could, in principle, associate the area σ with the size of a critical cellular component, for example, the cell nucleus, a nucleosome, a gene, or DNA.

> ### ■ Example
> A uniform, parallel beam of 4-MeV protons is used for irradiation in a cell survival experiment. The data for the survival fractions found at several dose levels are given in Table 13.1. The fluence-to-dose conversion factor for the protons is 1.53×10^{-12} Gy m^2.
>
> a) Show that the survival decreases exponentially with dose. What is the mean lethal dose?
> b) What is the LD$_{50}$ for the cells?
> c) What is the proton fluence for 22% survival of the cells?

Table 13.1 Survival fraction S/S_o at different doses D for example in the text.

D (Gy)	S/S$_o$
0.2	0.72
0.5	0.45
1.0	0.22
1.5	0.088
2.0	0.040

d) What is the target size for radiation lethality, based on a single-target, single-hit survival model?

Solution

a) Exponential survival means that the data satisfy an equation of the form Eq. (13.5). The relation between survival and dose would then be

$$\ln\left(\frac{S}{S_o}\right) = -\frac{D}{D_o}, \qquad (13.9)$$

which can be written as

$$D_o = \frac{-D}{\ln(S/S_o)}. \qquad (13.10)$$

Using the first and last data points from Table 13.1, we find, respectively, that

$$D_o = \frac{-0.2 \, \text{Gy}}{\ln 0.72} = 0.609 \, \text{Gy} \qquad (13.11)$$

and

$$D_o = \frac{-2.0 \, \text{Gy}}{\ln 0.04} = 0.621 \, \text{Gy}. \qquad (13.12)$$

The other three points in Table 13.1 give similar results, showing that the survival is exponential. We take the mean lethal dose to be $D_o = 0.63$ Gy, which is the average for the five values from Table 13.1.

b) The LD$_{50}$ is the dose that kills one-half of the cells. Applying Eq. (13.5) with the mean lethal dose just found, we write

$$\frac{S}{S_o} = 0.50 = e^{-D/0.63}, \qquad (13.13)$$

giving $D = 0.44$ Gy for the LD$_{50}$. Note that the relationship between the LD$_{50}$ and the mean lethal dose is like that between the half-life and the mean life of a radionuclide: $0.44/0.63 = \ln 2$ (to within roundoff).

c) The fluence-to-dose conversion factor was introduced in Eq. (13.4). We are given $k = 1.53 \times 10^{-12}$ Gy m^2. That is, the dose conversion is

1.53×10^{-12} Gy per unit fluence (i.e., Gy per proton m^{-2}). From Table 13.1 and Eq. (13.4), the fluence for 22% survival (1.0 Gy) is

$$\varphi = \frac{D}{k} = \frac{1.0 \text{ Gy}}{1.53 \times 10^{-12} \text{ Gy m}^2} = 6.54 \times 10^{11} \text{ m}^{-2}, \quad (13.14)$$

or 6.54×10^7 cm^{-2}.

d) According to Eq. (13.6), the target size is

$$\sigma = \frac{k}{D_0} = \frac{1.53 \times 10^{-12} \text{ Gy m}^2}{0.63 \text{ Gy}} = 2.4 \times 10^{-12} \text{ m}^2. \quad (13.15)$$

On a more convenient distance scale (1 µm = 10^{-6} m) for cellular dimensions, $\sigma = 2.4$ µm^2. A circle with this area has a radius of 0.87 µm, which is of subnuclear size for many cells.

While useful and instructive, such target models for cell killing are, at best, idealized approximations to reality. For one thing, the individual cells in a population are not identical. In addition to variations in size and shape, they are generally in different phases of the mitotic cycle, in which the radiosensitivity is different. (Synchronization can be achieved to some extent.) Also, no account is taken in the model of possible cell repair mechanisms that come into play in response to the radiation. The "target" itself within the cell is purely phenomenological. Experiments demonstrate clearly, though, that the cell nucleus is much more sensitive than the cytoplasm for radiation-induced cell lethality. Evidence indicates that chromosomal DNA and the nuclear membrane are probably the primary targets for cell killing. The reader is referred to the textbook by Hall (1994) in the Bibliography for more detailed information and references.

13.5
Multi-Target, Single-Hit Model

We treat next a somewhat more general model, which leads to a survival curve with a shoulder at low doses (Figure 13.2). In the multi-target, single-hit model, each cell is assumed to possess n identical targets of cross section σ. Death results when all n targets of a cell are struck at least once. As before (Eqs. (13.1) and (13.7)), the average number of hits in a given cell target at a dose level D is

$$\mu = \varphi \sigma = \frac{D}{D_0}. \quad (13.16)$$

The probability that a given target receives no hits is expressed by Eq. (13.3), and so the probability that it is hit at least once is

$$\Pr(X \geq 1) = 1 - \Pr(X = 0) = 1 - e^{-D/D_0}. \quad (13.17)$$

The probability that all n targets of a given cell are struck at least once is

$$[\Pr(X \geq 1)]^n = (1 - e^{-D/D_o})^n, \qquad (13.18)$$

which is the probability that the cell is killed. The probability that the cell survives is, therefore,

$$\frac{S}{S_o} = 1 - (1 - e^{-D/D_o})^n, \qquad (13.19)$$

according to the multi-target, single-hit model.

With $n=1$, Eq. (13.19) becomes identical with Eq. (13.5), describing the exponential survival of single-target, single-hit theory. Otherwise, the survival probability curve has a shoulder at low doses. The solid curve in Figure 13.3 shows a plot of Eq. (13.19) for $D_o = 0.90$ Gy and $n=3$. Each cell has three targets, which must all be hit in order to cause its death. Most cells survive low doses, since it is unlikely that a given cell will have multiple struck targets. As the dose is increased, cells accumulate targets that are hit, and the killing becomes more efficient. The survival curve then bends downward, becoming steeper with the increased efficiency, and then straightens out on the semilog plot. For the straight portion, most surviving cells have only the one remaining unstruck target. The response of the remaining population to additional radiation then becomes that of single-target, single-hit survival. The overall survival curve thus begins with zero slope at zero dose (Problem 13.10), bends over through a shoulder, and then becomes a straight line with slope $-1/D_o$ at high doses.

The dashed line in Figure 13.3 is extrapolated from the straight portion of the survival curve from high dose back to low dose. It intersects the ordinate at the value $n=3$, which is the number of targets in a cell. This result can be predicted from the

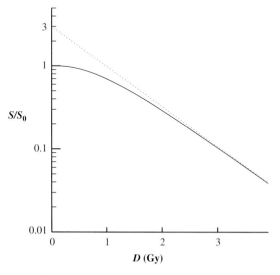

Figure 13.3 Plot (solid curve) of multi-target, single-hit model, Eq. (13.19), for $D_o = 0.90$ Gy and $n=3$.

equation that describes the model. At high doses, the exponential term in Eq. (13.19) is small compared with unity. Using the binomial expansion, $(1-x)^n \cong 1-nx$, for small x, we write in place of Eq. (13.19) with D large

$$\frac{S}{S_o} \cong 1 - (1 - n\,e^{-D/D_o}) = n\,e^{-D/D_o}. \tag{13.20}$$

This relation describes a straight line with slope $-1/D_o$ on the semilog plot. At $D=0$, the line represented by Eq. (13.20) intersects the ordinate at $S/S_o = n$, which is called the extrapolation number. The extrapolation number is thus equal to the number of targets in the multi-target, single-hit model. It provides a measure of the width of the shoulder of the survival curve. As a rule, reflecting the phenomenological nature of the model, observed extrapolation numbers are usually not integral. In the model, the existence of the shoulder is explained by the accumulation of hits in a cell before it is killed. A shoulder could also be explained by the action of repair processes set up in the cell in response to the radiation – as well as by other mechanisms not considered in the model.

A number of different target models for cell survival have been investigated. For instance, cell death can be attributed to striking any m of n targets in a cell ($m \leq n$), either once or a specified larger number of times. The different targets within a cell can also have different cross sections, σ. A variety of multi-target, multi-hit models provide cell survival curves with different detailed structures.

We have dealt with statistical aspects of cell survival only in terms of the random interaction of radiation with targets. Statistics is also important for the practical matters of experimental data collection and evaluation. The observed survival values at different doses, such as those in column 2 of Table 13.1, have associated error bars, which we have not discussed. These arise from various sources. A number of individual cells from a stock culture are counted for irradiation and controls, and then seeded into dishes for incubation and cloning. The ratio of the number of colonies and the initial number of cells, called the plating efficiency, is observed. This number is subject to experimental errors from cell counting as well as fluctuations in the conditions of handling and treating the cell colonies. Different samples are prepared for irradiation at different dose levels. The optimum seeding of the number of cells per dish to be irradiated is done in such a way as to result in countable numbers of colonies for good statistical reliability, but not so many as to cause a merging of subsequent colonies into one another. Irradiations at different doses can also be repeated several times to gain added statistical significance. The reader is again referred to the excellent book by Hall (1994) in the Bibliography for additional information.

13.6
The Linear–Quadratic Model

One is not limited to purely phenomenological modeling of cell killing. For example, a relationship is known to exist between certain kinds of radiation-induced chromosome changes and cell killing. Cells that undergo exchange-type aberrations do not survive. Such alterations require two separate chromosome breaks. When the dose is relatively small, the two breaks, if they occur, are likely to be caused by the same particle, such as a single secondary electron produced by the radiation. The probability of an exchange-type aberration is then proportional to the number of tracks per unit volume, and hence to the dose. The resulting survival curve is essentially linear on a semilog plot. At high doses, the two breaks can, in addition, be caused by two different, independent particles. The probability for this mode is proportional to the square of the dose, and its effect is to bend the survival curve downward. The expression for cell survival in this linear–quadratic model is then

$$\frac{S}{S_o} = e^{-\alpha D - \beta D^2}, \qquad (13.21)$$

where α and β are constants, which depend on the radiation and type of cells. When compared with Figure 13.2, it is apparent that individual particles of high-LET radiation, with its linear response at low dose, have a high probability of producing both chromosome breaks. This happens to a much lesser extent with low-LET radiation, which is characterized by a shoulder in the survival curve at low dose. The second chromosome break usually requires a second particle.

The two components of cell killing in the linear–quadratic model are consistent with evidence from quantitative studies of chromosome aberrations. The dose at which the linear and quadratic components contribute equally to cell killing occurs when $\alpha D = \beta D^2$, or when $D = \alpha/\beta$. Unlike the single-hit target models, which produce linear survival curves at large doses, the linear–quadratic model survival curve continues to bend downward.

Problems

13.1 Is the production of chromosome aberrations a deterministic or a stochastic effect of radiation? Explain.

13.2 What are some reasons for the uncertainties in risk estimates for stochastic effects of radiation at low doses?

13.3 a) Show that D_o in Eq. (13.5) is the average dose needed to kill a cell.
 b) Show that the ratio of D_o and LD_{50} is ln 2.

13.4 The fraction of surviving cells in a certain experiment is given by $S/S_o = e^{-0.75D}$, where D is the dose in Gy.
 a) What is the mean lethal dose?
 b) What is the survival probability for a dose of 1.0 Gy?
 c) What dose leaves a surviving fraction of 0.0010?
 d) What is the LD_{50} for the cells?

13.5 In an experiment in which cell survival is exponential, the survival fraction from a dose of 1.55 Gy is 0.050.
 a) What is the mean lethal dose?
 b) Write an equation giving the survival fraction as a function of dose.

Table 13.2 Data for problems in the text.

D (Gy)	S/S$_o$
1.0	0.648
2.0	0.269
3.0	0.0955
4.0	0.0308
5.0	0.0106
6.0	0.0036
7.0	0.0012

13.6 Cell survival data are fit with a multi-target, single-hit model, having $D_o = 1.4$ Gy and an extrapolation number $n = 4$. What fraction of the cells survive a dose of
   a) 1.0 Gy?
   b) 5.0 Gy?
   c) 10 Gy?
   d) Make a semilog plot of the surviving fraction as a function of dose.
13.7 a) What dose in the last problem results in 25% survival?
   b) What is the LD$_{50}$?
13.8 Repeat Problem 13.6 for $D_o = 1.4$ Gy and $n = 2$. Why are the survival levels lower than before at the same doses?
13.9 Repeat Problem 13.6 for $D_o = 2.2$ Gy and $n = 4$. Why are the survival levels higher than before at the same doses?
13.10 Show that the slope of the survival curve (13.19) is zero at zero dose.
13.11 Plot the survival data shown in Table 13.2. Based on a multi-target, single-hit model, write an equation that describes the data.
13.12 In a certain study, cell survival is found to be described by a single-target, single-hit model, $S/S_o = e^{-1.3D}$, where D is in Gy. At a dose of 2.0 Gy, what is the probability that, in a given cell, there are exactly
   a) no hits?
   b) 4 hits?
   c) 10 hits?
   d) What is the most likely number of hits?
13.13 Fit a multi-target, single-hit model to the cell survival data in Table 13.3.
   a) Find the slope at high doses.
   b) What is the extrapolation number?
   c) Write an equation that describes the specific data in Table 13.3.
13.14 The survival of cells exposed to photons in an experiment is described by the multi-target, single-hit function

$$\frac{S}{S_o} = 1 - (1 - e^{-1.35D})^{2.8},$$

where D is in Gy. The dose per unit fluence is 4.72×10^{-16} Gy m^2.
   a) Calculate the surviving fraction for a dose of 1.5 Gy.

13 Dose–Response Relationships and Biological Modeling

Table 13.3 Data for problems in the text.

Dose (Gy)	Surviving fraction
0.10	0.992
0.25	0.934
0.50	0.727
1.00	0.329
2.00	0.0460
3.00	0.00575
4.00	0.00071

b) What is the LD_{50} for the cells?
c) What photon fluence is required for a dose of 2.0 Gy?
d) What is the diameter of a cellular target, assumed to be circular?
e) Give one or more reasons why the extrapolation number is not necessarily an integer.

13.15 a) In Problem 13.6, what fraction of cells survive a dose of 0.60 Gy?
b) What is the average number of hits in a given cell target at 0.60 Gy?
c) What is the average number of struck targets in a given cell at 0.60 Gy?
d) What fraction of the cells have exactly three struck targets at 0.60 Gy?

13.16 Repeat the last problem for a dose of 1.5 Gy.

13.17 a) For the model in Problem 13.6, find the distribution of the number of hits in a given cell target at a dose of 1.0 Gy.
b) Find the distribution of the number of struck targets per cell at 1.0 Gy.
c) From (b), determine the cell survival fraction at 1.0 Gy and compare with that calculated from Eq. (13.19).

13.18 In a single-target, multi-hit model of cell survival, each cell contains a single target that, when struck m or more times, produces cell killing. Show that the survival of cells as a function of dose D is given by

$$\frac{S}{S_o} = e^{-D/D_o} \sum_{n=0}^{m-1} \frac{1}{n!} \left(\frac{D}{D_o}\right)^n,$$

where D_o is defined by Eq. (13.6).

13.19 a) Make a semilog plot of the survival curve in the last problem for $D_o = 0.92$ Gy and $m = 4$.
b) What is the survival fraction for a dose of 1.15 Gy?
c) Find the LD_{50}.
d) Is this model tantamount to the multi-target, single-hit survival model? Explain.

13.20 A cell population receives a total of N hits. Let ϱ be the "hit" probability—that is, the probability that a given cell receives a given hit, all hits in all cells being equally probable. The probability that a given cell gets exactly h of the N hits is which of the following?

a) $\varrho^h \frac{N!}{h!(N-h)!}$;
b) $\varrho^h(1-\varrho)^{N-h} \frac{1}{N!}$;
c) $e^{-h/N}$;
d) $\varrho^h(1-\varrho)^{N-h}$;
e) $\varrho^h(1-\varrho)^{N-h} \frac{N!}{h!(N-h)!}$.

13.21 When exposed to neutrons, the cell line in Problem 13.14 is found to have exponential survival, described by

$$\frac{S}{S_o} = e^{-1.82D},$$

with D in Gy. The *relative biological effectiveness* (RBE) of the neutrons (relative to the photons) is defined as the ratio of the photon and neutron doses that result in the same degree of cell killing.
  a) Calculate the photon and neutron doses that result in a survival level $S/S_o = 0.0010$.
  b) What is the RBE for $S/S_o = 0.0010$?
  c) Does the RBE depend upon the dose?

13.22 a) In the last problem, calculate the photon and neutron doses that result in 90% survival.
  b) What is the RBE for 90% survival?
  c) Suggest a hypothesis to explain why the neutron RBE should be larger at the higher level of survival, as in this problem, than in the last problem?

13.23 As a general rule, illustrated by the last two problems, the RBE for densely ionizing radiation (e.g., neutrons and alpha particles) increases as the dose gets smaller.
  a) Discuss the implications of such a finding for radiation protection, where one needs to assess the potential risks of low levels of radiation to workers and to members of the public.
  b) Is the larger RBE at smaller doses only an artifact – due to the shoulder of the photon response curve approaching zero slope as the dose approaches zero?

13.24 A linear–quadratic model is used to fit cell survival data with $\alpha = 0.080$ Gy^{-1} and $\beta = 0.025$ Gy^{-2}.
  a) Make a semilog sketch of the survival curve out to a dose of 12 Gy.
  b) At what dose are the linear and quadratic components of cell killing equal?

13.25 Fit the survival data in Table 13.2 to a linear–quadratic function.
  a) What are the numerical values of α and β?
  b) At what dose are the linear and quadratic contributions to cell killing the same?

13.26 Repeat the last problem for the data in Table 13.3.

13.27 How is the ratio α/β in Eq. (13.21) expected to behave with increasing LET?

14
Regression Analysis

14.1
Introduction

In many situations, a correlation might exist between two or more variables under study. For example, the weight of a male over 20 years of age presumably depends to some extent on his height. To consider the relationship more precisely, one could randomly sample a population of such men and, on the basis of the sample data, try to arrive at a mathematical expression for estimating weight based on height. To treat this formally, one can designate the weight Y as the *response variable* and the height X as the *independent variable*. A model, such as the following, can then be considered to relate the two quantities mathematically:

$$Y = \beta_0 + \beta_1 X + \varepsilon. \qquad (14.1)$$

Here β_0 and β_1 are unknown parameters, to be determined from the sample data. The term ε is the error due to a number of possible factors, such as the choice of the model and uncertainties in the measurements.

In this chapter we shall principally explore simple linear regression models like Eq. (14.1), which are linear in both the response and independent variables and the β parameters. While more complicated functions can be employed, the linear model expressed by Eq. (14.1) is often adequate to relate the two variables over some limited range of values. Our goal will include estimating the β parameters and their variances. We shall also determine for the response variable the variance of a predicted value, a mean value, and a future value. Regression analysis refers to the study of how one or more independent variables $(X_1, X_2, \ldots)$ relate to and enable the prediction of a dependent variable (Y). We will examine goodness of fit and also the inverse regression of obtaining the value of X when given Y.

14.2
Estimation of Parameters β_0 and β_1

We assume in the following discussion that we have pairs of variables (X_i, Y_i), $i = 1, 2, \ldots, n$. Each pair is independently obtained, and the value of X_i is measured without error or has such small error, compared to that of the response variable Y_i, that it is negligible. We also assume the following model:

$$Y_i = \beta_0 + \beta_1 X_i + \varepsilon_i, \tag{14.2}$$

where ε_i has mean zero, constant variance σ^2, and $\text{Cov}(\varepsilon_i, \varepsilon_j) = 0$ for $i \neq j$, that is, the errors ε_i are independent. If we plot the pairs (X_i, Y_i), we would want to select the straight line (Eq. (14.2)) that minimizes the errors. Figure 14.1 shows a hypothetical situation where (X_i, Y_i) are plotted, a straight line is drawn, and the errors are shown. A convenient way to define the error in Eq. (14.2) is to write

$$\varepsilon_i = Y_i - (\beta_0 + \beta_1 X_i), \tag{14.3}$$

and then, if the model (Eq. (14.2)) fits exactly, all $\varepsilon_i = 0$.

It is better to reduce the squared errors, ε_i^2, since errors can be either positive or negative. This method is called the *least squares method* of estimation. Thus, we want to find values b_0 and b_1 such that the sum of squared errors is minimized, that is, b_0 and b_1 are the values of β_0 and β_1 that minimize

$$S = \sum_{i=1}^{n} \varepsilon_i^2 = \sum_{i=1}^{n} [Y_i - (\beta_0 + \beta_1 X_i)]^2. \tag{14.4}$$

We can determine b_0 and b_1 by differentiating with respect to β_0 and β_1, setting the derivatives equal to zero, and solving

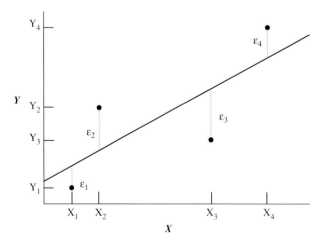

Figure 14.1 Plot of data pairs (X_i, Y_i) showing a straight line drawn through the data so that the errors ε_i are minimized.

14.2 Estimation of Parameters β_0 and β_1

$$\frac{\partial S}{\partial \beta_0} = \frac{\partial}{\partial \beta_0} \sum_{i=1}^{n} [Y_i - (\beta_0 + \beta_1 X_i)]^2 = -2 \sum_{i=1}^{n} [Y_i - (\beta_0 + \beta_1 X_i)] = 0 \quad (14.5)$$

and

$$\frac{\partial S}{\partial \beta_1} = \frac{\partial}{\partial \beta_1} \sum_{i=1}^{n} [Y_i - (\beta_0 + \beta_1 X_i)]^2 = -2 \sum_{i=1}^{n} [Y_i - (\beta_0 + \beta_1 X_i)] X_i = 0. \quad (14.6)$$

Now b_0 and b_1 may be substituted directly for β_0 and β_1 since the sum of squared errors is minimized, and Eqs. (14.5) and (14.6) can be rewritten as the so-called *normal equations*

$$\sum_{i=1}^{n} Y_i = b_0 n + b_1 \sum_{i=1}^{n} X_i \quad (14.7)$$

and

$$\sum_{i=1}^{n} X_i Y_i = b_0 \sum_{i=1}^{n} X_i + b_1 \sum_{i=1}^{n} X_i^2. \quad (14.8)$$

The solutions for the intercept b_0 and the slope b_1 are then (Problem 14.3)

$$b_0 = \frac{1}{n}\left(\sum_{i=1}^{n} Y_i - b_1 \sum_{i=1}^{n} X_i\right) = \bar{Y} - b_1 \bar{X}, \quad (14.9)$$

where $\bar{Y}$ and $\bar{X}$ are the sample means, and

$$b_1 = \frac{\sum_{i=1}^{n} X_i Y_i - (1/n) \sum_{i=1}^{n} X_i \sum_{i=1}^{n} Y_i}{\sum_{i=1}^{n} X_i^2 - (1/n)\left(\sum_{i=1}^{n} X_i\right)^2}. \quad (14.10)$$

The quantities in Eq. (14.10) have names used in many software programs. The term $\sum_{i=1}^{n} X_i^2$ is called the *uncorrected sum of squares*, and the term $(1/n)\left(\sum_{i=1}^{n} X_i\right)^2$ is the *correction for the mean* of the X_i's. The final term,

$$\sum_{i=1}^{n} X_i^2 - \frac{1}{n}\left(\sum_{i=1}^{n} X_i\right)^2, \quad (14.11)$$

is the *corrected sum of squares* or S_{XX}. Similarly, $\sum_{i=1}^{n} X_i Y_i$ is referred to as the *uncorrected sum of cross-products*, $(1/n) \sum_{i=1}^{n} X_i \sum_{i=1}^{n} Y_i$ is the *correction for the means*, and

$$\sum_{i=1}^{n} X_i Y_i - \frac{1}{n}\sum_{i=1}^{n} X_i \sum_{i=1}^{n} Y_i \quad (14.12)$$

is the *corrected sum of cross-products* of X and Y or S_{XY}. Another term, analogous to S_{XX}, is

14 Regression Analysis

Table 14.1 Equivalent computing formulas for S_{XY}, S_{XX}, and S_{YY}.

S_{XY}	S_{XX}	S_{YY}
$\sum_{i=1}^{n}(X_i-\bar{X})(Y_i-\bar{Y})$	$\sum_{i=1}^{n}(X_i-\bar{X})^2$	$\sum_{i=1}^{n}(Y_i-\bar{Y})^2$
$\sum_{i=1}^{n}(X_i-\bar{X})Y_i$	$\sum_{i=1}^{n}X_i(X_i-\bar{X})$	$\sum_{i=1}^{n}Y_i(Y_i-\bar{Y})$
$\sum_{i=1}^{n}X_i(Y_i-\bar{Y})$	$\sum_{i=1}^{n}X_i^2-\frac{1}{n}\left(\sum_{i=1}^{n}X_i\right)^2$	$\sum_{i=1}^{n}Y_i^2-\frac{1}{n}\left(\sum_{i=1}^{n}Y_i\right)^2$
$\sum_{i=1}^{n}X_iY_i-\frac{1}{n}\sum_{i=1}^{n}X_i\sum_{i=1}^{n}Y_i$	$\sum_{i=1}^{n}X_i^2-n\bar{X}^2$	$\sum_{i=1}^{n}Y_i^2-n\bar{Y}^2$
$\sum_{i=1}^{n}X_iY_i-n\bar{X}\bar{Y}$		

$$S_{YY} = \sum_{i=1}^{n}Y_i^2 - \frac{1}{n}\left(\sum_{i=1}^{n}Y_i\right)^2, \qquad (14.13)$$

which will be used later. These expressions are handy when using a calculator. Table 14.1 summarizes equivalent ways of computing S_{XY}, S_{XX}, and S_{YY}. Using these expressions, we can now write that

$$b_1 = \frac{S_{XY}}{S_{XX}}. \qquad (14.14)$$

We can also write our estimated equation as

$$\hat{Y} = b_0 + b_1 X \qquad (14.15)$$

and, since $b_0 = \bar{Y} - b_1\bar{X}$,

$$\hat{Y} = \bar{Y} + b_1(X-\bar{X}). \qquad (14.16)$$

Note that, if we set $X = \bar{X}$, then $\hat{Y} = \bar{Y}$, which means the point $(\bar{X}, \bar{Y})$ falls on the regression line.

■ *Example*
Known amounts of uranium were measured using a Geiger–Mueller counter, resulting in the data given below. Sources of random error other than counting statistics are assumed to be negligible, and the variability is considered constant over the range of the data.

X (g U)	15	20	30	40	50	60
Y (net counts)	1305	1457	2380	3074	3615	4420

a) Use the least squares formulas to obtain the estimates b_0 and b_1. Plot the corresponding line and the data points on a graph.

b) Suppose a new sample of 45 g U were counted. What would you predict the number of counts to be? Would you have confidence in this predicted value? Why or why not?

c) Suppose a new sample of 75 g U were counted. What would you predict the number of counts to be? Would you have confidence in this predicted value? Why or why not?

Solution

a) The least squares formulas give us $b_1 = S_{XY}/S_{XX}$ (recall Eq. (14.14) and the formulas for S_{XY} and S_{XX} given by Eqs. (14.12) and (14.11), respectively) and $b_0 = \bar{Y} - b_1 \bar{X}$. The values of the terms in S_{XY} and S_{XX} are computed to be $\sum_{i=1}^{n} X_i Y_i = 689\,025$, $\sum_{i=1}^{n} X_i = 215$, $\sum_{i=1}^{n} Y_i = 16\,251$, and $\sum_{i=1}^{n} X_i^2 = 9225$, giving $S_{XY} = 689\,025 - (215)(16\,251)/6 = 106\,697.5$ and $S_{XX} = 9225 - (215)^2/6 = 1520.8333$. Therefore, $b_1 = 1520.8333/106\,697.5 = 70.157$ and $b_0 = (16\,251/6) - (70.157)(215)/6 = 194.532$. Hence, the least squares regression line is given by

$$\hat{Y} = 194.532 + 70.157\,X. \tag{14.17}$$

The predicted values of Y for each value of X are

X	15	20	30	40	50	60
$\hat{Y}$	1247	1598	2299	3001	3702	4404

Note that we carry more significant digits in the model and round after the calculations. Plotting the pairs $(X, \hat{Y})$ and connecting the points gives the predicted line. Note that, since this is a linear fit to the data, we can simply plot the two extreme points and connect them using a straight line. Figure 14.2 shows the data and fitted regression line.

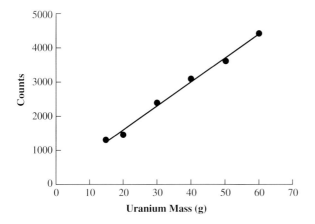

Figure 14.2 Data and fitted regression line from the example.

b) Using Eq. (14.17), $\hat{Y} = 194.532 + 70.157(45) = 3352$ counts. One could be confident of this predicted value, since the value $X = 45$ is internal to the empirical data set.

c) Again using Eq. (14.17), $\hat{Y} = 194.532 + 70.157(75) = 5456$. This time, the value of X is outside the set of empirical data, and so one might have less confidence in this predicted value. Our confidence in the predicted value is diminished, since we do not know the behavior of Y beyond $X = 60\,\text{g}$, unless we are certain that the assumed linear properties of our model hold.

14.3
Some Properties of the Regression Estimators

We may write $e_i = Y_i - \hat{Y}_i$ for $i = 1, 2, \ldots, n$ and note that $\sum_{i=1}^{n} e_i = 0$ by referring to the first normal equation given in Eq. (14.5). This is one way to check our arithmetic, since this sum should always equal zero. Using the data in the previous example, we have

X_i	15	20	30	40	50	60
Y_i	1305	1457	2380	3074	3615	4420
$\hat{Y}_i$	1247	1598	2299	3001	3702	4404
$e_i = Y_i - \hat{Y}_i$	58	−141	81	73	−87	16

We see that $\sum_{i=1}^{6} e_i = \sum_{i=1}^{6} (Y_i - \hat{Y}_i) = 0$. The e_i's are called *residuals*, and $\sum_{i=1}^{n} e_i^2$ is the *minimum sum of squared residuals*, as required by the least squares principle. We can rewrite Eq. (14.5) as

$$\sum_{i=1}^{n} (Y_i - b_0 - b_1 X_i) = \sum_{i=1}^{n} e_i = 0 \tag{14.18}$$

and Eq. (14.6) as

$$\sum_{i=1}^{n} (Y_i - b_0 - b_1 X_i) X_i = \sum_{i=1}^{n} e_i X_i = 0, \tag{14.19}$$

which provides some useful identities.

The residuals can be used to estimate the variance σ^2 mentioned with the assumptions in Eq. (14.2). Some things to note from Eq. (14.2) and the assumptions stated there are that

$$E(Y_i) = E(\beta_0 + \beta_1 X_i + \varepsilon_i) = \beta_0 + \beta_1 X_i + E(\varepsilon_i) = \beta_0 + \beta_1 X_i. \tag{14.20}$$

Therefore, the mean value of the distribution of Y_i for a corresponding X_i is $\beta_0 + \beta_1 X_i$. Also,

$$\text{Var}(Y_i) = E[(Y_i - \beta_0 - \beta_1 X_i)^2] = E[(\varepsilon_i)^2] = \sigma^2. \tag{14.21}$$

Hence, the variance of Y_i is constant, regardless of the value of X_i. Also,

$$E\{[Y_i - E(Y_i)][Y_j - E(Y_j)]\} = E(\varepsilon_i \varepsilon_j) = 0, \tag{14.22}$$

by our assumption of zero covariance of the random errors, and, hence, the Y_i's have zero covariance as well. Note that, so far, we have not stated any distributional assumptions regarding the error ε_i. Later on, when we want to make inferences, we will assume that the errors are normally distributed. The normal distribution with zero covariance implies that the errors ε_i and the observations Y_i are independent.

There is a very famous theorem called the *Gauss–Markov theorem*, which states some important properties of the estimators b_0 and b_1.

Gauss–Markov Theorem

Under the conditions of model (14.2), the least squares estimators b_0 and b_1 given by Eqs. (14.9) and (14.10) are unbiased and have minimum variance among all unbiased linear estimators.

Thus, $E(b_0) = \beta_0$ and $E(b_1) = \beta_1$ and, among all linear estimators, these have the smallest variance. Note that b_0 and b_1 are linear estimators – that is, they are linear combinations of the Y_i. To show this, consider the equation

$$b_1 = \frac{\sum_{i=1}^n (X_i - \bar{X})(Y_i - \bar{Y})}{\sum_{i=1}^n (X_i - \bar{X})^2}, \tag{14.23}$$

which we can write as (Problem 14.5)

$$b_1 = \frac{\sum_{i=1}^n (X_i - \bar{X}) Y_i}{\sum_{i=1}^n (X_i - \bar{X})^2} = \sum_{i=1}^n a_i Y_i, \tag{14.24}$$

where $a_i = (X_i - \bar{X}) / \sum_{i=1}^n (X_i - \bar{X})^2$.

Similarly, using Eq. (14.9), it is easy to show that b_0 is also a linear function of the Y_i's (Problem 14.6). The residuals are a natural moment estimator for σ^2. Recall the assumption that $E(\varepsilon_i^2) = \sigma^2$ and the function we minimized was $S = \sum_{i=1}^n \varepsilon_i^2 = \sum_{i=1}^n [Y_i - (\beta_0 + \beta_1 X_i)]^2$. Replacing β_0 and β_1 by the estimators b_0 and b_1, we have $S' = \sum_{i=1}^n e_i^2 = \sum_{i=1}^n (Y_i - b_0 + b_1 X_i)^2$. Most texts refer to S' by SSE, which stands for *sum of squares for error*. We note that $E(S) = n\sigma^2$; hence, SSE is a natural estimator for σ^2 and, in fact, $E(\text{SSE}) = (n-2)\sigma^2$. Thus, $\text{MSE} = \text{SSE}/(n-2)$ is an unbiased estimator of σ^2. MSE stands for *mean square error*, since it is a form of average (or mean) of the squared error terms. We shall not prove that $E(\text{SSE}) = (n-2)\sigma^2$, but simply comment that the form of the multiplier $(n-2)$ is due to the fact that we have two constraints on the residuals, $\sum_{i=1}^n \varepsilon_i = 0$ and $\sum_{i=1}^n \varepsilon_i X_i = 0$. The term $(n-2)$ is often referred to as the *degrees of freedom*.

Next we shall determine the variance of our estimators b_0 and b_1. Using the assumption of independence of the Y_i's and recalling that we showed b_1 to be a linear function of Y_i's, we find

$$\text{Var}(b_1) = \sum_{i=1}^n a_i^2 \text{Var}(Y_i) = \sum_{i=1}^n a_i^2 \sigma^2 = \sigma^2 \sum_{i=1}^n a_i^2, \tag{14.25}$$

where $a_i = (X_i - \bar{X})/S_{XX}$. It can be shown (Problem 14.7) that $\sum_{i=1}^{n} a_i^2 = 1/S_{XX}$, so that

$$\text{Var}(b_1) = \frac{\sigma^2}{S_{XX}} = \frac{\sigma^2}{\sum_{i=1}^{n} (X_i - \bar{X})^2}. \tag{14.26}$$

Since $b_0 = \bar{Y} - b_1 \bar{X}$ is a linear function of the Y_i's, it can be shown (Problem 14.8) that

$$\text{Var}(b_0) = \frac{\sigma^2 \sum_{i=1}^{n} X_i^2}{n S_{XX}}. \tag{14.27}$$

Normally σ^2 is unknown and we replace σ^2 with MSE in Eqs. (14.26) and (14.27) to obtain estimates for Var(b_1) and Var(b_0).

■ **Example**
Calculate the following quantities using the data from the previous example:

a) SSE;
b) MSE;
c) $\widehat{\text{Var}}(b_0)$ and standard error;
d) $\widehat{\text{Var}}(b_1)$ and standard error.

Solution
(It is important to keep a number of digits until the final result and then round accordingly.)

a)
$$\text{SSE} = \sum_{i=1}^{n} (Y_i - b_0 - b_1 X_i)^2$$
$$= \sum_{i=1}^{n} Y_i^2 - b_0 \sum_{i=1}^{n} Y_i - b_1 \sum_{i=1}^{n} X_i Y_i \tag{14.28}$$

or

$$\text{SSE} = S_{YY} - S_{XY}/S_{XX}. \tag{14.29}$$

Either equation can be used, but if b_0 and b_1 have not been calculated, the second equation is more direct. We shall use both and see how they compare. Using Eq. (14.28) gives

$$\text{SSE} = 51\,544\,375 - (194.53145)(16\,251) - (70.157262)(679\,025)$$
$$= 42\,936.924. \tag{14.30}$$

Using Eq. (14.29) gives

$$\text{SSE} = 7\,528\,541.5 - (106\,697.5)^2/1520.8333$$
$$= 42\,935.5812. \tag{14.31}$$

The relative percent difference, $(1.3428/42\,935.5812) \times 100\% = 0.0031\%$, is small, but still indicates the need for carrying many digits. We will use $SSE = 42\,935.5812$.

b) $MSE = SSE/(n-2) = 42\,935.5812/(6-2) = 10\,733.8953$. So $\hat{\sigma}^2 = 10\,733.8953$ and $\hat{\sigma} = 103.6045$

c) $\widehat{Var}(b_0) = \hat{\sigma}^2 \sum x_i^2 / nS_{xx} = 10\,733.8953(9225)/6(1520.8333) = 10\,851.52705$ and the standard error is $\sqrt{\widehat{Var}(b_0)} = 104.170663$.

d) $\widehat{Var}(b_1) = \hat{\sigma}^2 / S_{xx} = 10\,733.8953/1520.8333 = 7.057904$ and the standard error is $\sqrt{\widehat{Var}(b_1)} = 2.656672$.

14.4
Inferences for the Regression Model

Up to this point we have not assumed any distribution for ε_i or, equivalently, Y_i. Now we shall assume that the $\varepsilon_i \sim N(0, \sigma^2)$ and, consequently, that $Y_i \sim N(\beta_0 + \beta_1 X_i, \sigma^2)$. We shall derive the maximum likelihood estimators in place of the least squares estimators. The likelihood is simply the joint density of the observations, but treated as a function of the unknown parameters β_0, β_1, and σ^2. Thus, the likelihood function is given by

$$L(\beta_0, \beta_1, \sigma^2) = \prod_{i=1}^{n} f(Y_i | \beta_0, \beta_1, \sigma^2) \tag{14.32}$$

$$= \prod_{i=1}^{n} \frac{1}{\sigma\sqrt{2\pi}} e^{-(1/2\sigma^2)(Y_i - \beta_0 - \beta_1 X_i)^2} \tag{14.33}$$

$$= \frac{1}{(\sigma^2)^{n/2} (2\pi)^{n/2}} e^{-(1/2\sigma^2) \sum_{i=1}^{n} (Y_i - \beta_0 - \beta_1 X_i)^2}. \tag{14.34}$$

Recall that we want to find the values of β_0, β_1, and σ^2 that maximize the likelihood function. Maximizing the likelihood function is equivalent to maximizing the log (natural) likelihood, so

$$\ln[L(\beta_0, \beta_1, \sigma^2)] = -\frac{n}{2}\ln(2\pi) - \frac{n}{2}\ln\sigma^2 - \frac{1}{2\sigma^2}\sum_{i=1}^{n}(Y_i - \beta_0 - \beta_1 X_i)^2. \tag{14.35}$$

Note that, with respect to β_0 and β_1, we want to find the values that minimize the sum of squares, but this is identical to what we did using least squares and hence the normal assumption leads to equivalent estimators between maximum likelihood and least squares. Thus, the maximum likelihood estimators $\hat{\beta}_0 = b_0$ and $\hat{\beta}_1 = b_1$. The unbiased and minimum variance properties carry over to the normal distribution situation. The estimators have the same variances and since these estimators are linear functions of the Y_i and the Y_i are normally distributed then so are b_0 and b_1. We can write

$$b_0 \sim N\left(\beta_0, \sigma^2/n \sum X_i^2 / S_{xx}\right) \tag{14.36}$$

and

$$b_1 \sim N(\beta_1, \sigma^2 / S_{xx}) \tag{14.37}$$

The MLE for σ^2 is easily obtained by differentiating Eq. (14.35) with respect to σ^2, setting the result equal to zero, and solving. Doing so we find

$$\hat{\sigma}^2 = \frac{1}{n} \sum_{i=1}^{n} (Y_i - b_0 - b_1 X_i)^2 = \frac{SSE}{n}. \tag{14.38}$$

It can be shown that SSE/σ^2, under the normality assumption, is $\chi^2_{(n-2)}$ (i.e., chi-squared with $(n-2)$ degrees of freedom). Recalling that the expected value of a chi-square random variable is equal to its degrees of freedom, we see

$$E\left[\frac{n\hat{\sigma}^2}{\sigma^2}\right] = E\left[\frac{SSE}{\sigma^2}\right] = (n-2) \tag{14.39}$$

or

$$E\hat{\sigma}^2 = \frac{(n-2)}{n} \sigma^2. \tag{14.40}$$

Thus, the maximum likelihood estimator for σ^2 is biased, but $[n/(n-2)]\hat{\sigma}^2 = SSE/(n-2) = MSE$ is unbiased for σ^2.

In most situations, σ^2 will be unknown and Eqs. (14.36) and (14.37) will not be useful. In this case, we use the following Student's t-statistics to make inferences regarding β_0 and β_1:

$$\frac{b_0 - \beta_0}{\sqrt{(MSE/n)\left(\sum X_i^2 / S_{xx}\right)}} \sim t_{n-2} \tag{14.41}$$

and

$$\frac{b_1 - \beta_1}{\sqrt{MSE/S_{xx}}} \sim t_{n-2}. \tag{14.42}$$

■ *Example*
Using the previous example and data obtain the following:

a) A 95% confidence interval for β_0.
b) A 95% confidence interval for β_1.
c) Test the hypothesis that $H_0: \beta_1 = 0$ versus $H_1: \beta_1 \neq 0$ at the $\alpha = 0.05$ level of significance.

Solution

a) Using the results from the previous examples and Eq. (14.41) we have that

$$\Pr\left(-t_{n-2,\alpha/2} < \frac{b_0 - \beta_0}{\sqrt{(MSE/n)(\sum x_i^2 / S_{xx})}} < t_{n-2,\alpha/2}\right) = 1 - \alpha.$$

Using Table A.5, we have $t_{4,0.025} = 2.776$; the standard error is

$$\sqrt{\frac{MSE}{n}\left(\frac{\sum x_i^2}{S_{xx}}\right)} = 104.1707$$

and we have

$$\Pr\left(b_0 - t_{n-2,\alpha/2}\sqrt{\frac{MSE}{n}\left(\frac{\sum x_i^2}{S_{xx}}\right)} < \beta_0 < b_0 + t_{n-2,\alpha/2}\sqrt{\frac{MSE}{n}\left(\frac{\sum x_i^2}{S_{xx}}\right)}\right)$$
$$= 1 - \alpha.$$

Substituting we find $\Pr[194.532 - 2.776(104.1707) < \beta_0 < 194.532 + 2.776(104.1707)] = 0.95$, or $\Pr(-94.6458 < \beta_0 < 483.7098) = 0.95$. Thus, the 95% confidence interval for β_0 is $(-94.6, 483.7)$.

b) Using Eq. (14.42) and our previous results, we have $\Pr[70.157 - 2.776(7.0579) < \beta_1 < 70.157 + 2.776(7.0579)] = 0.95$, or $\Pr(50.5643 < \beta_1 < 89.7497) = 0.95$.

c) The hypothesis $H_0: \beta_1 = 0$ versus $H_1: \beta_1 \neq 0$ is a two-sided test, and so we will compare the result of Eq. (14.42) to $t_{4,0.025} = 2.776$. Using Eq. (14.42), we find $t = (b_1 - \beta_1)/\sqrt{MSE/S_{xx}} = (70.157 - 0)/\sqrt{7.0579} = 26.4079$. Since this is greater than $t_{4,0.025} = 2.776$, we reject H_0 at the 5% level of significance.

Next we shall look at predicted values and their associated variability. The prediction equation is

$$\hat{Y} = b_0 + b_1 X, \tag{14.43}$$

where b_0 and b_1 are unbiased estimators. We know that $E(\hat{Y}) = \beta_0 + \beta_1 X$, which is an unbiased estimator of the value of Y at the given value of X. It can be shown that $\bar{Y}$ and b_1 are independently distributed (Problem 14.10). Since Eq. (14.43) implies that $\hat{Y} = \bar{Y} + b_1(X - \bar{X})$, then

$$\text{Var}(\hat{Y}) = \text{Var}(\bar{Y}) + (X - \bar{X})^2 \text{Var}(b_1) \tag{14.44}$$

$$= \frac{\sigma^2}{n} + \frac{\sigma^2 (X - \bar{X})^2}{S_{XX}} = \sigma^2 \left(\frac{1}{n} + \frac{(X - \bar{X})^2}{S_{XX}}\right). \tag{14.45}$$

We know that a linear combination of normal random variables is itself normally distributed. Because $\bar{Y}$ and b_1 are normally distributed, $\hat{Y}$ is also. Hence, $\hat{Y}$ has mean $\beta_0 + \beta_1 x$ and variance given by Eq. (14.45). Thus, if we want to place a confidence on the mean predicted value, we can use normal theory to write

$$\Pr\left[-z_{1-\alpha/2} < \frac{\hat{Y} - (\beta_0 + \beta_1 X)}{\sqrt{\sigma^2 \xi}} < z_{1-\alpha/2}\right] = 1 - \alpha \qquad (14.46)$$

and

$$\Pr[\hat{Y} - z_{1-\alpha/2}\sqrt{\sigma^2 \xi} < \beta_0 + \beta_1 X < \hat{Y} + z_{1-\alpha/2}\sqrt{\sigma^2 \xi}] = 1 - \alpha, \qquad (14.47)$$

where $\xi = [1/n + (X - \bar{X})^2 / S_{XX}]$. We will not know σ^2 in most cases and must estimate this value using the *mean square error* (MSE). The Student's t-distribution must be utilized in this case, resulting in a very similar looking confidence interval, namely,

$$\Pr[\hat{Y} - t_{n-2,\alpha/2}\sqrt{\text{MSE }\xi} < \beta_0 + \beta_1 X < \hat{Y} + t_{n-2,\alpha/2}\sqrt{\text{MSE }\xi}] = 1 - \alpha. \qquad (14.48)$$

Before moving to an example, we mention that predicting values within the region of our observed X_i's is called *interpolation*, while predicting values outside the region is called *extrapolation*. Interpolation is generally safe to do, in that we can see how well the observations follow our assumption of linearity. We have little knowledge of how the relationship between Y and X may vary, once we move outside this region. Our prediction should be satisfactory if we are reasonably sure that linearity can be assumed, but we will have no proof that the linear model is correct beyond what we have observed. Note also that the variance of our predicted value is a function of $(X - \bar{X})^2$, and we can see that the variance increases greatly as we extrapolate further outside the experimental region.

■ *Example*

Using the previous data, obtain the following when $X = 30$, 36, and 65:

a) the predicted value, and
b) the estimated variance of the predicted value.

Solution

a) The prediction equation is given by $\hat{Y} = 194.532 + 70.157X$, so we find

$$X = \quad 30 \quad 36 \quad 65$$
$$\hat{Y} = 2299.2 \ 2720.2 \ 4754.7$$

b) Using MSE to estimate σ^2 and Eq. (14.45), we have

$$\widehat{\text{Var}}(\hat{Y}) = 10\,733.8953\left[\frac{1}{6} + \frac{(X - 35.83333)^2}{1520.83333}\right], \qquad (14.49)$$

giving

$$X = 30 \quad 36 \quad 65$$
$$\widehat{\text{Var}}(\hat{Y}) = 2029.1 \quad 1789.2 \quad 7793.1$$

Note that the variance of the predicted value becomes larger as we move further from $\bar{X}$. The value $X = 65$ is outside the range of the data on which we built our regression model. A prediction in this region is an extrapolation, which may be inaccurate, unless we are confident that the linear relationship holds in this region.

Predicting *future observations* is different from predicting expected values of Y for a given X, as we just did. Let Y denote the new observation and $\hat{Y}$ denote our predicted value at some given value of X. We were predicting $E(Y) = \beta_0 + \beta_1 X$ previously, using $\hat{Y} = b_0 + b_1 X$. Now we wish to predict Y, a new random variable. This new Y should come from a distribution with mean $\beta_0 + \beta_1 X$, if our model holds. We know that $\hat{Y}$ is unbiased and, hence, this would be our best estimate of Y as well. We can, therefore, use $\hat{Y}$ to estimate both the future value of a random variable and its mean value. Estimating Y by $\hat{Y}$ incurs an error, $e = Y - \hat{Y}$. We can see that the expected value of e is zero, that is,

$$E(e) = E(Y - \hat{Y}) = E(Y) - E(\hat{Y}) = (\beta_0 + \beta_1 X) - (\beta_0 + \beta_1 X) = 0. \quad (14.50)$$

The variance of e is

$$\text{Var}(e) = \text{Var}(Y - \hat{Y}) = \text{Var}(Y) + \text{Var}(\hat{Y}) - 2\text{Cov}(Y, \hat{Y}). \quad (14.51)$$

The variance of a future observation is simply σ^2 (recall Eq. (14.21)) and the variance of $\hat{Y}$ is $\sigma^2 \xi$ (Eq. (14.45)). The covariance term is zero, since the new Y is not involved in the determination $\hat{Y}$. We find that

$$\text{Var}(e) = \sigma^2 + \sigma^2 \xi = \sigma^2 (1 + \xi). \quad (14.52)$$

Thus, the variability in predicting a future observation is considerably larger than the variability in predicting the mean value. We can carry the logic one step further by considering an estimation of the mean of k future observations of Y. We represent the mean of k future values by $\bar{Y}_k$, and we can see that $\bar{Y}_k$ comes from a distribution with mean value $\beta_0 + \beta_1 x$ and variance σ^2/k. We see, following the arguments given above, that $\hat{Y}$ is still the best predictor, and now the variance of the error is

$$\text{Var}(e_k) = \sigma^2 \left(\frac{1}{k} + \xi \right). \quad (14.53)$$

Note that $\text{Var}(e_k)$ approaches $\text{Var}(\hat{Y})$ as the number of observations k, used in determining $\bar{Y}_k$, increases (Problem 14.20).

■ *Example*

Using the previous example, obtain the following when $X = 30$, 36, and 65:

a) the predicted value of Y, and

b) the estimated variance of the predicted value.

Solution

a) The predicted future observation is given by $\hat{Y}$, as in the previous example, so that

$$
\begin{array}{cccc}
X & = & 30 & 36 & 65 \\
Y_{\text{new}} & = & 2299.2 & 2720.2 & 4754.7
\end{array}
$$

b) The variance of the prediction of a future observation is given by Eq. (14.52). Replacing σ^2 by MSE we find

$$
\begin{array}{cccc}
X & = & 30 & 36 & 65 \\
\widehat{\text{Var}}(e) & = & 12\,763.0 & 12\,523.1 & 18\,527.0
\end{array}
$$

The variance for the prediction of a new value at a given X is considerably larger than the variance for the expected value at a given X. It is clear that the distribution for a future predicted value is normal with mean $(\beta_0 + \beta_1 x)$ and variance given by Eq. (14.53), where $k = 1$. A $(1 - \alpha)100\%$ confidence interval is given by

$$\hat{Y} \pm z_{1-\alpha/2} \sigma \left(\frac{1}{k} + \xi \right)^{1/2} \tag{14.54}$$

for the mean of k future values at x.

Similarly, if σ^2 is unknown, then we must estimate it using MSE and Student's t-distribution. The corresponding confidence interval is

$$\hat{Y} \pm t_{n-2,\alpha/2} \left[\text{MSE} \left(\frac{1}{k} + \xi \right) \right]^{1/2}. \tag{14.55}$$

We state the following without proof, but refer the interested reader to R.G. Miller's text, Simultaneous Statistical Inference (Miller, 1981). To obtain simultaneous confidence curves for the whole regression function over its entire range, replace $t_{n-2,\alpha/2}$ with $(2F_{2,n-2,1-\alpha/2})^{1/2}$ in Eq. (14.55) for the mean of k future observations, or in Eq. (14.48) for a predicted value.

14.5
Goodness of the Regression Equation

In this section, we will take a different approach to examining the regression equation. This alternative approach is not so important for simple linear regression, but is helpful for more complex regression models. It consists of partitioning the variability we observe in Y among the various components of the regression model. To this end, consider the following identity:

$$(Y_i - \hat{Y}_i) = (Y_i - \bar{Y}) - (\hat{Y}_i - \bar{Y}). \tag{14.56}$$

14.5 Goodness of the Regression Equation

The residual $e_i = (Y_i - \hat{Y}_i)$ can, as suggested by this identity, be partitioned into two parts: (1) the deviation of Y_i from the mean and (2) the deviation of the predicted $\hat{Y}_i$ from the mean. We can rewrite (14.56) as

$$\sum_{i=1}^n (Y_i - \bar{Y})^2 = \sum_{i=1}^n (Y_i - \hat{Y}_i)^2 + \sum_{i=1}^n (\hat{Y}_i - \bar{Y})^2 + 2\sum_{i=1}^n (Y_i - \hat{Y}_i)(\hat{Y}_i - \bar{Y}). \quad (14.57)$$

The term on the left-hand side of Eq. (14.57) is called the *total sum of squares* (TSS); the first term on the right, the *sum of squared deviations* or the *sum of squares for error* (SSE) (can you see why?); the second term on the right, the *sum of squares due to regression* (SSR). The cross-product term is zero (Problem 14.13). In the abbreviated nomenclature, we have

$$\text{TSS} = \text{SSE} + \text{SSR}. \quad (14.58)$$

Each of these sums of squares has associated degrees of freedom. The term TSS, for instance, has $(n-1)$ degrees of freedom. One way of seeing this is that, even though there are n terms in the sum, there are only $(n-1)$ independent terms, since $\sum_{i=1}^n (Y_i - \bar{Y}) = 0$. The SSR has only one degree of freedom. To see this, recall that $\hat{Y}_i = \bar{Y} + b_1(X_i - \bar{X})$, so that[1]) $\hat{Y}_i - \bar{Y} = b_1(X_i - \bar{X})$ is a function of only one parameter, b_1. The degrees of freedom on the left-hand side of the equation must equal those on the right; hence, the degrees of freedom associated with SSE are $(n-2)$ due to two constraints, namely, $\sum_{i=1}^n (Y_i - \hat{Y}_i) = 0$ and $\sum_{i=1}^n X_i(Y_i - \hat{Y}_i) = 0$. These sums of squares and degrees of freedom are usually presented in what is called an *analysis of variance* (ANOVA) table, which is usually constructed as shown in Table 14.2. The column MS shows the sum of squares divided by its degrees of freedom. The last column indicates that $E(\text{MSR})$ is a function of σ^2 and β_1^2, and that $E(\text{MSE})$ is equal to σ^2 alone. One can show that SSR and SSE are distributed as χ^2 random variables, the ratio of which, divided by their respective degrees of freedom, has an F distribution (Eq. (6.98)). We can see that the ratio $F = \text{MSR}/\text{MSE}$ can be used to test whether $H_0 : \beta_1 = 0$ versus $H_1 : \beta_1 \neq 0$. If $\beta_1 = 0$, then the expected mean squares are both equal to σ^2 and their ratio should be unity. If $\beta_1 \neq 0$, then the MSR will be inflated and the ratio should be greater than unity. Hence, we can use the $F_{1,n-2}$ distribution to test the null hypothesis above, which we reject if $F > F_{1,n-2,1-\alpha}$.

Table 14.2 Typical ANOVA table for the simple linear regression model.

Source of variation	df	SS	MS	E(MS)[a]	F	p-value
Regression	1	SSR	SSR/1	$\sigma^2 + \beta_1^2 S_{xx}$	MSR/MSE	$\Pr(F_{1,n-2} > F)$
Error	$n-2$	SSE	SSE/$(n-2)$	σ^2		
Total	$n-1$	TSS				

a) Expected value of mean squares is not typically given in the ANOVA table. It is shown here to see why the F ratio is a test of $\beta_1 = 0$.

1) The X_i are fixed and have no uncertainty.

Example

Complete the ANOVA table using data from the previous example.

Solution

Determine SSR, SSE, and TSS and put them in the ANOVA table using the form given in Table 14.2. The TSS is simply $\sum_{i=1}^{n}(Y_i-\bar{Y})^2 = \sum_{i=1}^{n} Y_i^2 - n\bar{Y}^2 = 51\,544\,375 - 44\,015\,834 = 7\,528\,541$. The SSR is calculated as $\sum_{i=1}^{n}(\hat{Y}_i-\bar{Y})^2 = b_1^2 S_{xx} = (70.1573)^2(1520.8333) = 7\,485\,612$. The SSE is obtained by subtraction, SSE = TSS − SSR = $7\,528\,541 - 7\,485\,612 = 42\,929$. Of course, SSE = $\sum_{i=1}^{n}(Y_i - \hat{Y}_i)^2 = (58.10959)^2 + (-140.67671)^2 + (80.75068)^2 + (73.17808)^2 + (-87.39452)^2 + (16.03288)^2 = 42\,937.22027$. (Although the two methods used to obtain SSE do not agree exactly, there is only a 0.02% difference.) The ANOVA table can now be filled in.

ANOVA for simple linear regression example:

Source of variation	df	SS	MS	F^a	p-value[b]
Regression	1	7 485 612	7 485 612	697.49	1.222×10^{-5}
Error	4	42 929	10 732.25		
Total	5	7 528 541			

[a] $F = $ MSR/MSE has the F distribution with degrees of freedom equal to regression and error, respectively. In this case, $F \sim F_{1,4}$.
[b] p-value = $\Pr(F_{1,4} > F) = \Pr(F_{1,4} > 697.49) = 0.00001222$.

The p-value is very small in this case, indicating a highly significant linear regression coefficient. In the case where we have a single dependent variable, the above F test is equivalent to testing $H_0 : \beta_1 = 0$ versus $H_1 : \beta_1 \neq 0$. Eq. (14.42) shows that b_1, suitably normalized, has the Student's t-distribution. Recall that a Student's t random variable is defined as the ratio of a standard normal random variable and the square root of a χ^2 random variable divided by its degrees of freedom. The square of a Student's t random variable, then, is the ratio of the square of a standard normal random variable (which has a χ^2 distribution with one degree of freedom) and a χ^2 random variable divided by its degrees of freedom. This is exactly the definition of an F random variable. We earlier used the t-distribution to test $H_0 : \beta_1 = 0$ versus $H_1 : \beta_1 \neq 0$ and obtained $t \approx 26.408$, so that $t^2 = 697.38$ — nearly the same as our F value, with the difference accredited to rounding. Thus, the F value is equivalent to testing $H_0 : \beta_1 = 0$ versus $H_1 : \beta_1 \neq 0$ in the single-variable regression. We note that, when there are several independent variables $(X_1, X_2, \ldots, X_k)$, one can calculate an F value that is then equivalent to testing the hypothesis that each $\beta_i = 0$ for all $i = 1, 2, \ldots, k$ simultaneously. The alternative hypothesis in this case is that at least one β_i is not zero.

Another statistic often used in regression analysis is called R^2, and it represents the percent of variation in the dependent variable Y that is explained by the independent variable X. The definition of R^2 is

$$R^2 = \frac{\text{SSR}}{\text{TSS}} = \frac{\sum_{i=1}^{n}(\hat{Y}_i - \bar{Y})^2}{\sum_{i=1}^{n}(Y_i - \bar{Y})^2}. \tag{14.59}$$

This is often expressed as a percentage by multiplying by 100%. We have, from the ANOVA table in the last example, $R^2 = 7\,485\,612/7\,528\,541 = 0.9943$, or $R^2 = 99.43\%$. This says that the independent variable X explains 99.43% of the variation that occurs in Y. This indicates a very strong linear relationship between X and Y. The values of R^2 range from 0 to 1, where 1 implies a perfect linear relationship between the dependent and independent variables, that is, $Y = \beta_0 + \beta_1 X$ with no error. If one took repeat observations, then R^2 must be less than 1, unless all the repeat measurements yielded equivalent results at a given value of X. This would occur very infrequently, if at all, in any real experimental situation.

14.6
Bias, Pure Error, and Lack of Fit

The simple linear regression model is something we have assumed in our analysis so far, but it is an assumption that we can examine. Recall that the residuals, $e_i = Y_i - \hat{Y}_i$, reflect the adequacy of the linear model to describe the data. It is common, if not imperative, to plot the residuals to see if they appear random. Recall that, in our assumed linear model $Y_i = \beta_0 + \beta_1 X_i + \varepsilon_i$, the ε_i are assumed independent and have constant variance σ^2 with mean zero. The residuals should mimic these characteristics. We know that $\bar{e} = 0$ by the normal equations, and this implies that the e_i are correlated. The correlation imposed by this constraint will not be that important in our examination of the residuals, where we are looking for discrepancies from the assumed model. Figure 14.3 shows some plots that would generally be used in the regression analysis. There are no obvious patterns in the above plots, and nothing to make us suspect there is any violation of our assumptions. Figure 14.3a is frequency plot of the residuals from a very large data set. We would expect the distribution of the values of residuals from a good regression fit to be symmetrically distributed about zero, with most of the values close around zero, and to have fewer points at the extremes, as we see in the figure. Figure 14.3b is a scatter plot of the residuals against the independent variable from the first example. These residuals are fairly randomly distributed around zero, indicating a good regression fit. We would suspect some possible issues if the residuals increased in magnitude as X increased (nonconstant variance), or if the residuals were negative at the low and high values of X but positive for the middle values of X (possibly indicating a quadratic model in X). Figure 14.3c is a plot of the residuals against the predicted values of the dependent variable in the first example. Again, we see that the distribution of residual values is fairly random about zero, indicating a good regression fit.

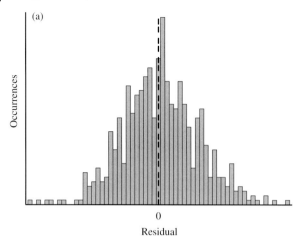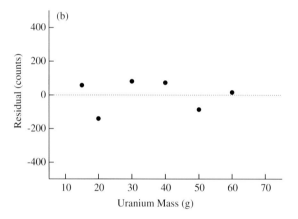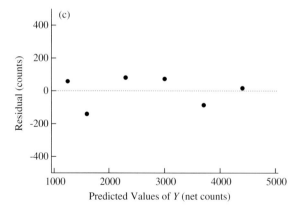

Figure 14.3 Plots of residuals from data fitted by linear regression: (a) frequency distribution of residuals from a large data set; (b) scatter plot of residuals as a function of uranium mass from the data in the first example; (c) scatter plot of residuals as a function of the predicted values for net counts in the first example.

14.6 Bias, Pure Error, and Lack of Fit

Recall that, by the normal equations, $\sum_{i=1}^{n} e_i \hat{Y}_i = 0$, so that the residuals and the predicted values are uncorrelated. This may not be the case with the actual observations, and so plotting the residuals against the observations might yield a nonrandom pattern, but that would still not be a violation of assumptions. We would suspect a nonconstant variance if the plot of the residuals against the predicted values shows an increase or decrease as $\hat{Y}$ increases. We might suspect that a quadratic or higher order model might be more appropriate if the pattern were curved. Figure 14.4 shows the general patterns that would cause concern, with (a) showing a nonconstant variance, (b) showing an inadequate model (using, in this case, a linear model to fit an exponential function), and (c) showing the pattern that would be considered acceptable. Figure 14.4d shows a situation where we know the time order in which the data were collected. Patterns such as those here (i.e., alternating groups of positive and negative residuals) suggest that there might be some learning effect or change in the process over time. This would imply a need to adjust the model for time, or for whatever factor might be changing with time. For example, temperature might be confounded with time, and beginning measurements might have been taken under cooler conditions. *Training* or *learning* is often confounded with time, chemicals can degrade over time, and so on. If the underlying cause for a time effect can be determined, then a new model incorporating this can be fit.

Another issue is that of systematic bias in the regression results. Let $\mu_i = E(Y_i)$ denote the value given by the "true" model, whatever it is, when $X = X_i$. Then we can write

$$Y_i - \hat{Y}_i = (Y_i - \hat{Y}_i) - E(Y_i - \hat{Y}_i) + E(Y_i - \hat{Y}_i) \tag{14.60}$$

$$= (Y_i - \hat{Y}_i) - [\mu_i - E(\hat{Y}_i)] + [\mu_i - E(\hat{Y}_i)] \tag{14.61}$$

$$= r_i + B_i, \tag{14.62}$$

where $r_i = (Y_i - \hat{Y}_i) - [\mu_i - E(\hat{Y}_i)]$ and $B_i = \mu_i - E(\hat{Y}_i)$. The quantity B_i is the bias at $X = X_i$. If the model is correct, then $E(\hat{Y}_i) = \mu_i$ and $B_i = 0$. If the model is not correct, then $E(\hat{Y}_i) \neq \mu_i$ and B_i takes on a nonzero value that depends on both the true model and the value of X. The quantity r_i is a random variable whose mean is zero, since $E(r_i) = E(Y_i) - E(\hat{Y}_i) - \mu_i + E(\hat{Y}_i) = 0$. It can be shown that $E(\sum_{i=1}^{n} r_i^2) = (n-2)\sigma^2$, based on the assumption that the Y_i's have constant σ^2 and that they are independent of each other. Then it can be shown that the expected value of the sum of squares for error is (Problem 14.14)

$$E(\text{SSE}) = E\left[\sum_{i=1}^{n}(Y_i - \hat{Y}_i)^2\right] = (n-2)\sigma^2 + \sum_{i=1}^{n} B_i^2, \tag{14.63}$$

and, since $\text{MSE} = \text{SSE}/(n-2)$,

$$E(\text{MSE}) = \sigma^2 + \frac{\sum_{i=1}^{n} B_i^2}{(n-2)}. \tag{14.64}$$

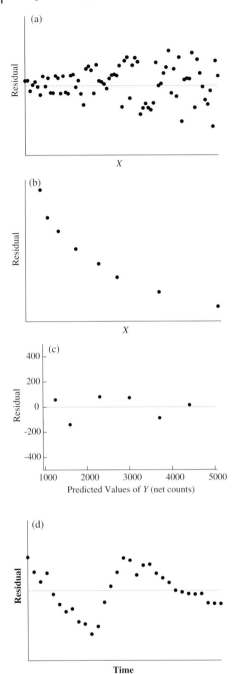

Figure 14.4 Plots of residuals showing patterns that might cause concern: (a) nonconstant variance that increases with X; (b) residuals from an inappropriate model, in this case a linear fit to data from exponential decay; (c) an acceptable pattern; (d) variance that changes with time.

Thus, the expected value of the MSE is inflated if the model is not correct. The bias, if large enough, will tend to increase our estimate of the population variance, and this will affect confidence intervals and tests of hypothesis accordingly. This increase in our estimate of the true variance leads to what is called a *loss of power* in a test of hypothesis. The loss of power is a reduction in the probability $((1-\beta)$, where β is the probability of a type II error) of detecting a true effect when one exists.

This can often be seen in a plot of the residuals versus the independent variable X in the case of a single predictor variable, but is more difficult to detect when there are multiple independent variables. One way of testing for model inadequacy is by using a prior estimate of σ^2 obtained by previous experiments. One can then compare the estimate of σ^2 obtained from the current model to that obtained in the past. If the current estimate is large compared to the past, then one might conclude that the current model is somehow inadequate and investigate possible reasons, amending the model accordingly. This can be done in another way if there are repeat observations taken at the same value of X. The variability we observe at each of these repeat points are estimates of the true variance, since the model, whatever it is, does not change at that given value of X. Thus, the variation at these repeat points is termed *pure error*, since it estimates only the random variation. It is important to understand what is meant by *repeat observations*. Repeat observations come by replicating the experiment at the same setting of X. For example, suppose we are interested in the amount of potassium in a human as a function of the weight of the person. A true repeat observation is obtained by measurement of [K] in two persons of the same weight. (Measuring the potassium concentration [K] in the tissue of one person at two points in time is a *reconfirmation*, which provides information about the variability in our measurement system, but not about variation in the values of [K] between people of the same weight.)

To analyze such data we need to use subscripts that identify particular values of X and a second subscript identifying the number of repeats at that value of X. For example, let $i = 1, 2, \ldots, m$ represent the number of distinct X values, and $j = 1, 2, \ldots, m_i$ be the number of repeat points at $X = X_i$. The contribution to the sum of squares pure error at the value $X = X_i$ is given by

$$\sum_{j=1}^{m_i} (Y_{ij} - \bar{Y}_i)^2 = \sum_{j=1}^{m_i} Y_{ij}^2 - m_i \bar{Y}_i^2. \tag{14.65}$$

Pooling the sum of squares across all the repeat points gives the sum of squares pure error,

$$SS_{pe} = \sum_{i=1}^{m} \sum_{j=1}^{m_i} (Y_{ij} - \bar{Y}_i)^2 \tag{14.66}$$

with associated degrees of freedom given by

$$df_{pe} = \sum_{i=1}^{m} (m_i - 1). \tag{14.67}$$

Then the pure error mean square is given by

$$\mathrm{MS_{pe}} = \frac{\mathrm{SS_{pe}}}{\mathrm{df_{pe}}}, \tag{14.68}$$

and this is an estimate of σ^2, regardless of whether the model is correct or not. Next we show that the SSE can be partitioned into two separate sums of squares, one being the *sum of squares pure error* and the other, what we will call the *sum of squares lack of fit*. We begin with an identity,

$$\sum_{i=1}^{m}\sum_{j=1}^{m_i} (Y_{ij} - \hat{Y}_i)^2 \equiv \sum_{i=1}^{m}\sum_{j=1}^{m_i} [(Y_{ij} - \bar{Y}_i) + (\bar{Y}_i - \hat{Y}_i)]^2. \tag{14.69}$$

Then it follows that Eq. (14.69) can be written as

$$\sum_{i=1}^{m}\sum_{j=1}^{m_i} (Y_{ij} - \hat{Y}_j)^2 \equiv \sum_{i=1}^{m}\sum_{j=1}^{m_i} (Y_{ij} - \bar{Y}_i)^2 + \sum_{i=1}^{m} m_i (\hat{Y}_i - \bar{Y}_i)^2$$

$$+ 2\sum_{i=1}^{m}\sum_{j=1}^{m_i} [(Y_{ij} - \bar{Y}_i)(\bar{Y}_i - \hat{Y}_i)]. \tag{14.70}$$

Note that the cross-product term vanishes (Problem 14.15). We can express Eq. (14.70) as

$$\mathrm{SSE} = \mathrm{SS_{pe}} + \mathrm{SS_{lof}}, \tag{14.71}$$

where "pe" stands for "pure error" and "lof" for "lack of fit." Let n denote the total number of observations so that $n = \sum_{i=1}^{m} m_i$, then we note that SSE has $(n-2)$ degrees of freedom associated with it, that $\mathrm{SS_{pe}}$ has $\sum_{i=1}^{m}(m_i-1)$ degrees of freedom associated with it, and we find by subtraction that the $\mathrm{SS_{lof}}$ has $(n-2) - \sum_{i=1}^{m}(m_i-1) = m-2$ degrees of freedom. One may consider that there are m distinct points in X on which to build the model, and that two of these are lost for the terms in our model, thus yielding $(m-2)$ degrees of freedom for lack of fit. The model's lack of fit can be tested by computing

$$F = \frac{\mathrm{SS_{lof}}/(m-2)}{\mathrm{SS_{pe}}/\sum_{i=1}^{m}(m_i-1)}, \tag{14.72}$$

which has an F distribution with $(m-2)$ and $\sum_{i=1}^{m}(m_i-1)$ degrees of freedom.

> ■ *Example*
> Suppose in the example examining weight and counts of ^{235}U we obtained three additional ingots that weighed 20, 40, and 60 g. The counts for these ingots are 1225, 3208, and 4480, respectively.
>
> a) Repeat the regression analysis using these additional points.
> b) Determine SSE, $\mathrm{SS_{pe}}$, and $\mathrm{SS_{lof}}$.
> c) Test whether there is significant lack of fit in the model.

Solution

a) $b_1 = S_{XY}/S_{XX} = 175\,707.2222/2355.5556 = 74.593$,

$$b_0 = \bar{Y} - b_1\bar{X} = 2784.889 - 74.593(37.222) = 8.373,$$

$$\text{TSS} = \sum_{i=1}^{n} Y_i^2 - \frac{1}{n}\left(\sum_{i=1}^{n} Y\right)^2 = 83\,171\,664 - \frac{1}{9}(25\,064)^2$$
$$= 13\,371\,208.89,$$

$$\text{SSR} = b_1 S_{XY} = 74.593(175\,707.2222) = 13\,106\,528.83,$$

$$\text{SSE} = \text{TSS} - \text{SSR} = 264\,680.06.$$

We can now fill in the ANOVA table:

Source of variation	df	SS	MS	F	p-value
Regression	1	13 106 528.83	13 106 528.83	346.63	3.2×10^{-7}
Error	7	264 680.06	37 811.44		
Total	8	13 371 208.89			

b) SSE = 264680.06.
We calculate the pure error contribution at each repeat point. If there are only two repeats, then $\sum_{j=1}^{m_i}(Y_{ij}-\bar{Y}_i)^2 = (1/2)(Y_{i1}-Y_{i2})^2$. Hence, we find $\text{SS}_{pe} = (1/2)(1457 - 1125)^2 + (1/2)(3074 - 3208)^2 + (1/2)(4420 - 4480)^2 = 65\,890$, with $(9 - 6) = 3$ degrees of freedom. Then SS_{lof} follows by subtraction: $\text{SS}_{lof} = \text{SSE} - \text{SS}_{pe} = 264\,680.06 - 65\,890 = 198\,790.06$, with $(m - 2) = (6 - 2) = 4$ degrees of freedom.

c) To test for lack of fit, we compute $F = [\text{SS}_{lof}/(m-2)]/[\text{SS}_{pe}/\sum_{i=1}^{m}(m_i-1)]$ $= (198\,790.06/4)/(65\,890/3) = 2.263$. We compare this to the upper 5% point on the $F_{4,3}$ distribution, which is $F_{4,3,0.05} = 6.59$, exceeding our calculated F, from which we conclude that there is no lack of fit with the linear model.

14.7
Regression through the Origin

The regression line may be known to pass through the origin in some situations. For example, the amount of precipitate Y resulting at a concentration X of a chemical reactant must be zero if X is zero, since there is no chemical reaction and, therefore, no precipitate. We would also expect the number of counts from a radiation detector to be zero if the amount of radioactive material is zero (but only in an environment with no background radiation). The model is the same in this situation, except that $\beta_0 \equiv 0$, so that the model is given by

$$Y_i = \beta_1 X_i + \varepsilon_i, \qquad (14.73)$$

where β_i is an unknown parameter, X_i is a known value of the independent variable, and ε_i is the random error term, which we assume to be normally distributed with mean 0 and variance σ^2. We assume, as before, that ε_i and ε_j are independent (implying that Y_i and Y_j are also independent).

The least squares or maximum likelihood estimation for β_1 is obtained by minimizing

$$Q = \sum_{i=1}^{n} (Y_i - \beta_1 X_i)^2 \qquad (14.74)$$

with respect to β_1. This leads to the following estimator:

$$b_i = \frac{\sum_{i=1}^{n} X_i Y_i}{\sum_{i=1}^{n} X_i^2}. \qquad (14.75)$$

An unbiased estimator of $E(Y) = \beta_1 X$ is given by

$$\hat{Y} = b_1 X. \qquad (14.76)$$

It can be shown, using the same arguments as those in Section 14.5, that an unbiased estimator for σ^2 is

$$\text{MSE} = \frac{\sum_{i=1}^{n}(Y_i - \hat{Y}_i)^2}{n-1} = \frac{\sum_{i=1}^{n}(Y_i - b_1 X_i)^2}{n-1}. \qquad (14.77)$$

Note that, in this case, the denominator is $(n-1)$, rather than $(n-2)$, since we are only estimating one parameter, β_1, thereby losing only one degree of freedom.

We can obtain the estimated variance and confidence interval for β_1, $E(Y)$, and a new observation Y_{new} using the techniques of Section 14.4. These are given in Table 14.3.

Table 14.3 Estimated variance and confidence interval for β_1, $E(Y)$, and a new observation, Y_{new}.

Parameter	Estimator	Variance of estimator	Estimated variance	Confidence interval[a]
β_1	$b_i = \frac{\sum_{i=1}^{n} X_i Y_i}{\sum_{i=1}^{n} X_i^2}$	$\frac{\sigma^2}{\sum_{i=1}^{n} X_i^2}$	$s_{b_1}^2 = \frac{\text{MSE}}{\sum_{i=1}^{n} X_i^2}$	$b_1 - ts_{b_1} \leq \beta_1 \leq b_1 + ts_{b_1}$
$E(Y)$	$\hat{Y} = b_1 X$	$\frac{\sigma^2 X^2}{\sum_{i=1}^{n} X_i^2}$	$s_{\hat{Y}}^2 = \frac{X^2 \text{MSE}}{\sum_{i=1}^{n} X_i^2}$	$\hat{Y} - ts_{\hat{Y}} \leq E(Y) \leq \hat{Y} + ts_{\hat{Y}}$
Y_{new}	$\hat{Y}_{\text{new}} = b_1 X$	$\sigma^2 \left(1 + \frac{X^2}{\sum_{i=1}^{n} X_i^2}\right)$	$s_{\hat{Y}_{\text{new}}}^2 = \text{MSE}\left[1 + \frac{X^2}{\sum_{i=1}^{n} X_i^2}\right]$	$\hat{Y}_{\text{new}} - ts_{\hat{Y}_{\text{new}}} \leq Y_{\text{new}} \leq \hat{Y}_{\text{new}} + ts_{\hat{Y}_{\text{new}}}$

a) Where $t = t_{n-1,\alpha/2}$.

■ *Example*
Varying amounts of a sample are analyzed to determine its activity concentration (in kBq g^{-1}). The following table relates the measured activity for a specified weight.

X = sample weight (g)	2.0	2.0	5.0	6.0	8.0	8.0
Y = activity (kBq)	42.8	45.1	109.4	129.4	176.3	175.1

a) Estimate β_1 and interpret what it represents, assuming $Y = \beta_1 X + \varepsilon$.
b) Estimate MSE for this model.
c) Obtain an estimate of Y_{new} and a 95% confidence interval when $X = 5.0$ g.

Solution
a) We find that $\sum_{i=1}^{n} X_i Y_i = 4309.4$ and $\sum_{i=1}^{n} X_i^2 = 19.7$, so that $b_1 = 4309.4/19.7 = 21.87513$ kBq g^{-1}.
b) MSE $= \sum_{i=1}^{n} (Y_i - \hat{Y}_i)^2/(n-1) = 1.576$ kBq^2.
c) When $X = 5.0$ g, $\hat{Y}_{new} = b_1 X = (21.87513$ kBq g$^{-1})(5.0$ g$) = 109.376$ kBq. We see from Table 14.3 that $s_{\hat{Y}_{new}}^2 = \text{MSE}[1 + X^2/\sum_{i=1}^{n} X_i^2] = 1.776$ kBq^2. The 95% confidence interval for Y_{new}, also from Table 14.3, is $\hat{Y}_{new} - ts_{\hat{Y}_{new}} \leq Y_{new} \leq \hat{Y}_{new} + ts_{\hat{Y}_{new}}$, which evaluates (using $t_{5,0.025} = 2.571$ from the Student's t-distribution table in Appendix A.5) to $109.376 - 2.571\sqrt{1.776} \leq Y_{new} \leq 109.376 + 2.571\sqrt{1.776}$ or $105.949 \leq Y_{new} \leq 112.803$ (in units of kBq).

Note that the assumption $\beta_0 = 0$ is a strong one. Fitting the full model, $Y = \beta_0 + \beta_1 X + \varepsilon$, is preferable in many situations where we cannot investigate the response near $X = 0$.

14.8
Inverse Regression

One might be interested in the value of X corresponding to a value of Y in some situations. For example, one might want to know the frequency of chromosome aberrations in blood cells corresponding to a whole-body dose above which radiation accident victims are referred for medical treatment, based on chromosome dicentric formation rate determined as a function of dose. The regression equation should be significant (i.e., $\beta_1 \neq 0$) for the inverse regression to be reasonable. A simple plot of the regression equation and its associated $(1 - \alpha) \times 100\%$ confidence interval for the true mean value illustrates the problem graphically. Assume that we are interested in the value of X corresponding to $Y = Y_0$. We represent $Y = Y_0$ by the horizontal line in Figure 14.5. It intersects the regression line at

$$Y = Y_0 = b_0 + b_1 \hat{X}_0, \tag{14.78}$$

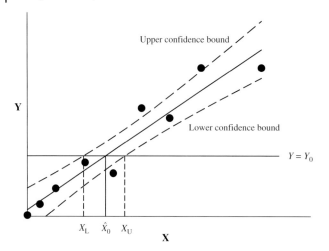

Figure 14.5 Graphical determination of fiducial limits X_L and X_U for the value of X corresponding to $Y = Y_0$.

giving a solution for $\hat{X}_0$,

$$\hat{X}_0 = (Y_0 - b_0)/b_1. \tag{14.79}$$

The line $Y = Y_0$ also intersects the 95% confidence band at

$$Y = Y_{X_L} + ts\sqrt{\left[\frac{1}{n} + \frac{(X_L - \bar{X})^2}{S_{XX}}\right]}, \tag{14.80}$$

where $Y_{X_L} = b_0 + b_1 X_L$, $t = t_{n-2,\alpha/2}$, and $s = \sqrt{MSE}$. Setting Eqs. (14.78) and (14.80) equal, canceling b_0 and rearranging the terms so that the square root is on one side only, squaring, and collecting terms yields a quadratic equation in X_L,

$$AX_L^2 + 2BX_L + C = 0, \tag{14.81}$$

where

$$A = b_1^2 - \frac{t^2 s^2}{S_{XX}}, \tag{14.82}$$

$$B = \frac{t^2 s^2 \bar{X}}{S_{XX}} - b_1^2 \hat{X}_0, \tag{14.83}$$

and

$$C = b_1^2 \hat{X}_0^2 - \frac{t^2 s^2}{n} - \frac{t^2 s^2 \bar{X}}{S_{XX}}. \tag{14.84}$$

We get exactly the same equation for X_U, so that X_L and X_U are the roots of the quadratic equation. Solving the equation and collecting terms, we find that

$$X_U, X_L = \hat{X}_0 + \frac{(\hat{X}_0 - \bar{X})g \pm (ts/b_1)\sqrt{(\hat{X}_0 - \bar{X})^2/S_{XX} + (1-g)/n}}{1-g}, \qquad (14.85)$$

where

$$g = \frac{t^2 s^2}{b_1^2 S_{XX}}. \qquad (14.86)$$

The values X_L and X_U are called *fiducial limits*. Small values of g indicate a significant regression coefficient β_1.

■ *Example*

Find the value of X and calculate the upper and lower 95% fiducial limits X_L and X_U when $Y = 2500$ counts using the initial example in Section 14.2 of detector response with mass of ^{235}U.

Solution

Using Eq. (14.1), $\hat{X}_0 = (2500 - 194.532)/70.157 = 32.86$ g. The value of g is calculated as $(2.776)^2 (42\,935.5812)^2/(70.157)^2 1520.8333 = 0.044$. The solution for X_L and X_U is given by Eq. (14.85),

$$X_U, X_L = 32.86$$
$$+ \frac{(32.86 - 35.83)0.044 \pm [(2.776)(207.21)/70.157]\sqrt{(32.86 - 35.83)^2/1520.8333 + (1 - 0.044)/6}}{1 - 0.044}$$

$$= 32.86 - \frac{0.13068}{0.956} \pm \left(\frac{3.33178}{0.956}\right) = 36.208, 29.238.$$

In this example, β_1 is highly significant and g is small. This allows inverse regression to be well determined. Figure 14.6 shows an example where the method given in Eq. (14.85) would give spurious results due to the regression not being significant. Plotting the graph and its confidence bands is a good practice and can be useful in avoiding issues evident in Figure 14.6. The inverse regression problem is also referred to as the *calibration problem*.

14.9 Correlation

Correlation is strongly related to regression, as we shall see. Correlation is appropriate when both X and Y are random variables. Correlation is a measure of the linear

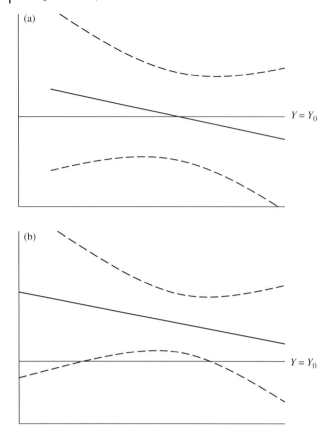

Figure 14.6 Spurious solutions for the upper and lower limits in the inverse regression problem: (a) the solution for X_U, X_L given by Eq. (14.85) has complex roots and line $Y = Y_0$ does not intersect the confidence intervals; (b) the solution for X_U, X_L has real roots, but both are on the same side of the regression line.

relationship between X and Y. The correlation coefficient, usually denoted by ϱ, is defined by

$$\varrho = \frac{\mathrm{Cov}(X, Y)}{\sqrt{\mathrm{Var}(X)\mathrm{Var}(Y)}}, \tag{14.87}$$

where $\mathrm{Cov}(X, Y)$ is the covariance between X and Y. It can be shown that $-1 \leq \varrho \leq 1$, where $\varrho = \pm 1$ indicates a perfect linear relationship between X and Y. If we have a random sample $(X_1, Y_1), (X_2, Y_2), \ldots, (X_n, Y_n)$ from the joint distribution $f_{X,Y}(x, y)$, then the quantity

$$r = \frac{\sum_{i=1}^n (X_i - \bar{X})(Y_i - \bar{Y})}{\sqrt{\sum_{i=1}^n (X_i - \bar{X})^2 \sum_{i=1}^n (Y_i - \bar{Y})^2}} = \frac{S_{XY}}{\sqrt{S_{XX} S_{YY}}} \qquad (14.88)$$

is called the *sample correlation coefficient*. Like ϱ, the sample correlation coefficient satisfies $-1 \le \varrho \le 1$. If we compare the equation for b_1 (Eq. (14.14)) and Eq. (14.88), it can be shown that (Problem 14.19)

$$b_1 = r\sqrt{\frac{\sum_{i=1}^n (Y_i - \bar{Y})^2}{\sum_{i=1}^n (X_i - \bar{X})^2}} = r\sqrt{\frac{S_{YY}}{S_{XX}}}. \qquad (14.89)$$

R.A. Fisher showed that the following transformation on r has an approximately normal distribution, namely,

$$z = \frac{1}{2}\ln\left(\frac{1+r}{1-r}\right) = \tanh^{-1}(r) \sim N\left(\tanh^{-1}\varrho, \frac{1}{n-3}\right), \qquad (14.90)$$

where $\tanh^{-1}$ is the inverse hyperbolic tangent. Equation (14.90) can be used to obtain confidence intervals on ϱ or to test the hypothesis that $H_0: \varrho = \varrho_0$ versus $H_1: \varrho \ne \varrho_0$. For example, $100(1 - \alpha)\%$ confidence intervals on ϱ are obtained by solving

$$\frac{1}{2}\ln\left(\frac{1+r}{1-r}\right) \pm z_{1-\alpha/2}\sqrt{\frac{1}{n-3}} = \frac{1}{2}\ln\left(\frac{1+\varrho}{1-\varrho}\right). \qquad (14.91)$$

Similarly, for testing $H_0: \varrho = \varrho_0$ versus $H_1: \varrho \ne \varrho_0$, we can compute

$$z = \left[\frac{1}{2}\ln\left(\frac{1+r}{1-r}\right) - \frac{1}{2}\ln\left(\frac{1+\varrho_0}{1-\varrho_0}\right)\right]\sqrt{n-3} \qquad (14.92)$$

and compare $|z|$ to $z_{1-\alpha/2}$ for a size α test. Of course, one can test one-sided hypotheses as well (e.g., $H_1: \varrho < \varrho_0$ or $H_1: \varrho > \varrho_0$).

> **Example**
> Suppose that a sample of size $n = 28$ is obtained on pairs (X, Y) and the sample correlation obtained is $r = 0.43$.
>
> a) Obtain a 90% confidence interval on ϱ.
> b) Test the hypothesis that $H_0: \varrho = 0.60$ versus $H_1: \varrho < 0.60$, using $\alpha = 0.05$.

Solution
a) Equation (14.91) yields

$$\frac{1}{2}\ln\left(\frac{1.43}{0.57}\right) \pm 1.645\sqrt{\frac{1}{25}} = \frac{1}{2}\ln\left(\frac{1+\varrho}{1-\varrho}\right) \quad (14.93)$$

$$(0.1309, 0.7889) = \frac{1}{2}\ln\left(\frac{1+\varrho}{1-\varrho}\right);$$

therefore, $0.1302 < \varrho < 0.6578$.

b) Using Eq. (14.92), we have

$$z = \left[\frac{1}{2}\ln\left(\frac{1.43}{0.57}\right) - \frac{1}{2}\ln\left(\frac{1.60}{0.40}\right)\right]\sqrt{25} = -1.1663. \quad (14.94)$$

We compare to $z_{0.05} = -1.645$, since this is a one-sided test, and reject if $z < z_{0.05} = -1.645$. Hence, we do not reject H_0.

The field of regression analysis is quite large and this chapter has explored some of the concepts for the simple linear regression case, that is, a single independent variable X. The extension to the multiple regression scenario, where we have p independent variables, $X_1, X_2, \ldots, X_p$, is straightforward (although the mathematics gets messy without the use of matrix theory). Those wanting a more thorough discussion of regression analysis should consult Draper and Smith (1998), Ryan (2009), or Neter et al. (2004).

Problems

14.1 A chemical engineer observes the following process yields at 10 corresponding temperatures:

Temp, °F (X)	100	110	120	130	140	150	160	170	180	190
Yield, g (Y)	52.0	58.6	60.0	62.6	65.4	64.1	72.6	72.4	82.4	83.8

a) Plot the data.
b) Fit the model, Yield $= \beta_0 + \beta_1$ Temp $+ \varepsilon$, and obtain estimates for β_0 and β_1.
c) Use the fitted model to plot a straight line through the data.
d) Test the hypothesis $H_0: \beta_1 = 0$ versus $H_1: \beta_1 \neq 0$ at the $\alpha = 0.05$ level of significance.

14.2 It was conjectured that the growth of pine trees was dependent on the amount of rainfall. A botanist studied the growth of seven pine trees, each initially 2 feet tall, for a 3-month period, where the trees were given different amounts of water ranging from 10 to 70 mm per month. Results were as shown:

Water (mm month^{-1})	10	20	30	40	50	60	70
Growth (cm)	2.72	2.85	2.41	2.68	2.98	3.36	3.09

a. Plot the data.
b. Fit a linear model to these data.
c. Obtain a 95% confidence interval on the slope, β_1.
d. Test $H_0: \beta_1 = 0$ versus $H_1: \beta_1 \neq 0$ at the $\alpha = 0.05$ level of significance.

14.3 Refer to Eqs. (14.7) and (14.8) and prove that the solution for b_0 and b_1 is given by Eqs. (14.9) and (14.10), respectively.

14.4 Show that $\sum_{i=1}^{n}(X_i-\bar{X})(Y_i-\bar{Y}) = \sum_{i=1}^{n}(X_i-\bar{X})Y_i$, where $\bar{X} = (1/n)\sum_{i=1}^{n}X_i$ and $\bar{Y} = (1/n)\sum_{i=1}^{n}Y_i$.

14.5 Use Problem 14.4 to show that Eq. (14.23) can be written as Eq. (14.24).

14.6 Recall that $b_0 = \bar{Y} - b_1\bar{X}$. Use this relationship and that of Eq. (14.24) to show that b_0 is a linear function of the Y_i's. That is, show that $b_0 = \sum_{i=1}^{n} a_i Y_i$ for some choice of a_i.

14.7 Show that if $a_i = (X_i-\bar{X})/S_{XX}$, then $\sum_{i=1}^{n} a_i^2 = 1/S_{XX}$.

14.8 Use Problem 14.6, where $a_i = [S_{XX}-n\bar{X}(X_i-\bar{X})]/nS_{XX}$, to show that Eq. (14.27) is true.
(Hint: Note that $S_{XX} + n\bar{X}^2 = \sum_{i=1}^{n} X_i^2$.)

14.9 The strength Y of concrete used for radiochemical storage tanks is related to the amount X of potash used in the blend. An experiment is performed to determine the strength of various amounts of potash. The data are given below.

X (%)	1.0	1.0	1.5	2.0	2.0	2.5	3.0	3.0
Y (ft lb)	103	104	103	111	110	118	119	113

a) Fit a linear model to these data with the amount of potash as the independent variable.
b) Construct the ANOVA table. Is the regression significant? (Test using $\alpha = 0.05$.)
c) Since there are repeat values of the independent variable, calculate the sum of squares pure error and the sum of squares lack of fit and test for lack of fit. (Test using $\alpha = 0.05$.)
d) Plot the data, the fitted line, and a 95% confidence interval for the mean value of the regression line using Eq. (14.48).

14.10 Consider reparameterizing the linear model to $Y_i = \alpha_0 + \beta_1(X_i-\bar{X}) + \varepsilon_i$ for $i = 1, 2, \ldots, n$.
a) Obtain the relationship between α_0 and β_0, where β_0 is the intercept in our original model (Eq. (14.1)).
b) Show that $\text{Cov}(\hat{\alpha}_0, b_1) = 0$.
(Hint: Since $\hat{\alpha}_0$ and b_1 are linear functions of the Y's, that is, $\hat{\alpha}_0 = \sum_i a_i Y_i$ and $b_1 = \sum_i c_i Y_i$, then all one needs to show is that $\sum_i a_i c_i = 0$.)

c) Obtain a test for the hypothesis $H_0: \alpha_0 = \alpha$ versus $H_1: \alpha_0 \neq \alpha$.

14.11 A teacher conjectured that a student's initial exam score might be a good predictor of the score on a second exam. The table below shows scores of 15 students on the two exams.

Student	1	2	3	4	5	6	7	8	9	10	11	12	13	14	15
Exam 1	95	88	76	77	78	85	78	80	76	83	81	87	76	83	82
Exam 2	94	91	86	87	87	86	86	84	86	88	89	88	86	88	90

a) Determine the correlation between the two exam scores.
b) Use Eq. (14.10) to determine the slope b_1, if Exam 2 scores are regressed as a function of scores from Exam 1.
c) Obtain a 95% confidence interval for the correlation coefficient ϱ.

14.12 The coating of a component for use in orbital satellites was tested to determine its degradation in the presence of energetic protons encountered in near-earth orbit. The coating integrity I, expressed as a dimensionless number between 0 and 1, is shown in the table below as a function of time t (in days) in the space environment.

t	I	t	I	t	I	t	I
10	0.998	60	0.994	110	0.987	160	0.983
20	0.998	70	0.991	120	0.985	170	0.982
30	0.996	80	0.991	130	0.988	180	0.982
40	0.997	90	0.988	140	0.985	190	0.981
50	0.994	100	0.989	150	0.984	200	0.978

a) Fit the linear model, $I = \beta_0 + \beta_1 t + \varepsilon$.
b) The coating's effectiveness is compromised when its integrity drops below 0.985. Estimate the time in orbit before this occurs.
c) Obtain the 95% confidence interval on the time estimate of part (b).

14.13 Show that the cross-product term in the total sum of squares, Eq. (14.57), is zero, that is,

$$2\sum_{i=1}^{n}(Y_i - \hat{Y}_i)(\hat{Y}_i - \bar{Y}) = 0,$$

so that we can write TSS = SSE + SSR, as given in Eq. (14.58).

14.14 Use the identity given in Eq. (14.62) and the fact that $E(\sum_{i=1}^{n} r_i^2) = (n-2)\sigma^2$ to show that $E(\text{SSE}) = (n-2)\sigma^2 + \sum_{i=1}^{n} B_i^2$, as given by Eq. (14.63).

14.15 Show that the cross-product term in Eq. (14.70), $\sum_{i=1}^{m}\sum_{j=1}^{m_i}(Y_{ij} - \bar{Y}_i)(\bar{Y}_i - \hat{Y}_i)$, equals zero, so that Eq. (14.71) holds.

14.16 Show that the variance for the estimator b_1, when the regression is through the origin, is given by $\text{Var}(b_1) = \sigma^2 / \sum_{i=1}^{n} X_i^2$.

14.17 Show that the variance of the estimator for $E(Y)$, when X is the value of the independent variable and regression is through the origin, is given by $X^2\sigma^2/\sum_{i=1}^{n} X_i^2$.

14.18 Show that the variance for the estimator of a new observation at $X = X_i$, when regression is through the origin, is given by $\text{Var}(e) = \text{Var}(Y_{new} - \hat{Y}) = \sigma^2(1 + X_i^2/\sum_{i=1}^{n} X_i^2)$.

14.19 Show that the slope b_1 can be expressed in terms of the sample correlation coefficient r, as indicated in Eq. (14.89).

14.20 Show that, from Eq. (14.53), $\lim_{k \to \infty} [\text{Var}(e_k)] = \text{Var}(\hat{Y})$.

14.21 The response Q of an ion chamber is tested as a function of pressure P in a hyperbaric chamber by taking 10 readings of a standard gamma source over a fixed time, with the results shown:

P (mmHg)	100	110	120	130	140	150	160	170	180	190
Q (nC)	52.0	58.6	60.0	62.6	65.4	64.1	72.6	72.4	82.4	83.8

a) Plot the data, with the response (Q) as the dependent variable.
b) Fit the model $Q = \beta_0 + \beta_1 P + \varepsilon$ and obtain estimates for β_0 and β_1.
c) Use the fitted model to plot a straight line through the data.
d) Test at the $\alpha = 0.05$ level the hypothesis $H_0: \beta_1 = 0$ versus $H_1: \beta_1 \neq 0$.
e) Calculate the residuals and plot them against P. Do the residuals appear random? Comment on any observations that raise issues with the assumption of randomness in the residuals.

14.22 The production of a particular mouse protein in response to a given radiation dose is postulated to be dependent on the dose rate $\dot{D}$. Irradiation of groups of mice of a particular strain with the same delivered dose, but at different dose rates, gave the following results:

$\dot{D}$ (rad h^{-1})	100	110	120	130	140	150	160
C (nmol l^{-1})	4.02	4.68	5.02	4.92	5.84	5.42	6.17

a) Plot the data, fit the model $C = \beta_0 + \beta_1 \dot{D} + \varepsilon$, and plot the line.
b) Obtain a 95% confidence interval for the slope parameter β_1.
c) A concentration of $C = 5.39$ nmol l^{-1} is of particular interest in assessing the mouse's metabolic response to ionizing radiation. Determine the dose rate $\dot{D}$ at which this protein concentration is expected and obtain the 95% confidence interval for this dose rate.

15
Introduction to Bayesian Analysis

15.1
Methods of Statistical Inference

Classical statistics defines the probability of an event as the limiting relative frequency of its occurrence in an increasingly large number of repeated trials. This "frequentist" approach to making statistical inferences was developed extensively in Chapter 7 under the heading of parameter and interval estimation. As an example, one might make measurements for the purpose of determining the true, but unknown, numerical value of the decay constant λ of a long-lived pure radionuclide. The true value can be estimated by repeatedly observing the number of disintegrations X in a fixed amount of time. The variability in the estimate of λ is evident from one set of measurements of the random variable X to another. Following classical procedures, the frequentist can provide an interval within which the true value of λ might lie with, for example, 95% confidence. The inference to be made is that 95% of the intervals so constructed by repeated sampling can be expected to contain the true value of λ. This is not to say, however, that the probability is 95% that any particular interval contains the true value. Such a probability statement would relate to a random variable, which the true value is not. This formal procedure for ascertaining the value of λ is embodied in the maximum likelihood estimation described in Section 7.9.

A different approach to statistical inference stems from the work of Thomas Bayes (1702–1761), whose theorem on conditional probability was presented in Section 3.5. The unknown quantity λ is assigned a prior probability distribution, based on one's belief about its true value. This assignment does not mean that λ is random, but only represents a statement of the analyst's belief about its true value. One can think of the true value of the parameter λ as being the realization of random variable Λ with a known distribution (which we call the prior distribution). This distribution is normally not the product of some realizable experiment on Λ, but rather is thought of as the belief of the experimenter's disposition to the true value before any data are collected. Observed data X are considered to be fixed information, which can be used to revise the probability distribution on λ. Given the data, what is the probability that the true value lies within a specified interval? The inference expresses subjectively a degree of belief in this probability. Furthermore, this probability is always conditional, depending on the information one has.

As an example to contrast the frequentist and Bayesian viewpoints, consider the experiment of tossing a fair coin 20 times and observing the number of heads. The interpretation of "fair" implies that the probability of heads occurring on any toss is 1/2. Suppose that the experiment ended with all 20 tosses resulting in heads. The classical statistician, if assuming the coin to be fair, would still consider the probability of heads to be 1/2. The Bayesian, on the other hand, would now consider that his belief about the true value of the probability for heads has shifted to some larger value.

In this chapter, we introduce the Bayesian approach. It is rapidly finding increasing use and importance today in a wide range of applications, including radiation protection (Martz, 2000).

15.2
Classical Analysis of a Problem

We begin by treating a specific problem by classical methods in this section and then analyzing the same problem by Bayesian methods in the next section.

We associate a random variable X with a large population of events that can result in one of only two possible outcomes: $X=1$ (which we term "success") or $X=0$ ("failure"). Our objective is to sample the population randomly in order to determine the probability p of success for the population. The frequentist considers the value of p to be fixed, but unknown. A random sample of size n is drawn from the population, and the number of successes Y is recorded:

$$Y = \sum_{i=1}^{n} X_i. \tag{15.1}$$

Since p is constant from draw to draw, Y has the binomial probability distribution,

$$\Pr(Y = y) = \binom{n}{y} p^y (1-p)^{n-y}. \tag{15.2}$$

Here $y = \sum x_i$, with the summation understood to run from $i=1$ to $i=n$. We can interpret this function as the probability $f(y|p)$ – that is, the probability for y successes given p. Showing the dependence on the x_i explicitly, we write in place of Eq. (15.2)

$$f(y|p) = f\left(\sum x_i | p\right) = \binom{\sum^n x_i}{} p^{\sum x_i}(1-p)^{n-\sum x_i}. \tag{15.3}$$

Regarding Eq (15.3) as a function of p for a given set of the x_i, we define the likelihood function (Section 7.9),

$$L\left(p | \sum x_i\right) = \binom{\sum^n x_i}{} p^{\sum x_i}(1-p)^{n-\sum x_i}. \tag{15.4}$$

15.2 Classical Analysis of a Problem

In the frequentist's view, this function tells us the likelihood of obtaining a sample with the value $y = \sum x_i$ for different values of p.

■ **Example**
A random sample of $n = 10$ events, drawn from a binomial population with unknown probability of success p, yields $\sum x_i = 4$.

a) Obtain the likelihood function.
b) Calculate the values of the function for $p = 0.0, 0.1, \ldots, 1.0$.
c) Plot the likelihood function and find the value of p that maximizes it.
d) What is the significance of the maximizing value of p?

Solution
a) With $n = 10$ and $\sum x_i = 4$, the likelihood function (15.4) becomes

$$L\left(p \mid \sum x_i = 4\right) = \frac{10!}{4!(10-4)!} p^4 (1-p)^{10-4}$$
$$= 210 p^4 (1-p)^6. \qquad (15.5)$$

b) Values of this function for the specified values of p are presented in Table 15.1. The largest value in the table occurs for $p = 0.4$.

c) The likelihood function (15.5) is plotted in Figure 15.1. To find its maximum, we differentiate L with respect to p and set the result equal to zero:

$$\frac{dL}{dp} = 210\left[4p^3(1-p)^6 - 6p^4(1-p)^5\right] = 0. \qquad (15.6)$$

The solution is $p = p_{\max} = 2/5$ ($=0.4$, exactly, which coincidently is one of the values assigned for Table 15.1).

d) This value, $p = p_{\max}$, is the value most likely to have produced the set of observations.

Table 15.1 Values of likelihood function, L, calculated as function of p from Eq. (14.5).

p	L
0.0	0.0000
0.1	0.0112
0.2	0.0881
0.3	0.2001
0.4	0.2508
0.5	0.2051
0.6	0.1115
0.7	0.0368
0.8	0.0055
0.9	0.0001
1.0	0.0000

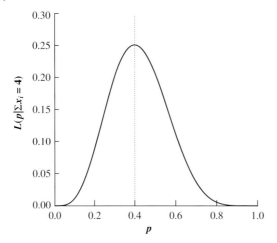

Figure 15.1 Plot of the likelihood function, Eq. (15.5).

The problem of determining p in this example has thus been treated in classical fashion by the method of maximum likelihood. As described in Section 7.9, the maximum likelihood estimator (MLE) has many desirable properties and is widely employed in statistics.

15.3
Bayesian Analysis of the Problem

The frequentist and Bayesian paradigms are quite different. In the Bayesian procedure, the unknown quantity p is the realization of the random variable P with an initial (prior) probability distribution, $g(p)$, of its own. This *prior distribution* for P is chosen on some basis (subjective) before a sample is drawn from the population. When the sample is taken, the data x_i are treated as fixed information. They are then used to make revisions and produce a *posterior distribution* on P. The end result is to provide a probability statement that the numerical value of P actually lies within a certain, specified interval. In contrast to the frequentist, the Bayesian analyst thus expresses a degree of "belief" in what the actual value of the parameter is.

To express these ideas formally, we set out to determine the probability function $f(p|\sum x_i)$, the posterior distribution on P given the sample $\sum x_i$. Applying Bayes' theorem, Eq. (4.137), with the prior distribution $g(p)$, we write for the posterior distribution

$$f\left(p\Big|\sum x_i\right) = \frac{f(p, \sum x_i)}{f(\sum x_i)} = \frac{f(\sum x_i|p)g(p)}{f(\sum x_i)}. \qquad (15.7)$$

■ **Example**

Treat the last example, in which $n = 10$ and $\sum x_i = 4$, by Bayesian methodology. In the absence of other information, let the prior distribution be uniform, namely,

$$g(p) = \begin{cases} 1, & 0 \le p \le 1, \\ 0, & \text{otherwise}. \end{cases} \quad (15.8)$$

a) Obtain the posterior distribution on p.
b) Plot the distribution.
c) Determine the mean of p based on the posterior distribution.
d) What is the significance of this mean?

Solution

a) As before, Eqs. (15.1)–(15.3) apply. Using Bayes' theorem (15.7) with the uniform prior $g(p) = 1$, we write for the posterior distribution

$$f(p|y) = \frac{f(y|p)g(p)}{f(y)} = \frac{\binom{n}{y} p^y (1-p)^{n-y} \times (1)}{\int_0^1 \binom{n}{y} p^y (1-p)^{n-y} \, dp}. \quad (15.9)$$

Note that the denominator, in which the integration is carried out over the range of p, is simply the marginal distribution (Section 4.4) on y. The binomial factor, which does not depend on p, cancels in the numerator and denominator. The remainder of the integral in the denominator can be evaluated through its relationship with the beta distribution. The probability density function for a random variable P, having the beta distribution with parameters $\alpha > 0$ and $\beta > 0$, is given by Eq. (6.111):

$$u(p; \alpha, \beta) = \begin{cases} \dfrac{\Gamma(\alpha+\beta)}{\Gamma(\alpha)\Gamma(\beta)} p^{\alpha-1}(1-p)^{\beta-1}, & 0 \le p \le 1, \\ 0, & \text{elsewhere}. \end{cases} \quad (15.10)$$

From Eq. (6.112), the mean and variance of P are

$$E(P) = \frac{\alpha}{\alpha+\beta} \quad (15.11)$$

and

$$\text{Var}(P) = \frac{\alpha\beta}{(\alpha+\beta+1)(\alpha+\beta)^2}. \quad (15.12)$$

Normalization of the beta probability density function (15.10) implies that

$$\int_0^1 p^{\alpha-1}(1-p)^{\beta-1}\,dp = \frac{\Gamma(\alpha)\Gamma(\beta)}{\Gamma(\alpha+\beta)}. \tag{15.13}$$

Comparison with the integral in Eq (15.9) shows that we can set (Problem 15.1)

$$\alpha = y+1 \quad \text{and} \quad \beta = n-y+1 \tag{15.14}$$

and write in place of Eq. (15.13)

$$\int_0^1 p^y(1-p)^{n-y}\,dp = \frac{\Gamma(y+1)\Gamma(n-y+1)}{\Gamma(n+2)}. \tag{15.15}$$

It follows from Eq. (15.9) that the posterior distribution on p given y is

$$f(p|y) = \begin{cases} \dfrac{\Gamma(n+2)}{\Gamma(y+1)\Gamma(n-y+1)} p^y(1-p)^{n-y}, & 0 \le p \le 1, \\ 0, & \text{elsewhere.} \end{cases} \tag{15.16}$$

The posterior distribution is thus the beta distribution with parameters α and β given by Eq (15.14). As in the last section, $n=10$ and $y=4$, and so the posterior distribution (15.16) becomes

$$f(p|y) = \begin{cases} \dfrac{\Gamma(12)}{\Gamma(5)\Gamma(7)} p^4(1-p)^6 = \dfrac{11!}{4!6!} p^4(1-p)^6 = 2310 p^4(1-p)^6, & 0 \le p \le 1, \\ 0, & \text{elsewhere.} \end{cases} \tag{15.17}$$

b) The distribution (15.17) is plotted in Figure 15.2. One sees that the shape of the posterior density on P is nearly the same as the shape of the likelihood function, indicating that the data have influenced our prior belief substantially, moving it from a uniform prior distribution to a beta distribution.

c) It follows from Eqs (15.14) and (15.11) that the mean is

$$E(P|y=4) = \frac{5}{5+7} = 0.4167. \tag{15.18}$$

d) The mean value given by Eq. (15.18) is now a weighted average of our prior belief (mean $=0.5$) and the data (mean $=0.4$). It will be shown later that the mean of the posterior distribution has a nice property with respect to estimators using a mean-square error criterion. It is also one way of comparing back to the classical estimator, which in this case is $\hat{p} = y/n = 0.40$.

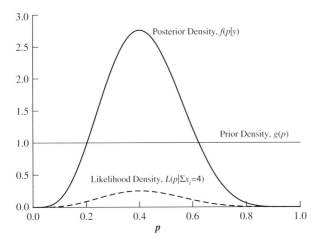

Figure 15.2 Plots of the prior, likelihood, and posterior densities.

The beauty of the Bayesian paradigm is that, once we have the posterior distribution, we know all that one needs to know about P, under the assumptions made. We can display the distribution, calculate its mean and variance, and compute any other quantities that are pertinent. We can calculate exact probability intervals for P, rather than confidence intervals that rely on repeated realizations of the sampling scheme. Thus, Bayesian statistics offers many features that the frequentist's does not.

15.4 Choice of a Prior Distribution

The major issue in the Bayesian methodology is the choice of the prior distribution. Its selection is necessary in order to apply Bayes' theorem and obtain the posterior distribution on the parameter, or parameters in a multivariate setting. The multiparameter situation entails some additional concepts with respect to distribution theory, but follows the same procedures that are employed for a single parameter. We shall investigate the single-parameter case only. The multivariate problem is discussed in other texts, for example, Press (1989).

To establish notation, we shall be interested in a random variable Θ, which has a prior distribution $g(\theta)$ and is a parameter in the sampling distribution $f(y|\theta)$. The joint distribution on (y, θ) can be expressed as

$$f(y, \theta) = f(y|\theta)g(\theta). \tag{15.19}$$

Suppose that we have drawn a sample that is characterized by θ, and that we have a prior distribution on θ that is independent of our sample. We can then use Bayes' theorem, Eq. (4.137), to derive a posterior distribution,

$$f(\theta|y) = \frac{f(\theta, y)}{f(y)} = \frac{f(y|\theta)g(\theta)}{f(y)}. \tag{15.20}$$

The marginal distribution $f(y)$ is obtained by integrating the joint distribution $f(\theta, y)$ over θ, and so

$$f(\theta|y) = \frac{f(y|\theta)g(\theta)}{\int f(y|\theta)g(\theta)d\theta}. \tag{15.21}$$

The function $f(y)$ does not depend on θ, and thus acts like a multiplicative constant in Eq. (15.20), specifying $f(\theta|y)$. Therefore, we may write the proportionality

$$f(\theta|y) \propto f(y|\theta)g(\theta), \tag{15.22}$$

which is sufficient to describe general properties, as we shall see. This important form of Bayes' theorem states that the posterior distribution is proportional to the product of the sampling and prior distributions. The statement (15.22) turns things around with respect to the conditioning symbol. It relates $f(\theta|y)$ to $f(y|\theta)$. The quantity $f(\theta|y)$ is of interest as expressing the probability distribution of θ given y (the data). The sampling distribution $f(y|\theta)$ expresses the likelihood of observing the data y given the value θ. As will become clearer in examples that follow, if we draw a random sample of size n from the population, then $f(y|\theta)$ is simply the likelihood function. The prior distribution $g(\theta)$, representing our knowledge or ignorance about the value of θ, is modified by the likelihood function to produce the posterior distribution. Thus, $f(\theta|y)$ in Eq (15.22) represents our state of knowledge about θ in light of the data.

As pointed out at the beginning of this section, the choice of the prior distribution is crucial. Yet, in most situations, nothing is known about this distribution, and so it must be selected subjectively. There are two basic ways to look at prior distributions. The "population" interpretation states that the prior distribution represents a population of possible parameter values from which the θ of current interest has been drawn. On the other hand, the "state of knowledge" interpretation says that we must express our knowledge about θ as if its value could be thought of as a random realization from the prior distribution. Usually, there is no relevant population of θ's from which the current θ has been drawn. This situation appears to present a dilemma, because the prior should reflect *some* things about θ. For example, it should include all possible values of θ. The *principle of insufficient reason*, first used by Laplace, states that if nothing is known about θ, then a uniform specification is appropriate. (We made this assumption in the last example by using Eq. (15.8).) As we shall see, however, the prior information is often outweighed by the information about θ that is contained in the sample. Since the prior distribution is a subjective choice, the domination of the sample information is desirable because it moves the inference from the subjective to the empirical.

■ **Example**
Return to the specific problem introduced through Eqs. (15.1)–(15.3) in Section 15.2, involving the binomial probability distribution. Choose the beta distribution with parameters α and β as the prior distribution on P.

a) Determine the posterior distribution on P.
b) Comment on how the parameters α and β can be interpreted.
c) What happens if $\alpha = \beta = 1$?
d) Find the mean of the posterior distribution.

Solution

a) We write for the sampling distribution from Eqs. (15.2) and (15.3)

$$f(y|p) = \binom{n}{y} p^y (1-p)^{n-y}. \tag{15.23}$$

The prior is given as the beta distribution, which from Eq (15.10) we write as

$$g(p) = \begin{cases} \dfrac{\Gamma(\alpha+\beta)}{\Gamma(\alpha)\Gamma(\beta)} p^{\alpha-1}(1-p)^{\beta-1}, & 0 \leq p \leq 1, \\ 0, & \text{elsewhere.} \end{cases} \tag{15.24}$$

Applying Bayes' theorem in the form of the proportionality (15.22), we write for the posterior distribution

$$f(p|y) \propto f(y|p)g(p) = \binom{n}{y} p^y(1-p)^{n-y} \frac{\Gamma(\alpha+\beta)}{\Gamma(\alpha)\Gamma(\beta)} p^{\alpha-1}(1-p)^{\beta-1}$$

$$\propto p^{y+\alpha-1}(1-p)^{n-y+\beta-1}, \tag{15.25}$$

where we have dropped all the multiplicative constant terms in the last step. Comparison with Eq (15.10) shows that the posterior distribution also has the form of a beta distribution with parameters $(y + \alpha)$ and $(n - y + \beta)$.

b) The dependence of the sampling distribution (15.23) on P is contained in the factors $p^y(1-p)^{n-y}$, where y is the number of successes and $(n-y)$ is the number of failures. The prior, $p^{\alpha-1}(1-p)^{\beta-1}$, given by Eq (15.24), is similar in form. It follows that $(\alpha - 1)$ and $(\beta - 1)$ may be thought of as the number of successes and failures, respectively, prior to sampling.

c) If $\alpha = \beta = 1$, then the (prior) beta distribution (15.24) becomes the uniform distribution. The posterior distribution (15.25) is beta, with parameters $(y + 1)$ and $(n - y + 1)$.

d) As shown in (a), the posterior is a beta distribution with parameters $(y + \alpha)$ and $(n - y + \beta)$. The mean is, by Eq. (15.11),

$$E(P|y) = \frac{\alpha + y}{\alpha + \beta + n}. \tag{15.26}$$

This example shows how one can select a rich prior distribution and can also choose values for the parameters of that prior distribution (called *hyperparameters*) to investigate special cases. In the example we saw that α and β can be interpreted as the prior number of successes and failures. Choosing $\alpha = \beta = 1$ yielded a uniform prior on p, which satisfies the *principle of insufficient reason*. One can also look at other choices of α and β to evaluate their effects on the posterior distribution and can get a good sense of how different choices can affect the posterior. The value of the posterior mean (15.26) lies between the sample proportion y/n and the prior mean $\alpha/(\alpha + \beta)$ (Problem 15.5). For fixed α and β, y and $(n-y)$ both become larger as the sample size increases, and $E(P|y) \cong y/n$. Thus, if the sample size is large enough, the choice of the hyperparameters will have little influence on the posterior. Conversely, if our sample is small, then the hyperparameters might make a considerable impact.

15.5
Conjugate Priors

The property that the posterior distribution follows the same parametric form as the prior distribution is called *conjugacy*. In the last example, the beta prior distribution for different values of α and β leads to a conjugate family for the binomial sampling distribution (or the binomial likelihood). Without discussing conjugate priors in any detail, we present several in Table 15.2 for specific sampling distributions.

> ■ **Example**
> Use the Poisson distribution to model the number of disintegrations X in a specified time t from a long-lived radioactive source.
>
> a) Obtain the posterior distribution by using the natural conjugate prior.
> b) Obtain the expected value of the posterior distribution.

Table 15.2 Natural conjugate priors.

Sampling distribution	Natural conjugate prior distribution
Binomial: $f(y\|p) \propto p^y(1-p)^{n-y}$	Beta: $g(p) \propto p^{\alpha-1}(1-p)^{\beta-1}$
Negative binomial: $f(y\|p) \propto p^r(1-p)^y$	Beta: $g(p) \propto p^{\alpha-1}(1-p)^{\beta-1}$
Poisson: $f(y\|\lambda) \propto e^{-\lambda} \lambda^y$	Gamma: $g(\lambda) \propto \lambda^{\alpha-1} e^{-\lambda\beta}$
Exponential: $f(y\|\lambda) \propto \lambda e^{-\lambda y}$	Gamma: $g(\lambda) \propto \lambda^{\alpha-1} e^{-\lambda\beta}$
Normal with known σ^2 and unknown μ: $f(y\|\mu) \propto e^{-(y-\mu)^2/2\sigma^2}$	Normal: $g(\mu) \propto e^{-(\mu-\theta)^2/2\tau^2}$
Normal with known μ and unknown σ^2: $f(y\|\sigma^2) \propto e^{-y^2/\sigma^2}/(\sigma^2)^{1/2}$	Inverse gamma:[a] $g(\sigma^2) \propto e^{-\beta/\sigma^2}/(\sigma^2)^{\alpha+1}$

[a] The inverse gamma is simply the distribution of $Y = 1/X$, where X has the gamma distribution. If the gamma distribution is $f(x; \alpha, \beta) = \beta^\alpha x^{\alpha-1} e^{-\beta x}/\Gamma(\alpha)$, then $Y = 1/X$ has the inverse gamma distribution with density given by $f(y; \alpha, \beta) = \beta^\alpha e^{-\beta/y}/\Gamma(\alpha)y^{\alpha+1}$.

Solution

a) Letting λ denote the decay constant, we write

$$f(x|\lambda) = \frac{(\lambda t)^x e^{-\lambda t}}{x!}. \tag{15.27}$$

From Table 15.2, the conjugate prior is the gamma distribution, which we write here as[1])

$$g(\lambda) \propto \lambda^{\alpha-1} e^{-\lambda \beta}. \tag{15.28}$$

From Bayes' theorem it follows that the (unnormalized) posterior distribution is

$$f(\lambda|x) \propto f(x|\lambda) f(\lambda) \propto \lambda^{x+\alpha-1} e^{-\lambda(t+\beta)}, \tag{15.29}$$

which we can identify as a gamma distribution with parameters $\alpha' = (x + \alpha)$ and $\beta' = (t + \beta)$.

b) The expected value of a gamma random variable with parameters α' and β' is α'/β' (Eq. (6.69)). Hence, the expected value of the posterior distribution is

$$E(\lambda) = \frac{x+\alpha}{t+\beta}. \tag{15.30}$$

(Note that the classical estimator of λ is $\hat{\lambda} = x/t$.)

It appears that one may interpret the parameter β as a prior timescale over which we observe the process and the parameter α is the number of occurrences observed prior to the sample. We note that if we choose $\alpha = 0$ and $\beta = 0$, then the posterior mean (15.30) converges to the classical estimator. This choice of parameter values is similar to having very little information about the process. Such priors are called *non-informative* and agree with the *principle of insufficient reason*.

15.6
Non-Informative Priors

We have already seen the application of a non-informative prior in our study of the binomial parameter P when we chose the uniform prior. Such distributions are also called *reference* prior distributions. The name *non-informative*, or *vague*, implies that the prior distribution adds little to the posterior distribution about information on the parameter. It lets the data set speak for itself. The idea is that no particular value of the

---

[1]) The following substitutions have been made in writing Eq. (15.28) from Eq. (6.68): $x \to \lambda$, $\lambda \to \beta$, and $k \to \alpha$.

prior distribution is favored over any other. For example, if Θ is discrete and takes on values θ_i, $i = 1, 2, \ldots, n$, then the discrete uniform distribution,

$$g(\theta_i) = \frac{1}{n}, \quad i = 1, 2, \ldots, n, \tag{15.31}$$

does not favor any particular θ_i over another. We say that $g(\theta_i)$ is a non-informative prior. Similarly, in the continuous case, if Θ is bounded, say $\theta \in [a, b]$, then we call the distribution,

$$g(\theta) = \frac{1}{b-a}, \quad a \le \theta \le b, \tag{15.32}$$

non-informative.

If the parameter space is infinite, for example, $\theta \in (-\infty, \infty)$, then the situation is less clear. We could write the non-informative prior as constant everywhere, $g(\theta) = c$, but this function (called an improper distribution) has an infinite integral and therefore cannot be a valid density function. However, Bayesian inference is still possible, provided the integral over θ of the likelihood $f(x|\theta)$ exists. As seen from Eq. (15.21), $g(\theta) = c$ cancels in the numerator and denominator. If the integral exists, then $f(\theta|x)$ is a proper posterior density.

A thorough discussion of non-informative prior distributions is beyond the scope of this text. Interested readers are referred to Carlin and Louis (1996), Box and Tiao (1973), Press (1989), and Martz and Waller (1982). Non-informative priors can be proscribed by attributes of the sampling distribution. If the sampling distribution is such that $f(x|\theta) = f(x-\theta)$, for instance, so that the density involves θ only through the term $(x - \theta)$, then θ is called a *location parameter*. In this case, if θ belongs to the whole real line, then the non-informative prior for a location parameter is

$$g(\theta) = 1, \quad -\infty < \theta < \infty. \tag{15.33}$$

If the sampling distribution is such that $f(x|\theta) = (1/\theta)f(x/\theta)$, where $\theta > 0$, then θ is called a *scale parameter*. Thus, the non-informative prior for a scale parameter is given by

$$g(\theta) = \frac{1}{\theta}, \quad \theta > 0. \tag{15.34}$$

One can also work with $g(\theta^2) = 1/\theta^2$ as well as Eq. (15.34) and use $f(x|\theta) = (1/\theta)f(x/\theta)$. Both of the prior distributions described by Eqs. (15.33) and (15.34) are improper priors, since the integrals of the densities are infinite. However, this fact is not as serious as it may seem if the sampling distribution and the prior combine in a form that is integrable.

■ *Example*

The random variable X has the exponential distribution with parameter λ, where $0 < \lambda < \infty$.

a) Show that $\theta = 1/\lambda$ is the scale parameter.
b) Obtain the posterior distribution on λ by using the non-informative prior.

c) Obtain the posterior distribution on λ by using the gamma prior with parameters α and β (i.e., the conjugate prior in Table 15.2).
d) Compare the posterior distributions provided by (b) and (c).

Solution

a) The sampling distribution is given by

$$f(x|\lambda) = \lambda e^{-\lambda x} = \frac{1}{1/\lambda} e^{-x/(1/\lambda)}. \tag{15.35}$$

The distribution has the form $(1/\theta)f(x/\theta)$, where $\theta = 1/\lambda$ is the scale parameter.

b) The non-informative prior for a scale parameter is, from Eq. (15.34),

$$g(\lambda) = \frac{1}{1/\lambda} = \lambda, \quad \lambda > 0. \tag{15.36}$$

From Eq (15.22), we can write for the posterior density

$$f(\lambda|x) \propto f(x|\lambda)g(\lambda) = \lambda e^{-\lambda x} \lambda = \lambda^2 e^{-\lambda x}. \tag{15.37}$$

This function has the form of a gamma density with $\alpha' = 3$ and $\beta' = x$.[2] In this case, even though the prior density is improper, the resulting posterior density is proper.

c) Given that the prior is gamma with parameters α and β, we have

$$f(\lambda|x) \propto f(x|\lambda)g(\lambda) = \lambda e^{-\lambda x} \lambda^{\alpha-1} e^{-\lambda \beta} = \lambda^\alpha e^{-\lambda(x+\beta)}, \tag{15.38}$$

which has the form of a gamma density with parameters (Problem 15.8)

$$\begin{aligned} \alpha' &= \alpha + 1, \\ \beta' &= x + \beta. \end{aligned} \tag{15.39}$$

d) Comparing the two posterior distributions, we see that choosing $\alpha = 2$ and $\beta = 0$ is equivalent to selecting a non-informative prior. In this case, the gamma distribution contains the non-informative distribution as well as many others, thus presenting a rich variety of priors.

Consider next a normal population with unknown mean μ and known standard deviation σ. The sampling distribution is given by

$$f(x|\mu) = \frac{1}{\sqrt{2\pi}\sigma} e^{-(1/2\sigma^2)(x-\mu)^2}, \quad -\infty < x < \infty, \tag{15.40}$$

where we have omitted writing σ in the conditional part, since it is known. It is apparent that μ is a location parameter, because $f(x|\mu)$ depends on μ only through the

---

[2] In the definition (6.68) of the gamma density, we can interchange the roles of x and λ, writing $f(\lambda; k, x) \propto x^k \lambda^{k-1} e^{-\lambda x}$. Dropping the scale factor x^k, we write this as proportional to the gamma density on λ with parameters α' and β': $f(\lambda; \alpha', \beta') \propto \lambda^{\alpha'-1} e^{-\lambda \beta'}$. Comparison with Eq. (6.68) shows that the parameters in Eq. (15.37) are $\alpha' = 3$ and $\beta' = x$.

term $(x - \mu)$. In this situation, if we want to use a non-informative prior on μ, we would use

$$g(\mu) = 1, \quad -\infty < \mu < \infty. \tag{15.41}$$

■ **Example**

Let $X_1, X_2, \ldots, X_n$ denote a random sample from a normal population with unknown mean μ and known standard deviation σ.

a) Assuming a non-informative prior, determine the posterior distribution on μ.
b) Choose the natural conjugate prior and determine the posterior distribution on μ.
c) Compare the two posterior distributions obtained in (a) and (b) when the uncertainty in the unknown mean is very large. If the uncertainty in the mean is not large, what is the effect of a large sample on the posterior distribution?

Solution

a) The posterior distribution is proportional to the product of the sampling distribution and the prior. Assuming a non-informative prior on μ, we use Eqs. (15.40) and (15.41) to write

$$f(\mu | x_1, x_2, \ldots, x_n) \propto e^{-(1/2\sigma^2) \sum_{i=1}^{n} (x_i - \mu)^2}. \tag{15.42}$$

To obtain the distribution on μ, we first add and subtract $\bar{x}$ within the quadratic term. We then express the sum from $i=1$ to $i=n$ in the exponent as

$$\sum (x_i - \mu)^2 = \sum (x_i - \bar{x} + \bar{x} - \mu)^2 \tag{15.43}$$

$$= \sum (x_i - \bar{x})^2 + 2 \sum (x_i - \bar{x})(\bar{x} - \mu) + \sum (\bar{x} - \mu)^2$$

$$= \sum (x_i - \bar{x})^2 + n(\bar{x} - \mu)^2, \tag{15.44}$$

since $\sum (x_i - \bar{x}) = 0$. Thus, the posterior distribution (15.42) becomes

$$f(\mu | x_1, x_2, \ldots, x_n) \propto e^{-(1/2\sigma^2) \sum (x_i - \bar{x})^2} e^{-[1/(2\sigma^2/n)](\mu - \bar{x})^2}$$

$$\propto e^{[1/(2\sigma^2/n)](\mu - \bar{x})^2}. \tag{15.45}$$

The first exponential term that appears here is a constant with respect to μ and can, therefore, be ignored in determining the proportional posterior distribution on μ. The remaining exponential term shows that the posterior on μ is normal with mean $\bar{x}$ and variance σ^2/n.

b) Table 15.2 shows that the natural conjugate prior is normal with mean θ and variance τ^2. Thus, the posterior is proportional to the product

$$f(\mu|x_1, x_2, \ldots, x_n) \propto e^{-(1/2\sigma^2)\sum(x_i-\mu)^2} e^{-(1/2\tau^2)(\mu-\theta)^2}. \quad (15.46)$$

As before, adding and subtracting $\bar{x}$ in the exponent of the first term on the right, we find that

$$f(\mu|x_1, x_2, \ldots, x_n) \propto e^{-[1/(2\sigma^2/n)](\mu-\bar{x})^2} e^{-(1/2\tau^2)(\mu-\theta)^2}. \quad (15.47)$$

We can expand and collect terms in the exponents. By completing the square and omitting multiplicative factors that do not depend on μ, we find that (Problem 15.9)

$$f(\mu|x_1, x_2, \ldots, x_n) \propto e^{-(1/2\tau_1^2)(\mu-\theta_1)^2}, \quad (15.48)$$

where

$$\theta_1 = \frac{(1/\tau^2)\theta + (n/\sigma^2)\bar{x}}{(1/\tau^2) + (n/\sigma^2)} \quad (15.49)$$

and

$$\frac{1}{\tau_1^2} = \frac{1}{\tau^2} + \frac{n}{\sigma^2}. \quad (15.50)$$

This result shows that the posterior distribution on μ is normal with mean θ_1 and variance τ_1^2.

c) To compare the results from (a) and (b), we first consider the case of very large τ^2, implying great prior uncertainty about the value of μ. Then, from Eqs (15.49) and (15.50),

$$\theta_1 \cong \bar{x} \quad \text{and} \quad \frac{1}{\tau_1^2} \cong \frac{n}{\sigma^2}. \quad (15.51)$$

This limiting case, $\tau^2 \to \infty$, yields the same posterior (15.45) as the non-informative prior. On the other hand, if τ^2 is not large, but n is large, then n/σ^2 dominates and again the posterior density is close to that obtained by using the non-informative prior. In this instance, when we collect a large sample, it would be expected that the sample dominates the posterior, as it does.

15.7
Other Prior Distributions

There are many prior distributions that can be used in doing Bayesian analysis. Press (1989) discusses *vague, data-based,* and *g-priors*. Carlin and Louis (1996) discuss *elicited priors*, where the distribution matches a person's prior beliefs. We note that prior distributions simply serve as a weighting function and, as such, one can imagine a large variety of them.

Prior distributions should satisfy certain criteria. The principle of insufficient reason would suggest a uniform or non-informative prior. On the other hand, selecting the natural conjugate argues that the selection of the sampling distribution is as subjective as the selection of the prior. Box and Tiao (1973) argue that the choice of the prior is not important as long as the information from the sample dominates the information contained in the prior. As we have pointed out, the main objection to the Bayesian analysis is the selection of the prior distribution. Much consideration and thought should go into its choice.

15.8
Hyperparameters

The prior distribution itself often depends on certain parameters that are conveniently referred to as *hyperparameters*. For example, the beta prior (15.24) is a function of p with hyperparameters α and β. Once a functional form for a prior distribution has been chosen, the values of its hyperparameters need to be selected (unless the prior is non-informative). To illustrate one method for determining the values of the hyperparameters, we consider the screening for radioactivity of a collection of soil samples from a retired production site. Random samples are taken and counted under uniform conditions. A given sample is classified as "high" or "low," depending on the magnitude of an observed count rate. We expect the sampling for "high" or "low" to follow the binomial model, and hence we chose the beta prior distribution as we did with Eq. (15.10). While we do not know the values of α and β directly, we might have a practical feel for the mean and standard deviation of the prior distribution. This information can be used to determine α and β. For instance, let us estimate that the prior mean is 0.040 and that the standard deviation is 0.025 in appropriate units (e.g., counts per second). This choice of distributional characteristics reflects a guess about the magnitude of mean, with uncertainty reflected in the relatively large value of the standard deviation. The assumed mean and variance satisfy Eqs. (15.11) and (15.12). In this case, then,

$$E(P) = \frac{\alpha}{\alpha + \beta} = 0.040 \tag{15.52}$$

and

$$\text{Var}(P) = \frac{\alpha \beta}{(\alpha + \beta + 1)(\alpha + \beta)^2} = (0.025)^2. \tag{15.53}$$

Solution for the two unknowns gives $\alpha = 2.42$ and $\beta = 58.0$. In practical terms, our prior is saying that, if we checked about 60 soil samples, we would expect to see 2 or 3 with "high" readings. Use of Eqs. (15.52) and (15.53) provides a way to determine values for the hyperparameters α and β.

In the next section, we discuss how one can use Bayesian analysis to make inferences. Additional sections will address applications of the inference method to the binomial probability, the Poisson rate, and the mean of a normal population with known variance.

15.9 Bayesian Inference

The Bayesian inference is embodied in the posterior distribution. All that one needs to know formally about the random variable is contained in the posterior density or its cumulative distribution. Knowledge of either or both is sufficient for answering most questions, but using them leaves comparison with frequentist methods somewhat unclear. Bayesians have been adept at finding good comparative measures, and we shall discuss these for point estimators, interval estimators, and hypothesis testing.

To consider a point estimate for a random variable Θ, we look at any summary entity, such as the mean, median, or mode of the posterior density. The mode is the value that maximizes the posterior density, which is proportional to the product of the sampling distribution and the prior. If we choose a flat prior, then the maximization of the posterior is equivalent to maximizing the sampling distribution or the likelihood function. Hence, the mode and the maximum likelihood estimator will be the same when the prior is flat. Because of this fact, the mode is often called the *generalized maximum likelihood estimator*. The mode is most appropriate if the density is two-tailed. In the case of a single-tail distribution, like the exponential, the mode may be the leftmost point, which is not very appropriate or informative.

The mean is often used to measure central tendency. The mean of the posterior can thus serve as a measure of centrality as well. Based on the posterior distribution, we can evaluate the accuracy of an estimator, $\hat{\theta}(y)$, by looking at its variance $E_{\theta|y}[(\Theta - \hat{\theta}(y))^2]$, where the subscript indicates that the expectation is taken with respect to the posterior distribution on Θ. Letting $\mu = E_{\theta|y}[\Theta]$ represent the posterior mean, we can write

$$E_{\theta|y}[(\Theta - \hat{\theta}(y))^2] = E_{\theta|y}[(\Theta - \mu + \mu - \hat{\theta}(y))^2] \tag{15.54}$$

$$= E_{\theta|y}[(\Theta - \mu)^2] + 2(\mu - \hat{\theta}(y))E_{\theta|y}[\Theta - \mu] + (\mu - \hat{\theta}(y))^2$$
$$= E_{\theta|y}[(\Theta - \mu)^2] + (\mu - \hat{\theta}(y))^2. \tag{15.55}$$

We see that the posterior variance of $\hat{\theta}(y)$ is equal to the variance of the posterior distribution plus the square of the difference between the estimator and the posterior mean. Therefore, to minimize the posterior variance of $\hat{\theta}(y)$, we should choose $\hat{\theta}(y)$ such that the last term in Eq (15.55) vanishes, that is, choose $\hat{\theta} = \mu$. Thus, μ minimizes the posterior variance of all estimators. The mean is a common measure of centrality. However, we note that highly skewed distributions will tend to have mean values that are large, due to the influence of long tails, and thus can vary significantly from the middle of the distribution.

The median is the value $\tilde{\theta}$ for which $\Pr_{\Theta|y}(\Theta \leq \tilde{\theta}) \geq 1/2$ and $\Pr_{\Theta|y}(\Theta \geq \tilde{\theta}) \geq 1/2$. (For continuous densities, the median is unique. For discrete distributions, it can take on infinitely many values.) Also, $\tilde{\theta}$ minimizes $E_{\Theta|y}(|\Theta - \tilde{\theta}|)$. Simply put, the median is the value that splits the distribution in half. The major issue with the median is the fact that finding an analytical representation for it is usually difficult. For a unimodal, symmetric distribution (such as the normal), the mean, median, and mode all coincide.

The Bayesian's version of the frequentist's confidence interval is the *credible interval*. Credible intervals are simply probabilistic intervals taken directly from the posterior distribution. If $F_{\theta|y}(\theta)$ is the cumulative distribution function of the posterior distribution, then we can find values a and b such that

$$\Pr(a < \Theta < b|y) = F_{\theta|y}(b) - F_{\theta|y}(a) = 1 - \alpha. \tag{15.56}$$

The interval (a, b) is a $100(1 - \alpha)\%$ "credibility interval" for θ. The values of a and b are not unique. That is, there are many possible choices that will yield a $100(1 - \alpha)\%$ credibility interval. The usual way to determine their values is to specify equal probabilities to the left and right of a and b, respectively. We choose a such that $\Pr_{\theta|y}(\Theta < a) = \alpha/2$ and b such that $\Pr_{\theta|y}(\Theta > b) = \alpha/2$. This partitioning can always be done when Θ is continuous. If Θ is discrete, however, one cannot always find values of a and b that capture exactly $100(1 - \alpha)\%$ of the distribution.

The symmetric form for determining a and b will not always yield the shortest interval. The *highest posterior density* (HPD) method seeks to avoid this defect. The HPD credible set C can be defined by writing

$$C = \{\theta \in \Theta : f(\theta|y) \geq k(\alpha)\}, \tag{15.57}$$

where $k(\alpha)$ is the largest constant that satisfies $\Pr(C|y) \geq 1 - \alpha$. This statement means that C contains all the values of θ for which the posterior density is greater than or equal to some constant $k(\alpha)$, where this constant is dependent upon the choice of α. Figure 15.3 indicates the idea behind the HPD credibility interval. The value $k(\alpha)$ cuts the posterior density function in two places, where $\theta = a$ and $\theta = b$. We move $k(\alpha)$ up

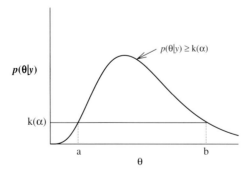

Figure 15.3 Graphical representation of the HPD credible interval of size $(1 - \alpha)$.

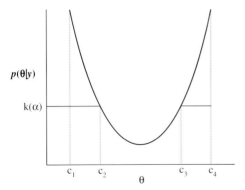

Figure 15.4 Graphical representation of example in which HPD credible interval method can yield two disjoint intervals.

and down until the integral over the interval from a to b is equal to $(1 - \alpha)$. The interval (a, b) is then the HPD interval.

For most well-behaved posterior densities, the method works well. However, if the density is single-tailed or multimodal, one can obtain some strange credible intervals. For example, if the posterior density is U shaped, Figure 15.4 shows that the HPD method would give two disjoint intervals. Again, we move $k(\alpha)$ up or down until the area under the posterior density function equals $(1 - \alpha)$. For the posterior density plotted in Figure 15.4, we see that the end points c_1 and c_4 must be included and, in addition, the line $f(\theta|y) = k(\alpha)$ cuts the graph at the points c_2 and c_3. Thus, there are two disjoint intervals (c_1, c_2) and (c_3, c_4) that make up the HPD interval. Although this problem is not common, the HPD interval is normally solved by numerical techniques on a computer. This procedure is computationally more intensive than using the symmetric form. We use the symmetric form throughout, but note that the HPD interval will, in general, be shorter than the symmetric interval.

The last inferential area to be discussed is hypothesis testing, which is notably different in the frequentist and Bayesian settings. The Bayesian paradigm allows for different hypotheses or models in a very natural way. In the usual hypothesis testing model, a null hypothesis H_0 and an alternative H_1 are stated. If the null and alternative hypotheses specify single values, for example, H_0: $\theta = \theta_0$ versus H_1: $\theta = \theta_1$, then the hypotheses are described as *simple versus simple*. If one hypothesis specifies a single value while the other gives an interval, for example, H_0: $\theta = \theta_0$ versus H_1: $\theta > \theta_1$, then they are referred to as *simple versus composite*. Composite versus composite hypotheses are possible as well, for example, H_0: $\theta \leq \theta_0$ versus H_1: $\theta > \theta_1$. Initially, we shall consider *simple versus simple* hypotheses. One can think of these equally well as two competing models, but we shall keep the hypothesis testing format. We let $T(y_1, y_2, \ldots, y_n)$ denote some appropriate test statistic based on a random sample of size n. By Bayes' theorem (3.44), the posterior probability of H_0 given the observed data T is

$$\Pr(H_0|T) = \frac{\Pr(T|H_0)\Pr(H_0)}{\Pr(T|H_0)\Pr(H_0) + \Pr(T|H_1)\Pr(H_1)}, \quad (15.58)$$

where $\Pr(H_0) + \Pr(H_1) = 1$ is the sum of the prior probabilities for H_0 and H_1. Similar posterior probability of H_1 is

$$\Pr(H_1|T) = \frac{\Pr(T|H_1)\Pr(H_1)}{\Pr(T|H_0)\Pr(H_0) + \Pr(T|H_1)\Pr(H_1)}. \tag{15.59}$$

The posterior odds of H_0 (compared with H_1) are given by the ratio of these quantities:

$$\frac{\Pr(H_0|T)}{\Pr(H_1|T)} = \left[\frac{\Pr(T|H_0)}{\Pr(T|H_1)}\right]\left[\frac{\Pr(H_0)}{\Pr(H_1)}\right], \tag{15.60}$$

which is just the product of the likelihood odds and the prior odds of H_0. If the ratio (15.60) is greater than unity, then we accept H_0 over H_1; otherwise, we reject H_0 in favor of H_1.

The *Bayes factor* is defined as the ratio of the posterior odds of H_0 and the prior odds of H_0. Thus, with the help of Eq (15.60) we have

$$\text{Bayes factor} = \frac{[\Pr(H_0|T)/\Pr(H_1|T)]}{[\Pr(H_0)/\Pr(H_1)]} = \frac{\Pr(T|H_0)}{\Pr(T|H_1)}, \tag{15.61}$$

which depends only on the data. When testing a *simple versus simple* hypothesis, we see that the last expression is the ratio of the likelihood obtained from the two competing parameters θ_0 and θ_1. Equation (15.60) shows that, if the hypotheses are simple and the prior probabilities are equal, then the Bayes factor equals the posterior odds of H_0, which equal the likelihood odds of H_0.

With the *simple versus composite* hypothesis (e.g., H_0: $\theta = \theta_0$ versus H_1: $\theta \neq \theta_0$), we need to define prior densities on θ_0 and θ. Under H_0, the prior density is degenerate with all its mass at the point θ_0. That is,

$$\Pr\nolimits_{H_0}(\Theta = \theta_0) = 1. \tag{15.62}$$

Under H_1, the prior on Θ can be any other prior, such as a conjugate or non-informative. We can then write for the probability

$$\Pr(T|H_i) = \int f(T|\theta_i, H_i) g_i(\theta_i) d\theta_i, \quad i = 0, 1. \tag{15.63}$$

Under H_0, the integral is simply the function $f(T|\theta_0, H_0)$. Under H_1, the integral represents the average likelihood over the prior density. Note that if the prior is non-informative and improper, then the integral might not exist, and so care must be taken. This circumstance furnishes one argument against using improper priors. Given that $\Pr(T|H_0)$ and $\Pr(T|H_1)$ are defined, the posterior odds and the Bayes factor can be calculated. The results given earlier follow for this *simple versus composite* situation.

Table 15.3 Comparison of frequentist and Bayesian inference methods.

Inference	Frequentist	Bayesian
Point estimate	Maximum likelihood estimate	Mean, median, or mode of posterior distribution
Interval estimate	Confidence interval	Credible, or highest posterior density, interval
Hypothesis testing	Likelihood ratio test	Ratio of posterior odds, or Bayes factor

Table 15.3 compares the frequentist and Bayesian inference methods just discussed. In the next three sections, we shall apply the Bayesian method to the binomial, Poisson, and normal mean parameters.

15.10
Binomial Probability

We return to the drawing of a random sample of size n from a Bernoulli population where the probability of success is p. When the prior is the beta distribution with parameters α and β, the posterior on P is given by Eq. (15.25). We can determine the mode by differentiating with respect to p in Eq (15.25), setting the result equal to zero, and solving for p (Problem 15.13). We find that

$$\text{Mode of } P = \frac{y + \alpha - 1}{n + \alpha + \beta - 2}. \quad (15.64)$$

In part (c) of the example in Section 15.4, we showed that, when $\alpha = \beta = 1$, the prior is equal to the uniform distribution. Because the prior is flat, the mode should then equal the maximum likelihood estimator, y/n. This result is borne out by Eq. (15.64).

The median is difficult to use in a straightforward manner, because there is no simple algebraic form that describes it. One needs actual numerical values for the terms n, y, α, and β. Using these and tables of the cumulative binomial distribution, one can determine the median.

The mean of the posterior was shown through Eq. (15.55) to be an optimal estimator in that it minimizes the squared error. With $\alpha' = y + \alpha$ and $\beta' = n - y + \beta$ in Eq. (15.25), the mean is

$$E(P) = \mu = \frac{y + \alpha}{n + \alpha + \beta}. \quad (15.65)$$

Recall that we can think of the prior as telling us the number of successes α and failures β in a sample of size $(\alpha + \beta)$. The mean simply incorporates the prior and data in a natural way in Eq. (15.65), giving an estimator that is the ratio of the total number of successes and the total sample size.

Example

A collection of air monitor filters from a building is to be checked for large particulates, defined for control purposes as having dimensions that exceed a certain control value. Assume that the number Y of such filters that check positive for large particles in a sample of size n from the total filter population has the binomial distribution with probability p of selection.

a) In a random sample of $n = 120$ filters, $y = 4$ are found to check positively. Using a uniform prior on p, find the posterior distribution. Obtain its mode, mean, and median.

b) Past history suggests that the average value of p is 0.010 with a standard deviation of 0.007. Find the prior on p that would yield these values, assuming a beta prior. Obtain the posterior distribution and its mode, mean, and median.

c) It is hypothesized that $p = 0.010$ is the historical value for p. Compare with the alternative hypothesis that $p = 0.040$ by using the posterior odds and the Bayes factor. Due to a lack of knowledge about the likelihood of one hypothesis over the other, we shall assume that $\Pr(H_0: p = 0.01) = \Pr(H_1: p = 0.04) = 1/2$.

Solution

a) From our previous examples, the posterior distribution is given by Eq. (15.25), with $\alpha = \beta = 1$ for the uniform prior. With $n = 120$ and $y = 4$, the mode (15.64) is

$$\text{Mode of } P = \frac{4 + 1 - 1}{120 + 1 + 1 - 2} = 0.033. \qquad (15.66)$$

The mean of the posterior distribution is, from Eq. (15.65),

$$E(P) = \frac{4 + 1}{120 + 1 + 1} = 0.041. \qquad (15.67)$$

The median $\tilde{p}$ is defined by writing

$$\frac{\Gamma(122)}{\Gamma(5)\Gamma(117)} \int_0^{\tilde{p}} p^4 (1-p)^{116} \, dp = \frac{1}{2}. \qquad (15.68)$$

The factor outside the integral is $\Gamma(n + \alpha + \beta)/[\Gamma(y + \alpha)\Gamma(n - y + \beta)]$. Numerical solution of Eq. (15.68) gives the result $\tilde{p} = 0.038$.

b) Past history suggests the mean $\mu = 0.010$ and standard deviation $\sigma = 0.007$. From Eqs. (15.11) and (15.12),

$$\mu = \frac{\alpha}{\alpha + \beta} = 0.010 \qquad (15.69)$$

and

$$\sigma^2 = \frac{\alpha \beta}{(\alpha + \beta + 1)(\alpha + \beta)^2} = (0.007)^2. \qquad (15.70)$$

Solving for α and β, we find $\alpha = 2.01$ and $\beta = 199$. Combining these values (rounded to the nearest integer) with the data yields a beta posterior with parameters $\alpha' = y + \alpha = 4 + 2 = 6$ and $\beta' = n - y + \beta = 120 - 4 + 199 = 315$. The mode (15.64) of the posterior is

$$\text{Mode of } P = \frac{4 + 2 - 1}{120 + 2 + 199 - 2} = \frac{5}{319} = 0.016. \tag{15.71}$$

The mean of the posterior is

$$E(P) = \frac{\alpha'}{\alpha' + \beta'} = \frac{6}{6 + 315} = 0.019, \tag{15.72}$$

and the median is

$$\tilde{p} = 0.018. \tag{15.73}$$

c) In the *simple versus simple* hypothesis, the posterior odds are given by the product of the likelihood odds of H_0 and the prior odds of H_0, as expressed by Eq. (15.60). The test statistic T in that equation is the binomial variable y here:

$$f(y|p) = \binom{n}{y} p^y (1-p)^{n-y}. \tag{15.74}$$

Under H_0

$$\Pr(y|H_0) = \Pr(y|p = 0.01) = \binom{120}{4}(0.01)^4(0.99)^{116}, \tag{15.75}$$

and under H_1

$$\Pr(y|H_1) = \Pr(y|p = 0.04) = \binom{120}{4}(0.04)^4(0.96)^{116}. \tag{15.76}$$

We are given $\Pr(H_0) = \Pr(H_1) = 1/2$, and so

$$\frac{\Pr(H_0|y)}{\Pr(H_1|y)} = \frac{(0.01)^4(0.99)^{116}}{(0.04)^4(0.96)^{116}} = 0.139. \tag{15.77}$$

Since the ratio is less than unity, we would accept H_1 in favor of H_0. Since $\Pr(H_0) = \Pr(H_1) = 1/2$, the posterior odds are equal to the Bayes factor, and both are equal to the likelihood ratio. In the above case, we see that the odds are roughly 6 to 1 in favor of the alternative hypothesis that $p = 0.04$.

15.11
Poisson Rate Parameter

We next consider the Bayesian method applied to the important Poisson rate parameter λ. We assume that, during a given time t, a number of events x occur from some process. In addition, we assume that the natural conjugate prior distribution for λ is the gamma distribution. Making the replacements, $x \to \lambda$, $k \to \alpha$, and $\lambda \to \beta$ in Eq. (6.68), we write for the prior

$$g(\lambda) = \frac{\beta^\alpha}{\Gamma(\alpha)} \lambda^{\alpha-1} e^{-\lambda\beta}, \quad \lambda > 0, \quad \alpha, \beta > 0. \tag{15.78}$$

With the application of Bayes' theorem from Eq (15.22), we can express the posterior distribution as proportional to the product of the likelihood function and the prior. Thus,

$$f(\lambda|x) \propto \lambda^{x+\alpha-1} e^{-\lambda(t+\beta)}. \tag{15.79}$$

We see that the posterior is of the gamma family. The parameters α and β can be interpreted as the number of prior events α in a time period β. Thus, the posterior density on λ is given by

$$f(\lambda|x) = \frac{(t+\beta)^{x+\alpha}}{\Gamma(x+\alpha)} \lambda^{x+\alpha-1} e^{-\lambda(t+\beta)}. \tag{15.80}$$

It can be shown (Problem 15.14) that the mode of $f(\lambda|x)$ is

$$\text{Mode of } \lambda = \frac{x+\alpha-1}{t+\beta} \tag{15.81}$$

and the mean is

$$E(\lambda) = \frac{x+\alpha}{t+\beta}. \tag{15.82}$$

One sees that if $(t+\beta)$ is large, then the mean and mode have nearly the same value.

The Bayesian inference for the Poisson parameter λ has been illustrated nicely by Martz (2000) in a study of scrams that occur at nuclear power stations. A scram produces a rapid decrease in the reactivity of a reactor in order to make it subcritical in response to some transient event, unplanned or otherwise, that could lead to loss of control. The frequency, or rate, of unplanned scrams is an important measure of how well a facility performs. The Nuclear Regulatory Commission has published for 66 licensed power reactors in the United States the number X_i of unplanned scrams and the total number of hours T_i that a reactor i was critical during the year 1984. The objective of the study is to determine the (unknown) scram rates λ_i (number of unplanned scrams per 1000 h of operation) from the published data. We shall assume that the X_i are described by Poisson distributions with parameters $\lambda_i T_i$ (Problem 15.15). If the rates appear to be different from reactor to reactor, then the λ_i might be considered as independent, to be estimated individually for each plant. On the other hand, the scram rates might have a commonality among different reactors by virtue of the standard operating and regulatory conditions under which all of the facilities are constrained. Estimations that reflect any such common aspects would, of course, be desirable.

Martz chooses a special type of gamma prior distribution with $\alpha = 1/2$ and $\beta = 0$, called the Jeffreys non-informative prior (Box and Tiao, 1973). Although this is an

15.11 Poisson Rate Parameter

Table 15.4 Unplanned scram data for two reactors in 1984 (Martz, 2000).

Reactor, i	Number of scrams, x_i	Operating time, t_i ($\times 10^3$ h)
1	6	5.5556
2	9	7.3770

improper prior, the posterior is a proper gamma distribution with parameters $\alpha' = x_i + 1/2$ and $\beta' = t_i$. Martz also considers an empirical gamma prior, suggested by the data collected on all 66 plants in the Commission report, with $\alpha = 1.39$ and $\beta = 1.21$.

■ *Example*
Data for two of the reactor facilities in Martz's study are given in Table 15.4, showing the numbers of unplanned scrams and the operating times in 10^3 h. Use both the Jeffreys ($\alpha = 1/2$, $\beta = 0$) and the empirical gamma ($\alpha = 1.39$, $\beta = 1.21$) priors.

a) Obtain the resulting two posterior densities for each facility.
b) Determine the mode and mean estimates for the four posteriors and compare them with the MLE.
c) Obtain a 95% confidence interval on the rates and symmetric 95% credible intervals from the posterior densities for reactor 1.

Solution
a) With a Poisson distribution for the X_i in time T_i with rate λ_i and a gamma prior with parameters α and β, we showed by Eq. (15.29) that the posterior density is also gamma with parameters $\alpha' = x_i + \alpha$ and $\beta' = t_i + \beta$. For reactor 1 and the Jeffreys prior, we find $\alpha' = 6 + 1/2 = 6.5$ and $\beta' = 5.5556 + 0 = 5.5556$. This result, together with the parameters for the other three posterior gamma distributions, is shown in Table 15.5 (Problem 15.16).
b) The posterior mode for λ is $(\alpha' - 1)/\beta'$ (Eq. (15.81)), and the mean is α'/β' (Eq. (15.82)). The MLE, obtained by maximizing the likelihood with respect to λ, is given by $\lambda_{MLE} = x_i/t_i$. Using these estimates, we obtain the results summarized in Table 15.6 (Problem 15.17).

Table 15.5 Parameters α' and β' for the gamma posterior densities for example in the text.

Reactor, i	Jeffreys prior ($\alpha = 1/2$, $\beta = 0$)		Empirical prior ($\alpha = 1.39$, $\beta = 1.21$)	
	α'	β'	α'	β'
1	6.5	5.5556	7.39	6.7656
2	9.5	7.3770	10.39	8.5870

Table 15.6 Mode, mean, and MLE for example in the text.

Reactor, i	Jeffreys prior		Empirical prior		MLE
	Mode	Mean	Mode	Mean	
1	0.99	1.71	0.94	1.09	1.08
2	1.15	1.29	1.09	1.21	1.22

c) To obtain a 95% confidence interval on the rate parameter, one can employ techniques described elsewhere (e.g., Miller and Freund, 1965). The MLE is x_i/t_i, and the $(1-\alpha)100\%$ confidence interval on λ is

$$[\lambda_L, \lambda_U] = \left[\frac{\chi^2_{2x_1,\alpha/2}}{2t_1}, \frac{\chi^2_{2x_1+2,1-\alpha/2}}{2t_1}\right], \quad (15.83)$$

where $\chi^2_{\nu,\theta}$ is the θ percentile of the chi-square distribution with ν degrees of freedom. Using the reactor 1 data from Table 15.4, we find that

$$[\lambda_L, \lambda_U] = \left[\frac{\chi^2_{12,0.025}}{2(5.5556)}, \frac{\chi^2_{14,0.975}}{2(5.5556)}\right] = \left[\frac{4.404}{11.11}, \frac{26.117}{11.11}\right]$$
$$= [0.396, 2.351]. \quad (15.84)$$

The Bayesian symmetric credible interval is obtained from the posterior distribution on λ, and is such that $\Pr(\lambda_L \leq \lambda \leq \lambda_U | x) = 1 - \alpha$. If we impose symmetry, then $\Pr(\lambda \leq \lambda_U) = \Pr(\lambda \geq \lambda_L) = 1 - \alpha/2$. It can be shown (Problem 15.20) that the posterior distribution of the transformed random variable $2T\lambda$ given x is a chi-square distribution with $(2x+1)$ degrees of freedom. Thus, the corresponding symmetric $100(1-\alpha)\%$ two-sided Bayesian credible interval on λ is given by

$$[\lambda_L, \lambda_U] = \left[\frac{\chi^2_{2x+1,\alpha/2}}{2t_1}, \frac{\chi^2_{2x+1,1-\alpha/2}}{2t_1}\right]. \quad (15.85)$$

Thus,

$$[\lambda_L, \lambda_U] = \left[\frac{\chi^2_{13,0.025}}{2(5.5556)}, \frac{\chi^2_{13,0.975}}{2(5.5556)}\right] = \left[\frac{5.009}{11.11}, \frac{24.736}{11.11}\right]$$
$$= [0.451, 2.226]. \quad (15.86)$$

Comparing the lengths of the frequentist (Eq. (15.84)) and Bayesian (Eq. (15.86)) intervals, we see that the Bayesian interval is shorter by $[(1.955 - 1.775)/1.995] \times 100\% = 9.02\%$.

For the Bayesian, there is little point in considering hypothesis testing. Having the posterior distribution, one can compute probability statements for the values that the parameters take on. To complete the comparison with the classical approach, we consider the simple versus simple hypothesis that H_0: $\lambda = 2$ versus H_1: $\lambda = 1$ for the reactor 1 plant. Recall from Eq. (15.60) that the posterior odds are equal to the product of the likelihood and prior odds of H_0. If we assume that H_0 is just as likely to occur as H_1, then the prior probabilities are the same and the posterior odds equal the likelihood odds. Thus, Eq. (15.60) yields

$$\frac{\Pr(H_0|T)}{\Pr(H_1|T} = \frac{\Pr(T|H_0)}{\Pr(T|H_1)} = \frac{(\lambda_0 t)^x \, e^{-\lambda_0 t}/x!}{(\lambda_1 t)^x \, e^{-\lambda_1 t}/x!} = \left(\frac{\lambda_0}{\lambda_1}\right)^x e^{-(\lambda_0 - \lambda_1)t}. \tag{15.87}$$

Substitution of the numerical values gives for the posterior odds ratio

$$\left(\frac{2}{1}\right)^6 e^{-2(2-1)(5.5556)} = 0.247. \tag{15.88}$$

Since the ratio is less than unity, we reject H_0 in favor of H_1. We note that, since the prior odds are equal, the posterior odds and the Bayes factor are equal. Here we see that the posterior odds in favor of H_1 are roughly 3 to 1.

The more common hypothesis testing is the *simple versus composite*, where H_0 specifies the value and H_1 is of the form $>$, $<$, or $\neq$. In this case, the ratio of the posterior density of H_0 compared with that of H_1 is

$$\frac{\Pr(H_0|T)}{\Pr(H_1|T)} = \frac{\Pr(T|H_0)}{\Pr(T|H_1)} \frac{\Pr(H_0)}{\Pr(H_1)} = \frac{\Pr(H_0)}{\Pr(H_1)} \frac{\Pr(T|H_0, \lambda_0)}{\int f(T|H_1 \lambda_1) g(\lambda_1) d\lambda_1}, \tag{15.89}$$

where $g(\lambda_1)$ denotes the prior density of λ_1 under H_1. Consider H_0: $\lambda = 2$ versus H_1: $\lambda \neq 2$, and assume that $\Pr(H_0) = \Pr(H_1) = 1/2$. The likelihood for reactor 1 under H_0 is Poisson with $\lambda = 2$, $t = 5.5556$, and $x = 6$. We take the Jeffreys prior for $g(\lambda_1)$ (i.e., gamma density with $\alpha = 1/2$ and $\beta = 0$). Thus,

$$\int_0^\infty f(T|H_1, \lambda_1) g(\lambda_1) d\lambda_1 = \int_0^\infty \frac{(\lambda_1 t)^x \, e^{-\lambda_1 t}}{x!} \frac{\lambda_1^{-1/2}}{\Gamma(1/2)} d\lambda_1$$

$$= \frac{t^x}{x! \Gamma(1/2)} \int_0^\infty \lambda_1^{x-1/2} e^{-\lambda_1 t} d\lambda_1 \tag{15.90}$$

$$= \frac{t^x}{x! \Gamma(1/2)} \frac{\Gamma(x+1/2)}{t^{x+1/2}} = \frac{\Gamma(x+1/2)}{x! \Gamma(1/2) t^{1/2}}. \tag{15.91}$$

The last integral in Eq (15.90) is a gamma function, which can be evaluated with the help of Eq. (6.70) (Problem 15.22). Given $x = 6$ and $t = 5.5556$, we find from Eqs. (15.89) and (15.91) that (Problem 15.23)

$$\frac{\Pr(H_0|t)}{\Pr(H_1|T)} = \frac{1/2}{1/2} \times \frac{(2 \times 5.5556)^6 \, e^{-2(5.5556)}/6!}{\Gamma(6+1/2)/[6!\Gamma(1/2)(5.5556)^{1/2}]} = 0.408. \tag{15.92}$$

Again, we would reject H_0 in favor of H_1. In this case, we had to determine a weighted average value for $\Pr(T|H_1)$ since λ was not specified in H_1 and we used Jeffreys

15.12
Normal Mean Parameter

The normal distribution plays a central role in statistics and no less a role in Bayesian analysis. The distribution is characterized by its mean μ and variance σ^2. In our analysis, we shall be concerned with the mean value alone, assuming that we have a good idea of the variance. (Generally, both μ and σ^2 should be considered; see Press (1989).) To simplify matters further, we initially treat the case in which a single data point is observed.

We let x denote a single observation from a normal distribution parameterized by an unknown mean θ and known variance σ^2. We write the likelihood function as

$$f(x|\theta) = \frac{1}{\sqrt{2\pi}\sigma} e^{-(1/2\sigma^2)(x-\theta)^2}. \tag{15.93}$$

We select the prior distribution on Θ to be the conjugate prior, which from Table 15.2 we can write as a normal distribution with mean μ_0 and variance τ_0. We also assume that these two hyperparameters are known. By combining the likelihood and the prior and using Eq. (15.22), it follows that the proportional posterior distribution on θ is

$$f(\theta|y) \propto e^{-(1/2)[(x-\theta)^2/\sigma^2 + (\theta-\mu_0)^2/\tau_0^2]}. \tag{15.94}$$

The exponent can be rewritten with the help of the identity

$$A(\theta - a)^2 + B(\theta - b)^2 = (A + B)(\theta - \theta_0)^2 \\ + (A^{-1} + B^{-1})^{-1}(a - b)^2, \tag{15.95}$$

in which A, B, a, and b are constants and $\theta_0 = (A + B)^{-1}(Aa + Bb)$. (The identity can be shown by expanding and collecting terms on the left and then completing the square for terms involving θ^2 and θ (Problem 15.24).) Substituting $A = 1/\sigma^2$, $a = x$, $B = \tau_0^2$, and $b = \mu_0$ into Eq. (15.95), we find that the exponent in Eq (15.94) becomes (Problem 15.25)

$$-\frac{1}{2}\left[\frac{(x-\theta)^2}{\sigma^2} + \frac{(\theta-\mu_0)^2}{\tau_0^2}\right] = -\frac{1}{2}\left[\left(\frac{1}{\sigma^2} + \frac{1}{\tau_0^2}\right)(\theta - \theta_0)^2 + (\sigma^2 + \tau_0^2)^{-1}(x - \mu_0)^2\right], \tag{15.96}$$

in which

$$\theta_0 = \left(\frac{1}{\sigma^2} + \frac{1}{\tau_0^2}\right)^{-1}\left(\frac{x}{\sigma^2} + \frac{\mu_0}{\tau_0^2}\right). \tag{15.97}$$

The second term in brackets on the right-hand side of Eq. (15.96) does not depend on θ, and so it can be dropped from the exponential term of the proportional posterior expression (15.94). We are left with

$$f(\theta|y) \propto e^{-(1/2)(1/\sigma^2 + 1/\tau_0^2)(\theta - \theta_0)^2}, \tag{15.98}$$

showing that the posterior distribution on θ is normal with mean θ_0 and variance

$$\tau_1^2 = \left(\frac{1}{\sigma^2} + \frac{1}{\tau_0^2}\right)^{-1}. \tag{15.99}$$

The results can be summarized in some meaningful ways. First, the reciprocal of the variance is often called the precision. Thus, we see from Eq (15.99) that the posterior precision, $1/\tau_1^2$, is equal to the sum of the prior precision $1/\tau_0^2$ and the data precision $1/\sigma^2$. Second, the posterior mean value (15.97) is the weighted average of the prior mean, μ_0, and the data mean, x (single value since $n=1$), with weights proportional to the respective precisions. Another way to express the relationship is to say that θ_0 is equal to the prior mean adjusted toward the observed x. That is, we can rewrite Eq (15.97) in the form

$$\theta_0 = \mu_0 + (x - \mu_0) \frac{\tau_0^2}{\sigma^2 + \tau_0^2}. \tag{15.100}$$

We can also rewrite Eq (15.97) as an expression of the data shrunk toward the prior mean,

$$\theta_0 = x - (x - \mu_0) \frac{\sigma^2}{\sigma^2 + \tau_0^2}. \tag{15.101}$$

These descriptions provide simple ways to think of how the prior mean and the data mean combine to yield the posterior mean.

We next take a random sample $X_1, X_2, \ldots, X_n$ of size n from a normal population with unknown mean θ and known variance σ^2. As before, we choose the normal prior with mean μ_0 and variance τ_0^2, and assume that these hyperparameters are known. The posterior density satisfies

$$f(\theta|x_1, x_2, \ldots, x_n) \propto g(\theta) f(x_1, x_2, \ldots x_n|\theta) = g(\theta) \prod_{i=1}^{n} f(x_i|\theta) \tag{15.102}$$

$$\propto e^{-(1/2\tau_0^2)(\theta - \mu_0)^2} \prod_{i=1}^{n} e^{-(1/2\sigma^2)(x_i - \theta)} = e^{-(1/2)\left[(1/\tau_0^2)(\theta - \mu_0)^2 + (1/\sigma^2) \sum_{i=1}^{n}(x_i - \theta)^2\right]}. \tag{15.103}$$

Earlier we used the technique of adding and subtracting $\bar{x} = \sum x_i/n$ within each squared term in the summation on the right. The reason for introducing $\bar{x}$ is that it is sufficient for θ, the mean value. That is, the sample mean carries all of the information about the population mean. As we did with Eqs. (15.43)–(15.44), we write

$$\sum_{i=1}^{n} (x_i - \theta)^2 = \sum_{i=1}^{n} (x_i - \bar{x} + \bar{x} - \theta)^2 = \sum_{i=1}^{n} (x_i - \bar{x})^2 + n(\bar{x} - \theta)^2. \tag{15.104}$$

The summation on the far right does not involve θ, and hence can be dropped from the proportional density (15.103). We are left with

$$f(\theta|x_1 x_2, \ldots x_n) \propto e^{-(1/2)[(1/\tau_0^2)(\theta-\mu_0)^2+(n/\sigma^2)(\bar{x}-\theta)^2]}. \tag{15.105}$$

Again, like Eqs. (15.94)–(15.97), we can write here

$$\frac{1}{\tau_0^2}(\theta - \mu_0)^2 + \frac{n}{\sigma^2}(\bar{x} - \theta)^2 = \left(\frac{1}{\tau_0^2} + \frac{n}{\sigma^2}\right)(\theta - \theta_2)^2 + \left(\tau_0^2 + \frac{\sigma^2}{n}\right)^{-1}(\bar{x} - \mu_0)^2, \tag{15.106}$$

where

$$\theta_0 = \left(\frac{1}{\tau_0^2} + \frac{n}{\sigma^2}\right)^{-1}\left(\frac{\mu_0}{\tau_0^2} + \frac{n\bar{x}}{\sigma^2}\right). \tag{15.107}$$

Neglecting the terms that do not involve θ, we see that the posterior distribution (15.105) is normal with mean θ_0 and variance $\tau^2 = (1/\tau_0^2 + n/\sigma^2)^{-1}$.

Often it is desired to analyze frequency data, such as numbers of counts, by using standard normal theory procedures. When the count numbers are large, the normal can give a good approximation to the Poisson distribution. In applying the normal theory procedures, the variance needs to be constant. It turns out that, by analyzing the square root of the counts rather than the counts themselves, the variance is stabilized and is approximated by the value 0.25. This result is discussed in Box, Hunter, and Hunter (1978). We use it in the following example.

■ **Example**

Repeated counts are made in 1-min intervals with a long-lived radioactive source. The results for $n = 100$ 1-min readings yield a sample with a mean of 848.37 counts and a variance of 631.41. For Poisson data, the square root transformation is a variance stabilizing transformation (recall that $\text{Var}(X) = \lambda = E(X)$). Therefore, as the mean changes, the variance also changes. In fact, the square root transformation yields a new random variable whose variance is approximately a constant equal to 1/4. A Poisson distribution with a mean this large is well approximated by a normal, and so the square root transformation actually improves the approximation. The mean and variance of the transformed data (replacing each count by its square root value) are $\bar{x} = 29.1200$ and $s^2 = 0.1862$. Use the transformed data and assume that $\sigma^2 = 1/4$. Assume that the prior distribution for the mean (on the square root data) has mean 30 and variance 1/2.

a) What is the posterior density on the mean?
b) Determine the mean, median, and mode for the posterior distribution.
c) Obtain a 95% symmetric credible interval on the mean.

Solution
a) The sample values are $n = 100$, $\sigma^2 = 1/4$, and $\bar{x} = 29.1200$, and the prior distribution has $\mu_0 = 30$ and $\tau_0^2 = 1/2$. Recalling Eq. (15.107) and the

sentence following it, we know that the posterior density on θ, the mean of the transformed data, is normal with mean

$$\theta_0 = \left(\frac{1}{1/2} + \frac{100}{1/4}\right)^{-1} \left(\frac{30}{1/2} + \frac{100(29.12)}{1/4}\right) = 29.1244 \qquad (15.108)$$

and variance

$$\tau^2 = \left(\frac{1}{1/2} + \frac{100}{1/4}\right)^{-1} = 0.0024876. \qquad (15.109)$$

We see that the posterior mean is numerically the same as the data mean. (The posterior mean is actually shifted a little to the right of the data mean.) As mentioned earlier, with increasing sample size, the data should have increasing influence, as comes out clearly in this example.

Also, the posterior variance is nearly the same as that ($\sigma^2/n = 0.0025$) of the sample mean. The effect of the prior variance is slight, as expected for a large sample.

b) For the normal density, the mean, median, and mode are identical, namely, 29.1244.

c) For the symmetric 95% credible interval for the normal posterior, we write, therefore,

$$\Pr\left(-1.96 \leq \frac{\Theta - \theta_0}{\tau} \leq 1.96\right) = 0.95. \qquad (15.110)$$

$$\Pr(29.0266 \leq \Theta \leq 29.2221) = 0.95. \qquad (15.111)$$

We note that the credible interval is on the square root of the mean of the original data. If we square each term inside the probability statement (15.111), the probability will not change. We then find that the credible interval in the original units is (842.54, 853.93). The original sample was generated randomly from a Poisson distribution with mean 850. The back-transformed estimate of the mean is $(29.1244)^2 = 849.25$, close to the true mean value.

To complete this section, we consider the *simple versus composite* hypothesis, $H_0: \theta = \theta_0$ versus $H_1: \theta = \theta_1 \neq \theta_0$, with $\Pr(H_0) = \Pr(H_1) = 0.5$. Taking a random sample $X_1, X_2, \ldots, X_n$ of size n, assume that $X|\theta \sim N(\theta, 1)$. We know that $\overline{X}$ is sufficient for the mean, θ, and that $\overline{X}|\theta \sim N(\theta, 1/\sqrt{n})$. Thus, the two sampling distributions can be written as

$$f_0(\overline{x}|H_0, \theta_0) = \left(\frac{n}{2\pi}\right)^{1/2} e^{-(n/2)(\overline{x}-\theta_0)^2} \qquad (15.112)$$

and

$$f_1(\overline{x}|H_1, \theta_1) = \left(\frac{n}{2\pi}\right)^{1/2} e^{-(n/2)(\overline{x}-\theta_1)^2}. \qquad (15.113)$$

Under H_0, the prior density on θ is a point distribution at $\theta = \theta_0$. Under H_1, we take the prior density on θ to be $N(\theta^*, 1)$. The posterior odds ratio (see Eqs. (15.60) and (15.63)) becomes

$$\frac{\Pr(H_0|X)}{\Pr(H_1|X)} = \frac{e^{-(n/2)(\bar{x}-\theta_0)^2}}{(1/\sqrt{2\pi})e^{-(2/n+2)^{-1}(\bar{x}-\theta^*)^2} \int_{-\infty}^{\infty} e^{-(n/2+1/2)(\theta-\mu)^2} d\theta}, \quad (15.114)$$

where

$$\mu = \left(\frac{n}{2} + \frac{1}{2}\right)\left(\frac{n\bar{x}}{2} + \frac{\theta^*}{2}\right). \quad (15.115)$$

The integrand is identical to the normal density except for the factor $[(n+1)/2\pi]^{1/2}$. Thus, the integral is equal to the reciprocal of this factor, and so the posterior odds ratio (15.114) is

$$\frac{\Pr(H_0|X)}{\Pr(H_1|X)} = \frac{e^{-(n/2)(\bar{x}-\theta_0)^2}}{(1/\sqrt{n+1})e^{-(1/2)[(n+1)/n](\bar{x}-\theta^*)^2}}. \quad (15.116)$$

Since the prior odds are equal, the Bayes factor reduces to the posterior odds.

■ **Example**
Ten film dosimeters are checked for possible exposure to radiation in a laboratory experiment. Calibration shows that the background densitometer reading for unexposed films should be $\theta_0 = 0.25$ (relative units). The mean reading for the $n = 10$ films is $\bar{x} = 0.65$. Assume that the prior for the film darkening is a normal distribution with mean $\theta^* = 0.45$ and standard deviation $\sigma = 1$. Let $H_0: \theta = \theta_0 = 0.25$ and $H_1: \theta \neq \theta_0$ be the hypotheses we wish to test and, for simplicity, let $\Pr(H_0) = \Pr(H_1) = 1/2$.

a) Find the posterior odds ratio.
b) Obtain the Bayes factor.
c) Which hypothesis should we accept?

Solution
a) Substituting the given information into Eq. (15.116), we compute

$$\frac{\Pr(H_0|\bar{x})}{\Pr(H_1|\bar{x})} = \frac{e^{-5(0.65-0.25)^2}}{(11)^{-1/2} e^{-1.1(0.65-0.45)^2/2}} = 1.523. \quad (15.117)$$

b) Since $\Pr(H_0) = \Pr(H_1) = 1/2$, the Bayes factor and the posterior odds ratio are equal.
c) Because the posterior odds are greater than unity, we accept $H_0: \theta = 0.25$ in favor of $H_1: \theta \neq 0.25$.

We have only scratched the surface concerning Bayesian analysis. For further reading, one can consult several books mentioned throughout this chapter and, in addition, the excellent introductory texts by Barry (1996) and by Sivia (1996).

Problems

15.1 Verify the relations (5.14).

15.2 Surface swipes are collected and counted at a nuclear facility in order to check for removable activity on various surfaces. A swipe sample result is declared to be "high" when the number of counts it produces exceeds an established control value. Random swipe samples are to be collected throughout the facility to determine the probability p that a given one will be "high." From a random sample of 450 swipe samples tested, 18 "high" ones were found.
   a) Write down the likelihood function for p.
   b) Sketch the likelihood function over the range $0 \leq p \leq 0.15$.
   c) Obtain the maximum likelihood estimate.
   d) Obtain the posterior distribution on p using the uniform prior.
   e) Find the mean of the posterior distribution and compare it to the maximum likelihood estimate. Are you surprised? Why or why not?

15.3 Refer to the last problem.
   a) Obtain the posterior distribution by using the beta prior with $\alpha = 10$ and $\beta = 290$.
   b) Calculate the mean value for this posterior distribution.

15.4 Suppose that T has an exponential distribution with parameter $\lambda > 0$ and $0 < T < \infty$. Let λ have a prior distribution that is exponential with parameter $c > 0$. Obtain the posterior density on λ.

15.5 Show that the expected value (15.26) lies between the sample proportion y/n and the prior mean $\alpha/(\alpha + \beta)$.

15.6 Provide a conjugate prior distribution for the unknown parameter.
   a) $X \sim$ binomial (n, p), n known.
   b) $X \sim$ negative binomial (r, p), r known.
   c) $X \sim$ normal (μ, σ), σ known.

15.7 State whether the following is a location family or a scale family.
   a) $X \sim$ uniform $(\theta - a, \theta + a)$.
   b) $X \sim$ gamma (α, β), α known.
   c) $X \sim N(\mu, \sigma)$, μ known.

15.8 Verify (15.39).

15.9 Show that Eqs. (15.48)–(15.50) follow from Eq. (15.47).

15.10 The negative binomial distribution was given by Eq. (5.70). Let $Y_1, Y_2, \ldots, Y_n$ be independent, identically distributed binomial variables with parameters (r, θ), with r known and θ having a beta (α, β) prior distribution. In this instance, Y is the number of *failures* that precede the rth *success*, and the probability function given by Eq (5.70) can be written as

$$p(y|r, \theta) = \binom{y + r - 1}{y} \theta^r (1 - \theta)^y.$$

The mean and variance are $E(Y) = r(1 - \theta)/\theta$ and $\text{Var}(Y) = r(1 - \theta)/\theta^2$.
   a) Show that the posterior distribution on θ is beta $\left(nr + \alpha, \sum_{i=1}^{n} y_i + \beta\right)$.

b) Show that the expected value of the posterior distribution is
$$E(\theta|y_1, y_2, \ldots, y_n) = \frac{nr + \alpha}{nr + \alpha + \sum y_i + \beta}.$$

c) Interpret the parameters α and β in the prior distribution on θ.

d) Write the likelihood function and show that the MLE for θ is
$$\hat{\theta} = \frac{nr}{nr + \sum y_i} = \frac{r}{r + \bar{y}}.$$

e) Compare the MLE and the mean of the posterior distribution as the number of experiments n gets larger.

f) Show that the MLE and the maximum of the posterior density on θ are the same when $\alpha = \beta = 1$ (and the prior is the uniform distribution).

15.11 In the last problem, $n = 10$ experiments were run with $r = 5$, and $\sum y_i = 40$ was observed ($i = 1, 2, \ldots, 10$).

a) Assume that the prior is beta with $\alpha = 2$ and $\beta = 2$. Calculate the posterior mean and the MLE for θ.

b) Write down the integral equations that need to be solved in order to obtain the 95% symmetric credible interval for θ.

c) Determine the Bayes factor for the simple versus composite hypothesis that $H_0: \theta = \theta_0 = 0.5$ versus $H_1: \theta \neq \theta_0$. Assume that both hypotheses are equally likely to occur and that the prior on θ under H_1 is uniform (0, 1), that is, beta with $\alpha = \beta = 1$. Which hypothesis do you accept?

(*Hint:* Since $\Pr(H_0) = \Pr(H_1) = 1/2$, the Bayes factor is equal to the posterior odds ratio. Show that, in general, the posterior odds ratio is

$$\frac{\Pr(H_0|y_1, y_2, \ldots, y_n)}{\Pr(H_1|y_1, y_2, \ldots, y_n)} = (1-\theta)^{\sum y_i} \theta_0^{nr} \left[\frac{\Gamma(nr+1)\Gamma\left(\sum y_i + 1\right)}{\Gamma\left(nr + \sum y_i + 2\right)} \right]^{-1},$$

then substitute the given data.)

15.12 Additional data for Problem 15.2 suggest that the prior is a beta distribution with mean 0.033 and standard deviation 0.010. Use this information to calculate the values of the hyperparameters, α and β, using Eqs. (15.11) and (15.12).

15.13 Show that the mode of the density,
$$g(p) = \frac{\Gamma(n + \alpha + \beta)}{\Gamma(y + \alpha)\Gamma(n - y + \beta)} p^{y+\alpha-1}(1-p)^{n-y+\beta-1},$$
is given by Eq. (15.64).

15.14 Show that the mode of the density,
$$f(\lambda) = \frac{\beta^\alpha \lambda^{\alpha-1} e^{-\lambda\beta}}{\Gamma(\alpha)},$$
is given by $(\alpha - 1)/\beta$. Use the result to show that Eq. (15.81) is correct.

15.15 Justify using the Poisson distribution as an appropriate model to describe the sampling of variability in the numbers X_i of unplanned reactor scrams among the different facilities, i, in Section 15.11.

15.16 Determine the values of the parameters α' and β' shown for the two reactor facilities and two priors in Table 15.5.

15.17 Calculate the entries in Table 15.6 for the two reactor facilities.

15.18 The random variable U has the density

$$h(u) = \frac{u^\alpha e^{-u}}{\alpha!}, \quad u, \alpha > 0,$$

where α is an integer.

a) If $Y = 2U$, show that the density on Y is given by

$$f(y) = \frac{y^\alpha e^{-y/2}}{2^{\alpha+1}\alpha!}, \quad y, \alpha > 0.$$

b) Show hat Y has a chi-square density with degrees of freedom $\nu = 2\alpha + 2$ by equating parameters in the density in part (a) to the chi-square density with ν degrees of freedom given by

$$f(\chi) = \frac{x^{\nu/2-1} e^{-\chi/2}}{\Gamma(\nu/2) 2^{\nu/2}}.$$

15.19 Show that

$$\int_{\lambda T}^{\infty} \frac{u^N e^{-u}}{N!} du = \sum_{k=0}^{N} \frac{e^{-\lambda T}(\lambda T)^k}{k!}.$$

15.20 Let X be a Poisson random variable with rate parameter λ and measurement time T. Then

$$\Pr(X \leq N|\lambda) = \sum_{k=0}^{N} \frac{e^{-\lambda T}(\lambda T)^k}{k!}.$$

Using the results from the last two problems, show that

$$\Pr(X \leq N|\lambda) = \Pr(\chi^2_{2(N+1)} > 2\lambda T|\lambda),$$

where $\chi^2_{2(N+1)}$ is interpreted as a random variable that is chi-square with $2(N+1)$ degrees of freedom.

15.21 The confidence interval given by Eq. (15.84) can be obtained in the following way. Suppose we are concerned about the mean $\mu = \lambda T$ of a Poisson process. Consider the hypothesis test of $H_0: \mu = \mu_0 = \lambda_0$ versus $H_1: \mu > \mu_0$. We observe the random variable X and find $X = n$. We can determine the exact significance level of the test corresponding to the observed value n by computing

$$\Pr(X \geq n|\mu = \mu_0 = \lambda_0 T) = \sum_{k=n}^{\infty} \frac{(\lambda_0 T)^k e^{-\lambda_0 T}}{k!}.$$

If we select λ_0 such that this probability is equal to $\alpha/2$, then this value of λ_0 is the *smallest* for which we would reject H_0 with probability $\alpha/2$. Thus, this value of λ_0 is λ_L. Similarly, if we turn the hypothesis around so that $H_0: \mu = \mu_0 = \lambda_0$ versus $H_1: \mu < \mu_0$, then the exact significance level of the test corresponding to an observed value n is given by

$$\Pr(X \le n | \mu = \mu_0 = \lambda_0 T) = \sum_{k=0}^{n} \frac{(\lambda_0 T)^k e^{-\lambda_0 T}}{k!}.$$

If we select λ_0 such that the value of this sum is equal to $\alpha/2$, then this value of λ_0 is the largest for which we would reject H_0 for H_1 with probability $\alpha/2$. This value of λ_0 is λ_U. The interval (λ_L, λ_U) thus forms a $(1-\alpha)100\%$ confidence interval for λ_0.

a) Verify Eq. (15.84) by showing that

$$\sum_{k=n}^{\infty} \frac{(\lambda_L T)^k e^{-\lambda_L T}}{k!} = \frac{\alpha}{2}$$

leads to the expression $\lambda_L = \chi^2_{2n,\alpha/2}/2T$.

b) Verify Eq. (15.84) by showing that

$$\sum_{k=0}^{n} \frac{(\lambda_n T)^k e^{-\lambda_n T}}{k!} = \frac{\alpha}{2}$$

leads to the expression $\lambda_U = \chi^2_{2n+2,1-\alpha/2}/2T$.

15.22 Show that Eq. (15.92) follows from (15.91).

15.23 Verify that Eq. (15.92) is correct.

15.24 Prove the identity (15.95).

15.25 Show that Eqs. (15.96) and (15.97) follow from Eq. (15.95).

15.26 A random sample of size $n = 10$ is drawn from a normal population with unknown mean μ and standard deviation $\sigma = 2$. The sample mean is $\bar{x} = 1.2$, and $\sum_{i=1}^{10}(x_i - \bar{x})^2 = 0.90$.

a) Write down the proportional likelihood function (ignoring the factor $(1/\sqrt{2\pi\sigma})^{10}$).

b) Calculate the likelihood function for $\mu = -1.0, -0.5, 0.0, 0.5, 1.0, 2.0,$ and 3.0.

c) Sketch the likelihood function.

d) Obtain the value of the maximum likelihood estimator analytically.

e) Determine the value of the proportional likelihood function at its maximum.

15.27 Assume in the last problem that the prior distribution on μ is $N(0, 1)$.

a) Show that the posterior distribution on μ is normal with mean $6/7$ and variance $2/7$.

b) Sketch the prior and posterior densities.

c) Obtain a 95% credibility interval on μ using the posterior density.

Appendix

Table A.1 Cumulative binomial distribution.

n	r	p									
		0.1	0.2	0.25	0.3	0.4	0.5	0.6	0.7	0.8	0.9
5	0	0.590	0.328	0.237	0.168	0.078	0.031	0.010	0.002	0.000	0.000
5	1	0.919	0.737	0.633	0.528	0.337	0.188	0.087	0.031	0.007	0.000
5	2	0.991	0.942	0.896	0.837	0.683	0.500	0.317	0.163	0.058	0.009
5	3	1.000	0.993	0.984	0.969	0.913	0.813	0.663	0.472	0.263	0.081
5	4	1.000	1.000	0.999	0.998	0.990	0.969	0.922	0.832	0.672	0.410
5	5	1.000	1.000	1.000	1.000	1.000	1.000	1.000	1.000	1.000	1.000
10	0	0.349	0.107	0.056	0.028	0.006	0.001	0.000	0.000	0.000	0.000
10	1	0.736	0.376	0.244	0.149	0.046	0.011	0.002	0.000	0.000	0.000
10	2	0.930	0.678	0.526	0.383	0.167	0.055	0.012	0.002	0.000	0.000
10	3	0.987	0.879	0.776	0.650	0.382	0.172	0.055	0.011	0.001	0.000
10	4	0.998	0.967	0.922	0.850	0.633	0.377	0.166	0.047	0.006	0.000
10	5	1.000	0.994	0.980	0.953	0.834	0.623	0.367	0.150	0.033	0.002
10	6	1.000	0.999	0.996	0.989	0.945	0.828	0.618	0.350	0.121	0.013
10	7	1.000	1.000	1.000	0.998	0.988	0.945	0.833	0.617	0.322	0.070
10	8	1.000	1.000	1.000	1.000	0.998	0.989	0.954	0.851	0.624	0.264
10	9	1.000	1.000	1.000	1.000	1.000	0.999	0.994	0.972	0.893	0.651
10	10	1.000	1.000	1.000	1.000	1.000	1.000	1.000	1.000	1.000	1.000
15	0	0.206	0.035	0.013	0.005	0.000	0.000	0.000	0.000	0.000	0.000
15	1	0.549	0.167	0.080	0.035	0.005	0.000	0.000	0.000	0.000	0.000
15	2	0.816	0.398	0.236	0.127	0.027	0.004	0.000	0.000	0.000	0.000
15	3	0.944	0.648	0.461	0.297	0.091	0.018	0.002	0.000	0.000	0.000
15	4	0.987	0.836	0.686	0.515	0.217	0.059	0.009	0.001	0.000	0.000
15	5	0.998	0.939	0.852	0.722	0.403	0.151	0.034	0.004	0.000	0.000
15	6	1.000	0.982	0.943	0.869	0.610	0.304	0.095	0.015	0.001	0.000
15	7	1.000	0.996	0.983	0.950	0.787	0.500	0.213	0.050	0.004	0.000
15	8	1.000	0.999	0.996	0.985	0.905	0.696	0.390	0.131	0.018	0.000
15	9	1.000	1.000	0.999	0.996	0.966	0.849	0.597	0.278	0.061	0.002
15	10	1.000	1.000	1.000	0.999	0.991	0.941	0.783	0.485	0.164	0.013
15	11	1.000	1.000	1.000	1.000	0.998	0.982	0.909	0.703	0.352	0.056

(*Continued*)

Table A.1 (Continued)

n	r	p									
		0.1	0.2	0.25	0.3	0.4	0.5	0.6	0.7	0.8	0.9
15	12	1.000	1.000	1.000	1.000	1.000	0.996	0.973	0.873	0.602	0.184
15	13	1.000	1.000	1.000	1.000	1.000	1.000	0.995	0.965	0.833	0.451
15	14	1.000	1.000	1.000	1.000	1.000	1.000	1.000	0.995	0.965	0.794
15	15	1.000	1.000	1.000	1.000	1.000	1.000	1.000	1.000	1.000	1.000
20	0	0.122	0.012	0.003	0.001	0.000	0.000	0.000	0.000	0.000	0.000
20	1	0.392	0.069	0.024	0.008	0.001	0.000	0.000	0.000	0.000	0.000
20	2	0.677	0.206	0.091	0.035	0.004	0.000	0.000	0.000	0.000	0.000
20	3	0.867	0.411	0.225	0.107	0.016	0.001	0.000	0.000	0.000	0.000
20	4	0.957	0.630	0.415	0.238	0.051	0.006	0.000	0.000	0.000	0.000
20	5	0.989	0.804	0.617	0.416	0.126	0.021	0.002	0.000	0.000	0.000
20	6	0.998	0.913	0.786	0.608	0.250	0.058	0.006	0.000	0.000	0.000
20	7	1.000	0.968	0.898	0.772	0.416	0.132	0.021	0.001	0.000	0.000
20	8	1.000	0.990	0.959	0.887	0.596	0.252	0.057	0.005	0.000	0.000
20	9	1.000	0.997	0.986	0.952	0.755	0.412	0.128	0.017	0.001	0.000
20	10	1.000	0.999	0.996	0.983	0.872	0.588	0.245	0.048	0.003	0.000
20	11	1.000	1.000	0.999	0.995	0.943	0.748	0.404	0.113	0.010	0.000
20	12	1.000	1.000	1.000	0.999	0.979	0.868	0.584	0.228	0.032	0.000
20	13	1.000	1.000	1.000	1.000	0.994	0.942	0.750	0.392	0.087	0.002
20	14	1.000	1.000	1.000	1.000	0.998	0.979	0.874	0.584	0.196	0.011
20	15	1.000	1.000	1.000	1.000	1.000	0.994	0.949	0.762	0.370	0.043
20	16	1.000	1.000	1.000	1.000	1.000	0.999	0.984	0.893	0.589	0.133
20	17	1.000	1.000	1.000	1.000	1.000	1.000	0.996	0.965	0.794	0.323
20	18	1.000	1.000	1.000	1.000	1.000	1.000	0.999	0.992	0.931	0.608
20	19	1.000	1.000	1.000	1.000	1.000	1.000	1.000	0.999	0.988	0.878
20	20	1.000	1.000	1.000	1.000	1.000	1.000	1.000	1.000	1.000	1.000

Tabulated values are $B(r; n, p) = \sum_{x=0}^{r} b(r; n, p) = \sum_{x=0}^{r} \binom{n}{x} p^x (1-p)^{n-x} = \Pr(X \leq r)$. Refer to Eq. (5.9).

Table A.2 Cumulative Poisson distribution.

r	μ											
	0.01	0.05	0.10	0.20	0.30	0.40	0.50	0.60	0.70	0.80	0.90	1.00
0	0.990	0.951	0.905	0.819	0.741	0.670	0.607	0.549	0.497	0.449	0.407	0.368
1	1.000	0.999	0.995	0.982	0.963	0.938	0.910	0.878	0.844	0.809	0.772	0.736
2	1.000	1.000	1.000	0.999	0.996	0.992	0.986	0.977	0.966	0.953	0.937	0.920
3	1.000	1.000	1.000	1.000	1.000	0.999	0.998	0.997	0.994	0.991	0.987	0.981
4	1.000	1.000	1.000	1.000	1.000	1.000	1.000	1.000	0.999	0.999	0.998	0.996
5	1.000	1.000	1.000	1.000	1.000	1.000	1.000	1.000	1.000	1.000	1.000	0.999
≥6	1.000	1.000	1.000	1.000	1.000	1.000	1.000	1.000	1.000	1.000	1.000	1.000

Table A.2 (Continued)

r	1.1	1.2	1.3	1.4	1.5	1.6	1.7	1.8	1.9	2.0	2.1	2.2
0	0.333	0.301	0.273	0.247	0.223	0.202	0.183	0.165	0.150	0.135	0.122	0.111
1	0.699	0.663	0.627	0.592	0.558	0.525	0.493	0.463	0.434	0.406	0.380	0.355
2	0.900	0.879	0.857	0.833	0.809	0.783	0.757	0.731	0.704	0.677	0.650	0.623
3	0.974	0.966	0.957	0.946	0.934	0.921	0.907	0.891	0.875	0.857	0.839	0.819
4	0.995	0.992	0.989	0.986	0.981	0.976	0.970	0.964	0.956	0.947	0.938	0.928
5	0.999	0.998	0.998	0.997	0.996	0.994	0.992	0.990	0.987	0.983	0.980	0.975
6	1.000	1.000	1.000	0.999	0.999	0.999	0.998	0.997	0.997	0.995	0.994	0.993
7	1.000	1.000	1.000	1.000	1.000	1.000	1.000	0.999	0.999	0.999	0.999	0.998
≥8	1.000	1.000	1.000	1.000	1.000	1.000	1.000	1.000	1.000	1.000	1.000	1.000

r	2.3	2.4	2.5	2.6	2.7	2.8	2.9	3.0	3.5	4.0	4.5	5.0
0	0.100	0.091	0.082	0.074	0.067	0.061	0.055	0.050	0.030	0.018	0.011	0.007
1	0.331	0.308	0.287	0.267	0.249	0.231	0.215	0.199	0.136	0.092	0.061	0.040
2	0.596	0.570	0.544	0.518	0.494	0.469	0.446	0.423	0.321	0.238	0.174	0.125
3	0.799	0.779	0.758	0.736	0.714	0.692	0.670	0.647	0.537	0.433	0.342	0.265
4	0.916	0.904	0.891	0.877	0.863	0.848	0.832	0.815	0.725	0.629	0.532	0.440
5	0.970	0.964	0.958	0.951	0.943	0.935	0.926	0.916	0.858	0.785	0.703	0.616
6	0.991	0.988	0.986	0.983	0.979	0.976	0.971	0.966	0.935	0.889	0.831	0.762
7	0.997	0.997	0.996	0.995	0.993	0.992	0.990	0.988	0.973	0.949	0.913	0.867
8	0.999	0.999	0.999	0.999	0.998	0.998	0.997	0.996	0.990	0.979	0.960	0.932
9	1.000	1.000	1.000	1.000	0.999	0.999	0.999	0.999	0.997	0.992	0.983	0.968
10	1.000	1.000	1.000	1.000	1.000	1.000	1.000	1.000	0.999	0.997	0.993	0.986
11	1.000	1.000	1.000	1.000	1.000	1.000	1.000	1.000	1.000	0.999	0.998	0.995
12	1.000	1.000	1.000	1.000	1.000	1.000	1.000	1.000	1.000	1.000	0.999	0.998
13	1.000	1.000	1.000	1.000	1.000	1.000	1.000	1.000	1.000	1.000	1.000	0.999
≥14	1.000	1.000	1.000	1.000	1.000	1.000	1.000	1.000	1.000	1.000	1.000	1.000

r	5.5	6.0	6.5	7.0	7.5	8.0	8.5	9.0	9.5	10.0	15.0	20.0
0	0.004	0.002	0.002	0.001	0.001	0.000	0.000	0.000	0.000	0.000	0.000	0.000
1	0.027	0.017	0.011	0.007	0.005	0.003	0.002	0.001	0.001	0.000	0.000	0.000
2	0.088	0.062	0.043	0.030	0.020	0.014	0.009	0.006	0.004	0.003	0.000	0.000
3	0.202	0.151	0.112	0.082	0.059	0.042	0.030	0.021	0.015	0.010	0.000	0.000
4	0.358	0.285	0.224	0.173	0.132	0.100	0.074	0.055	0.040	0.029	0.001	0.000
5	0.529	0.446	0.369	0.301	0.241	0.191	0.150	0.116	0.089	0.067	0.003	0.000
6	0.686	0.606	0.527	0.450	0.378	0.313	0.256	0.207	0.165	0.130	0.008	0.000
7	0.809	0.744	0.673	0.599	0.525	0.453	0.386	0.324	0.269	0.220	0.018	0.001
8	0.894	0.847	0.792	0.729	0.662	0.593	0.523	0.456	0.392	0.333	0.037	0.002
9	0.946	0.916	0.877	0.830	0.776	0.717	0.653	0.587	0.522	0.458	0.070	0.005
10	0.975	0.957	0.933	0.901	0.862	0.816	0.763	0.706	0.645	0.583	0.118	0.011
11	0.989	0.980	0.966	0.947	0.921	0.888	0.849	0.803	0.752	0.697	0.185	0.021
12	0.996	0.991	0.984	0.973	0.957	0.936	0.909	0.876	0.836	0.792	0.268	0.039

(Continued)

Table A.2 (Continued)

r	μ											
	5.5	6.0	6.5	7.0	7.5	8.0	8.5	9.0	9.5	10.0	15.0	20.0
13	0.998	0.996	0.993	0.987	0.978	0.966	0.949	0.926	0.898	0.864	0.363	0.066
14	0.999	0.999	0.997	0.994	0.990	0.983	0.973	0.959	0.940	0.917	0.466	0.105
15	1.000	0.999	0.999	0.998	0.995	0.992	0.986	0.978	0.967	0.951	0.568	0.157
16	1.000	1.000	1.000	0.999	0.998	0.996	0.993	0.989	0.982	0.973	0.664	0.221
17	1.000	1.000	1.000	1.000	0.999	0.998	0.997	0.995	0.991	0.986	0.749	0.297
18	1.000	1.000	1.000	1.000	1.000	0.999	0.999	0.998	0.996	0.993	0.819	0.381
19	1.000	1.000	1.000	1.000	1.000	1.000	0.999	0.999	0.998	0.997	0.875	0.470
20	1.000	1.000	1.000	1.000	1.000	1.000	1.000	1.000	0.999	0.998	0.917	0.559
21	1.000	1.000	1.000	1.000	1.000	1.000	1.000	1.000	1.000	0.999	0.947	0.644
22	1.000	1.000	1.000	1.000	1.000	1.000	1.000	1.000	1.000	1.000	0.967	0.721
23	1.000	1.000	1.000	1.000	1.000	1.000	1.000	1.000	1.000	1.000	0.981	0.787
24	1.000	1.000	1.000	1.000	1.000	1.000	1.000	1.000	1.000	1.000	0.989	0.843
25	1.000	1.000	1.000	1.000	1.000	1.000	1.000	1.000	1.000	1.000	0.994	0.888
26	1.000	1.000	1.000	1.000	1.000	1.000	1.000	1.000	1.000	1.000	0.997	0.922
27	1.000	1.000	1.000	1.000	1.000	1.000	1.000	1.000	1.000	1.000	0.998	0.948
28	1.000	1.000	1.000	1.000	1.000	1.000	1.000	1.000	1.000	1.000	0.999	0.966
29	1.000	1.000	1.000	1.000	1.000	1.000	1.000	1.000	1.000	1.000	1.000	0.978
30	1.000	1.000	1.000	1.000	1.000	1.000	1.000	1.000	1.000	1.000	1.000	0.987
31	1.000	1.000	1.000	1.000	1.000	1.000	1.000	1.000	1.000	1.000	1.000	0.992
32	1.000	1.000	1.000	1.000	1.000	1.000	1.000	1.000	1.000	1.000	1.000	0.995
33	1.000	1.000	1.000	1.000	1.000	1.000	1.000	1.000	1.000	1.000	1.000	0.997
34	1.000	1.000	1.000	1.000	1.000	1.000	1.000	1.000	1.000	1.000	1.000	0.999

Tabulated values are $P(r;\mu) = \sum_{x=0}^{r} p(x;\mu) = \sum_{x=0}^{r} \mu^x e^{-\mu}/x! = \Pr(X \le r)$. Refer to Eq. (5.27).

Table A.3 Cumulative normal distribution.

z	0.00	0.01	0.02	0.03	0.04	0.05	0.06	0.07	0.08	0.09
−3.3	0.0005	0.0005	0.0005	0.0004	0.0004	0.0004	0.0004	0.0004	0.0004	0.0003
−3.2	0.0007	0.0007	0.0006	0.0006	0.0006	0.0006	0.0006	0.0005	0.0005	0.0005
−3.1	0.0010	0.0009	0.0009	0.0009	0.0008	0.0008	0.0008	0.0008	0.0007	0.0007
−3.0	0.0013	0.0013	0.0013	0.0012	0.0012	0.0011	0.0011	0.0011	0.0010	0.0010

Table A.3 (*Continued*)

z	0.00	0.01	0.02	0.03	0.04	0.05	0.06	0.07	0.08	0.09
−2.9	0.0019	0.0018	0.0018	0.0017	0.0016	0.0016	0.0015	0.0015	0.0014	0.0014
−2.8	0.0026	0.0025	0.0024	0.0023	0.0023	0.0022	0.0021	0.0021	0.0020	0.0019
−2.7	0.0035	0.0034	0.0033	0.0032	0.0031	0.0030	0.0029	0.0028	0.0027	0.0026
−2.6	0.0047	0.0045	0.0044	0.0043	0.0041	0.0040	0.0039	0.0038	0.0037	0.0036
−2.5	0.0062	0.0060	0.0059	0.0057	0.0055	0.0054	0.0052	0.0051	0.0049	0.0048
−2.4	0.0082	0.0080	0.0078	0.0075	0.0073	0.0071	0.0069	0.0068	0.0066	0.0064
−2.3	0.0107	0.0104	0.0102	0.0099	0.0096	0.0094	0.0091	0.0089	0.0087	0.0084
−2.2	0.0139	0.0136	0.0132	0.0129	0.0125	0.0122	0.0119	0.0116	0.0113	0.0110
−2.1	0.0179	0.0174	0.0170	0.0166	0.0162	0.0158	0.0154	0.0150	0.0146	0.0143
−2.0	0.0228	0.0222	0.0217	0.0212	0.0207	0.0202	0.0197	0.0192	0.0188	0.0183
−1.9	0.0287	0.0281	0.0274	0.0268	0.0262	0.0256	0.0250	0.0244	0.0239	0.0233
−1.8	0.0359	0.0351	0.0344	0.0336	0.0329	0.0322	0.0314	0.0307	0.0301	0.0294
−1.7	0.0446	0.0436	0.0427	0.0418	0.0409	0.0401	0.0392	0.0384	0.0375	0.0367
−1.6	0.0548	0.0537	0.0526	0.0516	0.0505	0.0495	0.0485	0.0475	0.0465	0.0455
−1.5	0.0668	0.0655	0.0643	0.0630	0.0618	0.0606	0.0594	0.0582	0.0571	0.0559
−1.4	0.0808	0.0793	0.0778	0.0764	0.0749	0.0735	0.0721	0.0708	0.0694	0.0681
−1.3	0.0968	0.0951	0.0934	0.0918	0.0901	0.0885	0.0869	0.0853	0.0838	0.0823
−1.2	0.1151	0.1131	0.1112	0.1093	0.1075	0.1056	0.1038	0.1020	0.1003	0.0985
−1.1	0.1357	0.1335	0.1314	0.1292	0.1271	0.1251	0.1230	0.1210	0.1190	0.1170
−1.0	0.1587	0.1562	0.1539	0.1515	0.1492	0.1469	0.1446	0.1423	0.1401	0.1379
−0.9	0.1841	0.1814	0.1788	0.1762	0.1736	0.1711	0.1685	0.1660	0.1635	0.1611
−0.8	0.2119	0.2090	0.2061	0.2033	0.2005	0.1977	0.1949	0.1922	0.1894	0.1867
−0.7	0.2420	0.2389	0.2358	0.2327	0.2296	0.2266	0.2236	0.2206	0.2177	0.2148
−0.6	0.2743	0.2709	0.2676	0.2643	0.2611	0.2578	0.2546	0.2514	0.2483	0.2451
−0.5	0.3085	0.3050	0.3015	0.2981	0.2946	0.2912	0.2877	0.2843	0.2810	0.2776
−0.4	0.3446	0.3409	0.3372	0.3336	0.3300	0.3264	0.3228	0.3192	0.3156	0.3121
−0.3	0.3821	0.3783	0.3745	0.3707	0.3669	0.3632	0.3594	0.3557	0.3520	0.3483
−0.2	0.4207	0.4168	0.4129	0.4090	0.4052	0.4013	0.3974	0.3936	0.3897	0.3859
−0.1	0.4602	0.4562	0.4522	0.4483	0.4443	0.4404	0.4364	0.4325	0.4286	0.4247
−0.0	0.5000	0.4960	0.4920	0.4880	0.4840	0.4801	0.4761	0.4721	0.4681	0.4641
0.0	0.5000	0.5040	0.5080	0.5120	0.5160	0.5199	0.5239	0.5279	0.5319	0.5359
0.1	0.5398	0.5438	0.5478	0.5517	0.5557	0.5596	0.5636	0.5675	0.5714	0.5753
0.2	0.5793	0.5832	0.5871	0.5910	0.5948	0.5987	0.6026	0.6064	0.6103	0.6141
0.3	0.6179	0.6217	0.6255	0.6293	0.6331	0.6368	0.6406	0.6443	0.6480	0.6517
0.4	0.6554	0.6591	0.6628	0.6664	0.6700	0.6736	0.6772	0.6808	0.6844	0.6879
0.5	0.6915	0.6950	0.6985	0.7019	0.7054	0.7088	0.7123	0.7157	0.7190	0.7224
0.6	0.7257	0.7291	0.7324	0.7357	0.7389	0.7422	0.7454	0.7486	0.7517	0.7549
0.7	0.7580	0.7611	0.7642	0.7673	0.7704	0.7734	0.7764	0.7794	0.7823	0.7852
0.8	0.7881	0.7910	0.7939	0.7967	0.7995	0.8023	0.8051	0.8078	0.8106	0.8133
0.9	0.8159	0.8186	0.8212	0.8238	0.8264	0.8289	0.8315	0.8340	0.8365	0.8389
1.0	0.8413	0.8438	0.8461	0.8485	0.8508	0.8531	0.8554	0.8577	0.8599	0.8621
1.1	0.8643	0.8665	0.8686	0.8708	0.8729	0.8749	0.8770	0.8790	0.8810	0.8830
1.2	0.8849	0.8869	0.8888	0.8907	0.8925	0.8944	0.8962	0.8980	0.8997	0.9015
1.3	0.9032	0.9049	0.9066	0.9082	0.9099	0.9115	0.9131	0.9147	0.9162	0.9177
1.4	0.9192	0.9207	0.9222	0.9236	0.9251	0.9265	0.9279	0.9292	0.9306	0.9319
1.5	0.9332	0.9345	0.9357	0.9370	0.9382	0.9394	0.9406	0.9418	0.9429	0.9441

(*Continued*)

Table A.3 (*Continued*)

z	0.00	0.01	0.02	0.03	0.04	0.05	0.06	0.07	0.08	0.09
1.6	0.9452	0.9463	0.9474	0.9484	0.9495	0.9505	0.9515	0.9525	0.9535	0.9545
1.7	0.9554	0.9564	0.9573	0.9582	0.9591	0.9599	0.9608	0.9616	0.9625	0.9633
1.8	0.9641	0.9649	0.9656	0.9664	0.9671	0.9678	0.9686	0.9693	0.9699	0.9706
1.9	0.9713	0.9719	0.9726	0.9732	0.9738	0.9744	0.9750	0.9756	0.9761	0.9767
2.0	0.9772	0.9778	0.9783	0.9788	0.9793	0.9798	0.9803	0.9808	0.9812	0.9817
2.1	0.9821	0.9826	0.9830	0.9834	0.9838	0.9842	0.9846	0.9850	0.9854	0.9857
2.2	0.9861	0.9864	0.9868	0.9871	0.9875	0.9878	0.9881	0.9884	0.9887	0.9890
2.3	0.9893	0.9896	0.9898	0.9901	0.9904	0.9906	0.9909	0.9911	0.9913	0.9916
2.4	0.9918	0.9920	0.9922	0.9925	0.9927	0.9929	0.9931	0.9932	0.9934	0.9936
2.5	0.9938	0.9940	0.9941	0.9943	0.9945	0.9946	0.9948	0.9949	0.9951	0.9952
2.6	0.9953	0.9955	0.9956	0.9957	0.9959	0.9960	0.9961	0.9962	0.9963	0.9964
2.7	0.9965	0.9966	0.9967	0.9968	0.9969	0.9970	0.9971	0.9972	0.9973	0.9974
2.8	0.9974	0.9975	0.9976	0.9977	0.9977	0.9978	0.9979	0.9979	0.9980	0.9981
2.9	0.9981	0.9982	0.9982	0.9983	0.9984	0.9984	0.9985	0.9985	0.9986	0.9986
3.0	0.9987	0.9987	0.9987	0.9988	0.9988	0.9989	0.9989	0.9989	0.9990	0.9990
3.1	0.9990	0.9991	0.9991	0.9991	0.9992	0.9992	0.9992	0.9992	0.9993	0.9993
3.2	0.9993	0.9993	0.9994	0.9994	0.9994	0.9994	0.9994	0.9995	0.9995	0.9995
3.3	0.9995	0.9995	0.9995	0.9996	0.9996	0.9996	0.9996	0.9996	0.9996	0.9997

Tabulated values are $\Pr(Z < z) = F(z) = (1/\sqrt{2\pi}) \int_{-\infty}^{z} e^{-(1/2)t^2}\, dt$, as represented by the shaded area of the figure. Refer to Eq. (6.20).

Table A.4 Quantiles $\chi^2_{\nu,\alpha}$ for the chi-squared distribution with ν degrees of freedom.

ν \ α	0.005	0.010	0.020	0.025	0.050	0.100	0.200	0.250	0.300	0.500	0.700	0.750	0.800	0.900	0.950	0.975	0.980	0.990	0.995	0.999
1[a]	$0.0^4 393$	$0.0^3 157$	$0.0^3 628$	$0.0^3 982$	$0.0^2 393$	0.016	0.064	0.102	0.148	0.455	1.074	1.323	1.642	2.706	3.841	5.024	5.412	6.635	7.879	10.83
2	0.010	0.020	0.040	0.051	0.103	0.211	0.446	0.575	0.713	1.386	2.408	2.773	3.219	4.605	5.991	7.378	7.824	9.210	10.60	13.82
3	0.072	0.115	0.185	0.216	0.352	0.584	1.005	1.213	1.424	2.366	3.665	4.108	4.642	6.251	7.815	9.348	9.837	11.35	12.84	16.27
4	0.207	0.297	0.429	0.484	0.711	1.064	1.649	1.923	2.195	3.357	4.878	5.385	5.989	7.779	9.488	11.14	11.67	13.28	14.86	18.47
5	0.412	0.554	0.752	0.831	1.145	1.610	2.343	2.675	3.000	4.351	6.064	6.626	7.289	9.236	11.07	12.83	13.39	15.09	16.75	20.52
6	0.676	0.872	1.134	1.237	1.635	2.204	3.070	3.455	3.828	5.348	7.231	7.841	8.558	10.65	12.59	14.45	15.03	16.81	18.55	22.46
7	0.989	1.239	1.564	1.690	2.167	2.833	3.822	4.255	4.671	6.346	8.383	9.037	9.803	12.02	14.07	16.01	16.62	18.48	20.28	24.32
8	1.344	1.646	2.032	2.180	2.733	3.490	4.594	5.071	5.527	7.344	9.524	10.22	11.03	13.36	15.51	17.54	18.17	20.09	21.96	26.12
9	1.735	2.088	2.532	2.700	3.325	4.168	5.380	5.899	6.393	8.343	10.66	11.39	12.24	14.68	16.92	19.02	19.68	21.67	23.59	27.88
10	2.156	2.558	3.059	3.247	3.940	4.865	6.179	6.737	7.267	9.342	11.78	12.55	13.44	15.99	18.31	20.48	21.16	23.21	25.19	29.59
11	2.603	3.053	3.609	3.816	4.575	5.578	6.989	7.584	8.148	10.34	12.90	13.70	14.63	17.28	19.68	21.92	22.62	24.73	26.76	31.26
12	3.074	3.571	4.178	4.404	5.226	6.304	7.807	8.438	9.034	11.34	14.01	14.85	15.81	18.55	21.03	23.34	24.05	26.22	28.30	32.91
13	3.565	4.107	4.765	5.009	5.892	7.042	8.634	9.299	9.926	12.34	15.12	15.98	16.99	19.81	22.36	24.74	25.47	27.69	29.82	34.53
14	4.075	4.660	5.368	5.629	6.571	7.790	9.467	10.17	10.82	13.34	16.22	17.12	18.15	21.06	23.69	26.12	26.87	29.14	31.32	36.12
15	4.601	5.229	5.985	6.262	7.261	8.547	10.31	11.04	11.72	14.34	17.32	18.25	19.31	22.31	25.00	27.49	28.26	30.58	32.80	37.70
16	5.142	5.812	6.614	6.908	7.962	9.312	11.15	11.91	12.62	15.34	18.42	19.37	20.47	23.54	26.30	28.85	29.63	32.00	34.27	39.25
17	5.697	6.408	7.255	7.564	8.672	10.09	12.00	12.79	13.53	16.34	19.51	20.49	21.62	24.77	27.59	30.19	31.00	33.41	35.72	40.79

(*Continued*)

Table A.4 (Continued)

ν	α																		
	0.005	0.010	0.020	0.025	0.050	0.100	0.200	0.250	0.300	0.500	0.700	0.750	0.800	0.900	0.950	0.975	0.980	0.990	0.995	0.999
18	6.265	7.015	7.906	8.231	9.390	10.87	12.86	13.68	14.44	17.34	20.60	21.61	22.76	25.99	28.87	31.53	32.35	34.81	37.16	42.31
19	6.844	7.633	8.567	8.907	10.117	11.65	13.72	14.56	15.35	18.34	21.69	22.72	23.90	27.20	30.14	32.85	33.69	36.19	38.58	43.82
20	7.434	8.260	9.237	9.591	10.851	12.44	14.58	15.45	16.27	19.34	22.78	23.83	25.04	28.41	31.41	34.17	35.02	37.57	40.00	45.32
21	8.034	8.897	9.915	10.28	11.591	13.24	15.45	16.34	17.18	20.34	23.86	24.94	26.17	29.62	32.67	35.48	36.34	38.93	41.40	46.80
22	8.643	9.542	10.60	10.98	12.338	14.04	16.31	17.24	18.10	21.34	24.94	26.04	27.30	30.81	33.92	36.78	37.66	40.29	42.80	48.27
23	9.260	10.20	11.29	11.69	13.091	14.85	17.19	18.14	19.02	22.34	26.02	27.14	28.43	32.01	35.17	38.08	38.97	41.64	44.18	49.73
24	9.886	10.86	11.99	12.40	13.848	15.66	18.06	19.04	19.94	23.34	27.10	28.24	29.55	33.20	36.42	39.36	40.27	42.98	45.56	51.18
25	10.52	11.52	12.70	13.12	14.611	16.47	18.94	19.94	20.87	24.34	28.17	29.34	30.68	34.38	37.65	40.65	41.57	44.31	46.93	52.62
26	11.16	12.20	13.41	13.84	15.379	17.29	19.82	20.84	21.79	25.34	29.25	30.44	31.80	35.56	38.89	41.92	42.86	45.64	48.29	54.05
27	11.81	12.88	14.13	14.57	16.151	18.11	20.70	21.75	22.72	26.34	30.32	31.53	32.91	36.74	40.11	43.20	44.14	46.96	49.65	55.48
28	12.46	13.57	14.85	15.31	16.928	18.94	21.59	22.66	23.65	27.34	31.39	32.62	34.03	37.92	41.34	44.46	45.42	48.28	50.99	56.89
29	13.12	14.26	15.57	16.05	17.708	19.77	22.48	23.57	24.58	28.34	32.46	33.71	35.14	39.09	42.56	45.72	46.69	49.59	52.34	58.30
30	13.79	14.95	16.31	16.79	18.493	20.60	23.36	24.48	25.51	29.34	33.53	34.80	36.25	40.26	43.77	46.98	47.96	50.89	53.67	59.70

Refer to Eq. (6.87).
The first five values are represented here to three significant figures by indicating as a superscript the number of zeros after the decimal, for example, 0.0^3157 represents the number 0.000157.

Table A.5 Quantiles $t_{\nu,\alpha}$ that cut off area α to the right for Student's t-distribution with ν degrees of freedom.

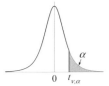

ν			α		
	0.100	0.050	0.025	0.010	0.005
1	3.078	6.314	12.706	31.821	63.657
2	1.886	2.920	4.303	6.965	9.925
3	1.638	2.353	3.182	4.541	5.841
4	1.533	2.132	2.776	3.747	4.604
5	1.476	2.015	2.571	3.365	4.032
6	1.440	1.943	2.447	3.143	3.707
7	1.415	1.895	2.365	2.998	3.499
8	1.397	1.860	2.306	2.896	3.355
9	1.383	1.833	2.262	2.821	3.250
10	1.372	1.812	2.228	2.764	3.169
11	1.363	1.796	2.201	2.718	3.106
12	1.356	1.782	2.179	2.681	3.055
13	1.350	1.771	2.160	2.650	3.012
14	1.345	1.761	2.145	2.624	2.977
15	1.341	1.753	2.131	2.602	2.947
16	1.337	1.746	2.120	2.583	2.921
17	1.333	1.740	2.110	2.567	2.898
18	1.330	1.734	2.101	2.552	2.878
19	1.328	1.729	2.093	2.539	2.861
20	1.325	1.725	2.086	2.528	2.845
21	1.323	1.721	2.080	2.518	2.831
22	1.321	1.717	2.074	2.508	2.819
23	1.319	1.714	2.069	2.500	2.807
24	1.318	1.711	2.064	2.492	2.797
25	1.316	1.708	2.060	2.485	2.787
26	1.315	1.706	2.056	2.479	2.779
27	1.314	1.703	2.052	2.473	2.771
28	1.313	1.701	2.048	2.467	2.763
29	1.311	1.699	2.045	2.462	2.756
∞	1.282	1.645	1.960	2.326	2.576

Refer to Eq. (6.96).

Table A.6 Quantiles $f_{0.95}(\nu_1, \nu_2)$ for the F distribution.

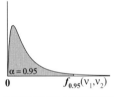

ν_2	ν_1											
	1	2	3	4	5	6	7	8	9	10	11	12
1	161.4	199.5	215.7	224.6	230.2	234.0	236.8	238.9	240.5	241.9	243.0	243.9
2	18.51	19.00	19.16	19.25	19.30	19.33	19.35	19.37	19.39	19.40	19.41	19.41
3	10.13	9.552	9.277	9.117	9.013	8.941	8.887	8.845	8.812	8.786	8.763	8.745
4	7.709	6.944	6.591	6.388	6.256	6.163	6.094	6.041	5.999	5.964	5.936	5.912
5	6.608	5.786	5.409	5.192	5.050	4.950	4.876	4.818	4.772	4.735	4.704	4.678
6	5.987	5.143	4.757	4.534	4.387	4.284	4.207	4.147	4.099	4.060	4.027	4.000
7	5.591	4.737	4.347	4.120	3.972	3.866	3.787	3.726	3.677	3.637	3.603	3.575
8	5.318	4.459	4.066	3.838	3.687	3.581	3.500	3.438	3.388	3.347	3.313	3.284
9	5.117	4.256	3.863	3.633	3.482	3.374	3.293	3.230	3.179	3.137	3.102	3.073
10	4.965	4.103	3.708	3.478	3.326	3.217	3.135	3.072	3.020	2.978	2.943	2.913
11	4.844	3.982	3.587	3.357	3.204	3.095	3.012	2.948	2.896	2.854	2.818	2.788
12	4.747	3.885	3.490	3.259	3.106	2.996	2.913	2.849	2.796	2.753	2.717	2.687
13	4.667	3.806	3.411	3.179	3.025	2.915	2.832	2.767	2.714	2.671	2.635	2.604
14	4.600	3.739	3.344	3.112	2.958	2.848	2.764	2.699	2.646	2.602	2.565	2.534
15	4.543	3.682	3.287	3.056	2.901	2.790	2.707	2.641	2.588	2.544	2.507	2.475
16	4.494	3.634	3.239	3.007	2.852	2.741	2.657	2.591	2.538	2.494	2.456	2.425
17	4.451	3.592	3.197	2.965	2.810	2.699	2.614	2.548	2.494	2.450	2.413	2.381
18	4.414	3.555	3.160	2.928	2.773	2.661	2.577	2.510	2.456	2.412	2.374	2.342
19	4.381	3.522	3.127	2.895	2.740	2.628	2.544	2.477	2.423	2.378	2.340	2.308
20	4.351	3.493	3.098	2.866	2.711	2.599	2.514	2.447	2.393	2.348	2.310	2.278
21	4.325	3.467	3.072	2.840	2.685	2.573	2.488	2.420	2.366	2.321	2.283	2.250
22	4.301	3.443	3.049	2.817	2.661	2.549	2.464	2.397	2.342	2.297	2.259	2.226
23	4.279	3.422	3.028	2.796	2.640	2.528	2.442	2.375	2.320	2.275	2.236	2.204
24	4.260	3.403	3.009	2.776	2.621	2.508	2.423	2.355	2.300	2.255	2.216	2.183
25	4.242	3.385	2.991	2.759	2.603	2.490	2.405	2.337	2.282	2.236	2.198	2.165
26	4.225	3.369	2.975	2.743	2.587	2.474	2.388	2.321	2.265	2.220	2.181	2.148
27	4.210	3.354	2.960	2.728	2.572	2.459	2.373	2.305	2.250	2.204	2.166	2.132
28	4.196	3.340	2.947	2.714	2.558	2.445	2.359	2.291	2.236	2.190	2.151	2.118
29	4.183	3.328	2.934	2.701	2.545	2.432	2.346	2.278	2.223	2.177	2.138	2.104
30	4.171	3.316	2.922	2.690	2.534	2.421	2.334	2.266	2.211	2.165	2.126	2.092
40	4.085	3.232	2.839	2.606	2.449	2.336	2.249	2.180	2.124	2.077	2.038	2.003
60	4.001	3.150	2.758	2.525	2.368	2.254	2.167	2.097	2.040	1.993	1.952	1.917
120	3.920	3.072	2.680	2.447	2.290	2.175	2.087	2.016	1.959	1.910	1.869	1.834
∞	3.841	2.996	2.605	2.372	2.214	2.099	2.010	1.938	1.880	1.831	1.789	1.752

Table A.6 (Continued)

v_2	v_1										
	13	14	15	16	17	18	19	20	21	22	23	24
1	244.7	245.4	246.0	246.5	246.9	247.3	247.7	248.0	248.3	248.6	248.8	249.1
2	19.42	19.42	19.43	19.43	19.44	19.44	19.44	19.45	19.45	19.45	19.45	19.45
3	8.729	8.715	8.703	8.692	8.683	8.675	8.667	8.660	8.654	8.648	8.643	8.639
4	5.891	5.873	5.858	5.844	5.832	5.821	5.811	5.803	5.795	5.787	5.781	5.774
5	4.655	4.636	4.619	4.604	4.590	4.579	4.568	4.558	4.549	4.541	4.534	4.527
6	3.976	3.956	3.938	3.922	3.908	3.896	3.884	3.874	3.865	3.856	3.849	3.841
7	3.550	3.529	3.511	3.494	3.480	3.467	3.455	3.445	3.435	3.426	3.418	3.410
8	3.259	3.237	3.218	3.202	3.187	3.173	3.161	3.150	3.140	3.131	3.123	3.115
9	3.048	3.025	3.006	2.989	2.974	2.960	2.948	2.936	2.926	2.917	2.908	2.900
10	2.887	2.865	2.845	2.828	2.812	2.798	2.785	2.774	2.764	2.754	2.745	2.737
11	2.761	2.739	2.719	2.701	2.685	2.671	2.658	2.646	2.636	2.626	2.617	2.609
12	2.660	2.637	2.617	2.599	2.583	2.568	2.555	2.544	2.533	2.523	2.514	2.505
13	2.577	2.554	2.533	2.515	2.499	2.484	2.471	2.459	2.448	2.438	2.429	2.420
14	2.507	2.484	2.463	2.445	2.428	2.413	2.400	2.388	2.377	2.367	2.357	2.349
15	2.448	2.424	2.403	2.385	2.368	2.353	2.340	2.328	2.316	2.306	2.297	2.288
16	2.397	2.373	2.352	2.333	2.317	2.302	2.288	2.276	2.264	2.254	2.244	2.235
17	2.353	2.329	2.308	2.289	2.272	2.257	2.243	2.230	2.219	2.208	2.199	2.190
18	2.314	2.290	2.269	2.250	2.233	2.217	2.203	2.191	2.179	2.168	2.159	2.150
19	2.280	2.256	2.234	2.215	2.198	2.182	2.168	2.155	2.144	2.133	2.123	2.114
20	2.250	2.225	2.203	2.184	2.167	2.151	2.137	2.124	2.112	2.102	2.092	2.082
21	2.222	2.197	2.176	2.156	2.139	2.123	2.109	2.096	2.084	2.073	2.063	2.054
22	2.198	2.173	2.151	2.131	2.114	2.098	2.084	2.071	2.059	2.048	2.038	2.028
23	2.175	2.150	2.128	2.109	2.091	2.075	2.061	2.048	2.036	2.025	2.014	2.005
24	2.155	2.130	2.108	2.088	2.070	2.054	2.040	2.027	2.015	2.003	1.993	1.984
25	2.136	2.111	2.089	2.069	2.051	2.035	2.021	2.007	1.995	1.984	1.974	1.964
26	2.119	2.094	2.072	2.052	2.034	2.018	2.003	1.990	1.978	1.966	1.956	1.946
27	2.103	2.078	2.056	2.036	2.018	2.002	1.987	1.974	1.961	1.950	1.940	1.930
28	2.089	2.064	2.041	2.021	2.003	1.987	1.972	1.959	1.946	1.935	1.924	1.915
29	2.075	2.050	2.027	2.007	1.989	1.973	1.958	1.945	1.932	1.921	1.910	1.901
30	2.063	2.037	2.015	1.995	1.976	1.960	1.945	1.932	1.919	1.908	1.897	1.887
40	1.974	1.948	1.924	1.904	1.885	1.868	1.853	1.839	1.826	1.814	1.803	1.793
60	1.887	1.860	1.836	1.815	1.796	1.778	1.763	1.748	1.735	1.722	1.711	1.700
120	1.803	1.775	1.750	1.728	1.709	1.690	1.674	1.659	1.645	1.632	1.620	1.608
∞	1.720	1.692	1.666	1.644	1.623	1.604	1.587	1.571	1.556	1.542	1.529	1.517

v_2	v_1									
	25	26	27	28	29	30	40	60	120	∞
1	249.3	249.5	249.6	249.8	250.0	250.1	251.1	252.2	253.3	254.3
2	19.46	19.46	19.46	19.46	19.46	19.46	19.47	19.48	19.49	19.50
3	8.634	8.630	8.626	8.623	8.620	8.617	8.594	8.572	8.549	8.526
4	5.769	5.763	5.759	5.754	5.750	5.746	5.717	5.688	5.658	5.628
5	4.521	4.515	4.510	4.505	4.500	4.496	4.464	4.431	4.398	4.365
6	3.835	3.829	3.823	3.818	3.813	3.808	3.774	3.740	3.705	3.669
7	3.404	3.397	3.391	3.386	3.381	3.376	3.340	3.304	3.267	3.230

(Continued)

Table A.6 (Continued)

v_2	v_1									
	25	26	27	28	29	30	40	60	120	∞
8	3.108	3.102	3.095	3.090	3.084	3.079	3.043	3.005	2.967	2.928
9	2.893	2.886	2.880	2.874	2.869	2.864	2.826	2.787	2.748	2.707
10	2.730	2.723	2.716	2.710	2.705	2.700	2.661	2.621	2.580	2.538
11	2.601	2.594	2.588	2.582	2.576	2.570	2.531	2.490	2.448	2.404
12	2.498	2.491	2.484	2.478	2.472	2.466	2.426	2.384	2.341	2.296
13	2.412	2.405	2.398	2.392	2.386	2.380	2.339	2.297	2.252	2.206
14	2.341	2.333	2.326	2.320	2.314	2.308	2.266	2.223	2.178	2.131
15	2.280	2.272	2.265	2.259	2.253	2.247	2.204	2.160	2.114	2.066
16	2.227	2.220	2.212	2.206	2.200	2.194	2.151	2.106	2.059	2.010
17	2.181	2.174	2.167	2.160	2.154	2.148	2.104	2.058	2.011	1.960
18	2.141	2.134	2.126	2.119	2.113	2.107	2.063	2.017	1.968	1.917
19	2.106	2.098	2.090	2.084	2.077	2.071	2.026	1.980	1.930	1.878
20	2.074	2.066	2.059	2.052	2.045	2.039	1.994	1.946	1.896	1.843
21	2.045	2.037	2.030	2.023	2.016	2.010	1.965	1.916	1.866	1.812
22	2.020	2.012	2.004	1.997	1.990	1.984	1.938	1.889	1.838	1.783
23	1.996	1.988	1.981	1.973	1.967	1.961	1.914	1.865	1.813	1.757
24	1.975	1.967	1.959	1.952	1.945	1.939	1.892	1.842	1.790	1.733
25	1.955	1.947	1.939	1.932	1.926	1.919	1.872	1.822	1.768	1.711
26	1.938	1.929	1.921	1.914	1.907	1.901	1.853	1.803	1.749	1.691
27	1.921	1.913	1.905	1.898	1.891	1.884	1.836	1.785	1.731	1.672
28	1.906	1.897	1.889	1.882	1.875	1.869	1.820	1.769	1.714	1.654
29	1.891	1.883	1.875	1.868	1.861	1.854	1.806	1.754	1.698	1.638
30	1.878	1.870	1.862	1.854	1.847	1.841	1.792	1.740	1.683	1.622
40	1.783	1.775	1.766	1.759	1.751	1.744	1.693	1.637	1.577	1.509
60	1.690	1.681	1.672	1.664	1.656	1.649	1.594	1.534	1.467	1.389
120	1.598	1.588	1.579	1.570	1.562	1.554	1.495	1.429	1.352	1.254
∞	1.506	1.496	1.486	1.476	1.467	1.459	1.394	1.318	1.221	1.000

Refer to Eq. (6.100). v_1: degrees of freedom in numerator; v_2: degrees of freedom in denominator.

Appendix | 435

Table A.7 Quantiles $f_{0.99}(\nu_1, \nu_2)$ for the F distribution.

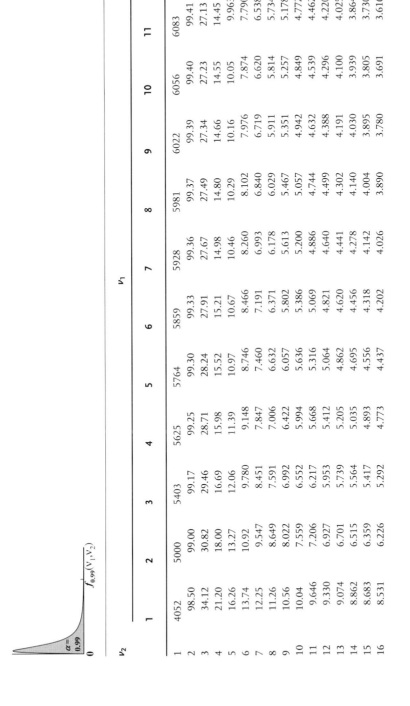

ν_2 \ ν_1	1	2	3	4	5	6	7	8	9	10	11	12
1	4052	5000	5403	5625	5764	5859	5928	5981	6022	6056	6083	6106
2	98.50	99.00	99.17	99.25	99.30	99.33	99.36	99.37	99.39	99.40	99.41	99.42
3	34.12	30.82	29.46	28.71	28.24	27.91	27.67	27.49	27.34	27.23	27.13	27.05
4	21.20	18.00	16.69	15.98	15.52	15.21	14.98	14.80	14.66	14.55	14.45	14.37
5	16.26	13.27	12.06	11.39	10.97	10.67	10.46	10.29	10.16	10.05	9.963	9.888
6	13.74	10.92	9.780	9.148	8.746	8.466	8.260	8.102	7.976	7.874	7.790	7.718
7	12.25	9.547	8.451	7.847	7.460	7.191	6.993	6.840	6.719	6.620	6.538	6.469
8	11.26	8.649	7.591	7.006	6.632	6.371	6.178	6.029	5.911	5.814	5.734	5.667
9	10.56	8.022	6.992	6.422	6.057	5.802	5.613	5.467	5.351	5.257	5.178	5.111
10	10.04	7.559	6.552	5.994	5.636	5.386	5.200	5.057	4.942	4.849	4.772	4.706
11	9.646	7.206	6.217	5.668	5.316	5.069	4.886	4.744	4.632	4.539	4.462	4.397
12	9.330	6.927	5.953	5.412	5.064	4.821	4.640	4.499	4.388	4.296	4.220	4.155
13	9.074	6.701	5.739	5.205	4.862	4.620	4.441	4.302	4.191	4.100	4.025	3.960
14	8.862	6.515	5.564	5.035	4.695	4.456	4.278	4.140	4.030	3.939	3.864	3.800
15	8.683	6.359	5.417	4.893	4.556	4.318	4.142	4.004	3.895	3.805	3.730	3.666
16	8.531	6.226	5.292	4.773	4.437	4.202	4.026	3.890	3.780	3.691	3.616	3.553

(Continued)

Table A.7 (Continued)

v_2	v_1											
	1	2	3	4	5	6	7	8	9	10	11	12
17	8.400	6.112	5.185	4.669	4.336	4.102	3.927	3.791	3.682	3.593	3.519	3.455
18	8.285	6.013	5.092	4.579	4.248	4.015	3.841	3.705	3.597	3.508	3.434	3.371
19	8.185	5.926	5.010	4.500	4.171	3.939	3.765	3.631	3.523	3.434	3.360	3.297
20	8.096	5.849	4.938	4.431	4.103	3.871	3.699	3.564	3.457	3.368	3.294	3.231
21	8.017	5.780	4.874	4.369	4.042	3.812	3.640	3.506	3.398	3.310	3.236	3.173
22	7.945	5.719	4.817	4.313	3.988	3.758	3.587	3.453	3.346	3.258	3.184	3.121
23	7.881	5.664	4.765	4.264	3.939	3.710	3.539	3.406	3.299	3.211	3.137	3.074
24	7.823	5.614	4.718	4.218	3.895	3.667	3.496	3.363	3.256	3.168	3.094	3.032
25	7.770	5.568	4.675	4.177	3.855	3.627	3.457	3.324	3.217	3.129	3.056	2.993
26	7.721	5.526	4.637	4.140	3.818	3.591	3.421	3.288	3.182	3.094	3.021	2.958
27	7.677	5.488	4.601	4.106	3.785	3.558	3.388	3.256	3.149	3.062	2.988	2.926
28	7.636	5.453	4.568	4.074	3.754	3.528	3.358	3.226	3.120	3.032	2.959	2.896
29	7.598	5.420	4.538	4.045	3.725	3.499	3.330	3.198	3.092	3.005	2.931	2.868
30	7.562	5.390	4.510	4.018	3.699	3.473	3.304	3.173	3.067	2.979	2.906	2.843
40	7.314	5.179	4.313	3.828	3.514	3.291	3.124	2.993	2.888	2.801	2.727	2.665
60	7.077	4.977	4.126	3.649	3.339	3.119	2.953	2.823	2.718	2.632	2.559	2.496
120	6.851	4.787	3.949	3.480	3.174	2.956	2.792	2.663	2.559	2.472	2.399	2.336
∞	6.635	4.605	3.782	3.324	3.018	2.802	2.639	2.511	2.407	2.321	2.248	2.185

Table A.7 (Continued)

v_2	v_1											
	13	14	15	16	17	18	19	20	21	22	23	24
1	6126	6143	6157	6170	6181	6192	6201	6209	6216	6223	6229	6235
2	99.42	99.43	99.43	99.44	99.44	99.44	99.45	99.45	99.45	99.45	99.46	99.46
3	26.98	26.92	26.87	26.83	26.79	26.75	26.72	26.69	26.66	26.64	26.62	26.60
4	14.31	14.25	14.20	14.15	14.12	14.08	14.05	14.02	13.99	13.97	13.95	13.93
5	9.825	9.770	9.722	9.680	9.643	9.610	9.580	9.553	9.528	9.506	9.485	9.466
6	7.657	7.605	7.559	7.519	7.483	7.451	7.422	7.396	7.372	7.351	7.331	7.313
7	6.410	6.359	6.314	6.275	6.240	6.209	6.181	6.155	6.132	6.111	6.092	6.074
8	5.609	5.559	5.515	5.477	5.442	5.412	5.384	5.359	5.336	5.316	5.297	5.279
9	5.055	5.005	4.962	4.924	4.890	4.860	4.833	4.808	4.786	4.765	4.746	4.729
10	4.650	4.601	4.558	4.520	4.487	4.457	4.430	4.405	4.383	4.363	4.344	4.327
11	4.342	4.293	4.251	4.213	4.180	4.150	4.123	4.099	4.077	4.057	4.038	4.021
12	4.100	4.052	4.010	3.972	3.939	3.909	3.883	3.858	3.836	3.816	3.798	3.780
13	3.905	3.857	3.815	3.778	3.745	3.716	3.689	3.665	3.643	3.622	3.604	3.587
14	3.745	3.698	3.656	3.619	3.586	3.556	3.529	3.505	3.483	3.463	3.444	3.427
15	3.612	3.564	3.522	3.485	3.452	3.423	3.396	3.372	3.350	3.330	3.311	3.294
16	3.498	3.451	3.409	3.372	3.339	3.310	3.283	3.259	3.237	3.216	3.198	3.181
17	3.401	3.353	3.312	3.275	3.242	3.212	3.186	3.162	3.139	3.119	3.101	3.084
18	3.316	3.269	3.227	3.190	3.158	3.128	3.101	3.077	3.055	3.035	3.016	2.999
19	3.242	3.195	3.153	3.116	3.084	3.054	3.027	3.003	2.981	2.961	2.942	2.925
20	3.177	3.130	3.088	3.051	3.018	2.989	2.962	2.938	2.916	2.895	2.877	2.859
21	3.119	3.072	3.030	2.993	2.960	2.931	2.904	2.880	2.857	2.837	2.818	2.801
22	3.067	3.019	2.978	2.941	2.908	2.879	2.852	2.827	2.805	2.785	2.766	2.749
23	3.020	2.973	2.931	2.894	2.861	2.832	2.805	2.781	2.758	2.738	2.719	2.702
24	2.977	2.930	2.889	2.852	2.819	2.789	2.762	2.738	2.716	2.695	2.676	2.659

(Continued)

Table A.7 (Continued)

ν_2	ν_1											
	13	14	15	16	17	18	19	20	21	22	23	24
25	2.939	2.892	2.850	2.813	2.780	2.751	2.724	2.699	2.677	2.657	2.638	2.620
26	2.904	2.857	2.815	2.778	2.745	2.715	2.688	2.664	2.642	2.621	2.602	2.585
27	2.871	2.824	2.783	2.746	2.713	2.683	2.656	2.632	2.609	2.589	2.570	2.552
28	2.842	2.795	2.753	2.716	2.683	2.653	2.626	2.602	2.579	2.559	2.540	2.522
29	2.814	2.767	2.726	2.689	2.656	2.626	2.599	2.574	2.552	2.531	2.512	2.495
30	2.789	2.742	2.700	2.663	2.630	2.600	2.573	2.549	2.526	2.506	2.487	2.469
40	2.611	2.563	2.522	2.484	2.451	2.421	2.394	2.369	2.346	2.325	2.306	2.288
60	2.442	2.394	2.352	2.315	2.281	2.251	2.223	2.198	2.175	2.153	2.134	2.115
120	2.282	2.234	2.192	2.154	2.119	2.089	2.060	2.035	2.011	1.989	1.969	1.950
∞	2.130	2.082	2.039	2.000	1.965	1.934	1.905	1.878	1.854	1.831	1.810	1.791

ν_2	ν_1											
	25	26	27	28	29	30	40	60	120	∞		
1	6240	6245	6249	6253	6257	6261	6287	6313	6339	6366		
2	99.46	99.46	99.46	99.46	99.46	99.47	99.47	99.48	99.49	99.50		
3	26.58	26.56	26.55	26.53	26.52	26.50	26.41	26.32	26.22	26.12		
4	13.91	13.89	13.88	13.86	13.85	13.84	13.74	13.65	13.56	13.46		
5	9.449	9.433	9.418	9.404	9.391	9.379	9.291	9.202	9.112	9.020		
6	7.296	7.280	7.266	7.253	7.240	7.229	7.143	7.057	6.969	6.880		
7	6.058	6.043	6.029	6.016	6.003	5.992	5.908	5.824	5.737	5.650		
8	5.263	5.248	5.234	5.221	5.209	5.198	5.116	5.032	4.946	4.859		
9	4.713	4.698	4.685	4.672	4.660	4.649	4.567	4.483	4.398	4.311		
10	4.311	4.296	4.283	4.270	4.258	4.247	4.165	4.082	3.996	3.909		

Table A.7 (Continued)

v_2	v_1									
	25	26	27	28	29	30	40	60	120	∞
11	4.005	3.990	3.977	3.964	3.952	3.941	3.86	3.776	3.69	3.602
12	3.765	3.750	3.736	3.724	3.712	3.701	3.619	3.535	3.449	3.361
13	3.571	3.556	3.543	3.530	3.518	3.507	3.425	3.341	3.255	3.165
14	3.412	3.397	3.383	3.371	3.359	3.348	3.266	3.181	3.094	3.004
15	3.278	3.264	3.250	3.237	3.225	3.214	3.132	3.047	2.959	2.868
16	3.165	3.150	3.137	3.124	3.112	3.101	3.018	2.933	2.845	2.753
17	3.068	3.053	3.039	3.026	3.014	3.003	2.92	2.835	2.746	2.653
18	2.983	2.968	2.955	2.942	2.930	2.919	2.835	2.749	2.66	2.566
19	2.909	2.894	2.880	2.868	2.855	2.844	2.761	2.674	2.584	2.489
20	2.843	2.829	2.815	2.802	2.790	2.778	2.695	2.608	2.517	2.421
21	2.785	2.770	2.756	2.743	2.731	2.720	2.636	2.548	2.457	2.360
22	2.733	2.718	2.704	2.691	2.679	2.667	2.583	2.495	2.403	2.305
23	2.686	2.671	2.657	2.644	2.632	2.620	2.535	2.447	2.354	2.256
24	2.643	2.628	2.614	2.601	2.589	2.577	2.492	2.403	2.31	2.211
25	2.604	2.589	2.575	2.562	2.550	2.538	2.453	2.364	2.27	2.169
26	2.569	2.554	2.540	2.526	2.514	2.503	2.417	2.327	2.233	2.131
27	2.536	2.521	2.507	2.494	2.481	2.470	2.384	2.294	2.198	2.097
28	2.506	2.491	2.477	2.464	2.451	2.440	2.354	2.263	2.167	2.064
29	2.478	2.463	2.449	2.436	2.423	2.412	2.325	2.234	2.138	2.034
30	2.453	2.437	2.423	2.410	2.398	2.386	2.299	2.208	2.111	2.006
40	2.271	2.256	2.241	2.228	2.215	2.203	2.114	2.019	1.917	1.805
60	2.098	2.083	2.068	2.054	2.041	2.028	1.936	1.836	1.726	1.601
120	1.932	1.916	1.901	1.886	1.873	1.860	1.763	1.656	1.533	1.381
∞	1.773	1.755	1.739	1.724	1.710	1.696	1.592	1.473	1.325	1.000

Refer to Eq. (6.100). v_1: degrees of freedom in numerator; v_2: degrees of freedom in denominator.

References

Altshuler, B. and Pasternack, B. (1963) Statistical measures of the lower limit of detection of a radioactive counter. *Health Phys.*, **9**, 293–298.

Anderson, V.L. and McLean, R.A. (1974) *Design of Experiments: A Realistic Approach*, Marcel Dekker, Inc., New York, NY.

Atkinson, A.C. (1980) Tests of pseudo-random numbers. *Appl. Stat.*, **29** (2), 164–177.

Barnett, V. and Lewis, T. (1994) *Outliers in Statistical Data*, 3rd edn, John Wiley & Sons, Ltd, Chichester, UK.

Barry, D.A. (1996) *Statistics: A Bayesian Perspective*, Duxbury Press, New York, NY.

Bickel, P.J. and Doksum, K.A. (1977) *Mathematical Statistics: Basic Ideas and Selected Topics*, Holden-Day, Inc., San Francisco, CA.

Bishop, Y.M.M., Fienberg, S.E., and Holland, P.W. (1975) *Discrete Multivariate Analysis: Theory and Practice*, MIT Press, Cambridge, MA.

Boise, J.D., Land, C.E., Shore, R.E., Norman, J.E., and Tokunaga, M. (1979) Risk of breast cancer following low-dose radiation exposure. *Radiology* **131**, 589–597.

Box, G.E.P. and Tiao, G.C. (1973) *Bayesian Inference in Statistical Analysis*, Addison-Wesley, Reading, MA.

Box, G.E.P., Hunter, W.G., and Hunter, J.S. (1978) *Statistics for Experimenters: An Introduction to Design, Data Analysis, and Model Building*, John Wiley & Sons, Inc., New York, NY.

Carlin, B.P. and Louis, T.A. (1996) *Bayes and Empirical Bayes Methods for Data Analysis*, Chapman & Hall, New York, NY.

Carter, L.L. and Cashwell, E.D. (1975) Particle-Transport Simulation with the Monte Carlo Method. TID-26607, Technical Information Service, U.S. Department of Commerce, Springfield, VA.

Cember, H. (1996) *Introduction to Health Physics*, 3rd edn, McGraw-Hill, New York, NY.

Currie, L.A. (1968) Limits for qualitative detection and quantitative determination. *Anal. Chem.*, **40**, 586–593.

Currie, L.A. (1984) Lower Limit of Detection: Definition and Elaboration of a Proposed Position for Radiological Effluent and Environmental Measurements. NUREG/CR-4007, U.S. Nuclear Regulatory Commission, Washington, DC.

DOE (1986) DOE/EH-0027: *Department of Energy Standard for the Performance Testing of Personnel Dosimetry Systems*, U.S. Department of Energy, Washington, DC.

DOE (2001) ANSI/HPS N13.11-2001: *Personnel Dosimetry Performance – Criteria for Testing*, Health Physics Society, McLean, VA.

Draper, N.R. and Smith, H. (1998) *Applied Regression Analysis*, 3rd edn, John Wiley & Sons, Inc., New York, NY.

Edwards, A.W.F. (1972) *Likelihood*, Cambridge University Press, London, UK.

Garthwaite, P.H., Jolliffe, I.T., and Jones, B. (2002) *Statistical Inference*, Oxford University Press, Oxford, UK.

Hall, E.J. (1994) *Radiobiology for the Radiologist*, 4th edn, J.B. Lippincott, Philadelphia, PA.

Hogg, R.V. and Craig, A.T. (1978) *Introduction to Mathematical Statistics*, 4th edn, Macmillan Publishing Company, New York, NY.

Hogg, R.V. and Tanis, E.A. (1993) *Probability and Statistical Inference*, 4th edn, Macmillan Publishing Company, New York, NY.

HPS (1996) HPS N13.30-1996: *Performance Criteria for Radiobioassay: An American National Standard*, Health Physics Society, McLean, VA.

Johnson, Norman L., Kotz, Samuel, and Balakrishnan, N. (1994) *Continuous Univariate Distributions*, Vol. 1, 2nd Edition, John Wiley and Sons, Hoboken, N.J.

Johnson, Norman L., Kemp, Adrienne W., and Kotz, Samuel (2005) *Univariate Discrete Distributions*, John Wiley and Sons, Hoboken, N.J.

Kalos, M. and Whitlock, P. (1986) *Monte Carlo Methods*, Wiley–Interscience, New York, NY.

Kennedy, W.J., Jr. and Gentle, J.E. (1980) *Statistical Computing*, Marcel Dekker, Inc., New York, NY.

Knuth, D.E. (1980) *The Art of Computer Programming, vol. 2: Seminumerical Algorithms*, 2nd edn, Addison-Wesley, Reading, MA.

Martz, H.F. (2000) Chapter 2: An introduction to Bayes, hierarchical Bayes, and empirical Bayes statistical methods in health physics, in *Applications of Probability and Statistics in Health Physics: Health Physics Society 2000 Summer School* (ed. T.B. Borak), Medical Physics Publishing, Madison, WI, pp. 55–84.

Martz, H.F. and Waller, R.A. (1982) *Bayesian Reliability Analysis*, John Wiley & Sons, Inc., New York, NY.

Metropolis, N. and Ulam, S.M. (1949) The Monte Carlo method. *J. Am. Stat. Assoc.*, **44**, 335–341.

Miller, Rupert G., Jr. (1981) *Simultaneous Statistical Inference*, 2nd ed, Springer, New York, NY.

Miller, I. and Freund, J.E. (1965) *Probability and Statistics for Engineers*, Prentice Hall, Inc., Englewood Cliffs, NJ.

Neter, J., Kutner, M.H., Nachtsheim, C.J., Christopher, J., and Li, W. (2004) *Applied Linear Statistical Models*, 5th edn, McGraw-Hill, New York, NY.

Newman, M.E.J. and Barkema, G.T. (1999) *Monte Carlo Methods in Statistical Physics*, Oxford University Press, New York, NY.

Parzen, E. (1960) *Modern Probability Theory and Its Applications*, John Wiley & Sons, Inc., New York, NY.

Press, S.J. (1989) *Bayesian Statistics: Principles, Models, and Applications*, John Wiley & Sons, Inc., New York, NY.

Roberson, P.L. and Carlson, R.D. (1992) Determining the lower limit of detection for personnel dosimeter systems. *Health Phys.*, **62**, 2–9.

Rossi, H.H. and Zaider, M. (1996) *Microdosimetry and Its Applications*, Springer Verlag, New York, NY.

Ryan, T.P. (2009) *Modern Regression Methods*, John Wiley & Sons, Inc., New York, NY.

Satterthwaite, F.G. (1946) An approximate distribution of estimates of variance components. *Biometrics*, **2**, 110–112.

Scheaffer, R.L. and McClane, J.T. (1982) *Statistics for Engineers*, Duxbury Press, Boston, MA.

Sivia, D.S. (1996) *Data Analysis: A Bayesian Tutorial*, Oxford University Press, New York, NY.

Sonder, E. and Ahmed, A.B. (1991) Background Radiation Accumulation and Lower Limit of Detection in Thermoluminescent Beta-Gamma Dosimeters Used by the Centralized External Dosimetry System. ORNL/TM-11995, Oak Ridge National Laboratory, Oak Ridge, TN.

Strom, D.J. and MacLellan, J.A. (2001) Evaluation of eight decision rules for low-level radioactivity counting. *Health Phys.*, **81**, 27–34.

Taylor, L.D. (1974) *Probability and Mathematical Statistics*, Harper & Row, New York, NY.

Turner, J.E. (1995) Atoms, Radiation, and Radiation Protection, 2nd edn, John Wiley & Sons, Inc., New York, NY.

Turner, J.E. (2007) *Atoms, Radiation, and Radiation Protection*, 3rd edn, Wiley-VCH Verlag GmbH, Weinheim, Germany.

Turner, J.E., Wright, H.A., and Hamm, R.N. (1985) A Monte Carlo primer for health physicists. *Health Phys.*, **48**, 717–733.

Turner, J.E., Bogard, J.S., Hunt, J.B., and Rhea, T.A. (1988) *Problems and Solutions in Radiation Protection*, Pergamon Press, Elmsford, NY. Available from McGraw-Hill, Health Professions Division, New York, NY.

Ulam, S.M. (1983) *Adventures of a Mathematician*, Charles Scribner's Sons, New York, NY.

Walpole, R.E. and Myers, R.H. (1989) *Probability and Statistics for Engineers and Scientists*, 4th edn, Macmillan Publishing Company, New York, NY.

Index

a

"absorbed fraction" 327
activity 15ff, 26ff, 85, 157ff, 177, 195, 199, 205ff, 213ff, 227ff, 231ff, 246f, 251ff, 266ff, 377, 402, 419
– becquerel (Bq) 16
– curie (Ci) 16, 19
– sample 92, 216, 217, 228
– specific 18, 19, 27, 213, 214
– TRUE 206, 217, 232, 234
– mean 217, 236
alpha decay 9
– alpha particle 4, 9, 144ff, 251ff, 276
– spectrum, alpha particle 9
analysis of variance (ANOVA) table 367
annual limits on intake 327
ANSI/HPS N13.30 248
anthropomorphic phantom 248, 327
atomic theory, semiclassical 3ff
attenuation coefficient 214, 317ff
– linear 214, 317, 319ff, 324, 334
– mass 318, 320, 321, 333
attenuation processes 320
– Compton scattering 3, 5, 12, 276, 277, 320, 321, 323ff
– pair production 3, 320
– photoelectric absorption 3, 12, 321, 323ff
– photoelectric effect 5, 320, 322
– photonuclear reaction 320
Auger electrons 276
avalanche 279
average life 17, 60
average, weighted 306, 392, 413, 415
Avogadro's number 18

b

Bayes factor 406ff, 409, 413, 418, 420
Bayes' Theorem 43
Bayesian statistics 393

Becquerel 4, 16
Bernoulli 22, 92, 172, 186, 192, 306, 407
– process 22, 92
– trial 92, 94, 112, 114, 186
Bernoulli distribution 92, 93, 192
beta decay 9, 12
– antineutrino 9
– beta particle 9, 12, 17, 77, 291
– spectrum, beta particle 9
beta distribution 154
bias, systematic 249, 371
binding energy 274, 325
binomial 93
– approximation to hypergeometric 107
– cumulative distribution 94, 96, 407
– distribution 15, 16, 22ff, 29, 78, 85, 91, 93ff, 99, 100, 104, 106, 114ff, 135, 136, 139ff, 159, 164, 216, 224, 407, 408
– normal approximation 135, 136
– Poisson approximation 100, 141
binomial distribution 15ff, 22ff
binomial series 24
blank, appropriate 248, 249, 253, 255
Bohr 4, 6, 11
Bragg-Gray chamber 300
bremsstrahlung 276, 325

c

calibration 164, 216ff, 231, 238, 249ff, 253, 272, 380, 418
calibration problem 380
Cauchy distribution 89, 123, 150, 157, 161, 192
cell killing 341ff
cell survival 341ff
– aberrations, exchange-type 347ff
– clonogenic 342
– curve 341ff
– probability 341ff

central limit theorem   124, 132, 160, 162, 172, 175, 181, 210, 216$f\!f$
Chadwick   8
characteristic function   192
characteristic X-ray   9, 276
Chauvenet's criterion   263$f\!f$
Chebyshev's inequality   76$f\!f$, 87$f\!f$, 96, 128, 158
chi-square distribution   142, 145$f\!f$, 160, 164, 168, 183, 184, 412
 – additivity, property of   148, 149
 – degrees of freedom   146$f\!f$
 – normal approximation   147
 – quantiles   146
 – relation to gamma distribution   146
 – relation to standard normal distribution   148
chi-square testing   145, 271, 281
chord length   300$f\!f$
 – Cauchy theorem   311
 – distribution, isotropic   300$f\!f$
classical laws   1$f\!f$
 – of electromagnetism   2
 – of gravitation   2
 – of mechanics   2
 – of motion   1$f\!f$
 – Newtonian mechanics   2
 – of physics   2
 – of thermodynamics   2
 – "ultraviolet catastrophe"   2
classical statistics   387
coefficient of variation (CV)   164
Compton   3, 5, 12, 276$f\!f$, 320$f\!f$, 327
 – distribution   276
 – edge   276
 – scattering   3, 5, 12, 276$f\!f$, 320$f\!f$
Compton scattering   3, 5, 12, 276, 277, 320, 321, 323$f\!f$, 327
 – attenuation processes   320
 – Klein-Nishina formula   324
conditional probability   21, 38$f\!f$, 44, 69$f\!f$, 87, 145, 319, 333, 387
 – Bayes' Theorem   43
 – definition   39
confidence interval   169
 – for difference in means   176$f\!f$, 196
 – for difference in proportions   181
 – for means   168
 – for Poisson rate   175
 – probable error   170, 195
 – for proportions   172
 – for ratio of variances   184
 – standard error   169
 – for variance   183

conjugate prior distribution   396$f\!f$, 399$f\!f$, 414, 419, 423
continuity correction factor   137$f\!f$, 141, 159
correlation   71, 353, 369, 381$f\!f$
 – coefficient   381
 – coeffcient, sample   381
correlation, coefficient of   71$f\!f$, 381$f\!f$
count   106, 144, 169, 176, 195, 199, 202$f\!f$, 210, 214$f\!f$, 231$f\!f$, 248$f\!f$, 266$f\!f$, 271, 281, 284$f\!f$, 317, 322, 402, 416
 – background   215$f\!f$
 – gross   204, 215, 216, 218, 219, 222, 226, 228, 232, 239, 267
counter   102, 144, 228$f\!f$, 231, 267, 271$f\!f$, 300, 314, 331
 – gas proportional   272, 274$f\!f$, 279
 – operation, chi-square testing of   281
 – scintillation   271
counter, gas proportional   272, 274$f\!f$, 279, 293, 300
 – dosimetry, applications in   300
 – microdosimetry, applications in   300
 – Rossi   300
 – spherical   300
 – tissue-equivalent   300
covariance   71$f\!f$, 87, 90, 200, 359, 365, 381
credible interval   404
critical value, L_C 231$f\!f$, 236, 243, 263, 268
cross section, macroscopic   317, 327, 329
 – mean free path, inverse   319, 327
 – neutron   327
cumulative distribution function   53$f\!f$, 57$f\!f$, 66, 84$f\!f$, 88, 121$f\!f$, 318, 404
cumulative normal distribution   125, 126, 128, 140, 147, 148

d

Davisson-Germer   5
de Broglie   5$f\!f$
 – momentum   7
 – wavelength   8
de Moivre, Abraham   132
dead time   284$f\!f$
dead time correction   271, 284$f\!f$
decay   1, 4, 7$f\!f$
 – constant   15, 17, 21, 23, 26$f\!f$, 60, 61, 143, 144, 161, 176, 205, 21, 223, 315, 319, 387
 – disintegration   12, 15$f\!f$, 46, 106, 198, 224, 253
 – exponential   15$f\!f$, 26, 315, 372
 – radioactive   1, 4, 7$f\!f$, 15$f\!f$, 22, 26, 29, 61, 99, 102, 140, 143$f\!f$, 214, 216, 223, 224, 231, 315, 322
 – rate   16, 17, 102

decay probability   16, 23, 225
decay time sampling   315
decision level, L_C 233, 241, 248ff, 270
delta theorem   210
derived air concentrations   327
derived quantity   199, 201, 202
detector   228, 232, 234, 236, 237, 250, 271ff, 275ff, 284, 285, 287, 290, 317, 322, 376
– "energy proportional"   272
– Fano Factor   273, 274, 277, 290
– linear response   278
– nonparalyzable   284ff, 290ff
– observed resolution of   274
– paralyzable   284ff, 291, 292
– resolution of   271ff, 290, 291
– response function of   272ff, 279
– scintillation   276ff, 280, 290, 291, 321
– semiconductor   274, 276, 280, 281, 290
– sodium iodide crystal scintillator   276, 277, 280, 281, 291, 320
discrete uniform distribution   91, 92, 167, 398
distribution   15ff, 51ff, 65ff
– Bernoulli   92, 93, 192
– beta   154ff, 161, 391, 392, 394, 395, 407, 420
– binomial   15, 16, 22ff, 29, 78, 85, 91, 93ff, 99, 100, 104, 106, 114ff, 135, 136, 139ff, 159, 164, 216, 224, 407, 408
– Cauchy   89, 123, 150, 157, 161, 192
– chi-square   142, 145ff, 160, 164, 168, 183, 184, 412
– conditional   70
– cumulative   53ff, 57ff, 66, 84ff, 88, 121ff, 318, 404
– discrete uniform   91
– exponential   142ff
– F   151ff
– gamma   142ff, 155, 159, 396, 397, 399, 410, 411
– Gaussian   124, 132
– geometric   110
– hypergeometric   106ff
– independence   66
– joint   65ff
– lognormal   153ff
– multivariate hypergeometric   109
– negative binomial   112
– normal   3, 15, 26, 89, 102, 124ff, 133, 135, 136, 139, 140, 147ff, 154, 157, 161, 164, 168, 169, 171, 173, 177, 189, 190, 215, 216, 221, 232, 242, 247, 248, 251, 258, 260ff, 269, 272, 290, 359, 361, 381, 414, 418
– Poisson   15, 26, 98ff, 114ff, 141, 143, 144, 160, 164, 166, 167, 176, 187, 191, 193ff, 204, 215, 216, 223ff, 231, 248, 251ff, 269, 273, 282, 283, 289, 290, 396, 410, 411, 416, 417, 421
– posterior   84, 390ff, 403ff, 422
– prior   84, 387, 390ff, 397ff, 406ff
– standard normal   124ff, 128, 131, 148ff, 157, 169, 173, 177, 201, 218, 232, 237, 243, 245, 247, 248, 263, 332
– Student's t   89, 123, 149ff, 161, 170, 171, 179ff, 183, 195, 247, 248, 257, 258, 264ff, 269, 364, 366, 368, 377
– uniform   91ff, 119ff, 156, 160, 167, 194, 295, 301, 308, 324, 328, 331, 335, 391, 392, 395ff, 402, 407, 408, 419, 420
– of values   15
distribution, probability   8, 51ff, 91, 92, 94, 109, 111ff, 119, 121, 122, 124, 137, 148, 165, 186, 283, 312, 313, 387, 388, 390, 394
– discrete   52, 67, 72, 91ff, 167, 398, 404
– cumulative   54, 55ff, 61, 62, 65, 66, 84, 85, 88, 94, 96, 97, 100, 102, 119ff, 136, 137, 148, 252, 313ff, 324, 331, 332, 334, 407
DOE Laboratory Accreditation Program (DOELAP)   255
dose   1, 242, 245, 255ff, 264, 266, 269, 270, 275, 293, 294, 317, 322ff, 334, 337ff, 378, 385
– absorbed, per unit fluence   322, 323, 325, 326, 334, 343, 345, 349
– committed   327
– LD_{50} 343, 344, 348ff
– mean lethal   343, 344, 348
– minimum   337
– model   327
– threshold   337
dose-response   1, 260, 337ff
– curve   338ff
– function   340ff
– relationship   338ff
dosimeter, thermoluminescence (TLD)   114, 159, 242, 255ff
dosimetry   122, 255, 260, 261, 269, 293, 300, 327, 328
– internal   327
– using gas-proportional counter   300

e

effects, biological   337
– deterministic   337
– and exposure, radiation   337
– genetic   337
– germ cells   337
– radiation induced   338
– severity and dose   337

– somatic cells   337
– stochastic   337
efficiency, counter   106, 210, 213, 214, 217, 219, 222, 225ff, 232, 234, 236, 237, 250, 277, 278, 291
Einstein   2, 5, 9, 11
– "God does not play dice."   11
– special theory of relativity   2
electron-hole pair   271, 276, 280
energy resolution   271ff, 276, 280, 281, 290, 291
error   163
– estimated   173, 174
– mean square, MSE 359ff, 366ff, 371, 373, 376ff
– random   163
– standard   168
– sum of squares for, SSE 359ff, 367, 368, 371, 374ff, 384
– systematic   163, 164, 193, 199, 249, 271
error in an estimation   174
error propagation   199
– analysis   279
– in confidence interval of mean   201
– in derived quantity   201
– in mean   201
– in standard error   201
error propagation formulas   202
– exponentials   203
– products and powers   202
– sums and differences   202
– variance of mean   203
error, systematic   164, 193, 249
estimate, interval   168, 169, 407
estimate, point   165, 168, 170, 177, 186, 403, 407
estimated error   173
estimation   145, 163ff, 231, 253, 354ff, 358ff, 365, 376, 387
– least squares method   354
estimator   165
– consistent   166
– efficient   166
– generalized maximum likelihood   403
– maximum likelihood   186ff, 197, 361, 362, 390, 403, 407, 422
– minimum variance unbiased   166
– pooled, for variance   178
– standard error   173
– unbiased   165
event   33
– complement   34
– exhaustive   41ff
– independent   21ff, 38ff, 322

– intersection of   34
– mutually exclusive   34ff, 41ff, 46, 47
– probability   36
– simple   33
– union of   33
expected value   27, 59ff, 63, 69, 72, 74, 85, 89, 92, 95, 99, 105, 165, 166, 189, 190, 192, 198, 200, 209, 224, 227, 229, 232, 267, 295ff, 314, 330, 331, 362, 365ff, 371, 373, 396, 397, 419, 420
experiment   29
exponential distribution   85, 142ff, 159, 160, 197, 198, 286, 289, 315ff, 320, 398, 419
– arrival time for first Poisson event   143
– memory   99, 145, 319, 320
– relation to gamma distribution   142
– relation to radioactive decay   142
extrapolation   364

f

F distribution   151
– degrees of freedom   151
– quantiles   151
– relation between upper and lower quantiles   152
false negative (type II) error   235
false positive (type I) error   233
Fano factor   273, 274, 277, 290
fiducial limits   378, 379
film, radiosensitive   255
finite population correction factor   107, 109, 116, 134
Fisher, R.A.   381
frequency, relative   302ff, 313, 387
frequentist   387ff, 393, 403ff, 407, 412
full width at half maximum (FWHM)   272ff, 290

g

gamma distribution   142ff, 155, 159, 396, 397, 399, 410, 411
gamma ray   9, 11, 12, 276ff, 291, 319ff, 327, 328, 341
gas multiplication   272, 279
Gauss, Carl Friedrich   132
Gauss-Markov theorem   359
generalized maximum likelihood estimator   403
geometric distribution   110
GM tube, self-quenching   290
goodness-of-fit   119, 353
Gosset, W.S.   151
gram atomic weight   18ff

h

half-life   12, 15*ff*, 19*ff*, 25, 27, 28, 62, 106, 140, 206, 207, 211, 213, 224, 225, 227, 333, 344
Heisenberg   5, 6
– quantum mechanics   5
– uncertainty principle   5*ff*, 12
highest posterior density (HPD)   404, 407
high-purity germanium (HPGe)   228, 276, 280, 281, 290
hypergeometric distribution   106
hypothesis   240
– alternative   240, 241, 243, 245*ff*, 270, 368, 408, 409, 414
– composite vs. composite   240
– null   240*ff*, 257, 258, 270, 367, 405
– simple vs. composite   240
– simple vs. simple   240
hypothesis testing   240

i

importance sampling   309
increment, independent   320
Independence Theorem   44
independent event   44
independent variable   304, 306, 308, 353, 368, 369, 373, 376, 382, 383, 385
inference, Bayesian   403
inference, statistical   84, 231, 366, 387
integral calculus, fundamental theorem   56
interpolation   364
ion pair   274*ff*, 281, 290
– average energy, W, to produce   275
ionization   270*ff*, 274*ff*, 279
– device   271
– potential   274

j

joint probability function   65*ff*, 186

k

Klein-Nishina formula   324

l

lack of fit   369, 371, 373*ff*, 383
laws of quantum physics   2
– definite   2
– statistical   2
L_C 231*ff*, 236, 243, 263, 268
– critical value   231
– decision level   248
L_D 237, 238, 250*ff*, 258*ff*, 266, 269, 270
– minimum detectable true net count number   237
– lower limit of detection, LLD   237
– minimum detectable amount, MDA   237, 248, 250*ff*, 269
LD_{50} 343, 344, 348*ff*
leakage, radiation   322
LET spectrum   300
lifetime, average   17, 60
likelihood function   186*ff*, 197, 262, 361, 388*ff*, 392, 394, 403, 410, 414, 419, 420, 422
linear energy transfer (LET)   294, 300, 327, 341, 348, 351
– distribution   294
– quality factor   327
– spectrum   300
lognormal distribution   153
lower limit of detection, LLD   237

m

marginal   66*ff*, 72, 84, 86, 87, 89, 391, 394
– density   66, 67, 69, 70, 84, 86, 87, 89
– distribution   67, 68, 72, 86, 391, 394
maximum likelihood estimator (MLE)   186*ff*, 197, 361, 362, 390, 403, 407, 422
Maxwell   4
mean   59*ff*
mean square error, MSE 359, 364, 392
mean, correction for   355
median   62*ff*, 86, 89, 161, 165, 292, 332, 334, 403, 404, 407*ff*, 416, 417
Medical Internal Radiation Dose (MIRD) Committee   327
memory   99, 145, 319, 320
microscope, scanning tunneling   10
Millikan   4
minimum detectable amount, MDA   237, 248, 250*ff*, 269
minimum detectable true activity, A_{II} 235*ff*, 241, 242, 258, 267
minimum detectable true net count number, L_D 237
minimum significant measured activity, A_I 231*ff*, 238*ff*, 266, 267
minimum variance unbiased estimator (MVUE)   166*ff*, 188, 189, 194
mode   403, 404, 407*ff*, 416, 417, 420
modeling, biological   338
– cross section   342
– extrapolation number   347, 349, 350
– for stochastic effects   338
– hits per target, average number of   342
– linear quadratic   347
– multi-target, multi-hit   347
– multi-target, single-hit   345, 346
– single-target, single hit   342
moment   189

- generating function about point b 192
- generating function for a sum of random variables   193
- generating function   191
- j^{th} 190
- j^{th}, of the sample   190
- method of   189
momentum   4ff, 11, 77, 335
Monte Carlo method   122, 293ff, 340
- to determine absorbed dose   294, 323, 325, 326, 334
- to determine dose equivalent   293
- to determine dose   293
- to determine LET distribution   294
- to determine shielding properties   293
- in dosimetry   293
- in energy losses   293
- in flight distances   293
- in neutron transport   293
- in photon transport   317, 319ff
- in radiation penetration   293
- in radiation physics   293
- Russian roulette   329
- in scattering angles   293
- splitting   329
multiplicative rule   44
multivariate hypergeometric distribution   109

n
negative binomial distribution   112, 113, 115, 117, 396, 419
net dosimeter reading   256
Newton   1
Neyman-Pearson Lemma   262
noninformative prior distribution   397ff, 410, 414
- Jeffreys   410ff
- location parameter   398
- reference   397
- scale parameter   398
- vague   397
nonparalyzable detector   284ff, 290ff
normal equations   355, 369, 371
normalize   61

o
outlier   263
overdispersion   253

p
pair production   3, 320
paralyzable detector   284, 285, 287
parameter   164
partitioning of counting times, optimum   222

permutation   23ff
photoelectric absorption   3, 12, 321, 323ff
photopeak   276
pivotal quantity   171
Planck   2
- constant   4, 6, 8, 9, 11
- quantum of action   2
Poisson   98
- cumulative distribution   100
- distribution   98
- process   99
precision   163ff
- double   295, 299
- limiting   221
- numerical, of the computer   299
- as reciprocal of variance   415
- roundoff, function of   299
predicting   364ff
- future observations   365
- mean of k future observations   365
- mean value   365
prior distribution   84, 387, 390ff, 397ff, 406ff
- conjugate   396ff, 399ff, 414, 419, 423
- data-based   401
- elicited   401
- empirical gamma   411
- g-priors   401
- hyperparameters   395, 396, 402, 414, 415, 420
- noninformative   397ff, 410, 414
- population   394
- principle of insufficient reason   394
- state of knowledge   394
probability   2, 36ff
- axioms of   37
- conditional   38ff
probability density function   55ff
- posterior   84ff
- prior   84ff
protection, radiation   338
pulse height   271ff, 276, 277, 300
pure error   369, 371, 373ff, 383
- mean square   374
- sum of squares   374
pure error mean square   374

q
quantum mechanics   5ff, 52, 53, 56

r
radiation, isotropic   89, 117, 122, 157, 300
radioactive decay   1, 4, 7ff, 15ff, 22, 26, 29, 61, 99, 102, 140, 143ff, 214, 216, 223, 224, 231, 315, 322

radiobioassay 248*ff*, 255, 269
radionuclides, short-lived 223
random number 293*ff*
– generator 92, 122, 294, 295, 302, 308, 316, 330
– seed 294
– sequence 295
randomness, test for 295
rate, count 106, 144, 204, 205, 210, 215*ff*, 226*ff*, 232, 236, 250, 267, 271, 284*ff*, 317, 318, 322, 402
– background 216, 217, 227
– gross 204, 205, 216, 218, 219, 222, 226, 228
– net 215, 217, 220*ff*, 227, 228, 232, 233, 236, 249, 267
rate, event 285
ratio of variances estimator 184
reconfirmation 373
region 240
– acceptance 240
– critical 240
– rejection 240
regression 259, 260, 269, 353*ff*
– inverse 353, 378*ff*
– linear 353*ff*
– through origin 376
regression analysis 353*ff*
relation between gamma and beta distributions 154
relative error 106
relativity, special theory of 2
repeat observations 373
residual 358
residual, minimum sum of squared 358
response variable 353*ff*
risk 327*ff*
– and acceptable radiation limits 340
– assessment 327
– estimation 338
Roentgen 3
Rule of Elimination 41
Rutherford 4

S
sample size estimation 174
sample space 29*ff*, 51, 82
– continuous 32, 33, 37
– discrete 32, 33, 36, 37, 51
– element, individual 29
– outcome 29
sampling 107
– decay times, from exponential distribution 315
– from known distribution 313

– importance 309
– stratified 308
– with replacement 107
– without replacement 107
sampling distribution 132*ff*, 149, 158, 159, 166, 168, 169, 171, 173, 194, 393*ff*, 398*ff*, 402, 403, 417
Satterthwaite's approximation 180
Schrödinger 5
scintillation 271
– counter 271
– photon 272
scintillation detector 276*ff*, 280, 290, 291, 321
– lanthanum bromide, cerium activated [LaBr(Ce)] 280
– sodium iodide, thallium-doped [NaI(Tl)] 228, 276, 277, 280, 281, 291, 320
scram 410
semiclassical physics 4
semiconductor detector 274, 276, 280, 281, 290
– cadmium zinc telluride [CdZnTe] 280, 281
– high-purity germanium (HPGe) 228, 276, 280, 281, 290
set 33
– empty 33
– null 33
smoothing techniques 306
specific activity 18, 19, 27, 213, 214
spectrometer, alpha particle 251*ff*
standard deviation 63*ff*, 95*ff*, 106, 124*ff*, 132*ff*, 144, 154, 156*ff*, 164, 168*ff*, 173, 193*ff*, 202, 210*ff*, 216*ff*, 222*ff*, 232*ff*, 242*ff*, 249, 264*ff*, 273, 291, 307, 315*ff*, 330, 402, 420, 422
– net count rate 217
standard error 168
standard error of the mean 133, 195
standard normal distribution 124*ff*, 128, 131, 148*ff*, 157, 169, 173, 177, 201, 218, 232, 237, 243, 245, 247, 248, 263, 332
statistic 165
– sufficient 415
– test 240
statistical inference, methods 387
– Bayesian 387*ff*, 398, 403*ff*, 410
– classical 387
– frequentist 387*ff*, 393, 403*ff*, 407, 412
stratified sampling 308
strong force, short range 327
Student's t-distribution 149
– degrees of freedom 149
– quantiles 150
– relation to Cauchy distribution 150

sum of cross products   355
– corrected   355
– uncorrected   355
sum of squares   355, 359, 361, 367, 371, 374, 383, 384
– corrected   355
– due to regression, SSR 367
– for error, SSE 359
– lack of fit   374, 383
– pure error   374, 383
– uncorrected   355
sum of squares due to regression, SSR 367
sum of squares for error, SSE 359
sum of squares lack of fit   374
sum of squares pure error   374
sum of squares, corrected   355
sum of squares, uncorrected   355
support   80
survival probability   16, 20, 21, 23, 26, 143, 322, 341, 342, 346, 348

t

Taylor series expansion   199
test   241
– most powerful   262
– one-sided   245
– one-tailed   245, 246
– power curve for   241
– power of   241, 242, 373
– significance level of   241
– size   241
– two-tailed   245, 246
test statistic   240
Theorem of Total Probability   41ff
Thomson   4
time   290
– dead   271, 284ff, 289ff
– real elapsed   290
– system live   290
transformations of random variables   77ff
transformation, radioactive   9, 16
transmutation   4
transport, photon   317
– linear attenuation coefficient   317
– mass attenuation coefficient   318
– Monte Carlo   122, 293ff, 340
– in shielding calculation   317ff
– under good geometry   52, 317, 318
tunneling   9

type I error   233ff, 241, 243, 249, 252, 257, 263, 266, 267, 269

u

unbiased estimator   165
uncertainty, systematic   249
underdispersion   253
uniform distribution   91ff, 119ff, 156, 160, 167, 194, 295, 301, 308, 324, 328, 331, 335, 391, 392, 395ff, 402, 407, 408, 419, 420
uniqueness property   192

v

variable, random   36
– continuous   36
– discrete   36
– expected value   27, 59ff, 63, 69, 72, 74, 85, 89, 92, 95, 99, 105, 165, 166, 189, 190, 192, 198, 200, 209, 224, 227, 229, 232, 267, 295ff, 314, 330, 331, 362, 365ff, 371, 373, 396, 397, 419, 420
– independent   304, 306, 308, 353, 368, 369, 373, 376, 382, 383, 385
– mean   59ff
– mode   403
– response   353ff
variance   63ff, 71ff, 84, 87, 89, 92, 94ff, 105, 107, 110, 113ff, 120, 135, 142, 145, 146, 151, 154ff, 159ff, 164ff, 170ff, 183ff, 189, 194ff, 208ff, 217, 222, 232, 240, 247, 248, 250, 253, 256ff, 262, 270, 273, 274, 282, 308, 309, 316, 317, 353, 354, 358, 359, 361, 364ff, 369, 371ff, 376, 377, 385, 391, 393, 400ff, 414ff, 419, 422
– interval estimate   183
– of a linear combination of variables   74
– pooled estimator for   178
Venn diagram   35
von Laue   5

w

wavelength   2, 4ff, 11, 291
wavelengths, distribution of   2
wave-particle duality   5
– photoelectric effect   5
– X-ray diffraction   5
whole-body count   251

x

X-rays   3, 5, 9, 164, 211, 276, 320, 334, 338, 340